Fundamentals of Ornamental Horticulture and Landscape Gardening

NIPA® GENX ELECTRONIC RESOURCES & SOLUTIONS P. LTD.
New Delhi-110 034

Fundamentals of Ornamental Horticulture and Landscape Gardening

AJAI KUMAR TIWARI
Senior Scientist
Floriculture and Landscaping
Directorate of Floricultural Research, MPKV College of Agriculture Campus
Shivaji Nagar, Pune - 411 005, Maharashtra

RAMESH KUMAR
Director
Directorate of Floricultural Research
MPKV College of Agriculture Campus
Shivaji Nagar, Pune - 411 005, Maharashtra

NIPA® GENX ELECTRONIC RESOURCES & SOLUTIONS P. LTD.
New Delhi-110 034

NIPA® GENX ELECTRONIC RESOURCES & SOLUTIONS P. LTD.

101,103, Vikas Surya Plaza, CU Block
L.S.C. Market, Pitam Pura, New Delhi-110 034
Ph : +91 11 27341616, 27341717, 27341718
E-mail: newindiapublishingagency@gmail.com
Web: www.nipabooks.com

For customer assistance, please contact
Phone: + 91-11-27 34 17 17
Fax: + 91-11- 27 34 16 16
E-Mail: feedbacks@nipabooks.com

ISBN: 978-81-19215-13-3

Composed and Designed by NIPA®.

Dedicated to

My beloved Parents

Preface

Horticulture including Floriculture is growing every year at a galloping speed and the sector has now created enough opportunities for economic growth for India by way of export and domestic turnover. Considering its growth potential in future is now termed as "Golden Revolution".Flowers have always held an aesthetic and mystical significance. Ornamental Horticulture is a branch of Horticulture which deals with commercial growing of flowers ornamental plants and beautification of surroundings. This subject assumes great importance in improving total environment and checks visual pollution by way of beautification. The flowers have been eulogised by poets and artists for generations not merely for their ornamental properties but for their versatile curative properties. The flowers have a unique impact on the human mind, the colour adds beauty, fragrance keeps the mind and body afresh but the harmony lies in the art of flowers arrangement. Human association with flowering plants is prehistoric. Man is always attracted by beautiful and fragrant flowers.

The Study of ornamental horticulture comprises of two parts i.e. the knowledge of growing of plants like annuals, shrubs, climber, trees, bulbous, succulents and cactus, shade loving plants, lawn etc. and their use in beautification programme. The information on above aspect is scattered in different book and a great difficulty was being felt by students, landscapers and scientist. In this context, a need of simple book on introduction, cultivation and management of ornamental plants is highly demanded by student of Horticulture and florists of India.This book "*Fundamentals of Ornamental Horticulture and Landscape Gardening*" is thus brought about mainly to fulfil the above requirement and is also the outcome of my teaching, research and field experiences of several years. This book describes in details collection identification, propagation, cultivation, nutritional & weed management, plant and flowers senescence management etc., of annuals, herbs, shrubs, climbers and ornamental trees. Since, beautification involves the principles of art and gardening styles. Therefore, bio aesthetic planning, basics of landscaping and flowers arrangement have also been discussed in details. Special care has been taken to present the subject matter in a simple and lucid style. Book further includes recent available information.The illustrations have been drawn from own specimens with photographs.

A number of text books, periodicals, bulletins etc. have been consulted during the preparation of this manuscript. I therefore, express my sincere gratitude to authors, scientists and publishers who have generated the aforesaid scientific literatures.

Since, this is the first edition of book and which has been published within a very short time. Hence, there may be possibility of several mistakes. I would therefore, feel pleasure if I get critical comments by the readers with regard to such mistakes and other limitations in the present edition.

The keen interest and the sincere efforts of the publisher of this book "*New India Publishing Agency, New Delhi*", needs a special appreciation for publication within a short time. Finally, I express my sincere thanks to colleagues and friends who always encouraged me to bring out this book.

Author

Contents

Glossary

Abaxial : Underside of leaf or side facing away from the stem is known as abaxial.

Abscission zone : Layer of weak, thin-walled cells that form across the base of the plant part where the break eventually occurs.

Abscission : The separation of a leaf, flower, or fruit from a plant as a result of natural structural and chemical changes.

Absorption : Intake of water and mineral elements by the root system.

Acervulus : A mass of closely clustered conidiophores and conidia not covered by fungal tissue, initially subcuticular or subepidermal but eventually exposed.

Achene : Small and dry fruit, fruit of compositeae family, composed of two combined.

Actinomycete : Any member of the order Actinomycetales; the name is often used to refer specifically to those species which form mycelium.

Acute : Sudden development of disease symptoms. 2. Denoted to less than 90 degrees.

Adaxial : The upper side of a leaf is known as the adaxial surface.

Adjuvants : Additives, as chemicals added to a preservative to enhance its effectiveness.

Adsorption : The process of attraction of atoms or molecules from an adjacent gas or liquid to an exposed solid surface. Such attraction forces (adhesion or cohesion) align the molecules into layers onto the existent surface.

Adventitious root : Root of a vascular plant that arises anywhere other than from the radical or the zone of lateral root formation.

Aerate : Loosening or puncturing the soil to increase water penetration.

Aerial roots : Roots that arise above the ground, from either the stem or the base of the plant.

Alliance : Tribe or group of related genera. Among orchids many of these genera can crossbreed with other genera.

Aggregate fruit : It is the product of a single flower having apocarps pistil (free carpels).

Air layering : Specialized method of plant propagation accomplished by cutting into the bark of the plant to induce new roots to form.

Alba : White form of a flower.

Alternate host : One of two kinds of plants on which a parasitic rust fungus must develop to complete its life cycle.

Alternation of generations : The alternation of a diploid,meiospore-producing generation, the sporophyte, with ahaploid,gamete-producing generation, the gametophyte, in the life history of a plant.

Ambe Bahar : Blossoming in the month of February to March as seen in citrus, guava.

Amphistomaty : A leaf having stomata on both surfaces.

Amyloid : Spore walls, spore ornamentation, hyphal walls, ascus tips, etc. are blue or grayish to blackish-violet in colour followed by staining in Melzer's reagent, presumably because of the presence of starch or a starch-like compound.

Analogy : Non-homologous similarity of structure resulting from similarity of function.

Anamorph : Asexual (usually conidial) stage in the life cycle of a fungus.

Anastomosis : Fusion of somatic hyphae.

Androecium (androecia) : The microsporangiate part of the flower; all the stamens collectively.

Annuals : Plants that complete their entire life cycle within one growing season. The plant germinates from seed, grows blooms and then produces seed and dies.

Antagonist : Agent or substance that counteracts the action of another.

Anther cap : In orchids, a cap of tissue that covers the clusters of pollen.

Anther : Cluster of one or more microsporangia in the Anthophyta, attached to a filamentto form a stamen.

Antheridium (antheridia) : The male gametangium, consisting of a single specialized cell or group of cells in which sperm are produced.

Anthocyanin : Class of water-soluble pigments usually found in the vacuoles of plant cells; anthocyanins are often blue, pink, or yellow.

Anthracnose : Any of various plant diseases, particularly those caused by fungi of the Melanconiales, in which discrete, dark-colored, necrotic lesions develop on the leaves, stems, and/or fruits.

Antibiosis : Association between two organisms that is detrimental to the vital activities of one of them.

Antibiotic : Substance used to inhibit the growth of micro-organisms, including bacteria and fungi.

Apomixis : Sexual development of embryo (seed) in the ovary without fertilization.

Apomorphy : A novel evolutionary trait that is unique to a particular species and all its descendants and which can be used as a defining character for a species or group in phylogenetic terms.

Apospory : Form of apomixes in which the embryo sac develops from a vegetative cell.

Apothecium : An ascus-bearing structure (Ascocarp) in which the ascus-producing layer (hymenium) is not covered by fungal tissue at maturity.

Appressorium : An enlargement on a hypha or germ tube that attaches itself to the host before penetration takes place.

Aquatic plants : Plants which grow in, live in, or live on the water.

Arboretum : Garden with a large collection of trees and shrubs cultivated for scientific or educational purposes.

Aropetal : Development of structures (such as spores) in succession from the base to apex.

Autotrophic : The ability to synthesize high-energy carbon compounds from inorganic raw materials using energy from sunlight or from certain inorganic chemical reactions.

Avoidance : First principle of plant disease control defined as avoiding the disease by planting at a time when or in area where the inoculum in ineffective due to environmental condition or rare or absent.

Axil : The location on a stem between the upper surface of a leaf or leafstalk and the stem from which it is growing.

Axillary bud : Meristem or bud, located on a stem just above the attachment of a leaf; ordinarily found in seed plants.

Back bulb : Pseudobulb that has completed its growth cycle and become dormant. It serves as a storage organ and may bear a dormant bud. Under favorable conditions, this bud may develop into a new plant if the pseudobulb is severed from the parent.

Bare root : Plants offered for sale which have had all of the soil removed from their roots.

Basipetal : Developing from apex to base, applied to inflorescence opposite to acropetal.

Bioaesthetic planning : Planning for beautifying a place with plants and animal.

Basipetal : Successive from apex to base.

Bedding plant : Plants (mainly annuals), nursery grown and suitable for growing in beds.

Biennial plants : Plant that grows, flowers, produces seeds or fruit and dies in two years.

Binomial nomenclature : The system for naming organisms developed by Carl Linnaeus, in which every organism has a generic name and a specific epithet.

Biennials : A plant that take two growing seasons to complete its life cycle. It germinates, grows leaves stems in the first year; produces flowers and fruit (seed) in the second, and then dies.

Bifoliate : Two leaves on a single pseudo bulb.

Biological control : The deliberate use by humans of one species of organism to eliminate or control another.

Bisexual : Presence of both male & female parts in the same flower.

Blight : Disease characterized by rapid and extensive death of plant foliage. A general term applied to any of a wide range of unrelated plant diseases (e.g., chestnut blight, fire blight, late blight, halo blight).

Bolting : Growth of an elongated stalk with flowers grown from within the main stem of a plant.

Bonsai : Japanese art of growing miniature trees and shrubs by extreme dwarfing.

Bract : A modified leaf often reduced in size and with little photosynthetic capability.

Bud : Undeveloped or embryonic shoot and normally occurs in the axil of a leaf or at the tip of the stem. Once formed, a bud may remain for some time in a dormant condition, or it may form a shoot immediately.

Bulb : The thickened underground storage organ of the group of perennials which includes daffodils and tulips.

Bullheads : Over size buds, normally having more petals than normal.

Calyx (calyces) : All of the sepals collectively; in a complete flower, the outermost whorl of parts.

Cambium : Thin ring of tissue within the stem, branch and trunk that continually forms nutrient and water-conducting vessels.

Cane : Long, narrow pseudobulb, usually having many joints or nodes.

Capsule : The seedpod of an orchid, often containing thousands, even millions, of seeds.

Carotenoid : A class of fat-soluble photosynthetic accessory pigments.

Caudicle : A slender stalk-like appendage of the pollinium or pollen mass.

Chloroplast : The intracellular organelle that carries out photosynthesis.

Chlorosis : The loss of chlorophyll from the tissues of a plant, resulting from microbial infection, viral infection, the action of certain phytotoxins, the lack of light, to magnesium or iron deficiency etc. Chlorotic tissues commonly appear yellowish.

Clamp connection : A curving outgrowth of a cell that, at cell division, acts as a bridge to allow passage of one of the products of nuclear division into the penultimate cell, thereby assuring maintenance of the dikaryotic condition in the members of the Basidiomycotina.

Cleistogamous : Used to describe a flower that self-pollinates, often without even opening the flower fully.

Cleistothecium : An ascocarp with the asci surrounded by fungal tissue and without regularly formed openings.

Clone : Asexually propagated offspring of a single parent plant. A clone is identical to its parents.

Cold frame : A wooden or concrete block box in which plants can grow or hold dormant during the cold winter months.

Column : The fused sexual organ of an orchid flower, found at top the lip.

Community Pot : Many tiny seedlings planted together in a single container before they are individually repotted.

Competition : A more or less active demand on the part of two organisms for sources of survival that rather inadequate to provide for all organisms present.

Complete fertilizer : A plant food which contains all three of the primary elements ie. nitrogen, phosphorus and potassium.

Compost : Organic soil amendment resulting from the decomposition of organic matter.

Conifer : Woody trees and shrubs that produce cones. Common conifers include pines, firs, spruce, juniper, redwood and hemlocks.

Corm : Short, solid, enlarged, underground stem from which roots grow. Corms are food-storage organs. They contain one bud that will produce a new plant.

Corolla : The corolla is the name for all the petals of a flower, the inner perianth whorl term used when this is not the same in appearance as the outermost whorl and is used to attract pollinators based on its bright color.

Cover crop : A crop which is planted in the absence of the normal crop to control weeds and add humus to the soil when it is ploughed in prior to regular planting.

Crosspollination : Transfer of pollen from the anther of a flower to the stigma of a flower of another plant.

Cross Protection : The protection conferred on a host by infection with one mild strain of a virus that prevents infection by a closely-related pathogenic strain.

Cross : The progeny that result from transferring pollen from one plant to the flower of another; the act itself.

Crown : The point at which a plants roots and top join (usually at soil level).

Cryo preservation : Preservation of seeds of pollen by immersing them in liquid nitrogen.

Cultivar : Specific cultivated plant grown from a single seed.

Cuticle : Waxy layer formed on the outer tangential walls of epidermal cells in the Kingdom Plantae.

Cytokinesis (cytokineses) : Cytoplasmic division in mitosis or meiosis.

Damping-off : A disease that results in the collapse and death of seedlings before or after they emerge from the soil (pre-emergence and post-emergence damping-off, respectively). Common causal agents include species of *Pythium* and *Phytophthora.*

Dead-heading : The removal of old blossoms to encourage continued blooming or to improve the appearance of the plant.

Deblossoming : Removal of flowers from the plant to reduce the crop load.

Decay : Gradual decomposition of dead organic matter.

Deciduous : Plant that sheds leaves once a year.

Defoliant : Chemical preparation used for causing leaf drop for a definite purpose.

De-greening : Process of decomposing the green pigment in fruit skins.

Desiccation : Dehydration or loss of water.

Determinate growth : Cessation of ontogeny when an organism reaches its final adult form.

Dibble stick : Pointed tool used to make holes in the soil for seeds, bulbs, or young plants.

Dichogamy : Absence of simultaneous maturation of stamens and stigmas in a flower.

Dichotomous : Type of branching where each axis bifurcates at the tip.

Dicliny : Condition in which male and female organs are separate in different flowers.

Dioecious : Having male (staminate) and female flowers (pistillate) on separate (different) unisexual plants e.g. date palm.

Dicot : A member of the Division Anthophyta, Class Magnoliopsida. This class is characterized by having two cotyledons, a feature shared with most other seed plants.

Dieback : Progressive death of shoots, branches and roots generally starting at the tip.

Diploid : Having a normal number of two sets of chromosomes; also known as 2N.

Diploid : Having two complete genomes; the chromosome condition resulting from syngamy and necessary for meiosis.

Disinfectant : Physical or chemical agent that frees a plant, organ, or tissue from infection.

Disinfest : To kill pathogens that have not yet initiated disease, but that occur in or on such in animate objects as soil, tools etc. or that occur on the surface of such plant parts as seed.

Disinfestant : Agent that kills or inactivates pathogens in the environment or on the surface of a plant or plant organ before infection takes place.

Dispersal : Spread of a pathogen within an area.

Dormancy : Rest period during which no vegetative growth occur.

Dormant oil : Oil sprayed on deciduous trees while they are dormant. Dormant oils are used to kill overwintering insects or insect eggs on plant bark.

Dormant : State of rest and reduced metabolic activity in which plant tissues remain alive but do not grow.

Double flower : Flower with many overlapping petals which gives it a very full appearance.

Double working : The practice of inserting the mutually compatible inter-stock piece.

Drip line : Circle which would exist if you drew a line below the tips of the outer most branches of a tree or plant.

Emasculation : Removal of immature anthers/male part from the flowers.

Embryo genesis : Morphological and physiological development of the embryo from the embryo less seeds.

Embryo : Rudimentary juvenile plant usually contained the seed.

Endocarp : Inner most layer of pericarp, usually hard in drupaceous fruits.

Endodermis (endodermis) : Layer of cells surrounding the vascular tissue in roots, stems, or leaves of many seed plants. The endodermis is the innermost layer of the cortex.

Endogenous : Something that originates from within an organism.

Endosperm : Nutritive material in seeds of the Anthophyta, formed from the triple fusion nucleus.

Epicarp : Outer most layer or exocarp of a fruit.

Epicotyl : Embryonic shoot of a seed plant.

Epidemiology : Study of sudden outbreak of disease.

Epidermis (epidermes) : Outermost cell layer of a plant.

Epigynous : Condition of a flower in which the sepals, petals, and stamens seemingly arise from the top of the ovary.

Epinasty : Downward curling of a leaf blade resulting from more rapid cell growth on the upper side of a petiole than on the lower side; often a hyperplastic symptom of plant disease.

Epiphyte : Plant which naturally grows upon another plant or other means of support above the earth's soil but is not parasitic; deriving it's needed moisture from the air.

Eradication : Control of plant disease by eliminating the pathogen after it is established or by eliminating the plants that carry the pathogen.

Erosion : Wearing away, washing away, or removal of soil by wind, water or man.

Espalier : Process of training a tree or shrub so its branches grow in a flat pattern.

Eukaryote : Organism with a membrane-bound nucleus, membrane-bound intracellular organelles and chromosomes.

Evaporation : Process by which water returns to the air. Higher temperatures speed the process of evaporation.

Evergreen : Evergreen plant is a plant having leaves all year round.

Exclusion : Second principle of plant disease control in which the pathogen is prevented from entering or establishing in the field or area where it does not exist earlier.

Exogenous : Something that originates from outside of an organism.

Explant : Plant part to be cultured in the process of micro propagation.

Ex-situ : Out of normal or natural habitat and away from the indigenous place of origin.

Exudates : Liquid discharges from plant organs followed by infection such as leaf exudates, root exudates, etc.

Eye : Undeveloped bud growth which will ultimately produce new growth.

Facultative parasite : Organism that normally lives as a saprophyte but under certain conditions can live as a parasite.

Fascicle : Group of needle leaves of pine or related conifers that are attached to a single lateral spur shoot.

Fertilizer : Organic or inorganic plant foods which may be either liquid or granular used to amend the soil in order to improve the quality or quantity of plant growth.

Filament : A structure consisting of many cells joined end-to-end, or a single, large multinucleate cell with the same thread-like appearance.

Firing of leaf : Yellowing, followed by rapid death of lower leaves, moving up the plant and giving the same appearance as if someone touched the bottom of the plants.

Floriferous : Used to describe a plant that flowers freely.

Flowers : Also known as a bloom or blossom is the reproductive structure found in flowering plants. The biological function of a flower is to mediate the union of male sperm with female ovum in order to produce seeds.

Foliar feeding : Fertilizer applied in liquid form to the plants foliage in a fine spray.

Foot candle : Measure of light useful in determining intensity of light for growing orchids; the illumination produced by a candle at a distance of one foot.

Forcing : Process of hastening a plants growth to maturity or bloom.

Frost : Condensation and freezing of moisture in the air.

Fungicide : Chemical or physical agent that kills the growth of fungi.

Fungus : Eukaryotic, heterotrophic organism that usually has thread like cells.

Gamete : Haploid reproductive cell that fuses with another gamete to produce a zygote.

Gametophyte : Haploid organism arising from meiospores; produces gametes by mitosis.

Gene : Unit that carries hereditary traits.

Genome : Complete set of chromosomes inherited as a unit from one parent.

Genus : Group of Plants that are classified together because of similar traits and an assumed common ancestry.

Germ cell : Reproductive cell which, when united with a germ cell of the opposite sex, develops into a new individual.

Germ tube : Short, juvenile hyphae arise from certain types of spores upon germination.

Germination by repetition : Formation of a secondary spores followed by germination of a spore.

Germination : Resumption of active metabolism by a spore or seed, rupture of its outer wall and beginning of renewed growth.

Germination : Spores of fungi and bacteria, give rise to a vegetative cell or hypha.

Girdle : Encircling of plant roots, stems, trunks or branches resulting in a constriction of the plant part, or a reduction of water and nutrient flow through the girdled plant part.

Girdling : Choking of a branch by a wire or other material, most often in the stems of woody plants that have been tied to tightly to a stake or support.

Graft : Unite a stem or bud of one plant to stem or root of another plant .

Grafting : Uniting of a short length of stem of one plant onto the root stock of a different plant. This is often done to produce a hardier or more disease resistant plant.

Gram stain : Important staining procedure of bacteria discovered by Danish scientist Christian Gram empirically in 1884. When bacteria are stained with certain basic dyes, some of them decolorized with organic solvents termed as Gram-negative species. Cells of Gram-positive species resist decolorization.

Grex : Flock or group, applied collectively to the offspring of a given cross.

Ground cover : Group of plants usually used to cover bare earth and create a uniform appearance.

Guard cell : One of a pair of cells that surround a stoma, an opening in the epidermis. By changing shape, the guard cells open and close the stoma.

Gummosis : General disorder of stone fruits, in which exudation and deposition of gum on the corner of fruit.

Guttation : Exudation of water from the ends of the vascular system at the margins of leaf.

Gymnosperm : Seed plants that are not members of the Anthophyta.

Gynoecium (gynoecia) : Megasporangiate part of the flower, consisting of one or more pistils made up of carpels.

Habitat : Place with a set of defined environment suitable for the growth of an organism.

Halophytes : Plants tolerant to high soil salinity conditions.

Haploid : Having one complete genome; the chromosome condition resulting from meiosis and necessary for syngamy.

Hardening off : Process of gradually acclimatizing greenhouse or indoor grown plants to outdoor growing conditions.

Hardiness : Ability of a plant to withstand low temperatures or frost, without artificial protection.

Hardpan : Impervious layer of soil or clay lying beneath the top soil.

Heading back : Cutting an older branch or stem back to a stub or twig.

Heeling in : Temporarily setting a plant into a shallow trench and covering the roots with soil to provide protection until it is ready to be permanently planted.

Heliophytes : Sun loving plant species which require intense light for normal growth.

Herbaceous : Plants that are mainly soft and succulent, forming little or no woody tissue.

Herbicide : Chemical that kills or retards plant growth. Herbicides may kill the entire plant; or they may kill only the above ground plant parts, leaving the roots alive.

Heterotrophic : Receiving nutrition by the ingestion or absorption of high-energy organic compounds produced by other organisms.

Hierarchy : Ordering of groups in which larger groups encompass sets of smaller groups.

Honeydew : Sticky secretion produced by sucking insects such as aphids.

Host : Plant gets harm while infected with pathogen as it grows and develop in/ on to the plant.

House Plants : Plant that is grown indoors in places such as residences and offices for decorative and health reasons such as indoor air purification.

Humus : Brown or black organic part of the soil resulting from the partial decay of leaves and other matter.

Humus : Stable end product of organic-matter decomposition that's believed to increase microbial activity in soil, improve soil structure, and enhance the root development of plants.

Hyaline : Transparent, translucent and or colourless.

Hybrid : Resulting progeny from the union of two different species (known as a primary hybrid), or of a species and a hybrid, or of two hybrids (known as a complex hybrid).

Hydroponics : The science of growing plants in mineral solutions or liquid, instead of in soil.

Hypha (hyphae) : Filament of the mycelium or vegetative body of a fungus or similar organism.

Hypocotyl : Embryo axis of a seed plant, connecting the radicle to the cotyledons and epicotyl.

Hypogynous : Condition of a flower in which the sepals, petals and stamens arise below the ovary.

Ikebana : Japanese art of flower arrangement.

Imbibitions : Absorption of fluid by a solid or colloid that results in swelling (Uptake of water, or other liquids, by the seed thus enabling germination).

Immunity : State of being absolute resistance in which plant remains free from disease.

Imperfect flower : Flower lacking stamens or pistils.

In vitro : In an artificial environment.

In vivo : Within a living organism.

Incompatibility : Sexual reproduction, it is the inability of the gametes to unite.

Incomplete flower : Flower lacking any of the four whorls or parts (sepals, petals, stamens, or pistils).

Incubation period : Period between penetration of a host by a pathogen and the first appearance of symptoms on the host.

Indeterminate growth : Continuation of ontogeny when an organism reaches its final adult form; characteristic of plants.

Indexing : Method to determine whether a given plant is infected by a virus. It involves the indicator plants sensitive to the virus and can produce symptoms.

Indicator host plant : Plant species that gives characteristic symptoms to a specific virus.

Indigenous : Native to a country or region.

Infection : Process by which pathogen get establish in to the host.

Inflorescence : Cluster of flowers and the manner in which the flowers are arranged.

Inoculation : Proper placement of pathogens or infecting propagules at a site where infection is possible.

Integument : Outer multicellular layer of the ovule, which develops into the seed coat.

Intercalary : Position of any thing (chlamydospores) in between apex and base of a given structure.

Intergeneric : Between or among two or more genera.

Internode : The portion of the stem, cane or pseudobulb located between two nodes or joints.

Interveinal Chlorosis : Yellowing in between leaf veins, but with the veins themselves remaining green. In grasses, this is called striping.

Intrageneric : Within a genus.

Introgression : Crossing of two species producing a fertile hybrid, which backcrosses for several generations with one of the parent species. In this way, genetic material and characteristics is transferred from one species to another.

Invasion : Spread of a pathogen through tissues of a diseased plant.

Irregular flower : Flower with bilateral symmetry.

Juvenile : Youngest plant, branch, bud and generally refers to the life stage of a nematode between the embryo and the adult; an immature nematode

Karyogamy : Fusion of the nuclei of two gametes; part of syngamy.

karyokinesis (karyokineses) : Nuclear division in mitosis or meiosis.

Keiki : Plantlet that grows high up on a cane, pseudobulb or flower stem. When it has developed sufficient roots system, it may be detached and grown separately.

Labellum : Third petal of an orchid flower, modified by evolution into a lip often used as an attractive landing platform for pollinators.

Latent infection : Infection unaccompanied by visible symptoms.

Latent period : 1. The period between infection and the appearance of first sign of the symptoms; 2. Elapsed time between phage infection or induction and lysis of bacterial cells; 3. Period after acquisition of virus by a vector before it becomes infective.

Lateral bud : A bud forming along the side of a stem or branch rather than at the end.

Lateral sepal : Term used to refer to the two lower most sepals that extend to the sides, versus the top most.

Lateral : In orchids, the lateral, or side, sepals are the lower pair of sepals.

Latex : Milky or clear, sometimes coloured, juice or emulsion of diverse compositions

Lath : Overhead structure of evenly spaced slats of wood or other materials used to create shade.

Layering : Method of propagation, by which a branch of a plant is rooted while still attached to the plant by securing it to the soil with a piece of wire or other means..

Leaching : Removal of salts and soluble minerals from the soil by flushing the soil with water.

Leaf mould : Partially decomposed leaf matter, used as a soil amendment.

Leaf spot : Self-limiting lesion on a leaf.

Leaf : Leaf is an above ground plant organ specialized for photosynthesis.For this purpose, aleaf is typically flat and thin.

Lesion : A localized area of diseased or damaged tissue.

Life cycle : A representation of all the stages of an organisms life from birth through reproduction.

Lignin : A polyphenolic substance, chemically related to tannins and flavonoids, that forms a ajor constituent of the secondary cell wall of vascular plant cells.

Lip : Orchid labellum.

Lipolysis : The breakdown of storage lipids (e.g. fats and oils) in living organisms.

Lithophyte : Plant that grows on rocks taking nourishment form mosses, litter, rain and their own dead tissues.

Loam : A rich soil composed of clay, sand and organic matter.

Lux : The lumen is the metric unit of light intensity and the term lux refers to the number of lumens per square metre of surface area and Klux stands for kilo lux or 1,000 lux.

Manse Bonsai : Very dwarf bonsai plants not exceeding a height of 5-20 cm.

Manure : Organic matter, excreted by animals, which is used as a soil amendment and fertilizer.

Mattocking : The operation of gradually removing the pseudo stem of a parent plant.

Mechanical inoculation : Method of transmitting the pathogen from infected to healthy plant through mechanical means; juice from diseased plants is rubbed on test-plant leaves that usually have been dusted with carborundum or some other abrasive material.

Medium : Potting material or mix of materials being used inside the orchid pot.

Medulla : Central part of an organ.

Meiosis (meioses) : Form of cell division of eukaryotes in which each of the four daughter cells has half as many chromosomes as the mother cell.

Mericlone : Generally exact copy of an original orchid plant made via the laboratory technique of meristem propagation; since it is a specific cultivar, it is designated by single quotes around its name.

Meristem : Undifferentiated tissue that is capable of developing into specialized tissue such as toot, leaf or stem. In orchids, a plant produced by culturing meristematic tissue.

Meristem : Group of undifferentiated cells that gives rise to mature cells and organs as a plant grows; also called a bud.

Microclimate : Environment immediately surrounding a plant; very localized climate conditions.

Micropropagation : Making new plants by any of the laboratory techniques, including meristem tissue propagation and sterile seed culture.

Mitochondrion (mitochondria) : Intracellular organelle that carries out oxidative respiration.

Mitosis (mitoses) : Division of the cell nucleus of eukaryotes producing daughter cells with the same chromosome number as the original cell.

Monocot : Member of the Division Anthophyta, Class Liliopsida. This class is characterized by having one cotyledon.

Monoecious : Condition in which staminate (male) and pistil late (female) flowers are found on the same plant but in different position.

Monophyletic group : Group of organisms consisting of a common ancestor and all its descendants.

Monopodial : Growth of orchid wherein a single vegetative shoot grows continually upward, such as in the central rosette of *Phalaenopsis* and the more vine like *Vanda*.

Mosaic : A common symptom induced in leaves by many plant viruses in which there is a pattern of dark green, light green and sometimes chlorotic areas. This pattern is often associated with the distribution of veins in the leaf. In monocotyledonous leaves it shows as stripes.

Mottle : Diffuse form of the mosaic symptom in plant leaves in which the dark and light green are less sharply defined. This term is frequently used interchangeably with mosaic.

Mowing : Operation consisting of trimming lawn grass at a certain height with the help of lawn mover.

Mucilage : Water-holding substance made of various complex carbohydrates, found in many plant groups.

Mulch : Protective covering that is spread on the ground around plants to inhibit evaporation of moisture, weed growth, control soil temperature, enrich the soil and prevent the dispersal of pathogens. It may be organic material such as leaves, peat, or wood chips, or inorganic material such as plastic sheeting.

Multifloral : Having more than one flower per inflorescence.

Mutation : A sudden departure from the parent type, due to a change in a gene or chromosome; an individual resulting from such a departure.

Mycelium (mycelia) : Mass of hyphae forming the body of a fungus.

Mycelium : Mass of hyphae, often used to denote all hyphae comprising a thallus.

Native plant : Plant that occurs and grows naturally in a specific region or locality.

Natural hybrid : Hybrid that occurs in the wild without the help of humans.

Necrosis : Plant tissue browns and dies. This is called necrosis. The tissue which has already died on a still living plant is called necrotic tissue.

Necrotroph : 1. An organism that kills part or all of another organism before deriving nutrients from it (usually applied to plant pathogens); 2. An organism that derives nutrients from dead plant or animal tissues, whether or not it is responsible for the death of those tissues.

Nocturnal : Refers to a plant which flowers opens at right.

Node : Part of a stem from which a leaf or new branch starts to grow.

Nomenclature : System of names, or the rules by which a system of names is formed.

Nonresupinate : In orchids, those plants whose flower lips are positioned uppermost relative to the inflorescence axis; the vast majority of orchid flowers are resupinate.

Obligate parasite : Parasite that in nature always grows and multiplies only on living organism.

Oblong : Spores are longer than broad with sides nearly parallel and with ends more or less flattened.

Offset : Side shoot used as a means of vegetative propagation.

Ontogeny : Developmental history of an individual, from its beginning as a single cell (zygote,meiospore, or asexual spore) or group of cells (gemma, soredium) to its mature multicellular structure.

Organic gardening : Method of gardening utilizing only materials derived from living things.

Organic matter : Substance derived from plant or animal material.

Organogenesis : The processes of forming roots and shoots from callus.

Osmotic pressure : Pressure exerted by the water-absorbing property of a solution.

Ovary : Portion of the pistil that contains the ovules.

Ovoid : Egg-shaped.

Ovule : Integumented megasporangium of a seed plant; develops into a seed.

Panicle : Type of flower inflorescence wherein the flowers are loosely arranged on a branching stem and open from the lowest or inner branches to the top.

Parallel venation : Arrangement of leaf vascular bundles in parallel along the long axis of a leaf; characteristic of monocot flowering plants.

Paraphyletic group : Group consisting of a common ancestor and some, but not all, of its descendants.

Paraphysis (paraphyses) : Hair-like structure associated with gametangia or sporangia; present in mosses, rockweeds and some other groups.

Parasite : Organism that derives its nourishment from another organism, with which it lives in close association and which is harmed by the association.

Parasitic plant : Plant which lives on and acquires it's nutrients from another plant. This often results in declined vigour or death of the host plant.

Parenchyma : Tissue formed by division of cells in three dimensions; also the unspecialized type of such tissue found in vascular plants.

Pathogenesis : Ability of pathogen to cause disease.

Pathogenic : Producing disease.

Pathogenicity : Capability of a pathogen to cause disease.

Peat moss : The partially decomposed remains of various mosses. This is a good, water retentive addition to the soil, but tends to add the acidity of the soil pH.

Peat : Partially degraded vegetable matter found in marshy areas. Peat is commonly used as a soil amendment.

Pedicel : Flowers having short floral stalk called pedicel.

Peduncle : Stem bearing a group of flowers.

Peloric : In orchids, a term used to describe an unusual and often beautiful (sometimes grotesque) condition where all three petals (instead of just one) attempt to fashion themselves into lip colors and/or shapes.

Perennial plant : Plant with a life cycle of three or more years. Herbaceous (non-woody) perennials include flowering plants and herbs, mainly. Woody perennials include trees and shrubs. Longevity depends on the plant and growing conditions.

Perfect flower : Flower with stamens and pistils, sepals and petals may or may not be present.

Perianth : Collectively, the calyx and the corolla.

Peristome : Ring of inward-directed teeth surrounding the opening of a moss capsule. The peristome teeth move in response to humidity, thus facilitating spore dispersal.

Perithecium : Closed ascocarp with a pore at the top, a true ostiole and a wall of its own.

Peritrichous : Flagella distributed more or less uniformly over the cell surface of bacterium.

Perlite : Mineral, which when expanded by a heating process forms light granuals. Perlite is a good addition to container potting mixes, to promote moisture retention while allowing good drainage.

Permineralization : Slow replacement of most or all of the organic material in a fossil with mineral crystals (usually silicate).

Pest : Person or thing that causes trouble, annoyance, discomfort, etc.; nuisance; specif., any destructive or troublesome insect, small animal, weed, etc.

Pesticide : Pesticide is any substance or mixture of substance intended for preventing, destroying, repelling or mitigating any pest. A pesticide may be a chemical substance, biological agent (such as a virus or bacterium), antimicrobial, disinfectant or device used against any pest.

Petal : Modified leaf, usually colored, forming a part of the corolla of a flower; occurs within the sepals (calyx).

Petiole : Stalk that attaches the leaf to the stem; a leafstalk.

pH : Measure of the acidity or alkalinity of a substance; a measure of the relative concentration of hydrogen ions and hydroxyl ions.

Phloem : Nutrient-conducting vessels found throughout the plant. Phloem vessels transport nutrients produced in the foliage down through the stems, branches, or trunk to the roots.

Photo morphogenetic : Influences of the light on growth and development of plants.

Photosynthesis : Process by which plants use the sun's light to produce food (carbohydrates).

Phylogeny : Genealogy of species; the pattern of ancestry and descent of species in evolution.

Pinching : Removing of the terminal shoots of plants.

Pith : Inner tissue of most stems and some roots.

Plasmogamy : Fusion of the cytoplasm of two gametes; part of syngamy.

Plastid : General name for chloroplasts, used especially for chloroplasts that are not green.

Pleiotropy : Phenomenon of a single major gone affecting more than one character.

Plesiomorphy : Character state already present in the ancestral species of a group under study.

Pod : Term used to refer to the seedpod or capsule.

Pollen tube : Structure formed by germinating pollen that absorbs nutrients, and in some groups carries the sperm cells to the eggs.

Pollen : Male gametophyte of seed plants, consisting of a generative cell containing the tube nucleus and generative nucleus and several (or no) prothallial cells.

Pollination : Transfer of pollen from the stamen (male part of the flower) to the pistil (female part of the flower), which results in the formation of a seed.

Pollinator : Agent of pollination.

Pollinia : Waxy pollen clumps or grains usually found in the anthers of most orchids; often yellow, distinct and found under the pollen cap of the column.

Pollinizer : Plant in fruit plantation which provides pollen for self-unfruitful varieties.

Polyembryony : Presence of more than one embryo seedlings in a ovule.

Polyphyletic group : Group of organisms not including the common ancestor of all members of the group.

Predisposition : Action of set of environments prior to penetration and infection that makes plant or plant parts susceptible to the pathogen.

Primary infection : First infection of the plant caused by a pathogen that emerging from a dormant stage (overwintering and oversummering propagules).

Primary inoculums : Overwintering and over summering propagules (resting structures) of pathogen that can cause primary infection on the host.

Primordium (primordia) : Group of cells that in the normal course of events will develop into a specific mature structure or organ.

Prokaryote : Organism without membrane-bound organelles, and with DNA organized in a single naked circular strand, rather than in chromosomes.

Proliferation : Rapid and repeated production of new cells, tissues or organs specifically, a hyperplastic symptom of plant disease in which organs continue to develop after they have reached the point beyond which they normally do not grow.

Propogation : Means of reproducing plants, such as by seeds, cutting , budding or grafting.

Propping : Practice of providing support to a plant with the help of crossed bamboo

Protection : Fourth principle of plant disease control in which a chemical, physical or biological barrier is created between host surface and pathogen.

Protoderm : Outermost tissue of root and shoot apex which develops into epidermis.

Pruning : Cutting and trimming of plants to remove dead or injured wood, or to control and direct the new growth of a plant.

Pseudobulb : Thickened portion of a stem, resembling a bulb but not being a true bulb, which is an underground modified bud.

Pure line : Progeny of a single homozygous, self-pollinated plant.

Pyriform : Pear-shaped.

Quarantine : Legal restriction on the movement of agricultural commodities for the purpose of exclusion, prevention or delay in the establishment of plant pests and diseases in area where they are not known to occur.

Raceme : Raceme is a type of inflorescence that is unbranched and indeterminate and bears pedicellate flowers along the axis.

Recombination : Formation of new combinations of genes as a result of the sexual process.

Relative humidity : Measurement of the amount of moisture in the atmosphere.

Resistance : Ability of an organism to exclude or overcome, completely or in some degree, the effect of a pathogen or other damaging factor.

Reticulate venation : Arrangement of leaf vascular bundles in a small network; characteristic of some Gnetophyta and dicot flowering plants.

Rhizoid : Hair like unicellular or filamentous absorbing and anchoring structure of organisms in the Kingdom Plantae. Root hairs are a type of rhizoid.

Rhizome : Underground stem usually horizontally oriented and sometimes specialized for food storage.

Rhizome : Stem that produces roots and above-ground organs such as stems, pseudobulbs and flowers. In epiphytic orchids it is usually found on the surface; in terrestrial orchids it may be underground.

Ring spot : Type of local lesion consisting of single or concentric rings of discoloration or necrosis, the regions between the concentric rings being green. The center of the lesion may be chlorotic or necrotic.

Root ball : Network of roots along with the attached soil, of any given plant.

Root bound : Condition which exists when a potted plant has outgrown its container. The roots become entangled and matted together and the growth of the plant becomes stunted. When repotting, loosen the roots on the outer edges of the root ball, to induce them to once again grow outward.

Rooting hormone : Powder of liquid growth hormone, used to promote the development of roots on a cutting.

Rot : Softening, discoloration and often disintegration of a succulent plant tissue as a result of fungal or bacterial infection.

Runners : Above ground, trailing stems that form roots at their nodes when they make contact with moist soil.

Sanitation : Principle of plant disease control involving removal and burning of infected plant parts and decontamination of tools, equipment, hands, etc.

Saprophyte : Plant or microorganism that obtains its nourishment from dead organic matter, such as most fungi and bacteria and a few non-photosynthetic flowering plants.

Scion : Short length of stem, taken from one plant which is then grafted onto the rootstock of another plant.

Sciophyte : Plant which grows well on or is adapted to shady sites.

Sclerotium : Hard, resistant, multicellular resting structure usually differentiated in a cortex, rind and medulla that under favorable conditions can germinate to produce mycelium or sexual or asexual fruiting bodies.

Secondary infection : Infection caused by inoculum produced as a result of a primary or a subsequent infection; an infection caused by secondary inoculum.

Secondary inoculums : Inoculum produced by infections that took place during the same growing season.

Seed : Multicellular structure containing the embryo of a seed plant, ordinarily with stored food, the whole protected by a seed coat.

Seedpod : Capsule bearing the seeds of an orchid.

Self-fertile : Ability of a cultivar to produce fruit with viable seeds following self pollination.

Self-pollination : Transfer of pollen from the anther of a flower to the stigma of the same flower or of another flower on the same plant.

Semi alba : White flower with a colored lip.

Seminal root : Root of a vascular plant that arises from the radicle or the zone of lateral root formation.

Sepal : Sepals in most flowers are green and lie under the more conspicuous petals.

Septate : Having cross walls.

Septum : Cross wall in a hypha or spore.

Serology : Branch of science dealing with properties and reactions of sera, particularly the use of antibodies in the sera to examine the properties of antigens.

Sexual phase : Portion of an organism's life cycle devoted to sexual reproduction.

Sheath : Flattened base of a leaf that wraps around a stem.

Sheath : Tubular structure surrounding some parts of the plant.

Sib cross : Method of seed propagation of plant wherein the pollen of one plant is placed on the stigma of another plant that was originally grown in the same seedpod as the first plant, therefore a cross pollination of siblings.

Sibling : Plant that is related to another orchid by virtue of having been produced from the same seedpod.

Species : Group of plants showing intergradation among its individuals and having in common one or more characteristics which definitely separate it from any other group; a kind of plant distinct from other plants.

Sphagnum : Bog moss which is collected and composted. Most peat moss is composed primarily of sphagnum moss. This moss is also packaged and sold in a fresh state, and used for lining hanging baskets and air layering.

Spike : Technically an unbranched flower stem with short stalked or stalkless flowers long inflorescence having multiple florets.

Splash : Petal of an orchid flower that modifies its petals by duplicating the coloring found on the lip; a type of peloric condition.

Spore : Discrete sexual or asexual reproductive unit, usually enclosed by a rigid wall, capable of being disseminated.

Saprophyte : Diploid organism arising from a zygote; produces meiospores by meiosis.

Spot : Symptom of disease characterized by a limited necrotic area, as on leaves, flowers and stems. Spots may be different in shape and size.

Spur shoot : Lateral shoot with short or nonexistent internodes, sometimes bearing reproductive structures or fascicles of leaves.

Spur : Hallow, sac-like or tubular extension of the base of the lip, usually bearing nectar.

Staking : Practice of driving a stake into the ground next to and as a support for a plant. When attaching the plant to the stake, be sure that it is tied loosely so it doesn't strangle the stem. When staking a potted plant, the stake should be set into the planter before the plant is added.

Stamen : Pollen-producing structure in a flower, consisting of a filament and an anther.

Stegmata : Thickened wall adjacent to the underlying sclerenchyma cells, with progressively thinner lateral walls and thin outer walls.

Sterigma : Spike-like structure that bears basidiospores on the basidium.

Sterilant : Substance used for sterilization of seeds, tools etc.

Sterile fungus : Fungus that is not known to produce any kind of spores.

Sterilization : Elimination of pathogens and other living organisms from soil, containers, etc. by means of heat, chemicals and radiation.

Stigma : Part of the pistil of a flower which is receptive to pollen.

Stipe : Stalk-like support of a pistil, or other small organ such as pollinium.

Stoma (stomata) : Opening in the epidermis of a plant controlled by two guard cells.

Stomates : Tiny pores located mainly on the underside of leaves. Oxygen, carbon dioxide, water vapor and other gases move in and out of the leaf through these pores.

Suberin : Complex of fatty substances present in the wall of cork tissue that waterproofs it and makes it resistant to decay.

Suberin : Water proofing material secreted by cells of endodermis or periderm.

Suberization : Deposition of suberin on the walls of plant cells and their subsequent conversion into cork tissue.

Sucker : Growth originating from the rootstock of a grafted plant, rather than the desired part of the plant.

Suppressive soil : Soil in which certain disease(s) fail to develop due to presence of antagonistic microorganisms in the soil.

Suscept : Plant that can be attacked by a given pathogen (a host plant).

Susceptibility : Inability of a host plant to resist the effect of a pathogen.

SWOT : Strength Weakness Opportunity Threat analysis.

Symbiosis : Mutual beneficial association between two or more disparate organisms.

Sympodial : One of the two forms of orchid growth (the other is monopodial), wherein each new growth arises from the rhizome of a previous growth, and each new growth is completely capable of bearing an inflorescence. (Sympodial means "with conjoined feet" and in biology is often used to refer to the outward morphology or mode of growth of organism).

Symptom : Visible sign of disease.

Symptomatology : Study of symptoms of disease and signs of pathogens for the purpose of diagnosis.

Synsepalum : Organ resulting from the fusion of the two lower sepals of a lady slipper orchid (*Paphiopedilum*). It is found below and behind the lip or pouch.

Systemic infection : Movement of pathogen / mycelium internally into the whole plant.

Systemic : Chemical which is absorbed directly into a plants system to either kill feeding insects on the plant, or to kill the plant itself.

Systemic : Pesticides or fungicides that are taken up by plant leaves and growths and then work from within the plant.

Tap root : Main, thick root growing straight down from a plant.

Taxon (taxa) : Group of organisms in a taxonomic hierarchy that have affinity to one another.

Tender plants : Plants which are unable to endure frost or freezing temperatures.

Tendril : Twisting, clinging, slender growth on many vines, which allows the plant to attach themselves to a support or trellis.

Terminal buds : Bud at the end of a stem or branch

Terrestrial : Growing in the ground or in the loose substrate atop the ground.

Tetraploid : Genetic aberration wherein the plant has twice as many chromosome sets as normal, often resulting in very vigorous, large plants and flowers.

Thallus (thalli) : Plant body that is formed of parenchyma or plectenchyma but is not well-differentiated into organs.

Thinning : Removing excess seedlings, to allow sufficient room for the remaining plants to grow. Thinning also refers to removing entire branches from a tree or shrub, to give the plant a more open structure.

Throat : Inner portion of a tubular orchid lip.

Tissue culture : Artificial propagation of plants via laboratory mericloning, also known as meristernming.

Tissue : Group of cells having similar structure performs a special function.

Topiary : Method of pruning and training certain plants into formal shapes such as animals.

Topsoil : Top layer of native soil. This term may also apply to good quality soil sold at nurseries and garden centers.

Toxin : Compound produced by a microorganism and being toxic to a plant or animal.

Tracheid : Water-cunducting cell of a vascular plant.

Translocation : Movement of materials within the plant.

Transpiration : Evaporation of water from plant tissue to the atmosphere.

Transplanting : Process of digging up a plant and moving it to another location.

Triple fusion nucleus : In the Anthophyta, the triploid nucleus that results from fusion of one sperm nucleus with both polar nuclei.

Tube nucleus : In seed plant pollen, the nucleus that directs growth of the pollen tube.

Tuber : Flat underground stem which stores food and plant energy and from which a plant grows.

Unifoliate : Bearing one leaf per growth.

Variegated : Leaves which are marked with multiple colours.

Variegation : Appearance of different coloured zones in the leaves and sometimes the stems of the plant.

Variety : Subdivision of a species that groups plants with a distinct form that is passed along to the progeny.

Vascular bundle : Strand of vascular tissue running through surrounding tissue such as parenchyma.

Vascular plant : Plant that produces vascular tissue.

Vascular tissue– Collectively, xylem and phloem.

Vector : Living agent that transmits a pathogen from an infected to an uninfected plant.

Vein banding : Symptom of virus-infected leaves in which tissues along the veins are darker green than other leaf area.

Vein clearing : Symptom of virus-infected leaves in which veinal tissue is lighter green than that of healthy leaves.

Velamen : Thick corky layer of cells covering the arial roots of epiphytic orchids, the function of whichis to absorb moisture from the atmosphere and attachments to supports.

Vermiculite : The mineral 'mica' which has been heated to the point of expansion. A good addition to container potting mixes, vermiculite retains moisture and air within the soil.

Vernalization : Method of inducing earlier flowering in plants by pre-treatment of seeds vigour, done either by withholding irrigation water or by spraying of chemicals or providing two temperature.

Virus : Infectious units composed of either RNA or DNA and enclosed by protein made protective coat.

Whorl : Three or more leaves, flowers or branches arranged in a circle at one point around the main stem.

Wilt : Disease symptom characterized by a loss of turgidity in a plant.

Wilting : Symptom characterized by loss of turgor pressure, which results in drooping of leaves, stems, and flowers.

Xanthophylls : Class of fat-soluble photosynthetic accessory pigments.

Xeric : Dry; applied to features of plants or places where they grow.

Xylem : Water-conducting vessels found throughout the plant. Xylem vessels transport water and minerals from the roots upward through the plant.

Xyloposlosis : Disorder associated with lack of compatibility between stock and scion,

Yellowing : Symptom characterized by the turning yellow of plant tissues that were once green.

Zygote : Diploid cell that is produced by fusion of ovum and sperm (syngamy).

CHAPTER 1

Identification and Use of Garden Implements

Garden Rakes

1. For use with moving debris.
2. Aids in the cleaning process.
3. Proper use is with two hands.

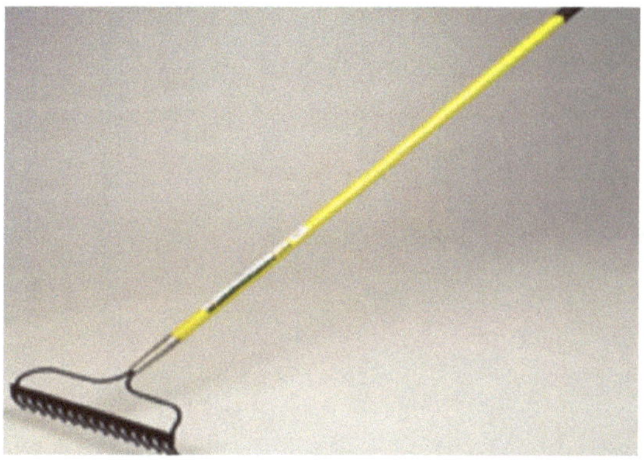

Fig. 1.1 : Steel Rake

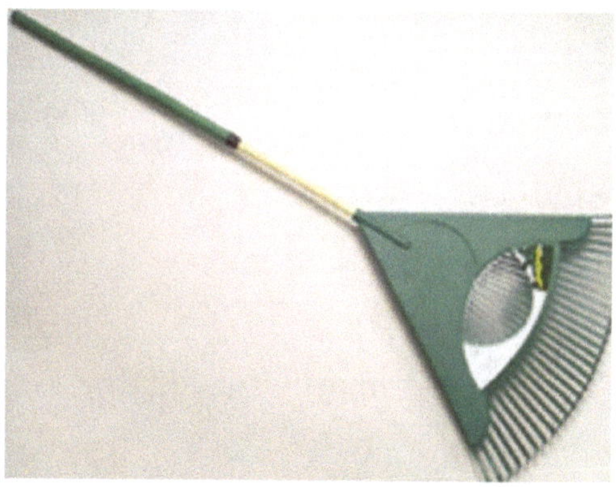

Fig. 1.2 : Leaf Rake

Pruners

Used for pruning of branches and suckers of plants.

There are two types of pruners

- Bypass
- Anvil

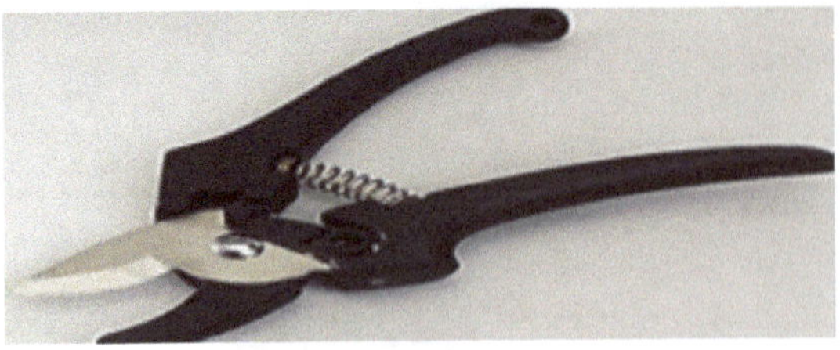

Fig. 1.3 : Anvil Pruner

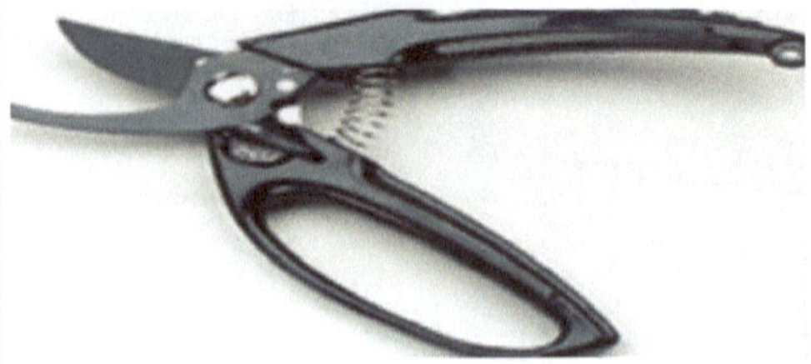

Fig. 1.4 : Bypass Pruner

Trowels

- Used to aid in the planting of medium to large plants.
- Moves small amount of dirt.
- For use with one hand while holding a plant in the other.

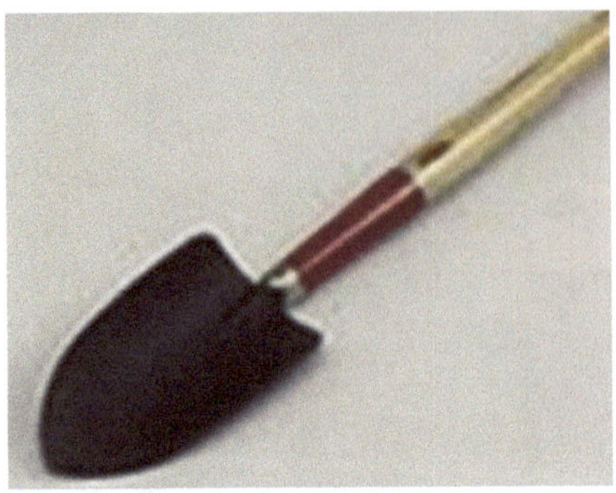

Fig. 1.5 : Trawel

Shovels

Two types of shovels

- Round point

  Used primarily for digging holes.

- Square point

- Better used for picking up trash and debris.
- The size is much larger and can pick up much more debris.
- The handle is much shorter with a different grip.

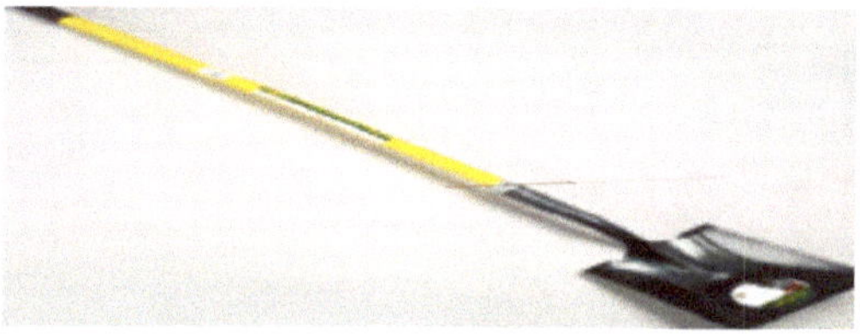

Fig. 1.7 : Square Point Shovel

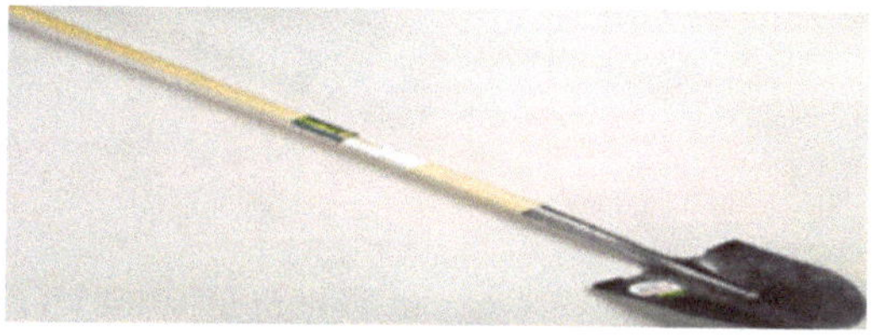

Fig. 1.8 : Round Point Shovel

Garden Hoe

1. Used to cultivate the soil.
2. Removes weeds.

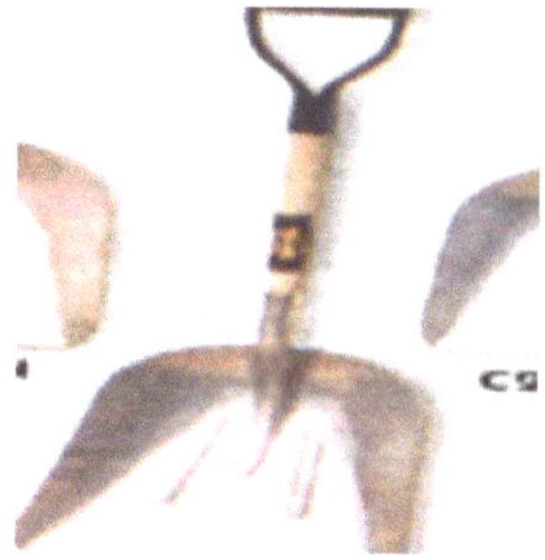

Fig. 1.9 :

Steel Head Garden Cultivator

- Cultivates the soil.
- Remove weeds.
- Allows oxygen to penetrate the soil.

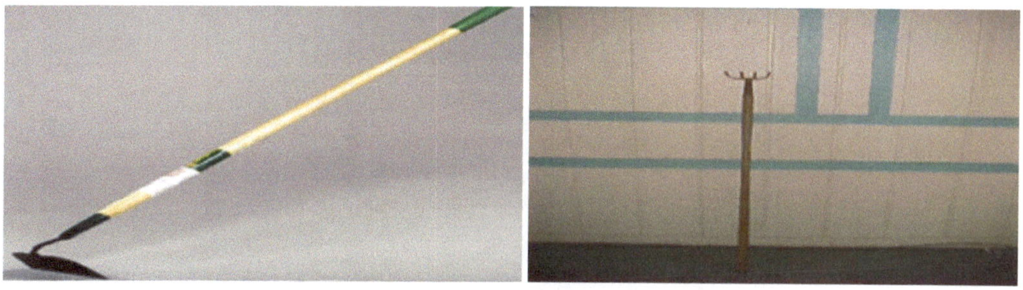

Fig. 1.10 : Hand cultivars

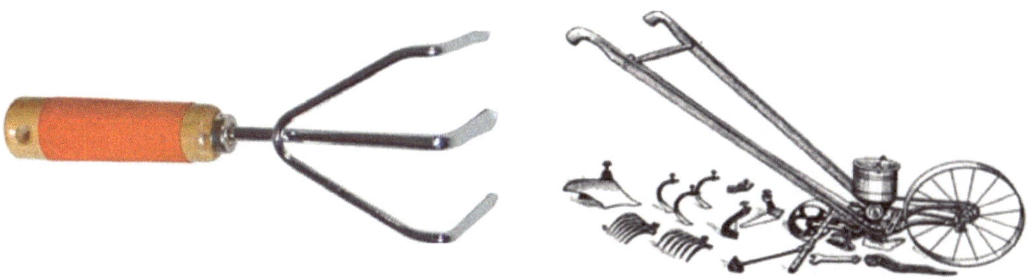

Fig. 1.11 : Hand cultivars

Wheelbarrow

- Used primarily to move heavy things from one place to another.
- Can be used to hold trash.
- Can also be used to make cement travel.

Fig. 1.12 :

Nozzles

- Used to water plants.
- Control direction of spray.
- Different ends allow different amounts of water.

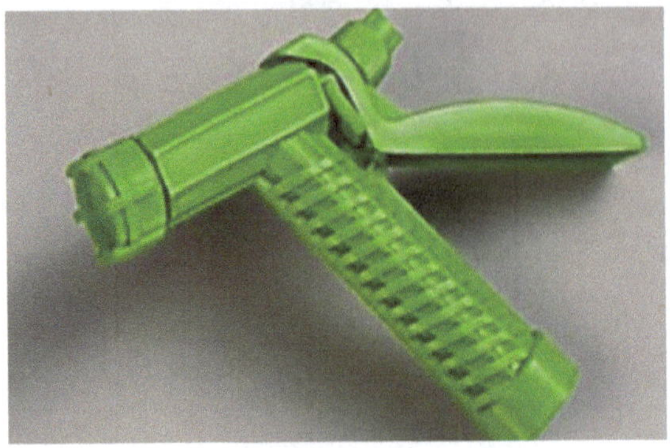

Fig. 1.13 :

Secateurs

Used to prune off branches of pencil thickness and making cutting etc.

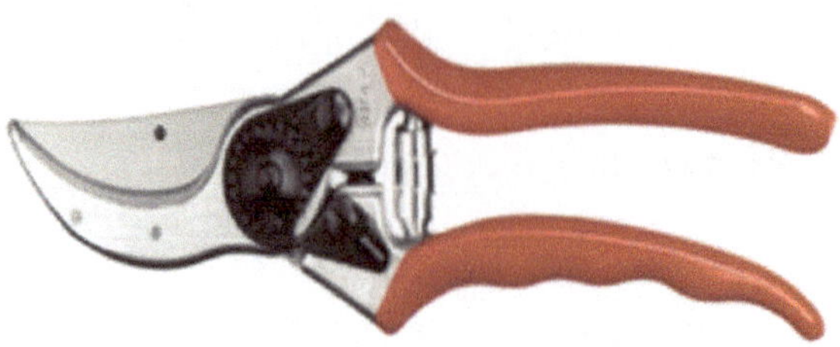

Fig. 1.14 :

Fig. 1.15 :

Tree Pruners

Used to prune off branches of trees at 3-4 meters height & 3-4 cm thick.

Fig. 1.16 :

Pruning Knife

Used for pruning of branches and suckers.

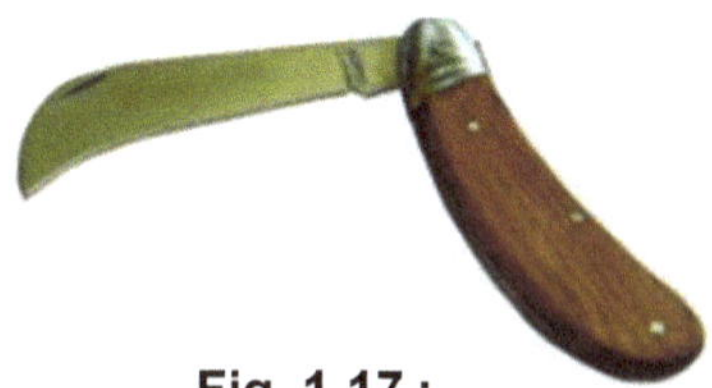

Fig. 1.17 :

Grafting and budding Knife

Used to perform grafting and budding operation.

Fig. 1.18 :

Hedge Shear

Used to cut the hedge and edge.

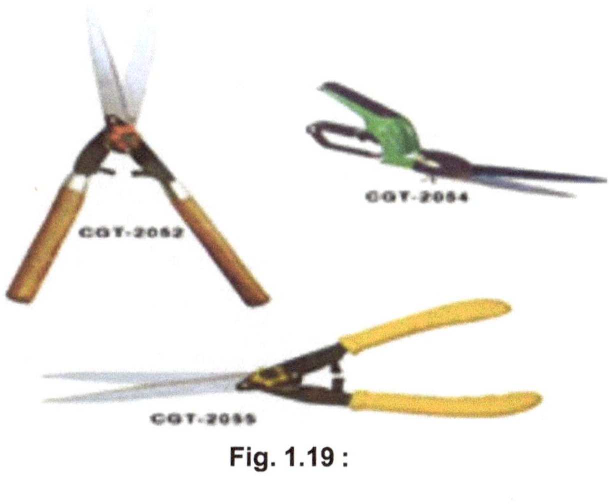

Fig. 1.19 :

Spade

Used to dug or turn the soil.

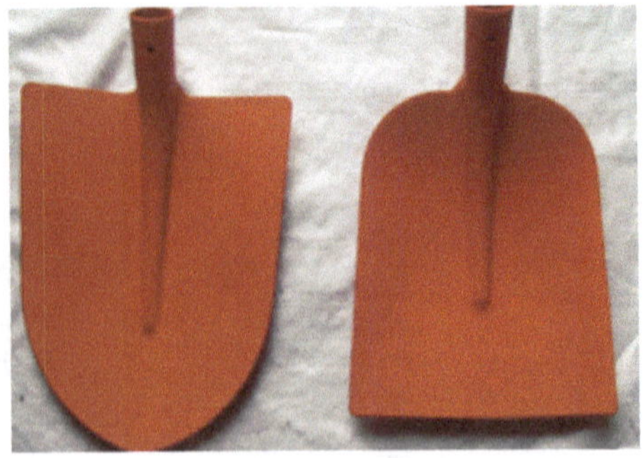

Fig. 1.20 :

Watering cane

Used to water the plants.

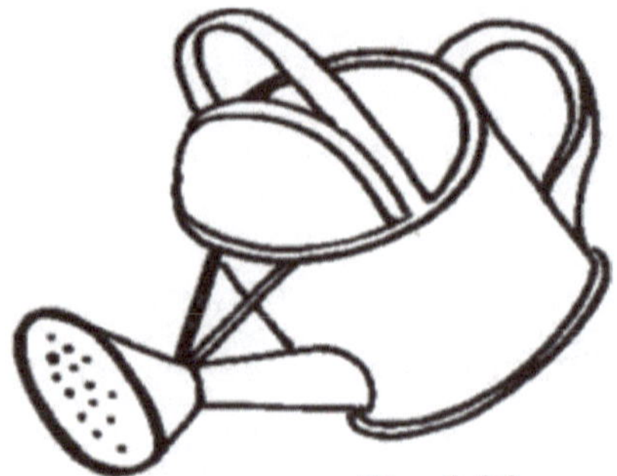

Fig. 1.21 :

Bill Hook

Used to cut the branches of trees.

Fig. 1.22 :

Planting Board

Used for proper placement of plants in dig.

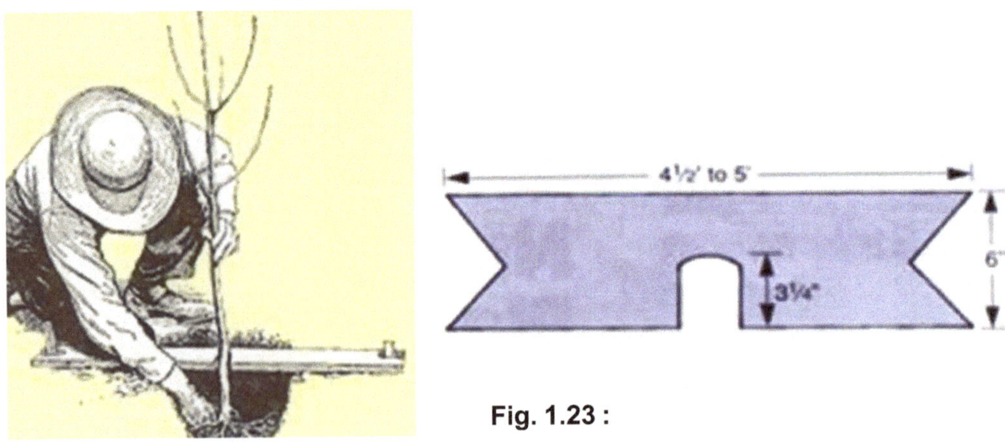

Fig. 1.23 :

Sprayer

Used to spray the pesticide and weedicide.

Fig. 1.24 : Knapsack Sprayer

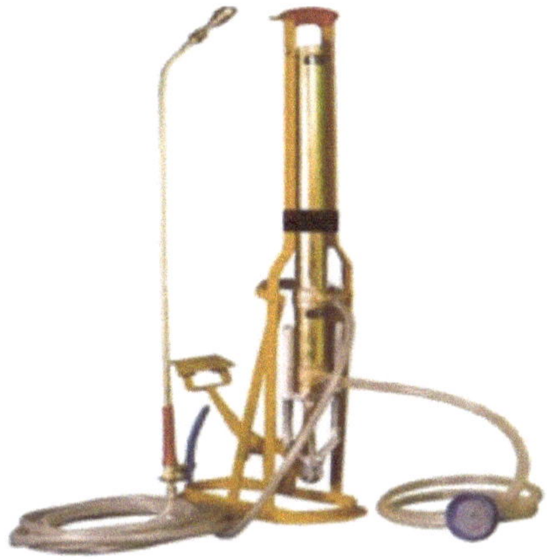

Fig. 1.25 : Foot sprayer

Hand Rotary duster

Used to dust the chemicals.

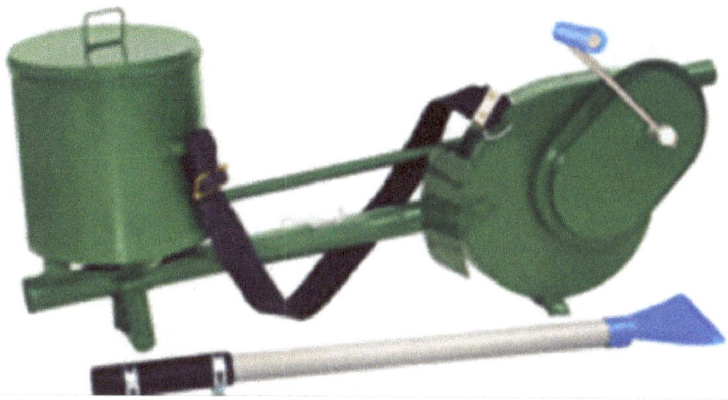

Fig. 1.26 :

CHAPTER 2

Classification and Identification of Ornamental Plants

Classifications and Nomenclature

Plants can be classified according to the following criteria: (1) botanical, (2) descriptive and (3) agricultural. Botanical classification is based on the morphological characteristics of plants as well as on their anatomy, physiology and DNA sequences. Descriptive classification is based on the environmental adaptation, growth habit and other observable features. In agriculture, plants can be broadly classified as either useful or useless. Those which are useful are called crops while those which are not useful are called weeds.

A common, well-defined system of crop classification is important in crop science and agriculture. Grouping of plants following established systems will simplify plant collection initiatives, research and breeding specialized development efforts. Having standardized botanical names will also facilitate efficient communication, dissemination and retrieval of scientific information. Additionally, the grouping of crops will indicate that these crops may have similar uses, adaptation, growth habits and methods of culture. Classifying crops by family is likewise important in intergeneric grafting. To students of crop science, a list of crops under the various agricultural classifications will be a useful reference in crop names and identification and as a review guide. To the farm manager himself, the same list will offer innumerable benefits in finding alternative crops for specific uses.

Despite these advantages, however, classifications of crops in agriculture vary from place to place and from one country to another due to differences in the primary usage of certain crops, intensity of care, extent of cultivation and other criteria under consideration. It is not uncommon to find crops having two or more classifications, used simultaneously, alternatively or sequentially. FAO(2010) in its report entitled "Crops Statistics - Concepts, Definitions and Classifications", expressed its difficulty in the collection and presentation of statistical data due to differences in the concepts, definitions, coverage and classifications" of crops which are grown in different countries.

Botanical Classification

Plants can be identified by observing certain distinguishing morphological characteristics. The process of categorizing plants is called Plant Classification and is nothing more than a systematic arrangement of plants into certain groups and even subgroups based on characteristics common to the group. This system of classifying plants is based on structural similarities (primarily flowers and fruits) and common ancestry (relationship by descent).Plant classification divides these groups in a hierarchical manner (from highest level to lowest level). One scheme commonly used is:

KINGDOM - (highest)

DIVISION - (next highest)

CLASS - (next)

ORDER - (next)

FAMILY - Family is highest group that is commonly used in classification reference because it is small enough to see or clearly distinguish among members of the group. Example is China Aster Family and Begonia Family,

GENUS - (next)

SPECIES - (next)

VARIETY - (sometimes)

CULTIVAR - (sometimes)

When we deal with genus, species, variety and cultivar, we are referring to groups or terms in the classification system that are commonly used in the nomenclature of plants. "Nomenclature" is nothing more than a system of naming plants using an international vocabulary primarily of Latin terms that have been standardized by experts in the field. These Latinized or formal names are referred to as the "Scientific" names or the "Botanical" names of the plants. Similar plants are placed in the same Genus. These individuals within genera (plural of genus) are called "Species" and differ from one another mainly in the appearance of their flowers. Only in rare cases is a variety name assigned; this means that the plant has some very special distinguishing feature which "botanists" feel is notable enough to be formally recognized in the scientific name. In general "botanist" tends to be less application-oriented than "horticulturist"; also horticulture is branch of agriculture dealing with certain specific types of plants - flowering and foliage plants for indoor and outdoor purposes, nursery crops etc. Botany is not branch of agriculture - plants involved may or may not have application to people.

If a plant has a Cultivar name, it is because that plant was isolated or manipulated by a "horticulturist" because it exhibits some characteristic different from the wild species type, such as flower difference, foliage difference, form or growth habit difference etc. So, cultivars are generally not found growing in the wild. The word "Cultivar" is a actually derived from Cultivated Variety.Below is the formal classification and nomenclature of a typical plant (Swiss Cheese Plant).

Kingdom - Plantae (Plant)
Division - Angiospermae (Angiosperm)
Class - Monocotyledoneae (Monocot)
Order - Arales
Family - Araceae (Arum Family - *Philodendron, Dieffenbachia,* Peace Lily)
Genus - Monstera
Species - deliciosa

The scientific name or botanical name is written: *Monstera deliciosa* - Like all scientific names, it must conform to a set of rigid rules with genus name beginning with a capital letter and the species name usually beginning with a small letter. Above system of Plant Classification is Known as "Binomial System" and gives each plant a *scientific name* using the genus and species. Both names are always underlined or italicized (commonly in books). Examples of scientific names:

Table 2.1 : Examples of Scientific names

Scientific Name	Common Name
Acer saccharinum	Silver maple
Acer platanoides	Norway maple
Tagetes erecta	African marigold
Tagetes patula	French marigold

When botanists group plants, they use flower parts as their primary guide because the flower is the least affected by growing conditions.

In the case of Croton: *Codiaeum variegatum pictum* with a variety name, it is also underlined and beginning with a small letter.In the case of the variegated wax plant, it is written: *Hoya carnosa 'variegata'* - with cultivar name in single quotes. The rule is that the cultivar name is always in single quotes, and may or may not be capitalized.Sometimes, a plant such as the common geranium is written as: *Pelargonium x hortorum* - in this case the "x" indicates that the plant is a hybrid, which is a plant produced from breeding two plants of the same genus, but of different species. Another type of hybrid - much rarer in occurrence, may arise from breeding two plants from completely different genera, such as *Fatshedera*, whose parents are in the *Fatsia* and *Hedera* genera. The scientific or botanical name of this plant would be written as *xFatshedera lizei*.These types of hybrid plants usually do not occur in the wild, but were created from sexual propagation.

Common names may vary a lot depending on country, region of country and even local pockets within regions. Ex. Swiss Cheese plant is also commonly known as Split-leaf *Philodendron*, but not so commonly known as Hurricane Plant. Ex. Strawberry begonia is neither a begonia nor a strawberry. However, the scientific name cannot be stressed enough because they are international in usage and cause less confusion. They are unique to one and only one plant and are subject to fewer changes because they are accepted internationally and can not change at the whim of anyone. Only experts in the field have the accepted authority to make changes in the scientific names.

Descriptive Classification of Crops

A. Classifications According to Mode of Reproduction

1. *Sexual-* plants that develop from a seed or spore after undergoing union of male and female gametes. Examples: palms and ferns.
2. *Asexual-* plants which reproduce by any vegetative means without the union of the sexual gametes or by apomixis. Examples: red mombin (sineguelas), breadfruit, mangosteen.

B. Classifications According to Mode of Pollination

1. *Self-pollinated plants* - the predominant mode of pollination in these plants is self-pollination in which both pollen and embryo sac are produced in the same floral structure or in different flowers but within the same plant. Examples: most pulses, okra, tobacco.
2. *Cross-pollinated plants* - pollen transfer in these plants is from the anther of one flower to the stigma of another flower in a separate plant, although

self-pollination may reach 5 percent or more. *Examples*: many grasses, avocado, grape, mango, many plants with unisexual or imperfect flowers.

3. *Often cross*-pollinated plants - these plants are largely self-pollinated but varying amounts of cross-pollination occur. Examples: Hollyhock

C. Classifications According to Life Span

1. *Annual*- a plant which live within a short period of time, for a few weeks or months, perpetuated by seed and which die soon after producing seeds; a plant which germinate, grow, flower, produce seed and die all in one season. Examples: candytuft Larspur Calendula etc.
2. *Biennial*- a plant which requires two growing seasons to complete its life cycle, the first for vegetative growth and accumulation of food reserves, and the second for the production of reproductive parts. It grows from seed, produces flower and seed and then dies in two growing seasons. Examples: bulb onion, cabbage, carrot, celery.
3. *Perennial*- a plant that lives indefinitely, including all trees and shrubs and many herbaceous plants with underground stems (e.g. corm, rhizome and tuber) like banana and clump-forming grasses. Perennial plants continue growing and produce seeds year after year, either from a single plant or, in herbaceous plants, from succeeding regrowth.

D. Classifications According to Growth Habit

1. *Herbs*- succulent plants with self-supporting stems. Examples: aglaonema, banana, dumbcane.
2. *Vines*- herbaceous climbing or twining plants without self-supporting stems. Examples: charantia Lathuyrus, Pea.
3. *Lianas*- woody climbing or twining plants which depend on other plants for vertical support to climb up to the top of the canopy. These climbers often form bridges between the forest canopies. Examples: Climbing bamboo, grape, Jade vine (*Strongylodon macrobotrys*), rattans, passion fruit.
4. *Shrubs*- small trees or tree-like plants, generally less than 5 meters in height but by other authorities it is restricted to small, erect, woody plants which produce several trunks from the base. Examples: Barbados cherry, pink jasmine (kamuning), Siam weed (hagonoy), santan, *Lantana*.
5. *Trees*- plants having erect and continuous growth with a large development of woody tissue with a single distinct stem or trunk, reaching a height of 5 meters or more. Examples: Neem, Cedar, Chinar etc.

E. Classifications According to Leaf Retention

1. *Evergreen* : Plants that maintain their leaves throughout the year. Abscissed leaves are continually replaced by new flushes. Examples: pines, banana, papaya, palms and most tropical plants.
2. *Deciduous* : Plants which naturally shed off or lose leaves annually for extended periods. Natural leaf shedding is pronounced in deciduous trees of temperate regions. In the India defoliation occurs during summer months in Fire Tree (*Delonix regia*) and red mombin or siniguelas. Many other plants exhibit partial defoliation during drought periods.

F. Classifications According to Ecological Adaptation or Habitat

1. *Aquatic, hydrophytes or hydrophytic plant* :A plant adapted to growing in water or waterlogged soil. It may grow entirely submerged, partly submerged or floating, or anchored to the ground in bogs, swamps, or beside the edges of ponds, lakes or streams. Examples: azolla, *Ipomea aquatica, Cyperus* spp., lotus (*Nelumbo nucifera*), water lily (*Nymphaea* spp.), mangrove species.
2. *Epiphyte or epiphytic plant* : A plant that grows above ground on another plant but is not parasitic, usually deriving only physical support from the host and obtaining nourishment from the air and other sources. Some have roots that take moisture and minerals leached from the canopy of trees and others catch rain and debris in special hollow leaves. The most common epiphytes belong to the pineapple (bromyliad), orchid, and fern families; also called air plantor tree dweller. Where a plant, e.g. strangler fig, initially grows as an epiphyte but later becomes rooted to the soil, it is especially called a *hemiepiphyte*.
3. *Halophyte or halophytic plant* : A plant that is able to grow in habitats excessively rich in salts or under saline conditions. Examples: nipa, talisay, bakawan (*Rhizophora mucronata*) and other mangrove species. Coconut, cashew and tamarind have varying levels of tolerance to saline conditions.
4. *Lithophyte or lithophytic plant* : A plant adapted to growing on rocks or in rocky terrain with little humus, absorbing nutrients from the atmosphere, rain and decaying matter which accumulate on the rocks. *Vanda, Ascocenda, Ascocentrum* and *Trudelia* orchids can be grown as lithophytes. *Dendrobium* has also been grown in pots filled with gravel or stone.
5. *Mesophyte or mesophitic plant* : A terrestrial plant which is adapted to moderate conditions for growth, i.e. not too dry and not too wet (e.g. corn and most commercially-grown crops).

6. *Parasite or parasitic plant* : a plant which grows on another plant from which it takes part or all nourishment e.g. *Cassytha, Loranthaceae* (mistletoe family), *Rafflesia; Neottia* and *Corallorhiza* orchids.

7. *Saprophyte or saprophytic crop* : Plants grows on decaying organic matter and has no green tissue.

8. *Sciophyte or Sciophyte plant* : A plant that is adapted to low light intensity or shade, e.g. most ferns and mosses, black pepper, cacao, coffee, lanzones, mangosteen, hot pepper, gingers and many orchids can tolerate or require shade.

9. *Terrestrial or land plant* : A plant which grows on land, rooting in the soil. It has aerial parts, collectively called *shoot* and an underground part called *root* which absorbs most of its water and nutrient needs from the soil. Most agricultural crops are terrestrial and are further sub classified into various groups such as halophytes, mesophytes, sciophytes and xerophytes depending on climatic and special adaptations.

10. *Xerophyte, xerophytic or xeric plant* : A plant which is adapted to conditions with little or no water. Examples: adelfa, bromyliads, euphorbias, cacti and many succulents.

G. Agricultural Classification of Plants

A. Crop vs. Weed

1. *Crop* : In agriculture, plants are classified as either crop or weed. A crop is useful plant, or a plant which is grown for any purpose. It is utilized by man directly or indirectly, raw or processed. Direct utilization is exemplified by the use of cereal grains as staple while they are indirectly consumed by man when they are first fed to livestock for meat, or processed into sugar for human consumption. These plants are intentionally grown or managed for various uses such as for food, spices, sugar, drinks, fiber, clothing, beverages, medicines, oils, tannins, rubber, ornamentation, fuels, construction materials, soil improvement, organic farming, landscaping and many more.

2. *Weed* : Unlike crop, a weed is an unintended plant or any plant which grows where it is not wanted. It competes with the intended crop for space, soil water, nutrients and light or becomes a nuisance in any manner. Many have allelopathic effect on other plants, both crop and weed, by releasing harmful allelochemicals which are considered plant-produced herbicides. Any plant can be a weed, depending on where it grows and how man is affected. Corn is a crop and commercially grown

world wide. But it may become a weed if it grows in a garden plot which is intended for the growing of leafy vegetables or any crop. A tree may also become a weed if it becomes invasive, prevents light penetration, and suppresses the growth of certain crops including pasture grasses.

B. Agronomic vs. Horticultural Crops

On the basis of tradition, extent of cultivation and intensity of culture, agricultural crops are classified into two main divisions: agronomic and horticultural.

1. *Agronomic crops are also called "field crops"* : They are mostly annual herbaceous plants that are grown under extensive or large-scale culture. The usable products are usually in high dry matter form. By tradition, cereals, seed legumes, root and tuber crops, sugar crops, latex and rubber crops, pasture and forage crops and fibre crops are classified under agronomic crops.

2. *Horticultural crops have been referred to as "garden crops"*: Included in this classification are those grown under any of the fields of horticulture such as olericulture or vegetable crops production, pomology or fruit crops production, and ornamental horticulture (floriculture and landscape horticulture), as well as spices and medicinal plants. Horticultural crops are annual and perennial plants which are grown under an intensive system of culture. Horticultural products are generally utilized with high moisture content and are therefore, highly perishable. From Janick (1972), these crops can be defined as "intensively cultured plants directly used by man for food, for medicinal purposes, or for aesthetic gratification". Intensive culture means a large input of capital, labour and technology per unit area of land.

However, there is no exact demarcation between agronomic and horticultural crops. No definite distinctions can be made of the different crops based on the area of cultivation and intensity of culture. The extent of cultivation varies from place to place and depends on such factors as the level of mechanization, adoption of technological advances, farm size, market stability and availability of capital. Changing market demands and prices may also cause shifts in the preferences of farmers and in the intensity of care to be adopted.Both agronomic and horticultural crops are further sub classified into specific groups according to the primary purpose of growing them or their intended major uses. Based on this criterion, a crop species may have different classifications. For example, corn (maize) is an agronomic crop if it is grown for its edible seeds which are harvested mature, or grown for silage; but it is horticultural if utilized primarily as vegetable as in the case of young corn and sweet corn. Another example is pineapple. Pineapple is generally classified as horticultural being commercially grown as a fruit crop but it becomes agronomic when the primary purpose of growing it is for the extraction of fibre which is used in making piña cloth. Papaya is a fruit crop but it can also be grown for vegetable or as a source of the industrial product papain. The basis of classification, therefore, should be "primary usage," disregarding the other minor uses.

A summary of the distinctions between agronomic and horticultural crops is further given in the following table from Janick (2005).

Table 2.2 : Differences between Horticulture and Agronomical crops

S.No.	Characteristic	Horticultural crops	Agronomic crops
1	Ultimate consumers	Human Typically	animals
2	Appearance	Important	Non-critical
3	Taste	Critical/subjective	Non-critical
4	Final product	Often consumed fresh in living state	Consumed processed or dried
5	Market type	Choice (non-commodity)	Commodity
6	Calories	Low density	High density
7	Vitamins and minerals	High	Low
8	Price per unit	High	Low
9	Total value per crop	Relatively low	High
10	Variability of cvs	High	Low
11	Breeding objectives	Quality, appearance	High yield, yield stability

C. Agricultural Classifications Based on Primary Uses

1. *Food Crops* : Plants grown primarily for the harvesting of any part which is used by man as food or processed into food product. This classification is a collective term for crops which are variously sub classified into smaller groups such as cereals, root and tuber crops, legume seed crops, sugar crops, beverage crops, fruit crops and vegetables.

2. *Non-food Crops* : Plants grown for the production of non-food products such as fibre, fodder, alcohol, tobacco, industrial oil, rubber, gums and resins, drugs, etc. or for ornamentation. Included in this classification are fibre crops, pasture and forage crops, rubber crops, latex and gum crops, dye and tannin crops, bio fuel crops, essential-oil crops, biocidal crops and most industrial crops.

3. *Staple Crops* : Plants grown for the harvesting of parts which are used as staple food. A staple food is one that is regularly consumed in such quantities as to form the basis of a traditional diet and from which people obtain a major proportion of their energy and nutrient requirements. In the India the top staple crop is rice followed by Wheat. Examples of other staple crops: banana and plantains, breadfruit, millet, white potato, sweet potato, yam.

4. *Cereal or Grain Crops* : Annual, herbaceous plants belonging to the grass family Gramineae (Poaceae) and some graminoids which are grown for their seeds or grains. The grains are harvested mature and utilized primarily as staple or feed or processed into carbohydrate-rich products. The word "cereal" is derived from Ceres, the ancient Roman goddess of harvest. The grain of cereals is technically a type of single, dry, indehiscent fruit called caryopsis, which has a hard outer pericarp fused to the seed coat. The endosperm predominantly consists of starch.Examples: corn, millet, rice, sorghum, wheat. According to FAO (2010) definition: Cereals are annual plants, generally of the gramineous family, yielding grains used for food, feed, seed and industrial purposes, e.g., ethanol.

5. *Legume Seed Crops or Pulses* : Leguminous plants which produce edible, protein-rich seeds. The seeds are utilized in the mature, dry form as food or feed or processed into various products. However, protein quality is inferior compared to meat because methionine content is generally insufficient. According to FAO (2010) definition; Pulses are annual leguminous crops yielding grains or seeds used for food, feed and sowing purposes. The denomination "pulses" should be limited to crops harvested for dry grain only, excluding, therefore, crops harvested green for forage, used for grazing or as green manure, and also crops harvested green for food (green beans, green peas, etc.), which are considered vegetables.

6. *Root and Tuber Crops or Tuberous Crops* : Plants with modified, swollen root or underground stem. Examples of modified roots are the tuberous and fleshy roots while the tuber and corm are examples of modified stems. A tuberous root is a thickened secondary root as in arrowroot, cassava, sweet potato and yam bean. A fleshy root is usually an enlarged primary root, as in carrot, ginseng (*Panax* spp.) and sugar beet. The upper portion on which secondary roots develop is hypocotyl or the first internode of the stem. In raddish, the fleshy root consists mainly of the hypocotyl. A tuber is an enlarged tip of an underground stem with leaves reduced to scales or scars subtending the auxillary buds, as in white potato and yam. The "eyes" represent buds in nodes, arranged in

spiral pattern from base to the apical end of the tuber. Aerial tubers are called tubercle. The corms are usually flattened from top to bottom with numerous roots at the lower part and a tuft of leaves on the upper. These crops are capable of producing high yields. However, they generally contain low amount of protein, minerals and vitamins than cereal crops. According to FAO (2010) definition:Roots and tubers grow generally as annual crops and yield roots, tubers, rhizomes, corms and stems which are used largely for human food, either as such or in processed form, but also for animal feed. In certain countries, they are used to manufacture starch and alcohol.

7. *Oil Seed Crops* : Plants grown for their seeds which are rich source of edible and industrial oil. The important oil seed crops in the world include soybeans, peanut, sunflower, oil palm, sesame and cotton.

8. *Sugar and Sweetener Crops* : Plants grown primarily for the production of sugar or other sweet-tasting products. Examples: sugarcane, sugar beet (a temperate crop), sweet sorghum, *Stevia*, corn, sweet potato, cassava, rice, many palms.

9. *Beverage Crops* : Plants which are sources of various drinks including fruit juices, tea, coffee, cocoa, toddy, beer and wine. They supply water which is essential to human nutrition. Some of these drinks also provide vitamins and minerals. Others have stimulating or relaxing effects. Examples: cacao, coconut, coffee, soursop, tea.

10. *Rubber Crops* : Plants grown for the production of latex which is processed into the industrial product called rubber. Rubber is an elastic hydrocarbon polymer constructed of isoprene units. It has widespread uses, from household to industrial products, with the main bulk in the transportation sector. Examples: para rubber tree (*Hevea brasiliensis*), Castilla rubber (*Castilla elastica*), Ceara rubber (*Manihot glaziovii*), guayule (*Parthenium argentatum*), Lagos Silk Rubber (*Funtumia elastica*).

11. *Latex and Gum Crops* : Plants grown for the collection of latex which is processed into chewing gum, in elastic rubber and other industrial products such as surgical tapes and dental supplies, insulation, splints, pipes, golf balls, waterproofing, adhesives, etc. Examples: chicle tree or chico (*Manilkara zapota*), gutta-percha (*Palaquium* spp.), balata (*Manilkara bidentata*), jelutong (*Dyera costulata*).

12. *Dye and Tannin Crops* : Plants grown as sources of tannin and colouring substances. Tannin is an aromatic, phenolic substance which is obtained from barks and other plant organs and variously used in tanning, medicines, dyeing, ink manufacture, etc. Examples: anatto or achuete,

Indian almond or talisay, indigo (*Indigofera tinctoria*), bakawan (*Rhizophora* and *Bruguiera*).

13. *Fibre Crops :* Plants grown as sources of fibre, a strong, thread-like material used in making textiles, rope, twine and similar materials. The fibre is extracted from the bark, leaves, or other organs including the husk of coconut. Examples: abaca, jute, kenaf, maguey and ramie.

14. *Pasture and Forage Crops :* Plants grown or managed as vegetable feed for grazing animals. They are classified as either native or improved species, grasses or legumes and may be fed fresh or dry or in processed form.

15. *Bio fuel Crops :* Plants grown for the production of fuel that is used as additive or replacement for petroleum products. The main biofuels are bioethanol, an alcohol derived from fermented sugar or starch and biodiesel from vegetable oils. Examples: sugarcane, cassava, corn, coconut, castor bean, Jatropha.

16. *Olericultural or Vegetable Crops :* Plants grown for their succulent and edible parts such as the roots, stems, leaves, young tops, fruits or seeds for use in culinary preparations either fresh or preserved in the fresh state. They are nearly all rich in vitamins A and C with high amounts of dietary fiber. According to FAO (2010) definition: Vegetables are plants cultivated both as field crops and garden crops, both in the open and under glass. Certain gramineous and leguminous plants which, if harvested for the dry grain, are classified among cereals and pulses, belong to this group as far as they are harvested green for the green grains and/or for the green pods (e.g., green maize, green peas, green beans, string beans, etc.). Moreover, only those vegetables which are cultivated principally for human consumption belong to this group. Consequently, vegetables grown principally for animal feed should be excluded, as should vegetables cultivated for seed. This group includes also melons and watermelons which some countries classify as fruit crops.

    Vegetables Classifications Based on Edible Part

    a. *Leafy Vegetables :* These crops are grown mainly for their leaves. They are rich in vitamins and minerals and they also look decorative.

    b. *Shoot Vegetables :* Plants grown primarily for their edible shoot, mainly the young, succulent stem.

    c. *Pod and Seed Vegetables :* Generally members of *Leguminosae* or *Fabaceae* family, these plants are grown for their young pods and seeds.

d. *Root and Bulb Vegetables :* Plants grown for their swollen underground roots and stems. Examples: carrot, potato, onion, raddish.

e. *Flower Vegetables :* Plants with edible flowers. Examples: horse raddish tree, katuray (*Sesbania grandiflora*), squash, rose, sunflower.

f. *Fruit Vegetables :* grown for their fleshy, succulent fruits. Examples: ampalaya, eggplant, tomato, peppers, melons.

Vegetable Classifications by Family

a. *Cole Crops or Crucifers :* These vegetable crops belong to the Cruciferae or Brassicaceae (Mustard) family, with edible leaves or heads. Examples: cabbage, cauliflower, mustard, pechay, Chinese cabbage, raddish.

b. *Cucurbits :* Belong to the Cucurbitaceae (Gourd and Squash) family. They are grown mainly for their fruits but some have edible young shoots and flowers; also called Vine Crops. Examples: bottle gourd, charantia, cucumber, luffa, melons, squash.

c. *Legume Vegetables :* Members of the Leguminaceal or Fabaceae (Bean) family. The seeds are rich in protein. Examples: bush sitao, katuray, kidney bean, lima bean, pea, pole.

d. *Lilies :* Members of the Liliaceae (Lily) family. Examples: asparagus, garlic, onion.

e. *Solanaceous Crops :* Belong to the Solanaceae (Nightshade or Eggplant) family and are also called *Fruit Vegetables*. Examples: eggplant, tomato, peppers, white potato.

f. *Mushrooms :* These are edible fungi belonging to the division Basidiomycota (club fungi). The edible part commonly consist of an upright stalk and an umbrella-shaped cap.

17. *Pomological or Fruit Crops and Nuts :* Plants grown primarily for their edible fruits or closely related structures which, as a rule, are consumed raw. Fruits borne on trees are called tree fruits. Fruits borne on low-growing plants such as shrubs, vines, lianas and some herbs are called small fruits (e.g. grape, passion fruit, pineapple, strawberry).Fruits are often important sources of vitamin C and many contain carotene. They are high in cellulose but usually have little fat or protein and little to no starch. According to FAO (2010) definition: Fruit crops are those yielding fruits and berries which generally are characterized by their sweet taste and their high content of organic acid and pectin. Apart from strawberries, all fruits and berries are permanent crops, mainly trees, bushes and shrubs, but also vines and palms.

18. *Spice Crops* : According to FAO (2010) definition. "Spices are plants which, in one or the other of their components (rhizome, bark, fruits, berries, seeds, etc.), contain strongly flavoured and aromatic substances and for that reason are used mainly as condiments. Most of them are perennial. Spices are rich in essential oils which, in addition to be used in the food industry, are also used in cosmetic and medicinal preparations. The nutritive value of spices is insignificant, but their commercial value is high".

19. *Essential-oil Crops* : Plants grown for the extraction of essential oils which are volatile, aromatic substances for perfumery and other uses.Examples: citronella, eucalyptus, peppermint, sampagita.

20. *Ornamental Crops* : Plants which are grown primarily for decoration or landscaping or to be appreciated because of their attractive flowers or foliage. They are further classified as cutflowers, cut foliage, turfgrasses, groundcovers, hedges, accents, specimen plants, avenue trees, screens, topiaries, fillers and others. Floricultural crops are valued for their attractive flowers, foliage ornamentals for their leaves. Ornamrental crop may further be classify as :

a. *Lawn or Turf Grasses* : Grasses grown for aesthetic purpose in the landscape or for any outdoor recreational use. They are usually maintained at a low height.

b. *Cutflowers* : Plants grown for their attractive flowers with long shelf life. Examples: anthurium, chrysanthemum, gladiolus, orchids, rose.

c. *Cutfoliage* : Plants grown for their attractive foliage which are cut for floral decoration.

d. *Edge Crops* : Short statured plants grown to serve as barrier between the lawn and garden, to highlight gardens, or to create stand-alone gardens; also called border plants.

e. *Groundcovers* : Low-lying, aesthetically appealing plants grown in the landscape primarily to suppress weed growth and to control, retard or prevent soil erosion by covering and binding loose, bare soil. It is oftenly used *en masse* to produce a carpeting effect. Examples: cucharita (*Alternanthera versicolor*), *Cuphea*.

f. *Hedges* : Plants grown at the edges of pathways or boundaries and continuously pruned to knee-high height or upper but below eye level. Examples: Duranta, hedge bamboo, Chinese holly, dwarf santan, papua.

g. *Accents*:Plants with showy features distinct from the rest of the other plants. It immediately attracts attention and becomes a focal item in the landscape garden and at the same time provides the contrast which gives attention to other plants.

h. *Specimens* : Plants having showy features, or with unique characteristics which make them pieces of conversation or botanical curiosity, or otherwise desired as collector's item. They are ideally planted in isolation rather than massed with other plants and easily become focal point in the landscape. Examples of potential specimens: queen of flowering trees (*Amherstia nobilis*), palms, Mussaenda.

i. *Screens* : Plants grown to serve as barrier against sun rays, to conceal certain parts of the landscape, or to obstruct view. Examples: shrubs, trellised vines.

j. *Shade crops* : Generally trees, shrubs, trellised vines and lianas which are grown mainly to provide shade singly or with supporting trellis.

k. *Avenue Trees* : Trees and shrubs grown, more or less equidistant, beside roads and streets. Palms are also used.

21. *Biocidal Crops* : Plants containing organic compounds with pesticidal or anti-microbial properties. The effective parts are either directly applied or seeped in water for foliar spray. Many have been commercially exploited by extracting the active ingredients. Examples: chrysanthemum (*Chrysanthemum cineriaefolium*), kayos (*Dioscorea hispida*), lagtang, makabuhay (*Tinospora crispa*), tobacco, tubli (*Derris elliptica*), neem.

22. *Industrial Crops* : Plants grown to provide materials for industrial processing and production of non-food products such as bio fuel, sugar, rubber, starch, industrial oil, aromatic compounds, steroids, medicinal drugs, organic pestides, tannins and dye. This is a special classification based on the method of processing and the nature of the product (non-food) and not on the part of the plant which is harvested and used as raw material. Based on these criteria, both agronomic and horticultural crops can be classified as industrial crops.

D. Special-Purpose Classifications of Crops

There are other groupings of agricultural crops or crop-epithets which are commonly used but can not be appropriately placed under either agronomic or horticultural classification. These classifications do not consider the direct benefit of the crop to the ultimate consumer, nor to the extent of cultivation and the intensity of care required. These classifications are used to refer to plants having special advantages to the farmer himself in relation to his farming practices.

1. *Main Crop* : Any crop which is intended by the grower to become his main source of revenue.

2. *Nursery crop* : A plant which is temporarily grown in the nursery and later planted in the field or garden or used for ornamental display when it reaches the proper age and size (e.g. most fruit crops and ornamental crops.
3. *Intercrop* : Any crop which is planted simultaneously with or before the flowering season of the main crop in intercropping.
4. *Filler Crop* : Any crop which is planted to fill a gap. In quincunx system of planting arrangement, the space at the centre of four hills of a main crop is commonly intended for a filler crop which is of a different variety or species.
5. *Relay Crop* : The crop which is planted after the flowering period or harvest of the main crop in relay cropping.
6. *Ratoon crop* : The crop consisting of the re-growth from shoots retained on the plants after harvest or from cut stalks of the previous crop (e.g. pineapple, sorghum, sugarcane).
7. *Cash crop* : Crop which is grown to generate income while the main crop is still in its vegetative stage of growth; any crop grown to generate cash rather than for subsistence.
8. *Catch crop* : Short maturing plant that is grown simultaneously with, or between successive plantings of a main crop to utilize residual fertilizer and soil moisture. It is often used as a green manure or to provide supplemental livestock feed; also called emergency crop.
9. *Nurse crop* : Crop which is grown to provide shade and increase humidity for the benefit of the main crop during its seedling and early stages of growth.
10. *Companion crop* : Any crop which is planted close to the main crop to complement the latter's growth and production, or to maximize utilization of space because they do not compete. Examples:
    a. Black pepper is planted with live *Gliricidia sepium* to serve as trellis.
    b. Lettuce acts as a living mulch to keep the soil moist while sunflowers to the south or west provide shade for the vegetable.
    c. Planting squash and onions will maximize production per unit area of land. These crops occupy different root zones, squash being deep rooted while onion is shallow rooted.
11. *Cover crop* : A crop grown mainly to control soil erosion, regulates soil temperature, control weeds and reduce evaporative losses. Leguminous

vines such as improved pasture and forage crops are excellent cover crops.

12. *Green manure crop* : Leguminous crop grown to be ploughed under the soil to increase organic matter and serve as organic fertilizer. Mungbean (mungo) is ideal for this purpose.

13. *Agroforest Crop* : Crop which is suited under a cropping system consisting of mixed agricultural and forest crops.

14. *Contour Hedgerow Crop* : Nitrogen-fixing trees and shrubs which are grown along contour lines in sloping lands. These crops are grown mainly to produce green manure and mulch, to serve as firebreak, to stabilize the soil and to control soil erosion. Examples: *Gliricidia sepium,Leucaena latisiliqua, Leucaena diversifolia,Flemingia macrophylla ,Desmodium renzonii,Calliandra calothyrsus,Senna spectabilis.*

15. *Trap crop or Decoy crop* : Plants grown to attract certain insect pests or parasites because they are favourite hosts. They act as decoys to lure pests away from the main crop. Also, they make pest control easier because the insects are concentrated on a few plants. Control measures will be concentrated to the infested plants including spraying of appropriate pesticides or these plants can be simply pulled out and destroyed together with the pests. Examples of trap crops and the insect pests that they attract:

    a. Basil (*Ocimum basilicum*)- green, loopy tomato caterpillar.

    b. Garlic (*Allium sativum*)- greenfly.

    c. Chive (*Allium schoenoprasum*)- greenfly and cutworms.

    d. Corn (*Zea mays*)- cotton bollworm.

    e. Pot Marigold (*Calendula officinalis*)- caterpillars and cutworms.

    f. Nasturtium (*Tropaeolum majus*)- aphids.

    g. Tarragon (*Artemisia dracunculus*)- loopy caterpillar.

    h. Tobacco (*Nicotiana tabacum*)- cotton bollworm.

16. *Insect pest repellent crop*: Plants grown along the borders and at strategic places in the farm to repel insect pests because of their strong aroma and anti-herbivory properties. Examples: Anise or Aniseed (*Pimpinella anisum*), Basil (*Ocimum basilicum*), Chive (*Allium schoenoprasum*), Garlic (*Allium sativum*), Marigold or Amarillo (*Tagetes* sp.), Marjoram (*Origanum majorama*), Nasturtium (*Tropaeolum majus*), Rosemary (*Rosmarinus officinalis*), Sesame (*Sesamum indicum*), Thyme (*Thymus vulgaris*).

17. *Natural enemy's attractant crop*: Flowering plants grown at strategic places in the farm to attract natural enemies of insect pests. Examples:

    a. Dill (*Anethum graveolens*)- can be grown to attract spiders, lacewings and parasitic wasps which eat or parasitize caterpillars, beetles and aphids.

    b. *Tagetes, Calendula* and Nasturtiums- attract hoverflies whose larvae feed on aphids.

Leaf Morphology and Plant Identification

Angiosperm Leaf Types :Leaf contributes many useful characters for identification at the family, genus and species levels. There are two basic types of leaf, Simple leaf and Compound leaf.

Simple Leaf vs. Compound Leaf

The position of the bud determines whether the leaf is simple or compound. In the case of the single leaf the bud is found in the axil of the leaf and stem. If the bud is located in the axil of a structure containing more than one leaf it is termed compound. Compound leaves may have from three to 1500 leaflets. Ex: *Acer* with three or *Albizia julibrissin* with 400 to 1500 leaflets.

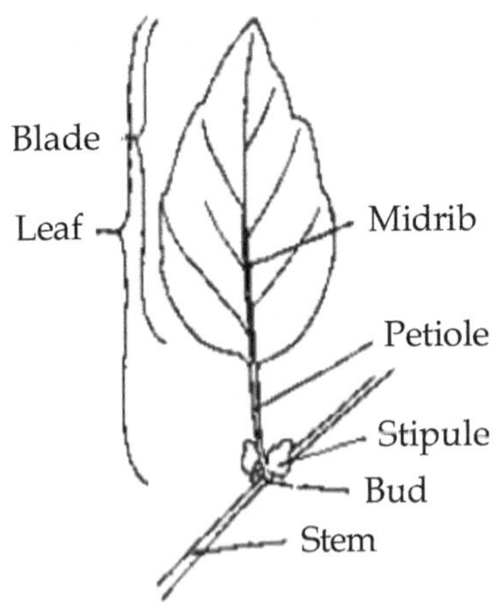

Fig. 2.1 : Simle Leaf

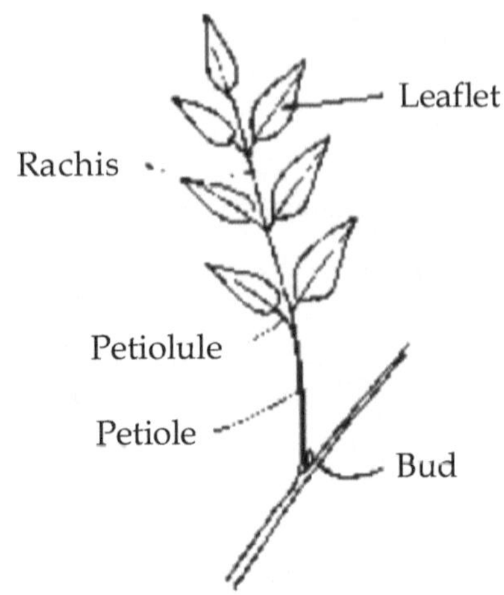

Fig. 2.2 : Compound Leaf

Gymnosperm Leaf Types

Cone-bearing or naked seeded plants often display different leaf types than those associated with angiosperm plants. Not all conifers (or cone-bearers) have evergreen foliage (exceptions include *Taxodium, Metasequoia, Larix,* and *Pseudolarix*).

Palmate
Ex: *Acanthopanax, Parthenocissus*

Odd Pinnate
Ex: *Acer negundo, Fraxinus*

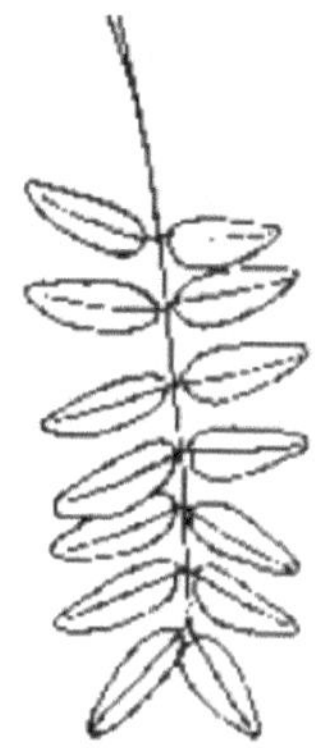

Even Pennate
Ex: *Gleditsia*

Fig. 2.3 : Variation in Compound Leaves

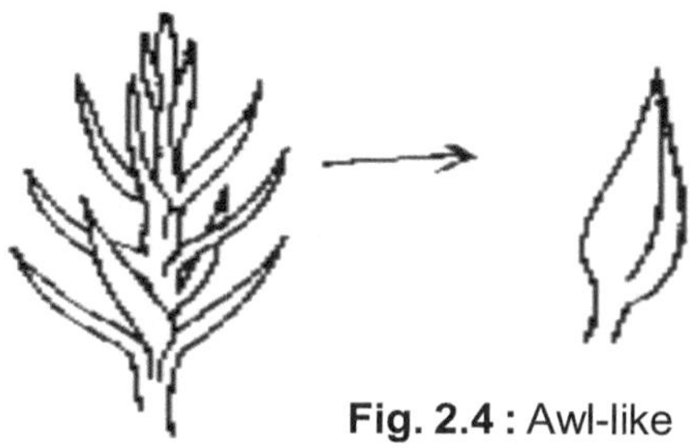

Fig. 2.4 : Awl-like

The needles (leaves) are shaped like an awl. They are usually very sharp to the touch.

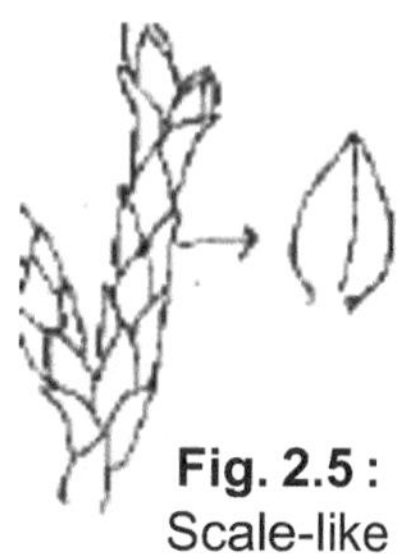

Fig. 2.5 : Scale-like

Many *Juniperus* (Junipers) exhibit awl-shaped foliage. This character is manifested in Juvenile forms of juniper; however, there are many species and cultivars (*Juniperus communis, J. Procumbens, J chinensis* 'Pyramidalis' to name a few) which possess the awl-like of needle foliage in youth and old age.

Scale-like foliage overlaps like the shingles on a roof or the scales on a fish. This type of foliage is relatively soft to the touch. *Thuja, Chamaecyparis, Cupressus, Calocedrus* and many *Juniperus* species exhibit this type of foliage.

Needle-like foliage is typical of several evergreen genera and species. In the genus *Pinus* the leaves (needles) are usually contained in fascicles of 2, 3, 2 and 3 or 5. Other species such as *Abies, Picea,Cedrus, Pseudotsuga,* and *Taxus* have the needles borne singly or in clusters along the stem. The needles may be relatively flat (2-sided) or angular (often quadrangular) in cross-section.

Arrangement of Leaves

Leaf arrangement is a very convenient character for use in identification keys and is often a constant feature in a family. The commonest arrangement, generally termed alternate, is one in which only one leaf is found at each node. Two leaves are found opposite each other at a node they are called opposite. Often the stem bears two sets of opposite leaves at right angles to each other resulting in a stem which has four angles and is square in cross-section. If more than two leaves occur at a node the arrangement is described as whorled. This arrangement must not be confused with the situation where the leaves are in irregular clusters, such an arrangement is called fasciculate. When leaves are whorled they occur all round the stem while fasciculate leaves are usually clustered on one side of the stem. The use of the four categories by the student allows him/her to categorize plants into groups and assists in eliminating many plants from consideration in the process of positive identification.

Opposite

Leaves and buds directly across from each other on the stem. Ex: *Acer, Lonicera, Deutzia, Viburnum.*

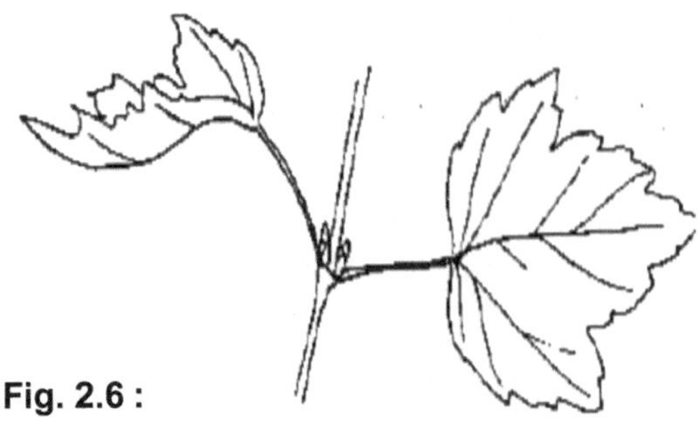

Fig. 2.6 :

Alternate : Leaves and buds are spaced in alternating fashion along the axis of the stem and seldom, if ever, are seated directly across from each other. Ex: *Betula,Fagus, Quercus, Celtis, Ulmus, Carya, Juglans.*

Fig. 2.7 :

Subopposite

Subopposite refers to a condition where the leaves and buds are not spaced sufficiently far apart to be considered alternate nor are they perfectly opposite, hence, the term subopposite. Ex: *Rhamnus cathartica, Cercidiphyllum japonicum, Chionanthus virginicus.*

Fig. 2.8 :

Whorled

Whorled refers to a condition when three buds and leaves (or more) are present at a node. Ex: *Catalpa,Hydrangea paniculata* 'Grandiflora', *Cephalanthus occidentalis.*

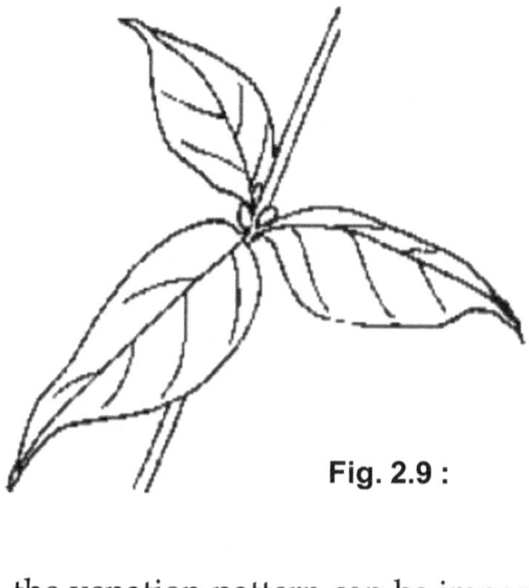

Fig. 2.9 :

Leaf Venation

In simple leaves, the venation pattern can be important in identification. Most dicotyledons have a pinnate venation pattern in which there is a midrib and a series of secondary veins to either side of this. In palmate or digitate venation three or more main veins arise from one point at the base of the lamina. Parallel venation is typical of many families of monocotyledons. The venation patterns found in simple leaves are paralleled by the arrangement of leaflets in compound leaves. When compound leaves are pinnate, the leaflets are arranged oneither side of the rachis. The end of the rachis may end in a pair of leaflets, in which case the leaf is described as even-pinnate or paripinnate, or in a single leaflet, when it is described as odd-pinnate or imparipinnate. In some groups of plants, such as *Acacia* and *Albizia* of the Fabaceae (Leguminosae), each leaflet is replaced by a rachis bearing small leaflets, i.e. the leaf is twice compound and is termed bipinnate. In pinnate leaves,each leaflet is termed a pinna, while in bipinnate leaves, the leaflets are called pinnules.A special and distinctive case of imparipinnate is the trifoliolate leaf which has three leaflets. This situation must not be confused with a group of three leaves, trifoliate, at a node. When the leaflets are arranged as in the fingers of a hand, the arrangement is called digitate or palmate and twice compound leaves with this arrangement are termed bidigitate or bipalmate. The base of the leaf stalk can be associated with a pair of small flattened appendages called stipules. Stipules can be small or large, attached firmly to the leaf stalk as in manyRosaceae, or falling off as or soon after the leaf unfolds as in *Ficus* (Moraceae). If stipules are associated with opposite leaves they may join-up across the stem forming aninter petiolar stipule. This is a distinctive feature of the Rubiaceae where such interpetiolar stipules may develop to the same size as the leaves, as in *Galium*, or be reduced to a line or linear scar. Stipules might be fused to surround the entire stem above the node and such a structure is called ochrea or stipular sheath. The base of apetiole or sessile leaf can surround or tightly wrap a part of the stem, and such a base is called sheath, while the free flat part of the leaf is called blade or lamina. In compound leaves, each leaflet may bear a pair of small stipules. Stems, leaves, leaflets and stipules can all be modified to form specialized structures. Tendrils are slender coiling organs. Sharp leaves are called spines, sharp shoots are called thorns and sharp outgrowths at the sides of stems or petioles are called prickles. In some plants, notably those adapted for extreme ecological conditions, leaves may be greatly reduced or lacking. The functions of the leaf can then be carried out by leaf-like structures called cladodes or cladophylls if they are modified stems, or phyllodes if they are modified leaf stalks. Cladodes are found in many Cactaceae such as *Opuntia* while phyllodes are typical of Australian species of *Acacia* introduced into Ethiopia, e.g. *Acacia melanoxylon*. Other extreme adaptations of leaves include the large petal-like bract, the spathe, typical of the Araceae. The leaves associated with inflorescences are usually reduced and/or modified and are refered to as bracts. Bracts may be colourful and petal-like as in *Bougainvillea* and Poinsettia, *Euphorbia pulcherima*, or highly reduced as phyllaries enclosing and surrounding the capitulum of the Asteraceae (Compositae) and the glumes enclosing the florets of Cyperaceae and Poaceae.

Both texture and the covering, indumentum, of leaves and other parts can assist in identification, particularly at the genus and species level. But these characters can also be useful at family level. Thick, leathery-textured leaves are typical of the families Balanitaceae and Salvadoraceae: star-shaped (stellate) hairs are a conspicuous feature of plants belonging to the Malvaceae, Tiliaceae and some genera of Solanaceae. Smell is another very useful character in identification. Some families are easily recognised by their smell, for example the Lamiaceae (Labiatae) and Apiaceae (Umbelliferae) contain many plants used as spices while the Burseraceae is the family which provides incense and myrrh. The juice or sap produced by a plant can be important for identification: plants of *Euphorbia* have milky sap while in the Cactaceae the sap is clear.

Pinnate

The leaf has a prominent central vein (often termed the mid rib) which extends from the base, where the petiole attaches to the blade, to the apex of the leaf. If the interveinal areas were removed the overall effect would be that of a fish bone. Pinnate venation occurs in the leaves of many plant types. The elm (*Ulmus*) and oak (*Quercus*) are classic examples.

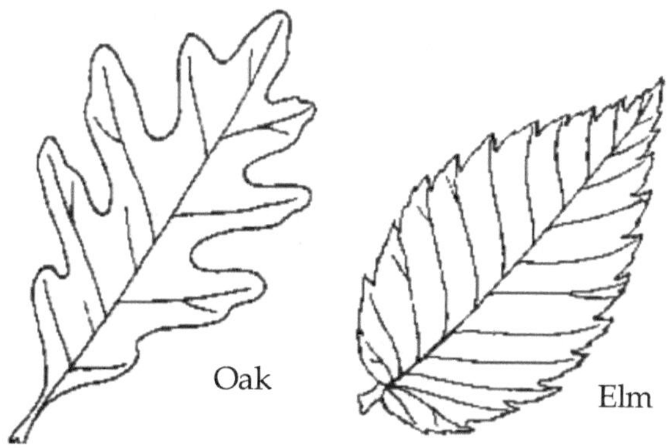

Fig. 2.10 :

Palmate

There are several main veins all of approximately equal size which extend from the base of the leaf to the apex of the lobe or margin of leaf. Ex: *Acer, Platanus, Cercis*

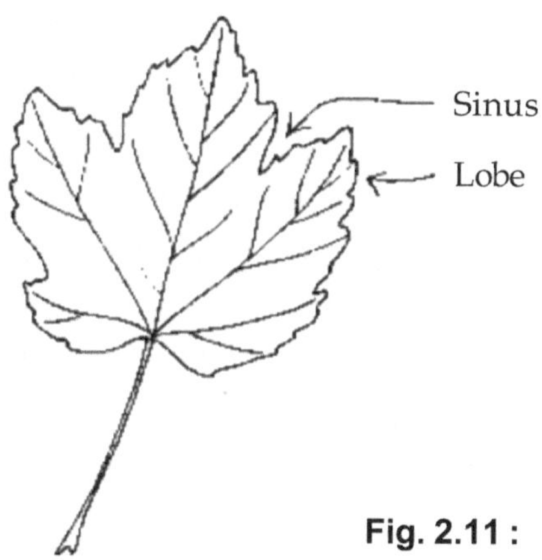

Fig. 2.11 :

Dichotomous

A very limited type of venation, the most familiar representative of which is *Ginkgo biloba*. The basal veins extend for a distance and then branch forming a "Y" type pattern.

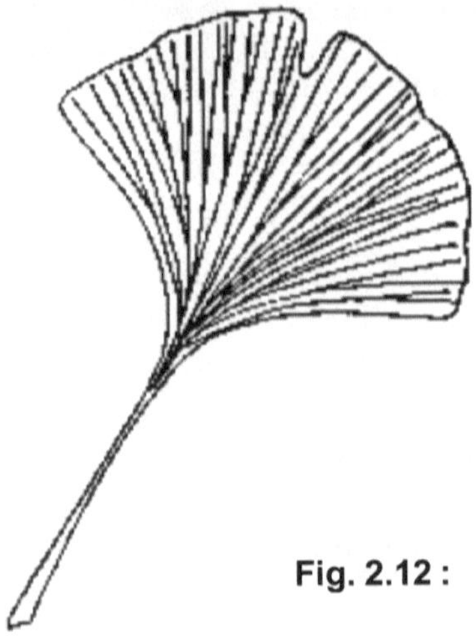

Fig. 2.12 :

Parallel : Typical of many monocotyledonous plants. The veins run essentially parallel to each other along the long axis of the leaf. Ex: *Ruscus, Danae*.

Inflorescence and Plant Identification

A flower may be solitary if it is produced by itself from the axil of a leaf or at the end of a branch. However, flowers are usually grouped together into inflorescences which can be either terminal and/or axillary. The stalk to an individual flower is termed a pedicel while that of an inflorescence is a peduncle. Leaves in an inflorescence are called bracts. They are usually smaller than the ordinary vegetative leaves but might also be large and coloured. Characters from inflorescence types are more important in identifying genera and species than families, but there are some families with distinctive inflorescence types. These are the capitulum, where many small flowers are densely grouped together to resemble a single flower (typical of the *Asteraceae* (*Compositae*) and *Dipsacaceae*); the umbel where the small flowers are all borne at the same level (typical for *Araliaceae*);several umbels can, in their turn, be arranged into a larger umbel, called compound umbel (typical of the Apiaceae (*Umbelliferae*)); the spadix with its thickened axis bearing densely crowded and reduced flowers inside the spathe (typical of the *Araceae*); the spikelet of a few to several very small and reduced, wind-pollinated flowers (typical of the *Cyperaceae* and *Poaceae*); and the fig (syconium) of unisexual flowers embedded in a swollen receptacle (typical of *Ficus*).

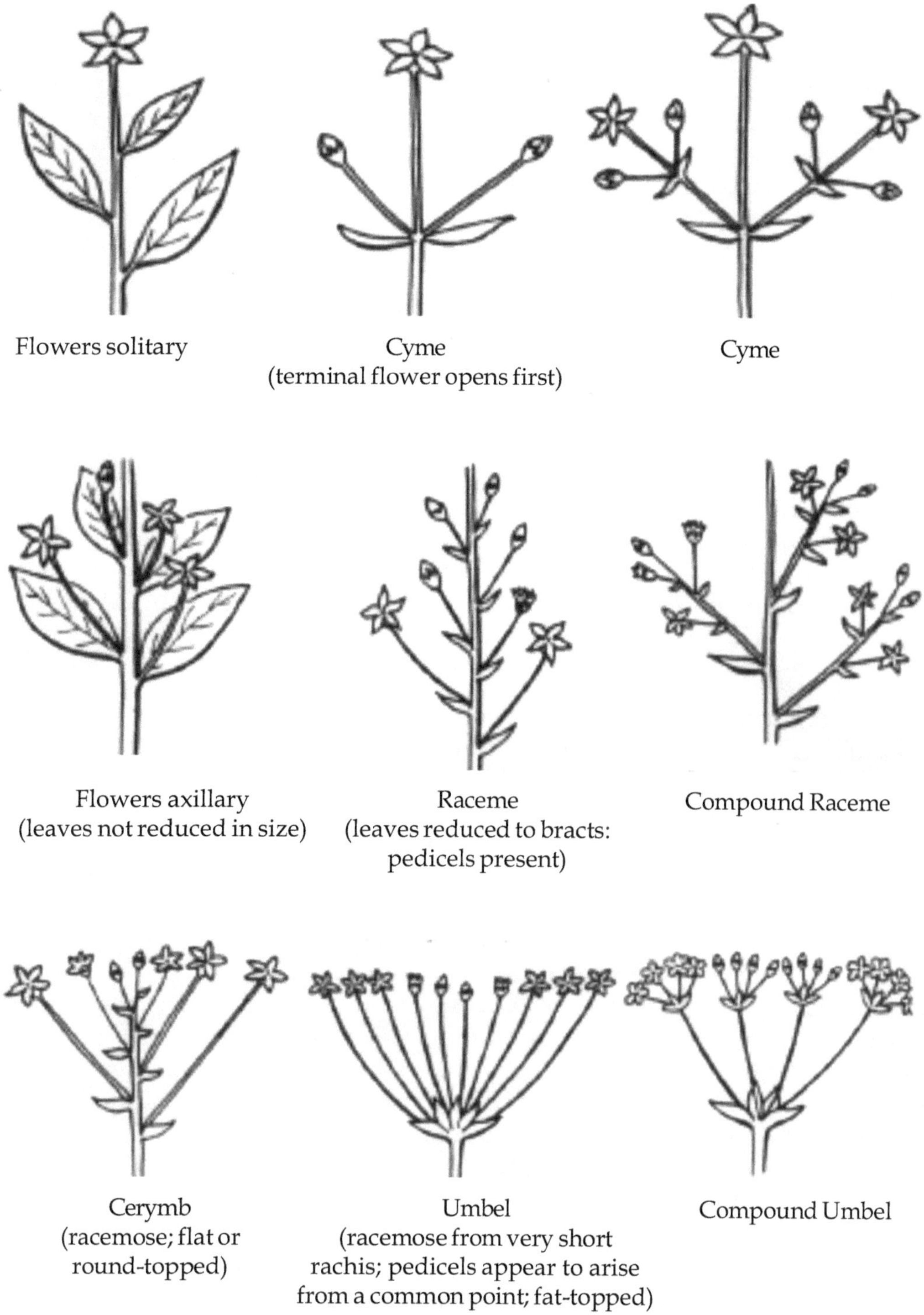

Fig. 2.13 : Inflorescence (Positions and Arrangements of Flowers)

Flowers

A plant may have three kinds of flowers, male female and bisexual (hermaphrodite). Neutral flowers can also be found in some plants; these are particularly frequent in ornamental plants which have been selected by man for their showy flowers. Monoecious plants have unisexual flowers of both kinds on the same plant. This situation is typical of some families including the Euphorbiaceae and Urticaceae. In dioecious plants an individual plant is either male or female. This situation can cause special problems in identification as the male and female plants may appear some what different from each other or one may find plants of only one sex which restricts the number of characters available for the identification exercise. Dioecious plants are characteristic of the Cucurbitaceae and Salicaceae and are common in families of woody plants more typical of temperate than tropical regions. Many botanists believe that the flower has been derived from a modified branch which has been greatly compressed and its leaves modified into the parts of the flower. Thus plants which have an elongated central axis, the receptacle or torus bearing numerous, spirally arranged parts are considered primitive. Such flowers are found in the Magnoliaceae, Annonaceae, Nymphaeaceae and Ranunculaceae. In the majority of flowering plants the receptacle is flattened or even concave with the parts arranged in whorls, each with a fixed and definite number of components.

Flowers are very important in for identification because they typically provide characters that are consistently expressed within a taxon (either at the family, genus, or species level). This is because floral characters are under strong genetic control and generally are not affected by changing environments. Certain floral characters may remain the same throughout a family or genus, while other characters are more variable and are used only to differentiate species. For example, floral symmetry, ovary position, type of placentation and kind of fruit, usually are used to differentiate families or genera, while petal color or shape and size of floral parts are more commonly used to distinguish species. Flowers arise from the apical portion of a stem in a region called the receptacle. They may be borne directly on a main stem axis or rachis (sessile) or on a slender stalk or stem called a pedicel. They usually consist of four whorls of parts that develop in the following series, from the outer whorl to the inner: sepals, petals, stamens and carpels.

Perianth (sepals + petals)

The accessory or sterile parts of a flower are called the perianth. If all the parts of the perianth look alike they are called tepals. This situation is found in some monocotyledonous families like the Liliaceae, Amaryllidaceae and Orchidaceae. In most dicotyledons the perianth is made up of two distinctly different whorls - an outer calyx composed of sepals which are usually green or dull-coloured and protect the flower in bud and an inner corolla composed of petals which are usually colourful and arranged to attract pollinators. The individual sepals and petals may be free from each other, then the flower is described as polypetalous or polysepalous, or united forming a sepal and/or petal tube, when the terms are gamopetalous/sympetalous or gamosepalous/synsepalous.Similarly all parts of the perianth may be fused to form a perianth tube. When the fused, tubular part of the flower appears to be formed from receptacle tissue carrying sepals, petals and stamens, it can be referred to as a hypanthium as found in Punica, the pomegranate. The receptacle can form a pad of tissue around the base of the ovary or from the base of the stamens. This is called a disc and may or may not be responsible for the production of nectar. The receptacle can also form additional structures within the flower, a gynophore is a stalk-like structure found under the ovary while an androphore is a column bearing only stamens (in male flowers) and an androgynophore is a column bearing both stamens and pistil. All these structures are found within the perianth and are always associated with a superior ovary. Other important characters of the perianth useful in identification of families are the position of the parts in the bud (aestivation), and their symmetry when the flower has opened. Contorted aestivation is having several parts in a whorl or close spiral with one margin overlapping and the other under lapping the adjacent structures, while in convolute one perianth part is rolled inside another and usually twisted apically. Imbricate is a general term for all arrangements where the parts overlap each other while in valvate, the margins of the adjacent parts touch only at the edges.

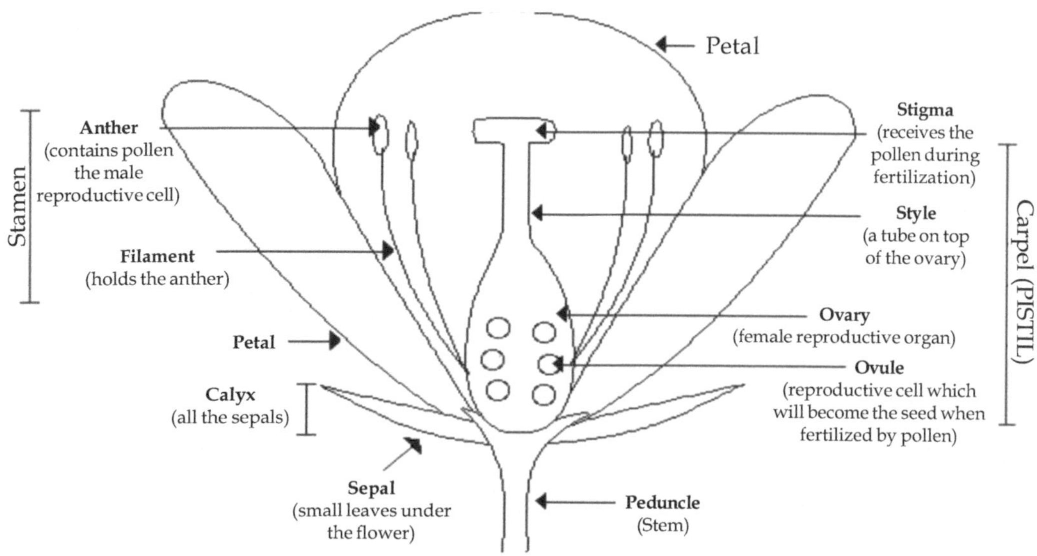

Fig. 2.14 :

A few flowers have no plane of symmetry (amorphic) or all the parts are spirally arranged (haplomorphic), but most can be divided along one or more lines into two equal and opposite halves.When a flower is only bilaterally symmetrical it is termed zygomorphic or irregular while actinomorphic or regular flowers are radially symmetrical.

Fig. 2.15 : Actinomorphic or radially symmetrical: regular flower

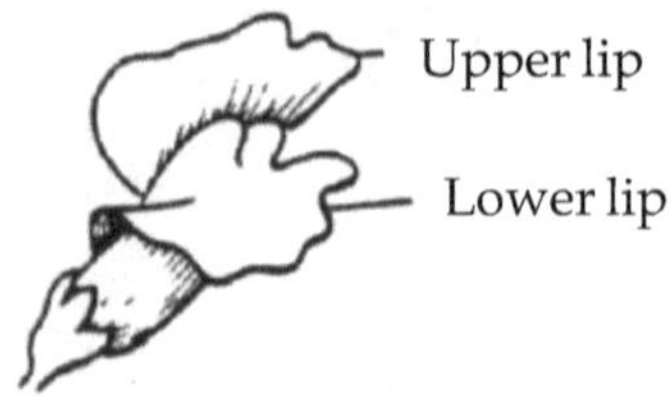

Fig. 2.16 : Zygomorphic or bilaterally symmetrical: irregular flower

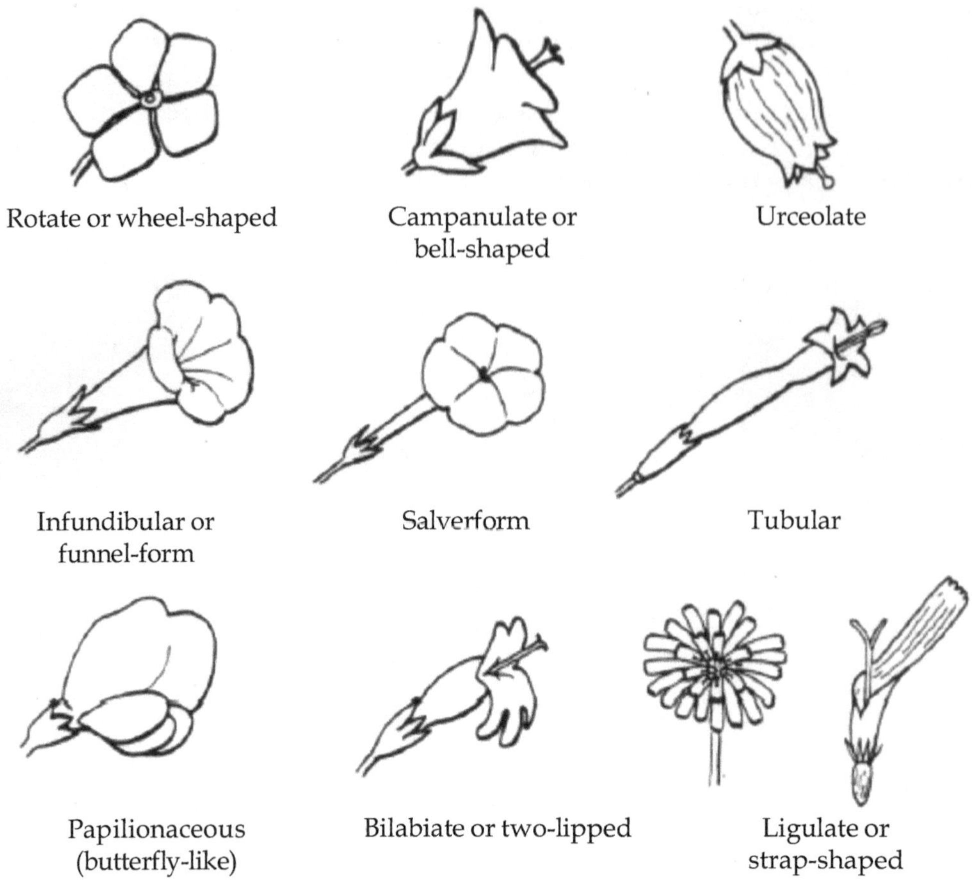

Fig. 2.17 : Corolla Shapes

Androecium

The androecium is made up of the stamens which produce the male spores (pollen grains) which contain the male gametes. Typically a stamen consists of a stalk -the filament - bearing an anther made up of 4 pollen sacs arranged in 2 thecae. The tissue between the thecae is called the connective. The filament may be attached to the base of the anther (basifixed) or the back of the anther. If the anther stays rigid it is termed (dorsifixed), but if it moves readily on the filament it is called versatile. Most the case open by longitudinal slits but in some families they open by transverse slits, or pores (Ericaceae and some Solanaceae and Viscaceae) or valves (Lauraceae). Stamens may have their filaments free or united. If united, this can be in groups as in Hypericum or to form a tube around the pistil as in the *Pisum* and *Vicia*. In the Asteraceae and *Lobelia* the filaments are free but the anthers are united (syngenesious). In epipetalous flowers the stamens are borne on the corolla and they may be sessile. The position of the stamens in relation to the perianth parts is important in identification: they may be antisepalous (opposite the sepals) or antipetalous (opposite the petals). When there are two whorls of stamens they may be diplostemonous with the outer whorl opposite the sepals and the inner opposite the petals, or obdiplostemonous in which the reverse arrangement is found. The stamens can be all the same length with the anthers borne at the same level or the lengths can vary. In the *Lamiaceae* (*Labiatae*) there are usually two long and two short stamens (didynamous) while in the *Brassicaceae* (*Cruciferae*) there are 4 long and 2 short stamens (tetradynamous).

Gynoecium

The gynoecium contains one or more pistils each made up of a basal ovary contain one or more ovules where the female gametes are produced and fertilization takes place. Pollen is caught on the stigma which is usually held by the style above the ovary in an appropriate position to receive pollen. The ovules develop into seeds while the ovary forms a true fruit (in a false fruit other plant parts are also included). The position of the ovary in relation to the other parts of the flower is important in both identification and classification of families.

These whorls of parts may develop in discrete cycles, or be more or less continuous and spirally arranged. The symmetry of flowers usually can be defined as radial (*actinomorphic*) or bilateral (*zygomorphic*). In a few groups the flowers are termed irregular (asymmetric) when the petals or sepals are dissimilar in form or orientation.

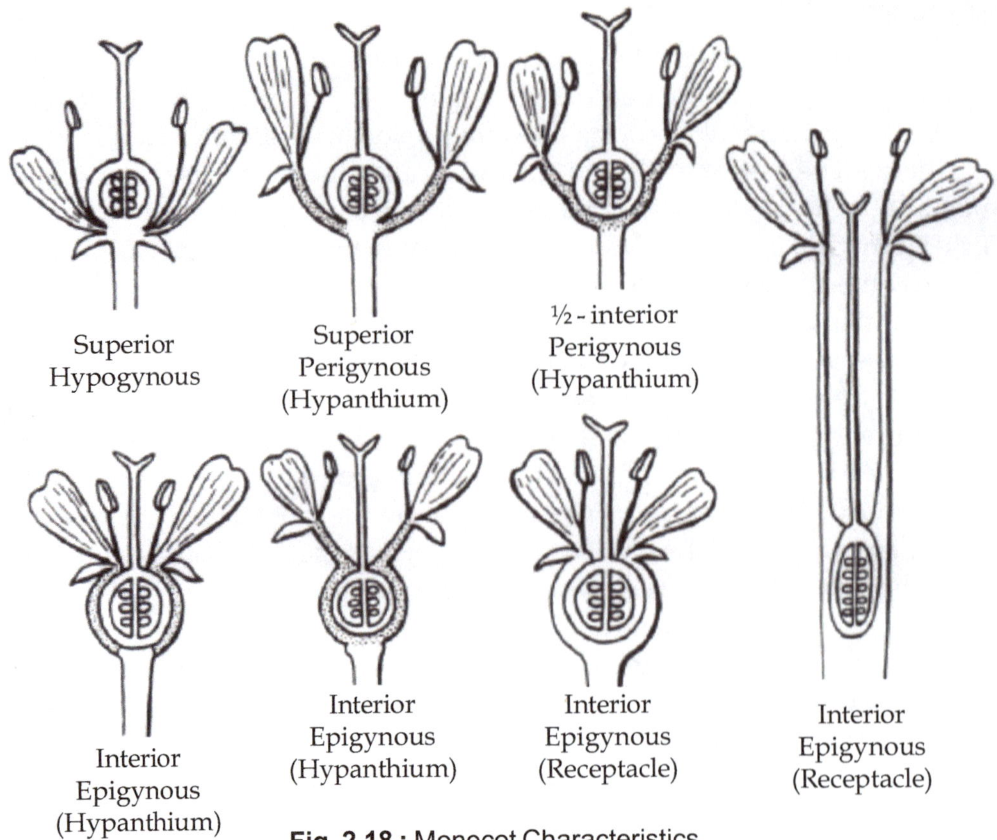

Fig. 2.18 : Monocot Characteristics

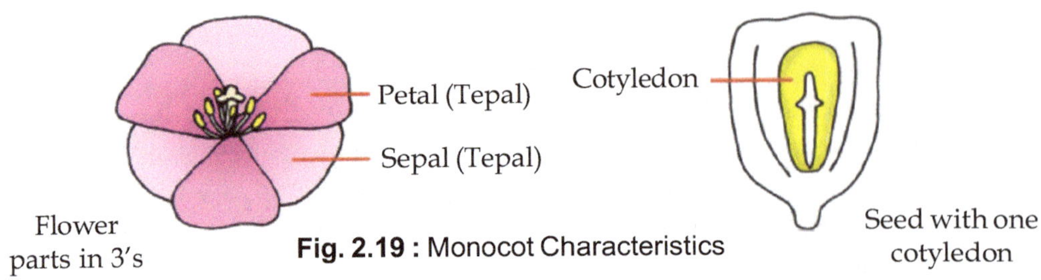

Fig. 2.19 : Monocot Characteristics

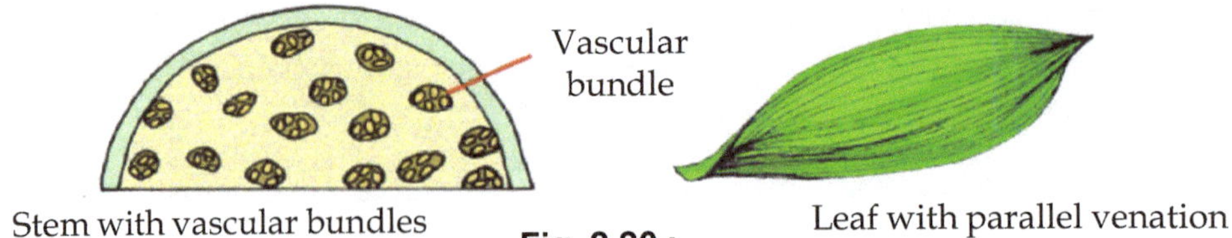

Fig. 2.20 :

Floral diversity is achieved by numerous variation patterns resulting from changes in : 1. Symmetry; 2. Numbers of each floral part; 3. Degree of fusion of the parts.

If same parts are fused, they are *connate*; if unlike parts are fused, they are *adnate*. Prefixes such as gamo-, sym-, and syn- denote connation, as in *gamopetaly* and *sympetaly* (fusion of petals to each other) and *syncarpy* (fusion of carpels). The prefix epi- refers to adnation, as in *epipetalous stamens* (stamens fused to the petals).

In floral morphology, the prefixes poly- and apo-represent the lack of fusion, as in *polypetaly* (separate petals) and *apocarpy* (separate carpels).

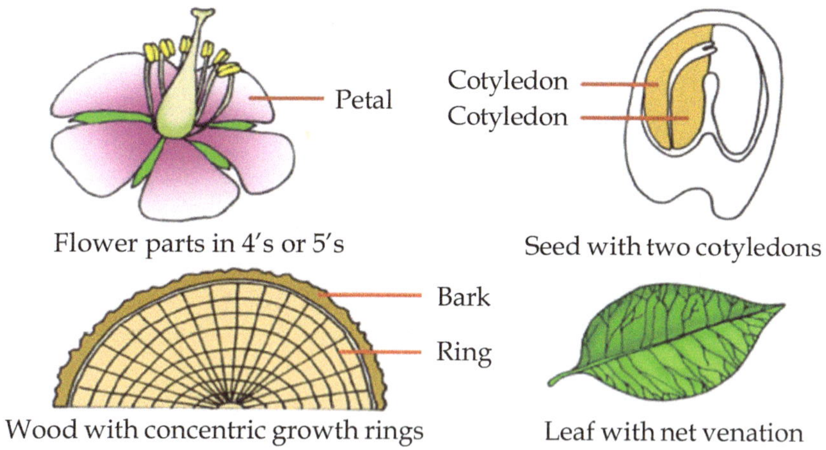

Fig. 2.21 : Dicot Characteristics

Fruits

Dehiscent, dry fruits include: capsule, follicle, legume, loment, schizocarp, silicle, and silique. A capsule is always derived from a compound pistil and splits open at maturity along one or more lines of dehiscence. A follicle is derived from a single carpel,has one or more seeds and splits open at maturity along the dorsal (front) line of fusion. Fruits of the Fabaceae are called legumes or loments and are derived from single carpels. A legume dehisces on both sides at maturity, while the loment breaks apart into one-seeded segments. A schizocarp splits apart at maturity into one-seeded parts(mericarps) that represent its separate carpels. Silicles and siliques are specialized capsules found only in the Brassicaceae; each of these has a membranous partition that separates the two locules and each is formed from a two-carpellate gynoecium. A silicle usually is less than two times longer than wide, while a silique usually is more than two times longer than wide. Fleshy fruits may be formed either from a simple or compound ovary and include: berry, drupe, pome, and sometimes follicle or capsule. A berry may have few or many seeds embedded in a pulpy matrix with a thin outer wall. A drupe is derived from a simple or compound ovary and has one or more seeds enclosed in a stony endocarp (theinner wall of the pericarp or ovary). A pome is formed from a compound pistil in an inferior ovary; the fleshy part of the pome is derived from the pericarp and hypanthium. Follicles, derived from single carpels and capsules, derived from compound gynoecia, are typically dry fruits, but may be fleshy in certain genera or families.

Fleshy fruits derived from several ovaries include the aggregate and multiple fruit types. In the aggregate fruit, the unicarpellate ovaries of a single flower are coherent at fruit maturity, as in blackberries and raspberries. Multiple fruits are formed by the coalescence of several flowers, as in mulberry or pineapple. Fruits may serve as the dispersal unit or they may dehisce, allowing the seeds to be dispersed. The unit of dispersal, either fruit or seed, may be transported by wind, water, or an animal. Fruit types generally are consistent within a genus and often within a family, there by providing useful characters for plant identification.

To be able to key flowering plants, it is important to become familiar with the nature of variation shown by all parts of a flower (sepals, petals, stamens, and carpels). To be able to key flowering plants, it is important to become familiar with the nature of variation shown by all parts of a flower (sepals, petals, stamens, and carpels). Learn the terms on the following diagram.

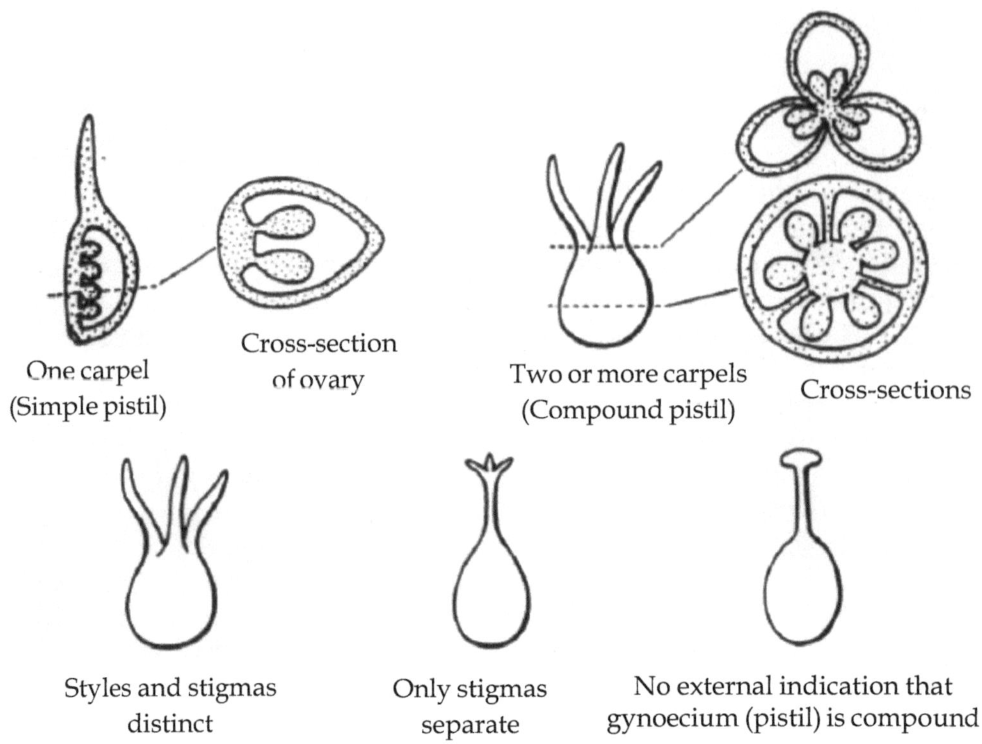

Fig. 2.22 : Carpels or Pistils

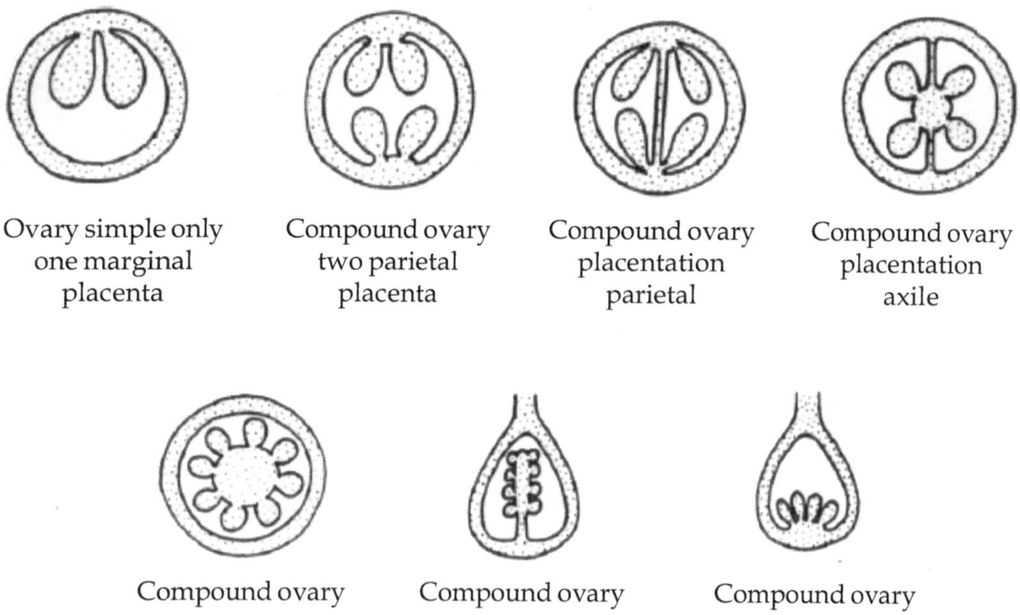

Fig. 2.23 : Type of Placentation

Leaf Shapes

The tremendous quantity of terminology related to leaf shapes can be confusing. Association of the following pictures with the terms will help to alleviate the burden of strict terminology. This also applies to leaf bases, margins, and apices.

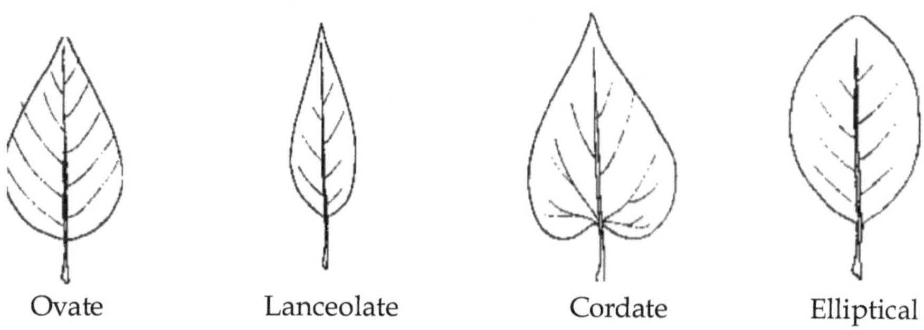

Fig. 2.24 : Various shapes of leaf

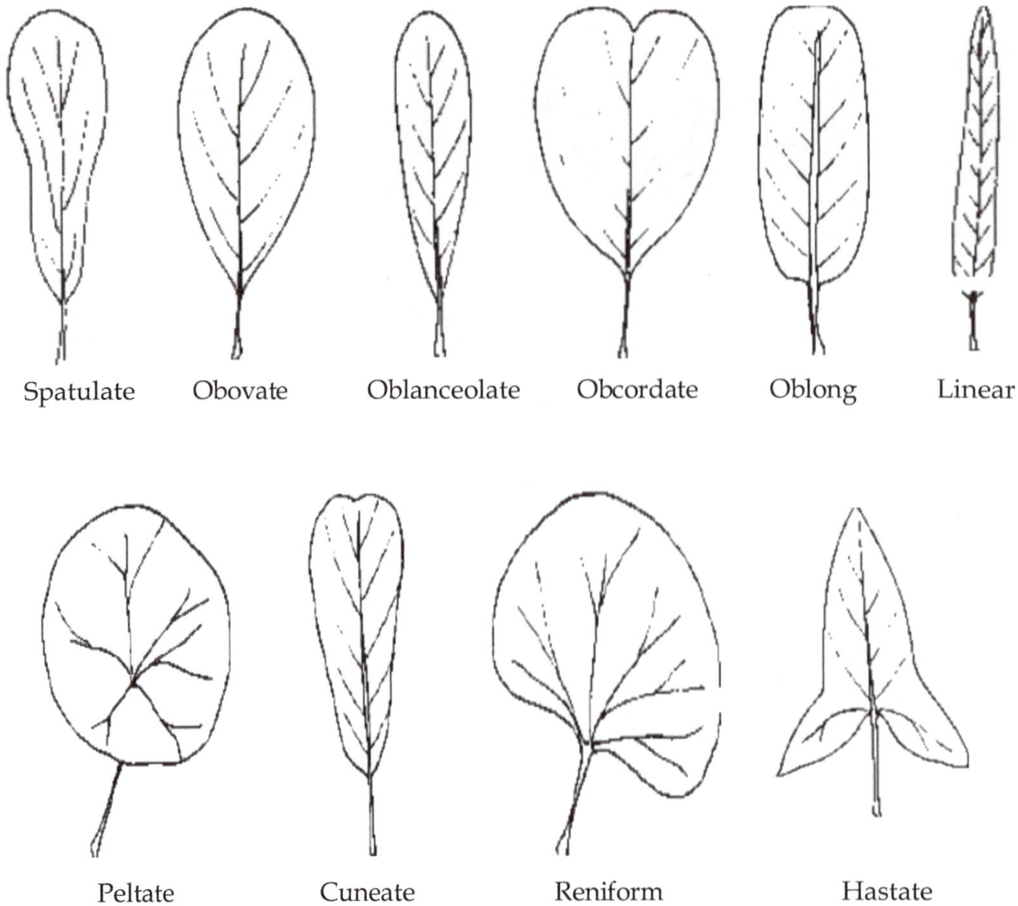

LEAF BASIS

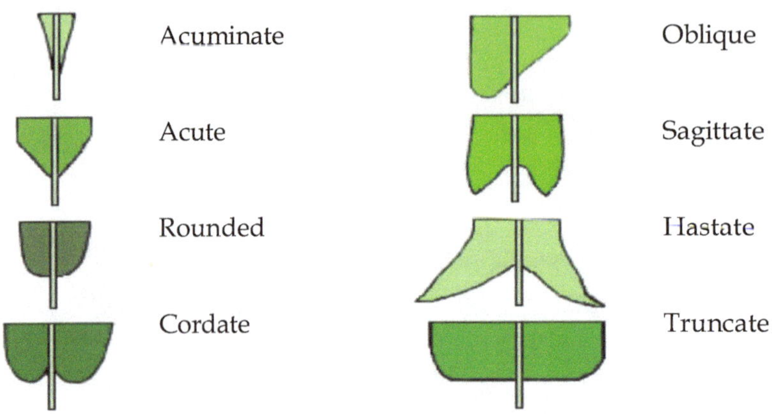

Fig. 2.25 : Various types of leaf basis

LEAF MARGINS

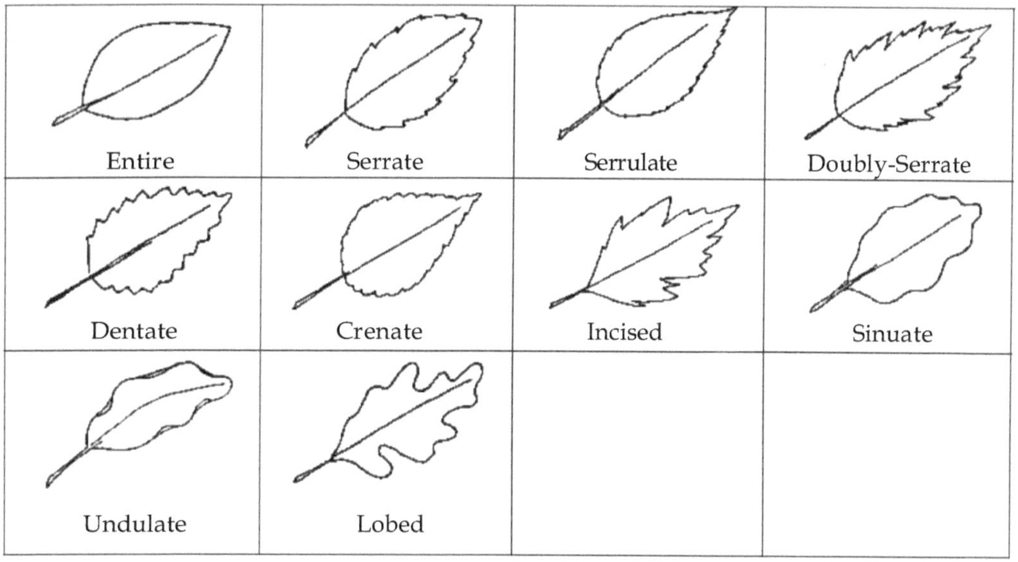

Fig. 2.26 : Various types of leaf margins

LEAF APICES

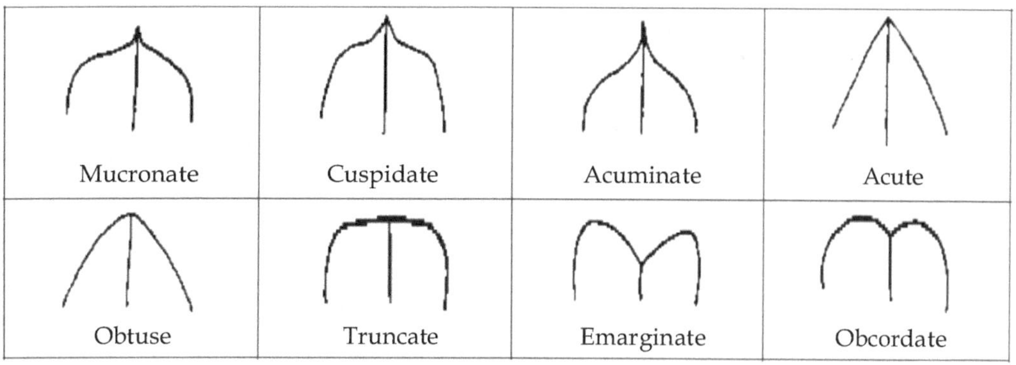

Fig. 2.27 : Various types of leaf apices

Guidelines for Completing Plant ID Sheets

Names

Family Name:

Origin:

Scientific Name:

Common Name:

Cultivar:

Environmental Requirements

1. Temperature

Cool (C) .... 10-15°C.

Medium (M) .... 15-20°C.

High (H) .... 21-26°C.

2. Light

Low (L) .... Minimum 0.50 Klux (recommended 0.70-1.5 Klux).

Medium (M) .... Minimum 1.0 Klux (recommended 2.0 Klux).

High (H) .... Minimum 2.0 Klux (recommended 5.0 Klux).

Very High (VH) Minimum 5.0 Klux (recommended 10+ Klux).

3. Moisture

Dry (D) Let dry completely between watering.

Moist (M) Keep uniformly moist but not wet.

Wet (W) Never let soil dry out.

4. Humidity

Low (L) Up to 40% RH.

Medium (M) 60-70% RH.

High (H) 70-80% RH.

5. Medium

Heavy (H) High in soil for wet conditions.

Medium (M) Well drained, moist conditions.

Light (L) Dry, sandy conditions.

Plant Characteristics

1. Plant Type

Tree (Single-stem, Multi-stem) (Tr).

Shrub (Shr).

Ground Cover (GC).

Vine (V).

2. Shape/Form

Upright (Up) Oval (Ov)

Spreading (Spr) Round (R)

Cascading or Weeping (Cas) Irregular (Ir)

Climbing (Cl) Pyramidal (Py)

3. Plant Size

Very Tall (VT) .... Greater than 3 meter

Tall (T) .... 2.5-3.0 meter

Medium (M) .... 2-2.5 meter

Short (S) .... 0.5-2.0 meter

Creeping (C) .... Shorter than 0.5 meter

4. Growth Rate

Slow (S)

Medium (M)

Fast (F)

5. Plant Texture

Fine (F), Medium (M), Coarse (C)

List of Characters

Plant Habit

1. Woody plants (excl. suffrutices)
2. Herbaceous plants (incl. suffrutices)
3. Aquatic plants, leaves floating or submerged
4. Climbers, lianas
5. Chlorophyll absent (parasites or saprophytes)
6. Bulb present (monocots only)
7. Milky juice present
8. Spiny stems or leaves
9. Tendrils present
10. Cladodes or phyllodes (modified branches or petioles)

Vestiture

11. Hairs glandular
12. Hairs stellate (also 2-armed, branched and tufted)
13. Hairs stellate (not 2-armed, branched and tufted)
14. Hairs 2-armed or t-shaped, non-glandular
15. Hairs branched
16. Hairs tufted, non-glandular
17. Hairs peltate or scale-like
18. Hairs vesicular or bladder-like
19. Hairs stinging
20. Cystoliths present (dicots only)

Leaves

21. Leaves opposite or verticillate
22. Leaves alternate (excl. distichous monocots)
23. Leaves distichous (monocots only)
24. Leaves equitant (e.g. Iris)
25. Leaves peltate
26. Leaves not compound
27. Leaves compound
28. Leaves pinnately compound (4 or more leaflets)
29. Leaves ternately compound (3 leaflets)
30. Leaves palmately compound (4 or more leaflets)
31. Venation pinnate or hardly visible in leaves or leaflets (incl. no. 30)
32. Venation invisible or leaves 1-nerved (monocots only)
33. Venation longitudinal in leaves or leaflets (incl. 3-nerved leaves)
34. Venation palmate in leaves or leaflets
35. Leaves or leaflets with entire margin (without any lobes or teeth)
36. Leaves or leaflets lobed or divided
37. Leaves or leaflets dentate, serrate, crenate, etc.
38. Epidermis of leaf papillose (dicots only)
39. Leaves with pellucid or glandular dots or lines
40. Leaves with a ligule
41. Stipules absent
42. Stipules present (only the scars may be left)

Inflorescences

43. Flowers solitary
44. Inflorescence a raceme, simple and monopodial
45. Inflorescence a spike, simple and monopodial
46. Inflorescence a corymb, simple and monopodial
47. Inflorescence an umbel, simple and monopodial
48. Inflorescence a fascicle, simple and monopodial
49. Inflorescence a head, simple and monopodial
50. Inflorescence simple and sympodial (cyme, dichasium etc.)
51. Inflorescence compound, sympodial or monopodial (panicle, thyrsus etc.)

Flowers

52. Flowers bisexual
53. Flowers unisexual
54. Flowers actinomorphic
55. Flowers zygomorphic or irregular
56. Receptacle small (ovary superior)
57. Receptacle enlarged, united with the ovary, totally or partially covering it
58. Receptacle enlarged, totally or partially free from the ovary (super. to inf.)
59. Receptacle enlarged, conical or hemispherical (ovaries superior)
60. Disc present (annular or separate glands)
61. Disc absent
62. Perianth segments 0
63. Perianth segments 1
64. Perianth segments 2
65. Perianth segments 3
66. Perianth segments 4
67. Perianth segments 5
68. Perianth segments 6
69. Perianth segments more than 6
70. Perianth of similar segments

71. Perianth of calyx and corolla
72. Sepals 0 (incl. a cupular calyx without lobes)
73. Sepals 1
74. Sepals 2
75. Sepals 3
76. Sepals 4
77. Sepals 5
78. Sepals more than 5
79. Sepals all free from one another
80. Sepals connate (at least two of them)
81. Sepals imbricate or contorted
82. Sepals valvate
83. Petals 0 (incl. a cupular corolla without lobes)
84. Petals 1
85. Petals 2
86. Petals 3
87. Petals 4
88. Petals 5
89. Petals 6
90. Petals 7
91. Petals 8
92. Petals 9
93. Petals 10
94. Petals more than 10
95. Petals all free from one another
96. Petals connate (at least two of them)
97. Petals imbricate
98. Petals contorted
99. Petals valvate
100. Corona present, or scales in the throat
101. Anther 1, fertile
102. Anthers 2, fertile
103. Anthers 3, fertile

104. Anthers 4, fertile
105. Anthers 5, fertile
106. Anthers 6, fertile
107. Anthers 7, fertile
108. Anthers 8, fertile
109. Anthers 9, fertile
110. Anthers 10, fertile
111. Anthers more than 10, fertile
112. Stamens the same number as and opposite the petals or alternate with the *sepals*
113. Flowers obdiplostemonous
114. Anthers dorsifixed or ventrifixed
115. Anthers basifixed
116. Anthers introrse
117. Anthers extrorse
118. Anthers opening by longitudinal slits
119. Anthers opening by apical pores
120. Anthers opening by valves
121. Anthers 1-locular at anthesis
122. Stamens free from the corolla
123. Stamens inserted on the corolla
124. Filaments not connate
125. Filaments connate into 1 tube or bundle
126. Filaments connate into separate bundles
127. Anthers with a conspicuous appendage
128. Pollen coherent into pollinia or wax-like masses
129. Staminodia present (in male or perfect flowers only)
130. Gynophore
131. Styles absent, stigmas sessile
132. Styles more than 1, free (carpels connate)
133. Style 1, or styles more or less connate (carpels free or connate)
134. Style gynobasic
135. Carpel 1

136. Carpels 2 (free or united)
137. Carpels 3 (free or united)
138. Carpels 4 (free or united)
139. Carpels 5 (free or united)
140. Carpels more than 5 (free or united)
141. Ovary 1-locular (incl. apocarpae)
142. Ovary 2-locular
143. Ovary 3-locular
144. Ovary 4-locular
145. Ovary 5-locular
146. Ovary with more than 5 locules
147. Ovule 1 in each loculus
148. Ovules 2 in each loculus
149. Ovules more than 2 in each loculus
150. Carpels free from each other or carpel 1
151. Ovules parietal, carpels connate
152. Ovules axile, carpels connate
153. Ovules attached to central axis or ventrally if carpels free or if carpel 1
154. Ovules attached to the apex of the ovary
155. Ovules attached to the base of the ovary

Fruit

156. Fruit a capsule (incl. pod, follicle etc.)
157. Fruit a nut (incl. achene, loment, schizocarp etc.)
158. Fruit fleshy (berry, drupe, pome)
159. Fruit with 1 seed
160. Fruit with 2 seeds
161. Fruit with more than 2 seeds
162. Fruits with wings
163. Fruits with hairs for dispersal by wind
164. Fruits with retrorsely barbed, hook-like or glochidiate hairs
165. Fruits with spines and hooks
166. Seeds with wings

167. Seeds with hairs
168. Arillus or aril-like organs present
169. Embryo straight
170. Embryo curved
171. Cotyledons folded or crumpled
172. Seeds without endosperm
173. Seeds with endosperm
174. Endosperm ruminate

Distribution

175. African
176. Asian
177. Australian and Oceanian
178. European
179. North American
180. South American

Identification of major herbaceous plant families

1. Amaryllidaceae (AMARYLLIS FAMILY)

- Herbaceous, perennial monocots.
- Stem with bulb at base.
- Leaves linear, often grass-like,with parallel veins.
- Inflorescence an umbel.
- Tepals 6, anthers 6.
- Hypanthium (fused tepals and base of stamens) often present.
- Ovary inferior.
- Fruit from 3 carpels.
- Seeds often black, many.

Examples: amaryllis (*Hippeastrum*), *Crinum*, snowdrops (*Galanthus*), belladonna-lily (*Amaryllis*), spider-lily (*Lycoris*), daffodil (*Narcissus*)

Apiaceae Carrot or Ginseng Family

- Mostly herbaceous.
- Aromatic, some very poisonous (oils, resins).

- Stems hollow
- Leaves alternate, often dissected or lobed, pinnate venation
- Leaf petiole broadened with sheath surrounding stem or base of leaf
- Flowers arranged in umbels or double umbels; small, white or yellow, many
- Petals 5, not fused, sepals reduced or absent
- Fruit a dry fruit that divides into 2 parts (schizocarp)

Apiaceae now includes Araliaceae, and these characters onlywork on the temperate traditional Apiaceae, not all genera in former Araliaceae. Examples : parsley (*Petroselinum*), dill (*Anethum*), hemlock (*Conium*),cilantro (*Coriandrum*), celery (*Apium*), Queen Anne's lace/carrot(*Daucus*), English Ivy (*Hedera*), ginseng (*Panax*), caraway (*Carum*),cumin (*Cuminum*), fennel (*Foeniculum*), *Schefflera*.

Apocynaceae Milkweed & Dogbane Family

- Leaves **opposite**, simple, pinnate venation.
- Leaf margin smooth.
- Stipules absent.
- Latex (milky sap) in all branches and leaves.
- Sepals 5, Petals 5, sometimes fused.
- Ovary superior.
- Anthers often fused, and sometimes fused with style head to a gynostegium pollen in pollinia in some species.
- Fruit usually with 2 separate carpels, developing into one-two dry capsular parts or berries.
- Seeds often with tufts of hairs at one end.

Asclepiadaceae is now included in Apocynaceae.Examples: dogbane (*Apocynum*), milkweed (*Asclepias*), rosy periwinkle(*Catharanthus*), oleander (*Nerium*), frangipani (*Plumeria*),hoya (*Hoya*).

Asteraceae Aster & Sunflower Family

- Herbaceous usually.
- Leaves variable, with pinnate venation.

- Inflorescence a head (involucre) with many flowers, with bracts surrounding it.
- Flowers small, either tubular or tongue-shaped (ligulate).
- Sepals absent.
- Petals fused, usually with 5 small lobes.
- Anthers fused.
- Ovary inferior.
- Fruit a dry nut (achene), often with hairs on top (pappus).

Examples: Echinacea (*Echinacea*), dandelion (*Taraxacum*), burdock (*Arctium*), mugwort/wormwood (*Artemisia*), chrysanthemum (*Dendranthema*), ox-eye daisy (*Leucanthemum*), asters (*Aster*, etc.), thistles (*Cirsium, Carduus*), sunflower (*Helianthus*), artichoke (*Cynara*).

Brassicaceae (Mustard Family)

- Herbaceous.
- With mustard oils.
- Leaves alternate, with pinnate venation.
- Leaf edge often dentate or lobed.
- Petals 4, not fused.
- Petals forming a cross + from above.
- Petals white, yellow, or pink.
- Stamens 6.
- Fruit a dry capsule with inner wall (silique).

These key characters only work for Brassicaceae in the traditional sense.Examples: mustard (*Sinapis*), garlic mustard, horse radish (*Armoracia*), cabbage/broccoli/brussels sprouts/kale collards/rutabaga/canola/black ustard/turnip (*Brassica*), arugula (*Diplotaxis*).

Caryophyllaceae Carnation & Pink Family

- Herbaceous.
- Leaves opposite , simple, with pinnate venation.
- Leaf edge smooth.
- Stems often with thickened nodes , at base of each leaf pair.
- Sepals 5, fused.
- Petals 5, not fused.

- Fruit a dry capsule opening at top.
- Seeds attached to central column inside capsule.
- Seeds many, black, often curved.

Examples: Carnation (*Dianthus*), *Agrostemma*, *Stellaria*, soapwort (Saponaria).

Cyperaceae Sedge Family

- Herbaceous monocot.
- Stems often as rhizomes and upright culms.
- Stems 3-sided, without nodes, solid, not hollow.
- Leaves linear, grass-like, with parallel veins, arranged at 3 angles (tristichous).
- Leaves sheathing at base.
- Inflorescence often divided into male and female parts, as spikelets on terminal branches.
- Flowers small, unisexual, sitting behind a bract.
- Sepals and petals absent (rarely present).
- Anthers 3, hanging free.
- Ovary superior, often inside abottle-shaped structure (perigynium).
- Fruit a small, 1-seeded nut.

Examples: sedge (*Carex*), papyrus (*Cyperus*)

Fabaceae Bean Family

- Mostly herbaceous.
- Leaves alternate, compound (with many small leaflets), sometimes with tendrils.
- Stipules at base of each leaf (variable in size).
- Corolla of 'butterfly-type', bilateral with 5 parts: banner/standard, wings, keel.
- Keel hidden between wings.
- Stamens and style hidden inside keel.
- Stamens 10 , 9 of these are fused.
- Fruit a bean (legume), a dry capsule without inner dividing walls, and with seeds attached to one side.
- Seeds splits in 2, nutrients stored in dicotyledons inside seed.

These characters work only for subfamily Faboideae. Examples: beans (*Phaseolus*), peas (*Vicia, Pisum*), licorice (*Glychyrriza*), soybean (*Glycine*), chickpeas, peanuts (*Arachis*), lentil (*Lens*), sweet pea (*Lathyrus*), carob (*Ceratonia*), alfalfa (*Medicago*), clover (*Trifolium*).

Lamiaceae (Mint Family)

- Herbaceous aromatic, with essential oils, often hairy with glands in or on leaves or glandular hairs.
- Leaves opposite and simple.
- Stipules absent.
- Stem usually quadrangular.
- Flowers in groups in leaf axils or in terminal spikes.
- Sepals fused.
- Corolla bilabiate (2-lipped mostly).
- Stamens 2 or 4.
- Fruits: 4 nutlets hidden inside calyx.

Examples: basil (*Ocimum*), mint (*Mentha*), sage (*Salvia*), thyme (*Thymus*), rosemary (*Rosmarinus*), lavender (*Lavandula*), catnip (*Nepeta*), beebalm (*Monarda*).

Poaceae Grass Family

- Herbaceous monocots.
- Stems often rhizomatous and/or with erect culms.
- Stems round, hollow, with nodes.
- Leaves linear, with parallel venation, sheathing, with ligule at top of sheath.
- Inflorescences with terminal and axillary spikelets.
- Flowers without sepals and petals.
- Stamens and ovary hidden inside bracts (called glumes, palea, and lemma), when flowering styles and anthers hanging out from spikelets.
- Spike (awn) often present on bracts.
- Fruit a one-seeded nut (caryopsis).

Examples: corn/maize (*Zea*), wheat (*Triticum*), barley (*Hordeum*), rye (*Secale*), oats (*Avena*), rice (*Oryza*), bamboo (many genera), Grass (*cynodon*).

Rosaceae Rose & Apple Family

- Herbaceous or woody, often shrubs or trees.
- Stipules at base of each leaf.
- Sepals 5, petals 5 free.
- Hypanthium (bottom in flower from fused bases from petals, sepals, and stamens).
- Stamens usually numerous.
- Fruit type variable, often pome (apple-like), drupelets (blackberry-like), drupe (stone fruit), dry capsules, or dry nutlets.

Examples: Rose (*Rosa*), apple (*Malus*), pear (*Pyrus*), blackberry/raspberry (*Rubus*), Lady's mantle (*Alchemilla*), hawthorn (*Crataegeus*), Strawberrie

Plantaginaceae ('Veronicaceae') SPEEDWELL FAMILY

- Herbaceous (rarely shrubs).
- Hairy plants, often aromatic.
- Leaves alternate or opposite.
- Flowers bilateral (often 2-lipped).
- Stamens 4, 2+2 together.
- Ovary superior.
- Seeds attached to center of fruit (axile placentation).
- Fruit a dry capsule.
- Seeds numerous.

Examples: plaintain (*Plantago*), eyebright (*Veronica*), snapdragon (*Antirrhinum*), foxglove (*Digitalis*), turtleheads (*Chelone*), chinese house (*Collinsia*).

Solanaceae Tomato, Pepper, & Potato Family

- Herbaceous in temperate areas,often woody in tropics.
- Leaves alternate, often lobed, often hairy, sometimes with prickles.
- Stipules absent.
- Sepals 5, fused.
- Petals 5, fused a little or a lot, corolla star-, trumpet-, or funnel shaped or tubular.

- Ovary superior.
- Anthers often fused, porate (pepper shaker-type).
- Berry, drupe, or capsule.
- Seeds many, attached to center of fruit (axile placentation).

Examples: potato/tomato (*Solanum,* incl. *Lycopersicon*), chili pepper/ sweet pepper (*Capsicum*), angel trumpet (*Brugmansia*), tobacco (*Nicotiana*), petunia (*Petunia*), tomatillo (*Physalis*), jimsonweed.

Identification and description of major house plants

1. Shrimp Plant *Beloperone guttata* (also *Justicia brandegeana*)

Family Acanthaceae.
250 Genera of dicots-herbs or shrubs-perfect flowers.

Temp. Medium

Light High

Moist. Dry

Prop. Cutting

Keep plants on dry side. Cut back 1/3 of plants in the spring.

2. Dragon Tree of Madagascar *Dracaena marginata*

Family Agavaceae
20 Genera of monocots - leaves mostly narrow

Temp. Medium

Light Medium

Moist Moist

Prop. Tip cutting

Pointed leaves, sensitive to fluoride.

3. Snake Plant *Sansevieria trifasciata*

Family Agavaceae (also found it listed in Liliaceae family in two references).

Temp. Cool to high

Light low to high

Moist. Dry to medium dry

Pests- Dis. Mealybug, root rot if too wet

Prop. Division, leaf cutting

Excellent low light plant; used for terrarium, dish garden, atrium; slow-growing; durable; does best in peat containing soil.

4. Giant Dumbcane *Dieffenbachia amoena*

Family	Araceae - Arum or Philodendron family. 13 genera of monocots - herbs -stemless or erect and climbing stems; inflorescence spadix usually subtended by a spathe; genera include *Anthurium* plus others listed.
Temp.	Medium to high
Light	Medium
Moist.	Moist
Pests-	Dis. Mealybug, root rot and stem rot
Prop.	Tip cutting, stem cutting

Can be tall (4.5 m); susceptible to cold (at 12^{0}C); produces oxalic acid crystals that are toxic to skin (causes inflamation).

5. Heart-leaf Philodendron *Philodendron scandens oxycardium*

Family	Araceae
Temp.	Medium to high
Light	Low to medium
Moist.	Medium
Pests	Dis. Mealybug, root rots
Prop.	Tip or nodal cuttings

Very tolerant to all indoor conditions including low light; most comon of all philodendrons.

6. Golden Pothos *Epipremnum aureum*

Moist.	Moist
Family	Araceae
Temp.	Medium
Light	Medium
Moist.	Medium
Pests	Dis. Mealybug, root rot
Prop.	Tip and nodal cuttings; tissue culture

Used for terrarium, atrium, hanging baskets; easy to grow.

7. Peace Lily or White Flag *Spathyphyllum clevelandii*

Family	Araceae
Temp.	Medium to high
Light	Low to medium
Moist.	Moist
Pests	Dis. Mealybug, root rot
Prop.	Tissue culture, division

Very tolerant to low light conditions; use well-drained medium.

8. Hawaiian Schefflera *Brassaia arboricola*

(also *B. actinophylla, Schefflera arboricola*)

Family	Araliaceae 80 genera of dicots - herbs, shrubs, trees, vines
Temp.	Cool to medium
Light	Medium to high
Moist.	Medium
Pests	Dis. Spider mite, scale, root rot
Prop.	Seed, cutting

More bushy and compact than *B. actinophylla*; attractive shrub.

9. English Ivy *Hedera helix*

Family	Araliaceae
Temp.	Cool to medium
Light	Medium to very high
Moist.	Medium
Pests	Dis. Spider mite, scale
Prop.	Tip cutting, nodal cuttings

Used as for hanging basket, dish gardens; terrariums; does poorly in low light

10. Norfolk Island Pine *Araucaria heterophylla*

Family	Araucariaceae
	Two genera of gymnosperms. Large pine-like trees; high light; won't take a freeze

Temp. Medium

Light High

Moist. Moist

Pests Dis. Spider mite

Prop. Tip cuttings, seed

Used for specimen, atrium, terrarium; seasonal use as Christmas tree

11. Iron Cross Begonia *Begonia masoniana*

Family - Begoniaceae
3 genera of dicots; herbs with lopsided leaves

Temp. High

Light Medium to high

Moist. Moist

Pests Dis. Mealybug, aphid, whitefly, root rot, powdery meldew

Prop. Leaf sections

Used for terrarium, pot plants; best in cool greenhouse, need high humidity; attractive foliage accent.

12. Silver Vase *Aechmea fasciata*

Family Bromeliaceae - Pineapple Family

44 genera of monocots; mostly epiphytic; stiff leaves which are colored toward base; leaves basal or rosette forming to hold water

Temp. Cool to medium

Light Medium to high

Moist. Dry

Prop. Tissue culture, division

Floral accent, pot plant; flowers last 4 -6 months, easy to grow

13. Mammillaria Cactus *Mammillaria albilanata*

Family Cactaceae - Cactus Family

50 to 150 genera of dicots; succulents found in drier regions of tropical areas; leaves reduced to spines; flowers showy and solitary

Temp. High

Light	Very high
Moist.	Dry
Pests	Mealybug, root rot
Prop.	Seed, division

Very attractive; used for dish garden and as a pot plant; keep dry during winter months.

14. Jade Plant *Crassula argentea*

Family	Crassulaceae - Orpine Family 35 genera of succulent herbs or under shrubs; annuals or perennials; includes *Sedum sempervivum, Kalanchoe*
Temp.	Cool, medium, high
Light	Medium to very high
Moist.	Dry
Pests	Mealybug, scale aphid, spider mite; root and stem rot
Prop.	Stem cuttings

Very tolerant of all conditions; prefers high light; sensitive to salty soil; used for bonsai, specimen.

15. Crown of Thorns *Euphorbia splendens (E. milii splendens)*

Family	Euphorbiaceae - Spurge Family 290 genera of dicots with often milky sap; herbs, shrubs and trees; frequently cactus like
Temp.	Medium to high
Light	High to very high
Moist.	Medium
Pests-	Mealybug, root rot
Prop.	Stem cuttings

Thorny; very good indoor plant for warm sunny areas

16. African Milk Tree *Euphorbia trigona*

Family	Euphorbiaceae
Temp.	Medium to high
Light	High

Moist. Dry

Pests Spider mite, mealybug, root rot

Prop. Stem cuttings

Prefers dry, high-light areas; durable; latex poisonous.

17. African Violet *Saintpaulia ionantha*

Family Gesneriaceae

Moist tropical herbs and creepers; leaves frequently colored above or below and hairy

Temp. Medium to high

Light Very high

Moist. Medium

Pests Spider mite, cyclamen mite, root and stem rot

Prop. Tissue culture, seed, leaf cuttings

Most popular flowering house plant; avoid chilling.

18. Swedish Ivy *Plectranthus australis*

Family Lamiaceae - Mint Family

180 genera of dicots with square stems; aromatic

Temp. Medium to high

Light Low to high

Moist. Moist

Pests Mealybug, spider mite, root rot

Prop. Tip cutting; nodal cutting

Very easy to grow, does not like chilling; sap stains skin and clothing orange.

19. Variegated Spider Plant *Chlorophytum comosum* 'Vittatum'

Family Liliaceae - Lily Family

335 genera of monocots; herbaceous perennials; flowers often showy; grow from rhizomes, corms or bulbs; includes lily, daylily, tulip, onion, daffodil, hyacinth.

Temp. Cool to high

Light Medium to high

Moist. Medium

Pests Scale, mealybug, spider mite

Prop. Division, stolons; tissue culture

Good for hanging basket; sensitive to fluoride damage.

20. Sprenger Asparagus *Asparagus densiflorus* 'Sprengeri'

Family Liliaceae

Temp. Cool to high

Light High to very high

Moist. Dry

Pests Spider mite

Prop. Seed

Used for hanging baskets and cut foliage; very durable; easy to grow; does best under high light conditions.

21. Weeping Fig *Ficus benjamina*

Family Moraceae - Mulberry Family
53 genera of dicots; trees and shrubs - rarely herbs; milky sap

Temp. Cool to high

Light Medium to very high

Moist. Medium

Pests- Spider mite, mealybug, scale

Prop. Tip cutting, air layering

Most common indoor tree; used as specimen, bonsai, pot plant; avoid drying, drafts, low lights; requires acclimatization; grow up to 70 feet.

22. Rubber Plant *Ficus elastica*

Family Moraceae

Temp. Cool to high

Light Medium to very high

Moist. Moist

Pests Scale, mealybug

Prop. Air-layering, tip cuttings; tissue culture

Many different cultivars; used as specimen, pot plant; can be used for exterior.

23. Cattleya Orchid *Cattleya sp.*

Family	Orchidaceae - Orchid Family 735 genera of monocots; herbaceous perennials; epiphytic, saprophytic, or terrestrial; stem often with pseudobulbs and aerial roots; one of the largest families of flowering plants
Temp.	High
Light	Very high
Moist.	Medium
Pests-	Scale, spider mite, mealybug
Prop.	Division, protocorm culture; seed

Flower initiation takes place at 2000-3500 fc; a large number of hybrids exists; used for flowering pot or corsages; slow growing.

24. Variegated Peperomia *Peperomia obtusifolia* 'Variegata'

Family	Piperaceae tropical herbs and vines; small flowers without petals or sepals
Temp.	Medium to high
Light	High to very high
Moist.	Medium to dry
Pests	Mealybug, spider mite, root and stem rot
Prop.	Tip cuttings, leaf cuttings

Used for terrarium, dish gardens.

25. Dallas Fern *Nephrolepis exaltata* 'Dallas'

Family	Polypodiaceae - Ferns (Pteridophytes)-Common Fern Family most common ferns; no distinct trunk; nonflowering plants reproducing from spores
Temp.	Medium to high
Light	High to very high
Moist.	Medium
Pests	Spider mite, mealybug, scale, whitefly, root rot
Prop.	Division, tissue culture

Used for hanging basket, pot plant; burns in direct sun; more compact than the Boston fern.

26. Staghorn Fern *Platycerium bifurcatum*

Family Polypodiaceae

Temp. Medium to high

Light Medium to very high

Moist. Medium

Pests Spider mite, scale, root rot

Prop. Spore, division

Used for hanging basket or slab; epiphytic; attractive fronds.

Table 2.3 : Identification of Plants used for Interiors

S.No.	Scientific Name	Family Name	Common Name
1	*Beloperone guttata*	Acanthaceae	Shrimp Plant
2	*Dracaena marginata*	Agavaceae	Dragon Tree of Madagascar
3	*Sansevieria trifasciata*	Agavaceae	Snake Plant
4	*Dieffenbachia amoena*	Araceae	Giant Dumbcane
5	*Philodendron scandens oxycardium*	Araceae	Heart leaf Philodendron
6	*Epipremnum aureum*	Araceae	Golden Pothos
7	*Spathyphyllum clevelandii*	Araceae	Peace Lily or White Flag
8	*Brassaia arboricola*	Araliaceae	Hawaiian Schefflera
9	*Hedera helix*	Araliaceae	English Ivy
10	*Araucaria heterophylla*	Araucariaceae	Norfolk Island Pine
11	*Begonia masoniana*	Begoniaceae	Iron Cross Begonia
12	*Aechmea fasciata*	Bromeliaceae	Silver Vase
13	*Mammillaria albilanata*	Cactaceae	Mammillaria Cactus
14	*Crassula argentea*	Crassulaceae	Jade Plant
15	*Euphorbia splendens*	Euphorbiaceae	Crown of Thorns
16	*Euphorbia trigona*	Euphorbiaceae	African Milktree
17	*Saintpaulia ionantha*	Gesneriaceae	African Violet
18	*Plectranthus australis*	Lamiaceae	Swedish Ivy
19	*Chlorophytum comosum* 'Vittatum'	Liliaceae	Variegated Spider Plant
20	*Asparagus densifloris* 'Sprengeri'	Liliaceae	Sprenger Asparagus

21	*Ficus benjamina*	Moraceae	Weeping Fig
22	*Ficus elastica*	Moraceae	Rubber Plant
23	*Cattleya spp.*	Orchidaceae	Cattleya Orchid
24	*Peperomia obtusifolia variegata*	Piperaceae	Variegated Peperomia
25	*Nephrolepis exaltata* 'Dallas'	Polypodiaceae	Dallas Fern
26	*Platycerium bifurcatum*	Polypodiaceae	Staghorn Fern

Table 2.4 : Plants for Interiors Listing by Plant Families

ACANTHACEAE	
Aphelandra squarrosa	Zebra Plant
Beloperone guttata	Shrimp Plant
Crossandra infundibuliformis	Crossandra
Fittonia minima	Miniature Silver Nerve Fittonia
Fittonia vershaffeltii 'Pearlei'	Pink Nerve Fittonia
Fittonia vershaffelti 'Argyroneura'	White Nerve Fittonia
Graptophylum pictum	Caricature Plant
Hemigraphis 'Exotica'	Purple Waffle Plant
Hypoestes sanguinolenta	Polka Dot Plant or Freckle Face
Pachystachys lutea	Lollipop Plant
Jacobinia carnea	Pink Lollipop Plant
Pseuderanthemum atropurporeum tricolor	
Sanchezia speciosa	Tiger Plant
AGAVACEAE	
Agave americana	Century Plant
Agave angustifolia marginata	Variegated Caribbean Agave
Agave attenuata	Dragon Tree Agave
Beaucarnea recurvata	Ponytail or Elephant-Foot Tree
Cordyline terminalis	Baby Doll Dracaena
Cordyline stricta	Cordyline
Dracaena craigii 'Compacta'	Compact Dracaena
Dracaena deremensis 'Bausei'	Striped Dracaena
Dracaena deremensis 'Janet Craig'	Janet Craig Dracaena
Dracaena deremensis 'Warneckii' Warneckii	Dracaena
Dracaena fragrans 'Massangeana'	Corn Plant
Dracaena godseffiana 'Florida Beauty'	Gold Dust Dracaena
Dracaena marginata	Dragon Tree of Madagascar
Dracaena marginata 'Tricolor'	Tricolor Dragon Tree

Dracaena sanderiana	Ribbon Plant
Dracaena thalioides	Lance Dracaena
Pleomele reflexa	Pleomele
Pleomele reflexa 'Variegata'	Song of India Pleomele
Sansevieria intermedia	Pygmy Bowstring
Sansevieria trifasciata	Snake Plant
Sansevieria trifasciata 'Golden Hahnii'	Golden Birdsnest Sansevieria
Sansevieria trifasciata laurentii Sansevieria	Variegated Snake Plant or Birdsnest
Yucca elephantipes	False Agave
AIZOACEAE	
Lithops leslii	Living Stones
AMARANTHACEAE	
Alternanthera versicolor	Joseph's Coat
Iresine herbstii 'Aureo-reticulata'	Clown Plant
Iresine lindenii Bloodleaf or	Achyranthus
Iresine lindenii formosa	Bloodleaf or Achyranthus
AMARYLLIDACEAE	
Clivia miniata	Kaffir Lily
Eucharis grandiflora	Eucharis Lily
Hippeastrum spp.	Amaryllis
Nerine bowdenii	Naked Lady Lily
APOCYNACEAE	
Mandevilla splendens	Alice DuPont Mandevilla
Mandevilla sanderi 'Rosea'	Rose Dipladenia
Nerium oleander	Oleander
ARACEAE	
Aglaonema commutatum maculatum	Silver Chinese Evergreen
Aglaonema commutatum 'Pseudo bracteum'	Golden Chinese Evergreen
Aglaonema commutatum 'Treubii'	Variegated Chinese Evergreen
Aglaonema crispum	Painted Droptongue
Aglaonema modestum	Chinese Evergreen
Anthurium scherzeranum	Flamingo Flower
Caladium spp.	Caladium
Diffenbachia amoena	Giant Dumbcane
Dieffenbachia 'Exotica'	Dumbcane
Dieffenbachia oerstedii 'Variegata'	Velvet Dumbcane

Dieffenbachia maculata	Spotted Dumbcane
Dieffenbachia maculata 'Rudolph Roehrs'	Gold Dieffenbachia
Epipremnum aureum	Golden Pothos
Epipremnum aureum 'Marble Queen'	Marble Queen Pothos
Monstera deliciosa 'Variegata'	Variegated Split-Leaved Philodendron
Monstera deliciosa	Split-Leaved Philodendron
Monstera guttiferum	Angel Winged Philodendron
Monstera friedrichsthalii	Windowleaf Philodendron
Philodendron bipennifolium	Horsehead Philodendron
Philodendron x 'Burgundy'	Purple Prince Philodendron
Philodendron cannifolium	Flask Philodendron
Philodendron hastatum	Spade-Leaf Philodendron
Philodendron scandens oxycardium	Heart-Leaf Philodendron
Philodendron micans	Velvet-Leaf Philodendron
Philodendron mortianum	Giant Philodendron
Philodendron panduraeforme	Fiddle-Leaf Philodendron
Philodendron selloum	Tree Philodendron
Philodendron squamiferum	Red Bristly Philodendron
Scindapsus pictus 'Argyraeus'	Satin Pathos
Spathiphyllum clevelandii	Peace Lily or White Flag
Syngonium auritum	Five Fingers Syngonium
Syngonium podophyllum	Arrow head or Nephthytis
Syngonium podophyllum 'White Butterfly'	White Butterfly Nephthytis
Zamioculcas zamiifolia	Z Plant
ARALIACEAE	
Brassaia actinophylla	Schefflera or Umbrella Tree
Brassaia arboricola	Hawaiian Schefflera
Dizygotheca elegantissima	False Aralia
Fatsia japonica	Japanese Aralia
Fatshedera x *lizei*	Tree Ivy, Botanical Wonder
Hedera canariensis	Algerian Ivy
Hedera canariensis variegata	Variegated Algerian Ivy
Hedera helix	English Ivy
Hedera helix 'Glacier'	Glacier Ivy
Hedera helix 'Golddust'	Golddust Ivy
Hedera helix 'Needlepoint'	Needlepoint Ivy
Hedera helix 'Scutifolia'	Sweetheart Ivy
Polyscias balfouriana	Balfour Aralia
Polyscias balfouriana marginata	Variegated Balfour Aralia

Polyscias fruticosa	Ming Aralia
Polyscias guilfoylei 'Victoriae'	Lace Aralia
Tupidanthus calyptratus	Tupidanthus
ARAUCARIACEAE	
Araucaria bidwillii	Monkey Puzzle, Bunya Bunya Tree
Araucaria heterophylla	Norfolk Island Pine
ARECACEAE	
Caryota mitis	Cluster Fishtail Palm
Caryota obtusa	Fishtail Palm
Chamaedorea elegans 'bella'	Neanthe Bella Palm
Chamaedorea seifrizii	Bamboo Palm
Chamaerops humilis	European Fan Palm
Chrysalidocarpus lutescens	Areca Palm, Butterfly Palm
Howea forsteriana	Kentia Palm
Phoenix roebelenii	Pigmy Date Palm, Dwarf Date Palm
Rhapis excelsa	Lady Palm
ASCLEPIADACEAE	
Ceropegia ampliata	Lantern Flower
Ceropegia woodii	String of Hearts
Hoya bella	Miniature Wax Plant
Hoya carnosa	Wax Plant
Hoya carnosa rubra	
Hoya carnosa 'Tricolor'	Tricolor Wax Plant
Hoya carnosa 'Variegata'	Variegated Wax Plant
Hoya compacta	Hindu Rope
Hoya compacta 'Variegata'	Variegated Wax Plant
Hoya keysii	Queensland Wax Plant
Hoya 'Silver Pink'	Silver Pink Hoya
Stapelia gigantea	Giant Toad Plant
Stephanotis floribunda	Stephanotis, Wedding Flower
ASTERACEAE	
Gynura aurantiaca	Velvet Plant
Gynura sarmentosa	Trailing Velvet Plant
Senecio macroglossus variegatus	Variegated Wax Ivy
Senecio mikaioides	Parlor Ivy, German Ivy
Senecio herreianus	String of Beads
Senecop rowleyanus	String of Pearls
Senecio serpens	Blue Chalksticks

BEGONIACEAE	
Begonia bowerii	Eyelash Begonia
Begonia 'Chantilly Lace'	
Begonia foliosa	
Begonia 'Lucerna'	Angelwing Begonia
Begonia masoniana	Iron Cross Begonia
Begonia rex	Rex Begonia
Begonia semperflorens 'Charm'	Charm Begonia
BIGNONIACEAE	
Radermachera sinica	China Doll
BROMELIACEAE	
Aechmea chantinii	
Aechmea rhodocyanea	Silver Urn Plant
Aechmea fulgens discolor	Coralberry
Ananas comosus	Pineapple
Ananas comosus variegatus	Variegated Pineapple
Billbergia nutans	Queen's Tears
Cryptanthus roseus pictus	Earth Star
Gusmania musaica	Mosaic Vase
Neoregelia carolinae 'Tricolor'	Tricolor Neoregelia
BUXACEAE	
Buxus microphylla japonica	Japanese Little Leaf Boxwood
CACTACEAE	
Astrophytum myriostigma	Bishop's Cap
Cephatocereus senilis	Oldman's Cactus
Cereus peruvianus 'Monstrosus'	Giant Club
Echinocactus grusonii	Golden Barrel
Epiphyllum ackermannii	Red Orchid Cactus
Epiphyllum cooperi	White Orchid Cactus
Ferocactus lastispinus	Devil's Tongue
Gymnocalycium mihanovichii	Plain Chin Cactus
Mammilaria celsian	
Mammilaria collinsii	
Mammilaria elongata	Golden Star
Opuntia basilaris	
Opuntia microdasys 'Albispina'	Bunny Ears
Opuntia microdasys rufida	Red Bunny Ears
Opuntia subulata	Eve's Pin Cactus

Rhipsalis cereuscula	Coral Cactus
Schlumbergera bridgesii	Christmas Cactus
Schlumbergera truncata	Thanksgiving Cactus
CELASTRACEAE	
Euonymus japonicus	Euonymus
Euonymus japonicus variegatus	Variegated Euonymus
CLUSIACEAE	
Clusia rosea	Fat Pork Tree
COMMELINACEAE	
Callisia elegans	Striped Inch Plant
Cyanotis kewensis	Teddy Bear Plant
Rheo spathaceae	Moses on a Raft
Setcreasea purpurea	Purple Heart
Tradescantia albiflora 'Albo-Vittata'	Giant White Inch Plant
Tradescantia fluminensis 'Variegata'	Variegated Wandering Jew
Tradescantia multiflora	Tahitian Bridal Veil
Tradescantia silamontana	White Velvet Tradescantia
Zebrina pendula	Wander Jew
CORNACEAE	
Aucuba japonica 'Variegata'	Gold Dust Tree
CRASSULACEAE	
Aeonium arboreum 'Atropurpureum'	Tree Aeonium
Aeonium hawarthii	Pinwheel
Crassula arborescens	Silver Dollar
Crassula argentea	Jade Plant
Crassula argentea 'Variegata'	Variegated jade Plant
Crassula flacata	Propeller Plant
Crassula tetragona	Miniature Pine Tree
Echeveria 'Doris Taylor'	Wooly Rose
Echeveria gilva	Green Mexican Rose
Echeveria pulvinata	Plush Plant
Graptopetalum paraguayense	Ghost Plant, Mother-of-Pearl
Gassulaceae huerina spp.	Dragon Flower
Kalanchoe beharensis	Velvet Leaf
Kalanchoe daigremonitana	Maternity Plant
Kalanchoe marmorata	Pen Wiper
Kalanchoe tomentosa	Panda Bear Plant

Kalanchoe tubiflora	Chandelier Plant
Pachyphytum oviferum	Moonstones
Sedum morganianum	Burro Tail
CYCADACEAE	
Cycas circinalis	Fern Palm
Cycas revoluta	Sago Palm
CYPERACEAE	
Cyperus alternifolius	Umbrella Plant
Cyperus papyrus	Papyrus
EUPHORBIACEAE	
Acalypha hispida	Chenille Plant, Foxtail
Acalypha wikesiana macafeana	Copper Leaf
Codiauem variegatum pictum	*'Bravo'* Croton
Euphorbia mammilaris	Indian Corncob
Euphorbia pseudocactus	
Euphorbia pulcherrima	Poinsettia
Euphorbia splendens prostrata	Crown of Thorns
Euphorbia tirucalli	Pencil Tree
Euphorbia trigona	African Milk Tree
Pedilanthus tithymaloides 'Variegatus'	Redbird Cactus, Devil's Backbone
FARBACEAE	
Mimosa pudica	Sensitive Plant
GESNERIACEAE	
Aeschynanthus lobbianus	Lipstick Plant
Aeschynanthus marmoratus	Black Pagoda
Chrysothemis folgens	Sunset Plant
Columnea microphylla	Miniature Lipstick Plant
Episcia cupreata	Flame Violet
Nautilocalyx lynchii	Purple Shrub Violet
Saintpaulia ionantha	African Violet
Sinningia speciosa	Gloxinia
Streptocarpus rexii	Cape Primrose
Streptocarpus saxorum	False African Violet
LAMIACEAE	
Coleus blumei	Coleus
Plectranthus australis	Swedish Ivy, Creeping Charlie
Plectranthus coleoides	'Marginatus' Candle Plant

Plectranthus oertendahii	Prostrate Charlie
Plectranthus purpuratus	Moth King
LEEACEAE	
Leea coccinea	West Indian Holly
LILIACEAE	
Aloe aristata	Lace Aloe
Aloe barbadensis	Aloe Vera, Medicine Plant
Aloe variegata	Tiger Aloe
Asparagus densiflorus 'Meyeri'	Plume Asparagus Fern
Asparagus densiflorus 'Sprengeri'	Sprenger Asparagus
Asparagus setaceus	Asparagus Fern
Aspidistra elatior	Cast Iron Plant
Aspidistra elatior 'Variegata'	Variegated Cast Iron Plant
Chlorophytum comosum	Spider Plant
Chlorophytum comosum 'Variegatum'	Inside-Out Spider Plant
Chlorphytum comosum 'Vittatum'	Variegated Spider Plant
Haworthia cuspidata	Star Window Plant
Haworthia fasciata	Fairy Washboard
Haworthia margaritifera	Pearl Plant
Haworthia reinwardtii	
MALVACEAE	
Abutilon x *hybridum*	Chinese Lantern
Abutilon x hybridum 'Souvenir de Bonn'	Variegated Flowering Maple
Abutilon megapotamicum	Hanging Chinese Lantern
Abutilon pictum 'Thompsonii'	Flowering Maple
Hibiscus rosa-sinensis	Chinese Hibiscus, Rose of China
Hibiscus rosa-sinensis cooperi	Variegated Hibiscus
MARANTACEAE	
Calathea clossonii	
Calathea insignis	Rattlesnake Plant
Calathea makoyana	Peacock Plant
Calathea picturata 'Argentea'	
Calathea roseo-picta	
Maranta leuconeura 'Erythroneura'	Red-Veined Prayer Plant
Maranta leuconeura kerchoveana	Prayer Plant
MORACEAE	
Ficus benjamina	Weeping fig
Ficus benjamina 'Exotica'	Javan Fig

Ficus deltoidea	Mistletoe Fig
Ficus elastica 'Decora'	Rubber Plant
Ficus elastica 'Honduras'	Variegated Rubber Plant
Ficus elastica 'Variegata'	Variegated Rubber Tree
Ficus lyrata	Fiddle Leaf Fig
Ficus palmeri	Mexican Blue Fig
Ficus petiolaris	Redvein Mexican Blue Fig
Ficus pumila	Creeping Fig
Ficus retusa nitida	Indian Laurel
Ficus rubignosa	Rusty Fig
Ficus rubignosa 'Variegata'	Variegated Rusty Fig
MUSACEAE	
Musa acuminata	Dwarf Banana
MYRSINACEAE	
Ardisia crispa	Coral Berry
NICTAGINACEAE	
Bougainvillea glabra	Bougainvillea
Bougainvillea glabra 'Raspberry Ice'	Variegated Bougainvillea
Pisonia umbellifera 'Variegata'	Bird Catcher Tree
OLEACEAE	
Jasminum sambac	Arabian Jasmine
Jasminum polyanthum	Pink Jasmine
ORCHIDACEAE	
Cattleya spp.	Lady of the Night Orchid
Cymbidium spp.	Cymbidium Orchid
Dendrobium spp.	Dendrobium Orchid
Epidendrum spp.	Fiery Reed Orchid
Miltonia spp.	Pansy Orchid
Oncidium spp.	Dancing Doll Orchid
Paphiopedilum spp.	Lady Slipper Orchid
Phalaenopsis spp.	Moth Orchid
Vanda spp.	Vanda Orchid
Vanilla planifolia	Vanilla Orchid
OXALIDACEAE	
Oxalis regnellii 'rubra alba'	Oxalis
Oxalis rubra	Pink Oxalis
PANDANACEAE	
Pandanus veitchii	Veitch Screw Pine

PASSIFLORACEAE

Passiflora caerulea	Passion Flower

PIPERACEAE

Peperomia astrid	
Peperomia caperata 'Emerald Ripple'	Emerald Ripple Peperomia
Peperomia griseoargentea	Silver Leaf Peperomia
Peperomia griseoargentea 'Blackie'	Dark Silver Leaf Peperomia
Peperomia incana	Felted Peperomia
Peperomia obtusifolia	Baby Rubber Plant
Peperomia obtusifolia variegata	Baby Rubber Plant
Peperomia rubella	Pepe Peperomia
Peperomia sandersii	Watermelon Peperomia
Peperomia scandens	Philodendron Peperomia
Peperomia scandens 'Variegata'	Variegated Philodendron Peperomia
Peperomia viridis	

PITTOSPORACEAE

Pittosporum tobira	Mock Orange
Pittosporum tobira 'Variegata'	Variegated Mock Orange

POACEAE

Oplismenus hirtellus 'Variegatus'	Ribbon Grass

PODOCARPACEAE

Podocarpus macrophylla	Buddhist Pine
Podocarpus macrophyllus 'Maki'	Japanese Yew Pine
Podocarpus gracilior	Fern Pine

POLYGONACEAE

Coccoloba latifolia	Sea Grape

POLYPODIACEAE

Adiatum cuneatum	Delta Maidenhair Fern
Adiatum mircophyllum	Maidenhair Fern
Alsophila australis	Australia Tree Fern
Asplenium nidus	Birdnest Fern
Asplenium bulbiferum	Mother Fern
Crytomium falcatum 'Rochefordianum'	Rochford Holly Fern
Davallia fejeensis	Rabbit's Foot Fern
Davallia trichomanoides	Squirrel's Foot Fern
Nephrolepis exaltata bostoniensis	Boston Fern
Nephrolepis exaltata 'Florida Ruffles'	Florida Ruffles Fern

Nephrolepis exaltata 'Fluffy Ruffles'	Dwarf Feather Fern
Nephrolepis exaltata 'Whitmanii'	Feather Fern
Pellaea rotundifolia	Button Fern
Phlebodium aureum	Hare's Foot Fern
Platycerium wilhelminae reginae	Staghorn Fern
Polystichum tsus-simense	Leather Fern
Pteris cretica	Table Fern
Pteris ensiformis 'Victoria'	Victoria Table Fern
Stenochlaena palustris	Climbing Fern
PORTULACACEAE	
Portulacaria afra	Elephant Bush
Portulacaria afra 'Variegata'	Variegated Elephant Bush
PRIMULACEAE	
Cyclamen percicum	Cyclamen
RUBIACEAE	
Gardenia jasminoides	Gardenia
Ixora spp.	Jungle Geranium
Coffea arabica	Coffee Plant
RUTACEAE	
Citrus mitis	Calamondin
SAXIFRAGACEAE	
Saxifraga stolonifera	Strawberry Begonia
Saxifraga stolonifera 'Tricolor'	Variegated Strawberry Begonia
Tolmiea menziesii	Piggy-back Plant
SELAGENELLACEAE	
Selaginella lepidophylla	Resurrection Plant
Selaginella kraussiana	Creeping Moss
STRELITZIACEAE	
Strelitzia nicolai	Giant Bird of Paradise
Strelitzia reginae	Bird of Paradise
URTICACEAE	
Helxine soleirolii	Baby's Tears
Pellionia pulchera	Satin Pellionia
Pilea cadierei	Aluminum Plant
Pilea depressa	Creeping Pilea
Pilea involucrata	Friendship Plant

Pilea microphylla	Artillery Plant
Pilea 'Moon Valley' Moon	Valley Pilea
Pilea 'Silver Tree'	Silver Tree Pilea
Pilea spruceana 'Norfolk'	Norfolk Pilea
VERBENACEAE	
Clerodendrum thomsoniae	Bleeding Heart Plant
VITACEAE	
Cissus adenopoda	Pink Cissus
Cissus antarctica	Kangaroo Vine
Cissus rotundifolia	Arabian Wax Cissus
Cissus quadrangula	Veldt Grape
Cissus rhombifolia	Grape Ivy
Cissus rhombifolia 'Ellen	Danica' Oak-Leaf Grape Ivy
ZAMIACEAE	
Zamia furfuracea	Cardboard Palm

Table 2.5 : Identification of annual flowers

Scientific Name	Common Name
Ageratum houstonianum	Ageratum
Erysimum linifolium	Alpine wallflower
Gomphrena globosa	Amaranth, Globe
Amaranthus 'Magic Fountain'	Amaranthus, Magic Fountain
Browallia speciosa	Amethyst flower,
Datura metal	Angel's Trumpet,
Agastache foeniculum	Anise hyssop,
Callistephus chinensis	Aster, China,
Gypsophila elegans	Baby's breath,
Nemophila menziesii	Baby-blue-eyes
Centaurea cyanus	Bachelor's button
Sutera corda	Bacopa
Impatiens balsamina	Balsam,
Diascia barberae	Barber's diascia
Bidens laevis	Begger's tick
Begonia semperflorens	Begonia, Wax
Begonia xargenteogutta	Begonia, Angelwings
Begonia tuberhybridacultorum	Begonia, Tuberous
Molucella laevis	Bells of Ireland

Ammi majus	Bishop's weed
Gaillardia pulchella	Blanket flower
Asclepias curassavica	Blood flower
Evolvulus glomeratus	Blue daze
Clerondendron ugandens	Blue butterfly bush
Browallia speciosa	Browallia
Browallia viscosa	Browallia,
Anchusa capensis	Bugloss,
Bidens laevis	Burmarigold,
Browallia speciosa	Bush violet
Schizanthus pinnatus	Butterfly flower
Brassica oleracea var. *acepola*	Cabbage, Flowering
Caladium bicolor; Caladium x hortulanum	Caladium
Calendula officinilalis	Calendula
Eschscholzia californica	California poppy
Iberis amara, I. umbellata	Candytuft (annual)
Iberis umbellata	Candytuft, Annual
Iberis umbellata	Candytuft, Globe
Osteospermum hyoseroides	Cape daisy
Anchusa capensis	Cape forget-me-not
Dimorphotheca sinuata	Cape marigold
Cynara cardunculus	Cardoon
Ricinus communis	Castor bean
Celosia crista	Celosia
Celosia argentea var. *cristata*	Celosia, Cockscomb
Celosia argentea var. *plumosa*	Celosia, Plumed
Acalypha hispida	Chenille plant
Cynoglossum amabile	Chinese forget-me-not
Cirsium	Cirsium
Cleome hasslerana	Cleome
Celosia crista	Cockscomb
Celosia argentea var. *plumosa*	Cockscomb, Plume
Coleus x *hybridus*	Coleus
Solenostemon pumilus; Coleus pumilus	Coleus, Trailing
Acalypha wilkesiana	Cooperleaf
Russelia equisentiformis	Coral plant
Coreopsis tinctoria	Coreopsis, Golden
Zea mays var. *Japonica*	Corn, Ornamental
Cosmos spp.	Cosmos
Cosmos bipinnatus	Cosmos, Garden

Cosmos sulphureus	Cosmos, Klondike
Cosmos sulphureus	Cosmos, Orange
Cosmos sulphureus	Cosmos, Sulphur
Cosmos bipinnatus	Cosmos, Tall
Nierembergia hippomanica	Cupflower
Dahlia hortensis	Dahlia hybrids
Arctosis hybrids	Daisy, African
Dyssodia tenuiloba	Daisy, Dahlberg
Rudbeckia hirta	Daisy, Gloriosa
Osteospermum hyoseroides	Daisy, South African
Datura metal	Datura
Dianthus armeria	Deptford pink
Dianthus chinensis	Dianthus
Datura innoxia	Downy thorn apple
Centaurea cineraria	Dusty miller
Senecio cineraria	Dusty miller
Bellis perennis	English daisy
Campanula isophylla	Falling stars
Scaevola aemula	Fan flower
Clarkia hybrids	Farwell-to-spring
Cuphea ignea	Firecracker plant
Odontonema strictum	Firespike
Nicotiana alata	Flowering tobacco
Linum grandiflorum	Flowering flax
Abutilon hybridum	Flowering maple
Nicotiana x sanderae	Flowering tobacco, Hybrid
Myosotis sylvatica	Forget-me-not
Pennisetum setaceum	Fountain grass, Annual
Pennisetum setaceum	Fountain grass, Crimson
Amaranthus tricolor	Fountain plant
Russelia equisentiformis	Fountain plant
Mirabilis jalapa	Four-o'clock
Digitalis purpurea	Foxglove, Annual
Hypoestes phyllostachya	Freckle face
Fuchsia x hybrida	Fuchsia
Gaillardia pulchella	Gaillardia
Polygonum orientale	Gardengate
Gazania rigens	Gazania
Pelargonium x *hortorum*	Geranium
Pelargonium peltatum	Geranium, Ivy

Gerbera jamesonii	Gerber daisy
Clarkia hybrids	Godetia
Dyssodia tenuiloba	Golden fleece
Gomphrena haageana	Gomphrena
Cuphea llavea	Heather, False,
Heliotropium arborescens	Heliotrope
Hibiscus rosa-sinensis	Hibiscus, Chinese
Hibiscus acetosella	Hibiscus, Red
Datura metal	Horn-of-plenty
Cynoglossum amabile	Hound's tongue
Dianthus chinensis x barbatus	Hybrid pinks
Impatiens wallerana	Impatiens
Impatiens 'New Guinea'	Impatiens, New Guinea
Campanula isophylla	Italian bellflower
Viola cornuta, V. tricolor	Johnny-jump-up
Amaranthus tricolor	Joseph's coat
Polygonum orientale	Kiss-me-over-the-garden-gate
Argyranthemum foeniculaceum	Lanceleaf marguerite
Lantana trifoliata	Lantana, Three leaved
Lantana montevidensis	Lantana, Weeping
Briza maxima	Large quaking grass
Delphinium spp.	Larkspur
Lantana trifoliata	Lavender popcorn
Plumbago auriculata	Leadwort, Cape
Eustoma grandiflorum	Lisianthus
Lobelia erinus	Lobelia
Amaranthus caudatus	Love-lies-bleeding
Malope trifida	Mallow
Tagetes erecta, T. patula	Marigold
Tagetes erecta	Marigold, African
Tagetes erecta	Marigold, American
Tagetes erecta	Marigold, Aztec
Tagetes patula	Marigold, French
Tagetes temurfolia	Marigold, Signet
Salvia farinacea	Mealycup sage
Melampodium paludosum	Melampodium
Cuphea hyssopifolia	Mexican heather
Tithonia rotundifolia	Mexican sunflower
Cuphea ignea	Mexican cigar plant
Asclepias curassavica	Milkweed, Annual

Mimulus x hybridus	Monkey flower
Convolulus tricolor	Morning glory, Dwarf
Portulaca grandiflora	Moss rose
Tropaeolum majus	Nasturtium
Omphalodes linifolia	Navelwort
Nemophila menziesii	Nemophila
Nicotiana alata	Nicotiana
Nierembergia hippomanica	Nierembergia
Nigella damascena	Nigella, Love-in-a-mist
Salpiglossis sinuata	Painted tongue
Viola x *wittrockiana*	Pansy
Pentas lanceolata	Pentas
Capsicum annuum	Pepper, Ornamental
Catharanthus roseus	Periwinkle, Madagascar
Strobilanthes dyerianus	Persian shield
Petunia x *hybrida*	Petunia
Calibrachoa xhybrida	Petunia, Trailing
Phlox drummondi	Phlox
Scabiosa atropurpurea	Pincushion flower
Dianthus chinensis	Pinks
Plectranthus argentatus	Plectranthus, Silver
Calceolaria crenatiflora	Pocketbook plant
Hypoestes phyllostachya	Polka dot plant
Portulaca grandiflora	Portulaca
Calendula officinilalis	Pot marigold
Nemesia strumosa	Pouch nemseia
Briza maxima	Quaking oats
Ammi majus	Queen Anne's lace, False
Cuphea llavea	Red cuphea
Ruellia elegans	Red ruellia
Clarkia hybrids	Rocky Mountain garland
Salvia coccinea	Sage, Bloody
Lantana camara	Sage, Red
Salvia argentea	Sage, Silver
Salvia coccinea	Sage, Texas
Lantana camara	Sage, Yellow
Salvia splendens	Salvia
Browallia speciosa	Sapphire flower
Scabiosa atropurpurea	Scabiosa
Salvia splendens	Scarlet sage

Limonium sinuatum	Sea lavender
Calceolaria crenatiflora	Slipperwort
Antirrhinum majus	Snapdragon
Angelonia angustifolia	Snapdragon, Summer
Euphorbia marginata	Snow-on-the-mountain
Cleome hasslerana	Spider flower
Plectranthus fosteri 'Marginatus'	Spur flower
Ipomopsis rubra	Standing cypress
Pentas lanceolata	Star cluster
Limonium sinuatum	Statice, Sea lavender
Limonium sinuatum	Statice
Helichrysum bracteatum	Strawflower
Anchusa capensis	Summer forget-me-not
Helianthus annuus	Sunflower
Lobularia maritima	Sweet alyssum
Dianthus barbatus	Sweet William
Amaranthus caudatus	Tassel flower
Cirsium	Thistle, Plume
Coreopsis tinctoria	Tickseed
Linaria maroccana	Toadflax
Gazania rigens	Treasure flower
Lavatera arborea; Lavatera trimestris	Tree mallow
Bellis perennis	True daisy
Datura metal	Trumpet flower
Diascia barberae	Twinspur
Diascia intergerrima	Twinspur
Alternanthera dentata	Upright calico plant
Verbena x *hybrida*	Verbena
Catharanthus roseus	Vinca
Viola cornuta, V. tricolor	Viola
Cheiranthus cheiri	Wallflower
Nicotiana sylvestris	White shooting stars
Ammobium alatum	Winged everlasting
Torenia fournieri	Wishbone flower
Brugmansia x candida	Yellow angel's trumpet
Zinna angustifolia	Zinnia, Creeping
Zinnia elegans	Zinnia
Zinnia haagenana	Zinnia, Mexican,
Zinnia elegans	Zinnia

Table 2.6 : Identification of flowering Trees and their classification on the basis of Uses

Botanical name (Family)	Local name	Growth Characteristics, Leaf Canopy characteristics
Acacia auriculiformis A. Cunn (Mimosaeae, Leguminosae)	Australian Wattle; Phyllode Acacia, Akashmoni	Quick growing, drought resistant, hardy, Nitrogen fixer. Not grazzed by cattle; Oblong and dense, Height - 16m; Flowering - June-Jan; Pro.-seed, sapling.
Acacia anilotica (mimosaeae, Leguminosae)	Babul, Wattles	Other species are: *A. decurrens* (green wattles); *A. dealbata* (silver wattles); *A. meamii* (black wattle).
Adina cordifolia Hook (Rubiaceae)	Keli-kadam, Haldu	Normally 15-20m, slow growing; Oblong, Leaves are acute at tip and heart-shaped at base; Pro- seeds.
Aegle mermelos (L.) (Rutaceae)	Wood apple , Sriphal , Bel	Slow growing, xerophytic, hardy. Grow well on OB dumps. Oblong. Leaves with 3 or 5 leaflets
Ailanthus excelsa Roxb (Simaroubaceae)	The tree of Heaven	Fast growing, 12-25m, shade tree.
Alstonia scholaris (Linn) R. Br. (Apocynaceae)	Devil tree, Chatim	Quick growing, hardy, 4-7 leaves in a whorled; produces latex, Round (umbrella shaped); oblong; Height: 10-15m; Flowering :Dec-March.
Albizzia lebbeck, Benth. (Mimosae, Leguminosae)	Silk flower or Parrot tree, Siris	Moderate growing; grown in dumps; drought resistant.Round/ spreading; Pro by Seeds. Other two species *A. odoratissima* (Black siris) and *A. procera* also common in ISM campus.
*Anthocephalus cadamba*Miq. (=A. indica A. Reich) (Rubiaceae)	The Kadamba, Kadamba	Quick growing, large; Spreading, Grow rapidly in first 6-8 yr, produce golden ball of flowers.
Anacardium occidentale L. (Anacardiaceae)	Cashewnut, Kaju	Slow growing, evergreen; round.
Azadirachta indica A. Juss, (=*Melia azadirachta* L) Meliaceae	Margosa tree, Nim, Neem	Medium size to large tree; Pinnately compound, Semi deciduous; Grow slowly on Mine dumps; Round/ oblong, spreading;. Blossom in the 1st-week of April; Good purifier of air. Seeds do not keep long vitality, sown immediately after collection.

Bauhunia purpurea Linn. (Cesalpineae)	The purple bauhinia or Mountain ebony, Devkanchan	Medium size tree, deep pink-rosy pink flowers; Leaves 10-14 cm long; Open branch 9-11 strong nerve radiating from base.
Bauhunia variegata Linn (Cesalpineae)	Variegated Bahunia Raktokchan, Kachnar	Small tree, Flowers wwhite with light yellow spot, Open branch11-15 strong nerve rediating from base;
Butea frondoasa Roxb, (=*Butea monosperma* L) Leguminosae	Flame of forest, Dhak, Palas	Slow growing; Medium size tree; Gives flowers onset of Holi. Oblong/ ovoid, In March gets covered with scarlet flowers; extremely resistant to draught.
Bombax malabaricum D.C. (= *Salmalia malabarica* Schoot; = *Bombax cieba*Linn.) (Bombacaceae)	Silk-cotton tree, Shimool, Simbal	Tall, wide spreading branches, wood is soft, leaves divided into 5-7 leaflets, flowers are bright crimson or red. Spreading. Often reaches 25m height, cuttings or seeds, beauty and brightness flower and shadiness of leaves.
Cassia seamea L. (Leguminosae Papilionaceae)	Iron wood tree, Chakundi	Fast growing, pinnately compound, dark green. Grow very well in subsided area, OB dumps. Oblong, Pro. by seeds, sapling; 10-12 m height.
Cassia fistula L. (Leguminosae, Papilionacaea)	Indian laburnum, Amaltash, Bandar lathi	Slow growing; Medium size tree; xerophytic; not eaten by goats or cattle; Fruits pods are long 1-2 ft and dark brown and remain hang for next flowing season.Round/ oblong; open branched Height: 10-12m; Pro. by seeds. Easily attacked by caterpillar; Gets covers with golden yellow flowers in April-May.
Callistemon lanceolatus Sweet. C. linearis L (Myrtaceae)	Lal bottle brash or Bottle brass plant	Handsome evergreen with pendulous branch. Propagate from seed.
Casuarina equisitifolia Frost. (casuarinaceae)	Beef-wood tree, Australian or Whistling pine, Jau, Casuarina	Tall tree, quick growing, evergreen. Oblong. Multiplied by seeds.
Delonix regia Rafin. (=*Poinciana regia* Bojer.)(Caesalpin eae)	The GulMohur, Krisnachura, Gulmohur	Quick growing. Evergreen, Middle size tree, 5-10 m tall. Propagated by seeds or cuttings; Spreading/ Flat topped; Open-branched spreading, irregular shaped. Pods are 1-1.5 feet long, black.

Dalbergia sissoo Roxb. Leguminosae Papilionaceae)	Sissoo, Shishum, Sisu	New leaves appear in mid Feb; Yield excellent timber; Grow very well in mining areas. Round; Height: 10-12m
Duranta repens L (verbinaceae)	Duranta	Shrub, 2-3m, quick growing, evergreen
Feronia elephantum (Rutaceae)	Elephant apple, Koth-bel	Big tree, evergreen.
Emblica officinalis Gaertn. (=*Phyllanthus emblica* L) (Euphorbiaceae	Gooseberry, Amla, Amloki	Medium size deciduous tree, quick growing.Oblong. Regenerated By seeds, cuttings.
Erythrina india Lam. Var. parcelli Papilionaceae (leguminosae).	The Indian Coral Tree	Flowering tree, *E. blakei*. - other flowering tree.
*Eucalyptus citrodora*Hook. Myrtaceae	Lemon scented gum (Eucalyptus tree)	Quick growing; Conical; Height: 15m or more; Reg.-seed, sapling.
Eucalyptus hybrid, Myrtaceae	Mysore gum	Quick growing; Conical; Height: 15m or more; Reg.-seed, sapling
Eugenia jambolana Lamk., (= *Syzygium cuminii* Skeels) Myrtaceae	Java Plum, Jambu, Jamun, Kala jam	Moderate growing; Middle size tree, 10-15m height; Oblong; multiplied by seeds, cuttings.
Ficus benghalensis L Moraceae	Banyan, Bot	Moderate growing; Spreading; Height 20m or sometimes more. Regeneration by Stem cutting, seeds.
Ficus infectoria, Roxb. (Moraceae)	Pakur, Pakar	Quick growing, evergreen, thick shady, sending down aerial roots. Round Small to big tree.
Ficus religiosa L. (Moraceae)	Peepal tree, Pipal, Aswatha	Vigorous seedlings can be seen on some older buildings; natural regeneration takes place by birds.Round/ oblong; Sacred tree; Reg: seeds or by cuttings; Grows slowly in early stages.
Ficus elastica Roxb. (Moraceae)	Indian rubber tree	Quick growing; Evergreen.

Ficus retusa Linn. (Moraceae)	Chilkan	Large spreading evergreen tree, glossy barks and green leaves, grow 30-35 feet large as banyan tree. Leaves alternate, Fruits are axially pair, yellowish when ripe.
Ficus cunea, Ham., (Moraceae)	Fig, Dumur,	Common tree.
Ficus glomerata Roxb.(Moraceae)	Fig tree, Gular, Jagyadumur	Common tree.
Ficus hispida (L), F. (Moraceae)	Kala-umbar, Kakdumur, Kagsha, Gobla	Tree, 10 m, quick growing, evergreen. Spreading.
Gliricidia sepium Walp Papilionaceae.	Gliricidia	Quick growing, medium size tree.
Gmelina arborea Roxb. (Verbenaceae)	White teak, Gamar	Moderate growing, wood is valuable, Planted in mining areas. Oblong. Reg by seeds, saplings; 10-12m height.
Grevillea robusta A. Cunn.(Proteaceae)	Silvery oak	Quick growing, hardy, suitable for mining areas. (Silvery colour in the back of leaves); Oblong. Good for dust and noise attenuation. Reg: by seeds. Very common in West Bokaro Coalfield areas.
Ixora coccinea L (Rubiaceae)	Rangan	Quick growing, Evergreen; Oblong, 6m, Multiplied By cuttings.
*Jacaranda mimosaefolia*D.Don. (Bignoniaceae)	Nil Gulmohar (E, H, B), The Jacaranda.	Native of Brazil. Quick growing; short lived. Introduced in some dumps but growth is not so attractive. Oblong; Open branched, fruit is flat, disc-like contains numerous papery winged seeds. Feathery leaves and blue colour flowers are added attraction.
Leucaena leucocephala L. Legiminosae (Momosae)	Subabool, Koo-babool	Fast growing and ranked first in mining areas, good fodder; grow very well in dumps. Oblong. Reg: by Seeds, saplings, 7-10m height.
*Lagrestroemia flos-reginae*Retz. (syn. L. speciosa Pers.)	The Queen's flower or pride of India; Jarul	Quick growing; medium size tree, well-known ornamental tree, grown as a avenue tree. Oblong; Shades all leaves in Feb- March, totally looks leafless and dull; Fruit - Woody capsule remain throughout the year.

Madhuca indica Gemel (=*Bassia latifolia* Roxb. (Sapotaceae)	The Butter tree or Mohwa tree, Mauha, Maul	Moderate growth, Oblong/round, Spreading crown.
Michelia champaca L (Magnoliaceae) *Magnolia stellata*	The Golden Champa, Swarnachampa, Champak.	Moderate height tree.
Mangifera indica L. (Anacardiaceae)	Mango, Am	Quick growing after 1yr. Round/ oblong; Height 10-15m; reg-seeds, stem cutting.
Mimusops elangi Linn. (sapotaceae)	Bakul, Bakul	Moderate growth, Round to oblong, By seeds, 4-10m height.
Melia azedarach L. (Meliaceae)	The Persian lilac, Bakayan, Ghoranim	Quick growing, grow very well in OB dumps and subsided area. Medium tree; Oblong to round; Height: 20m; Reg - seeds, saplings; confused with Neem (margosa tree).
Millingtonia hortensis (Bignoniaceae)	The Indian Cork tree akasneem, neem chameli, Jasmine tree	Tall, straight evergreen tree, deep green foliage.
Murraya exotica L. (= *M. paniculata* Jack) (Rutaceae)	Orange jessamine, Kamini	Shrub, 4-5 m, quick growing, evergreen; Round; Flowering seasons: June- Oct.
Murraya Koeningii Spreng. (Rutaceae)	Curry-leaf plant, Curry-pata	Small tree.
Nyctanthus arbotristis Linn (= *Jasminum multiflorum* Andr) (Oleaceae)	Night jasmine, Sheuli, Harsinghar	Shrub, 4-5m, quick growing, Oblong By seeds, cuttings; FOS- more or less throughout the year.
Pithecolobium dulce Benth. = Inga dulcis Willd. (Mimosae)	Manila tamarind, Vilayatiimli, Jilapi	Small to big tree, fast growing Reg: seeds.Spreading, oblong.
Plumeria acutifolia Poir (= *P. rubra* Linn) (Apocynaceae)	The temple tree, Gorurchampa, Gulancha, Golainchi	Small deciduous tree, Open. Flowering plant, hardy, pro: by stem cuttings.
Polyalthia longifolia, Var. pendulous Benth and Hook (Anonaceae)	Mast tree, Ashok, Debadaru. (another var is *P. longifolia, erectus*)	Quick growing, tall and majestic tree; branches do not spread out. In erectus var branches are spread out.Conical (Pyramidal); good shade or avenue tree; Height: 10-15m; Flowers in Feb and fruits ripens in Aug - September

Pterospermum acerifolium, Willd. (Sterculiaceae)	Moochkunda, Kanthachampa	Produce fragrant flower.Round, dense Medium to Big size tree. Dense canopy. Common in lower ground.
Pongamia pinnata, Pierrer, (= *Derris indica* (Lam) Bennett) (Leguminosae, Papilionaceae)	Indian beech, Karanj, Karanja	Quick growing; grow very well in subsided area and OB dumps. Round to oblong.
Peltophorum inerme {=syn.*P. ferrugineum* Benth;*P.ptero carpum* (DC.). Backer} (Cesalpineae, Leguminosae)	The copper Pod; The Rusty Shield-Bearer Radhachura	Deciduous, Quick growing; Shades all the leaves in Jan-Feb looks totally leafless and dull; Road- side plantation. Oblong;easiest method is through seeds; Blooms twice in a year: March-May and Sep.- Nov.
Pithecolobium dulce Benth. (= Inga dulcis Willd.) (Mimosae)	Manila tamarind, Vilayatiimli, Jilapi	Small to big tree, fast growing Spreading, oblong Reg: seeds.
Saraca indica Linn. (=*Saraca asoka* Roxb.) (Cesalpinaceae)	The asoka tree, Sita asoka, Asoka	Handsome small (5m) dense crown, foliage 7-paris of leaflets +1. Beautiful flowers.Oblong, Propagated from seeds. Beautiful sheltered plant.
Samanea saman Jacq. Mimoseae, Legumonosae	Rain tree	Big Tree, umbrella shaped canopy. Spreading.
*Spathodea campanulata*Beauv. (Bignoniaceae)	Tulip tree	Dry localities, ornamental flowering plant.
Streblus asper Lour. (Urticaceae)	Siamese rough bush Shaorah, Siora	Medium size tree.
Spondias pinnata	Indian Hogplum, Amra	Big tree.
Swietenia mahogoni L. (Meliaceae)	Mahogony tree	Other sp: are - S. macrophyllaS. microphylla.
Tamarindus indica Linn. (Cesalpineae)	The Tamarind tree Imli, Tentul	Tree, moderate growing, evergreen.Oblong/ spreading;
Tectona grandis Linn. (Verbenaceae)	Teak tree Segun Sagwan	Large tree, quick growing.Oblong, 25-30 m tall, by seeds.
Terminalia arjuna (Roxb) Weight & Arn.(Combretaceae)	Arjun	Large evergreen tree with smooth gray barks, quick growing Oblong, By seeds, cuttings.

Thevetia neriifolia Juss (*T. peruviana* K. Schum) (Apocynaceae).	Yellow oleander, Kolkey-phul, Holdey karabi	Small plants.
Ziziphus jujuba Lamk. (Rhamnaceae)	Kul, Ber	Tree, Quick growing (early); evergreen Round to oblong,
Z. oenoplia Mill	Siakul	Straggler shrub, 5m quick growing, evergreen, round. Other sp: Z. rugosa Lamk. Z. xylopyra wild.

CHAPTER 3

Factors Affecting Ornamental and Landscape Gardening

Climate and Ornamental Crops

There are many factors which affect the production of the ornamental crops. It may be internal or external factors. Climate and soils are the chief natural factors on which the success or failure of ornamental crops production is dependent. A knowledge of the effect of various soil and climatic conditions on ornamental crops production is very essential for every prospective to grower, climate includes several factors like temperature, relative humidity, rainfall, wind, hail, light whereas, soil covers such factors as moisture supply texture, pH etc.

Temperature: The rate of development of a plant (the leaf unfolding rate) is a function of temperature. As temperatures increase, the leaf unfolding rate increases. If the temperature exceeds the maximum temperature, plants experience stress and leaf unfolding rate declines. Given the high costs of fuel, many growers are tempted to reduce the greenhouse temperature to save money on fuel. Because the rate of crop development is dependent on temperature, reducing temperatures increases production time of all crops. However, some crops will be more severely delayed than others. Reducing temperatures from 20°C to 16°C increased time to flower of *Salvia farinacea* by 13 days. This same temperature drop only increased days to flower of pansy by four days. For crops that are severely delayed by lower temperatures, reducing the growing temperature may actually increase total fuel costs for producing that crop. Every plant has fairly well defined range of temperature to which it is tolerant and below which the plants of that variety are liable to be injured to a more or less extent.

i. *Minimum temperature*: Plants cease to grow at the onset of sufficiently cold weather, although how could it must because order for them to stop growing varies with the kind of plant and its condition. It is generally regarded that 0°C or below, the growth is suspended, above this it proceeds. The minimum temperature that on plant can endure may be much lower than that which another will tolerate. Hardiness to cold is related to the water content of the tissue. However, it is not the absolute water content but the form in which water is held by the tissues that brings about the hardiness. The water in the plant tissues is held in three forms:

   a. Free water

   b. Osmotically held water

   c. Bound or colloidally held water

   It is therefore, important for the grower not only to know what the minimum temperatures are in the region where he is to grow his crops but also the approximate minimum temperature that the particular plant or crops will with stand at different stages of growth.

ii. *Maximum Temperature* : Just like the minimum, there is maximum temperature which the plant will endure. This varies with the kind of plant and its conditions. Though the absolute maximum temperature for living protoplasm is closer to boiling point, for most of the higher plants, the lethal points lies somewhere between 43°C and 50°C. A number of growth processes are much retarded at temperature between these.

iii. *Optimum temperature* : It is therefore, obvious that somewhere between the minimum and maximum temperature that a plant or fruit tree will endure, lies the narrow range of temperature which is most congenial for its growth. This is known as optimum temperature. The cardinal temperature ranges for cool season crops are

   - Minimum : 0°C - 5°C
   - Optimum : 25°C - 30°C
   - Maximum : 30°C - 38°C

The cardinal temperature ranges for warm season crops are :

- Minimum : 15°C - 20°C
- Optimum : 30°C - 38°C
- Maximum : 45°C - 50°C

Biochemical processes are affected mainly by temperature. Any chemical reaction increases with increase in temperature. The rate of reaction may double or even be more for every 10°C increase in temperature. This rate of increase in reaction for every 10°C increase in temperature is called quotient 10 or temperature coefficient (Q10).

Q10 = Rate of reaction at (t+10)°C/ Rate of reaction at t°C

Where, t is temperature in °C.

Biochemical reactions are mediated by enzymes. These reactions increase with increase in temperature up to a limit beyond which the rate of reaction decreases. For many fruit crop plants processes Q10 is between 2 and 3 within a moderate temperature range. Q10 of 2 means the process doubles with increases in 10°C temperature. On the basis of temperature requirement ornamental plants may be grouped as under :

1. Plants For High Temperature Situations

These plants will withstand average temperatures between 26°C and 35°C. Remember that at approximately 35°C, plant metabolism stops for most plants and cooler night temperatures are necessary for survival. With average temperaturesover 38°C, few plants can be expected to survive.

Table 3.1 : List of Plants suitable for higher temperature

Scientific Name	Common Name
Agave americana	Century Plant
Agave angustifolia marginata	Variegated Caribbean Agave
Agave attenuata	Dragon Tree Agave
Agave victoriae-reginae	
Asparagus sprengeri	Sprenger Asparagus
Astrophytum myriostigma	Bishop's Cap
Bougainvillea glabra	Bougainvillea
Caryota mitis	Cluster Fishtail Palm
Caryota obtusa	Fishtail Palm
Cephatocereus senillis	Oldman's Cactus
Cereus peruvianus 'Mostrosus'	Giant Club
Coldiaeum variegatum pictum 'Bravo'	Croton
Crassula argentea	Jade Plant

Crassula argentea 'Variegata'	Variegated Jade Plant
Echinocactus grusonii	Golden Barrel
Ferocactus lastispinus	Devil's Tongue
Ficus benjamina	Weeping Fig
Ficus benjamina 'Exotica'	Weeping Fig
Ficus deltoides	Mistletoe Fig
Ficus rubignosa	Rusty Fig
Hibiscus rosa-sinensis	
Jasminum sambac	Arabian Jasmine
Kalanchoe beharensis	Velvet Leaf
Kalanchoe tomentosa	Panda Bear Plant
Lithops leslii	Living Stones
Pandanus veitchii	Veitch Screw Pine
Pittosporum tobira	Mock Orange
Pittosporum tobira 'Variegatum'	Variegated Mock Orange
Sedum morganianum	Burro Tail
Yucca elephantipes	False Agave

2. Plants for Cool Situations

These plants will tolerate average temperatures of 10°C to 18°C. They will also do well in normal interior temperature ranges but are good for placement near cooler windows and entryways.

Table 3.2 : List of Plants suitable for cool situations

Abutilon hybridum 'Fireball'	Chinese Lantern
Abutilon pictum 'Thompsonii'	Flowering Maple
Adiatum cuneatum	Delta Maidenhair Fern
Adiatum microphyllum	Maidenhair Fern
Ardisia crispa	Coral Berry
Asparagus meyeri	Plume Asparagus Fern
Asparagus plumosus	Plumosa Fern
Asparagus sprengeri	Sprenger Asparagus
Aspidistra elatior	Cast Iron Plant
Aspidistra elatior 'Variegata'	Variegated Cast Iron Plant
Asplenium nidus	Birdsnest Fern
Aucuba japonica variegata	Gold Dust Tres
Beloperone guttata	Shrimp Plant
Brassaia actinophylla	Schefflera or Umbrella Tree
Brassaia arboricola	Hawaiian Schefflera
Calthea clossonii	
Calthea insignis	Rattlesnake Plant

Calthea makoyana	Peacock Plant
Calthea picturata 'Argentea'	
Calthea roseo-picta	
Ceropegia woodii	String of Hearts
Chlorophytum comosum	Spider Plant
Chlorophytum comosum 'Variegatum'	Inside-Out Spider Plant
Chlorophytum comosum 'Vittatum'	Variegated Spider Plant
Cycas revoluta	Sago Palm
Cyclamen percicum	Cyclamen
Cyrtomium falcatum 'Rochefordianum'	Rochford Holly Fern
Davallia fejeensis	Rabbit's Foot Fern
Davallia trichomanoides	Squirrel's Foot Fern
Dizygotheca elegantissima	False Aralia
Dracaena deremensis 'Janet Craig'	Janet Craig Dracaena
Dracaena deremensis 'Warneckei'	Warneckei Dracaena
Dracaena fragrans massangeana	Corn Plant
Dracaena godseffiana 'Florida Beauty'	Gold Dust Dracaena
Dracaena marginata	Dragon Tree of Madagascar
Epiphyllum ackermannii	Red Orchid Cactus
Epiphyllum cooperi	White Orchid Cactus
Euonymus japonicus	Euonymus
Fatshedera lizei	Tree Ivy
Fatsia japonica	Japanese Aralia
Hedera canariensis variegata	Algerian Ivy
Hedera helix	English Ivy
Hedera helix 'Golddust'	Golddust Ivy
Hedera helix 'Needlepoint'	Needlepoint Ivy
Hedera helix 'Scutifolia'	Sweetheart Ivy
Helxine soleirolii	Baby's Tears
Hippeastrum spp.	Amaryllis
Hoya carnosa	Wax Plant
Hypoestes sanguinolenta	Polka-Dot Plant or Freckle Face
Leea coccinea	West Indian Holly
Maranta leuconeura 'Erythroneura'	Red-Veined Prayer Plant
Maranta leuconerua	kerchoveana Prayer Plant
Mimosa pudica	Sensitive Plant
Musa acuminata	Dwarf Banana
Nephrolepis exaltata 'Florida Ruffles'	Flordia Ruffles Fern
Nephrolepis exaltata 'Fluffy Ruffles'	Dwarf Feather Fern
Nephrolepis exaltata 'Whitemanii'	Feather Fern

Nephrolepis exaltata bostoniensis	Boston Fern
Oplismenus hirtelus 'Variegatus'	Ribbon Grass
Oxalis regnellii 'Rubra Alba'	Oxalis
Oxalis rubra	Pink Oxalis
Passiflora caerulea	Passion Flower
Pellaea rotundifolia	Button Fern
Pellionia pulchera	Satin Pellionia
Phlebodium aureum	Hare's Foot Fern
Pilea 'Moon Valley'	Moon Valley Pilea
Pilea 'Silver Tree'	Silver Tree Pilea
Pilea cadierei	Aluminum Plant
Pilea depressa	Creeping Pilea
Pilea involucrata	Friendship Plant
Pilea microphylla	Artillery Plant
Pilea spruceana 'Norfolk'	Norfolk Pilea
Pittosporum tobira	Mock Orange
Pittosporum tobira 'Variegatum'	Variegated Mock Orange
Platycerium wilhelminae reginae	Staghorn Fern
Plectranthus australis	Swedish Ivy or Creeping Charlie
Podocarpus macrophylla	Buddhist Pine
Podocarpus macrophylla maki	Japanese Yew Pine
Polystichum tsus-simense	Leather Fern
Pteris cretica	Table Fern
Pteris ensiformis 'Victoria'	Victoria Table Fern
Sansevieria trifasciata	Snake Plant
Sansevieria trifasciata laurentti	Snake Plant
Saxifraga stolonifera	Strawberry Begonia
Saxifraga stolonifera 'Tricolor'	Variegated Strawberry Geranium
Stenochlaena palustris	Climbing Fern
Tolmiea menziesii	Piggy-back Plant
Tradescantia albiflora 'Albo-Vittata'	Giant White Inch Plant
Tradescantia fluminensis 'Variegata'	Variegated Wandering Jew
Tradescantia multiflora	Tahitian Bridal Veil
Tupidanthus calyptratus	Tupidanthus
Zamia furfuraceae	Cardboard Palm
Zebrina pendula	Wandering Jew

Atmospheric Humidity : Higher humidity & higher temperature are favourable for Growth of certain ornamental corps. Higher humidity, being congenial for growth of fungus and bacteria, may be harmful to the plants.

Rainfall : The quantity of annual as well as its distribution plays an important role in the success and failure of ornamental plants. Excessive rain occurring in short periods is generally unfavourable to fruits as it leads to water logging rain at to blooming period may wash way pollens and thereby inhibit pollination. In low rainfall areas, cultivation of ornamental plants is difficult if adequate and cheap irrigations facilities are not available.

Plants With Low Water Requirements

These plants tolerate low water situations and are valuable for hard to reach or low water areas.

Table 3.3 : List of Plants with low water requirement

Scientific name	Common Name
Aechmea chantinii	
Aechmea fulgens discolour	Coralberry
Aechmea rhodocyanea	Silver Urn Plant
Aeonium arboreum 'Atropurpreum'	Tree Aeonium
Aeonium hawarthii	Pinwheel
Agave americana	Century Plant
Agave angustifolia	Variegated Caribbean Agave
Agave attenuata	Dragon Tree Agave
Agave victoriae-reginae	
Aloe aristida	Lace Aloe
Aloe barbadensis	Aloe Vera
Aloe variegata	Tiger Aloe
Ananas comosus	Pineapple
Ananas comosus variegatus	Variegated Pineapple
Asparagus densiflorus 'Myers'	Myers Asparagus
Asparagus densiflorua 'Sprengeri'	Sprenger Asparagus
Asparagus setaceus	Asparagus Fern
Aspidistra elatior	Cast Iron Plant
Aspidistra elatior 'Variegata'	Variegated Cast Iron Plant
Astrophytum myriostigma	Bishop's Cap
Beaucarnea recurvata	Ponytail or Elephant-Foot Tree
Bilibergia nutans	Queen's Tears
Brassaia arboricola	Hawaiian Schefflera
Cephatocereus senilis	Oldman's Cactus
Cereus peruvianus 'Mostrosus'	Giant Club
Ceropegia woodii	String of Hearts
Chamaedorea elegans 'Bella'	Neanthe Bella Palm
Chamaerops humilis	European Fan Palm

Chlorophytum comosum	Spider Plant
Chlorophytum comosum 'Variegatum'	Inside-Out Spider Plant
Chlorophytum comosum 'Vittatum'	Variegated Spider Plant
Cordyline terminalis	Baby Doll Dracaena
Crassula arborescens	Silver Dollar
Crassula argentea	Jade Plant
Crassula argentea 'Variegata'	Variegated Jade Plant
Crassula falcata	Propeller Plant
Crassula tetragona	Miniature Pine Tree
Cryptanthus roseus pictus	Earth Star
Cycas revoluta	Sago Palm
Dracaena craigii 'Compacta'	Compact Dracaena
Dracaena deremensis 'Bausei'	Stripped Dracaena
Dracaena deremensis 'Janet Craig'	Janet Craig Dracaena
Dracaena deremensis 'Warneckei'	Warneckei Dracaena
Dracaena fragrans massangeana	Corn Plant
Dracaena godseffiana 'Flordia Beauty'	Gold Dust Dracaena
Dracaena marginata	Dragon Tree of Madagascar
Dracaena marginata 'Tricolor'	Tricolor Dragon Tree
Dracaena sanderiana	Ribbon Plant
Dracaena thalioides	Lance Dracaena
Echeveria 'Doris Taylor'	Wooly Rose
Echeveria gilva	Green Mexican Rose
Echeveria pulvinata	Plush Plant
Echinocactus grusonii	Golden Barrel
Epiphyllum ackermannii	Red Orchid Cactus
Epiphyllum cooperi	White Orchid Cactus
Euphorbia mammilaris	Indian Corncob
Euphorbia tirucalli	Pencil Tree
Ferocactus lastispinus	Devil's Tongue
Graptopetalum paraguavense	Ghost Plant or Mother-of-Pearl
Gymnocalcycium mihanovichii	Plain Chin Cactus
Haworthia cuspidata	Star Window Plant
Haworthia fasciata	Fairy Washboard
Haworthia margaritifera	Pearl Plant
Haworthia reinwardtii	
Hoya 'Silver Pink'	
Hoya bella	Miniature Wax Plant
Hoya carnosa	Wax Plant
Hoya carnosa rubra	
Hoya carnosa tricolor	Tricolor Wax Plant
Hoya carnosa variegata	Variegated Wax Plant

Hoya compacta	Hindu Rope
Hoya keysii	
Huerina spp.	Dragon Flower
Kalanchoe beharensis	Velvet Leaf
Kalanchoe daigremontiana	Maternity Plant
Kalanchoe marmorata	Pen Wiper
Kalanchoe tomentosa	Panda Bear Plant
Kalanchoe tubiflora	Chandelier Plant
Lithops leslii	Living Stones
Mammilaria celsian	
Mammilaria collinsii	
Mammilaria elongata	Golden Star
Neoregelia carolinae tricolor	
Opuntia basilaris	
Opuntia microdasys 'Albispina'	Bunny Ears
Opuntia microdasys rufida	Red Bunny Ears
Opuntia subulata Eve's	Pin Cactus
Pachyphytum oviferum	Moonstones
Pandanus veitchii	Veitch Screw Pine
Pedilanthus tithymaloides 'Variegatus'	Redbird Cactus or Devil's Backbone
Phoenix roebelenii	Pigmy Date Palm or Dwarf DatePalm
Pleomele angustifolia honoriae	Narrow-Leaved Pleomele
Polyscias balfouriana	Balfour Aralia
Polyscias balfouriana marginata	Variegated Balfour Aralia
Polyscias fruticosa	Ming Aralia
Portulacaria afra	Elephant Bush
Portulacaria afra 'Variegata'	Variegated Elephant Bush
Rhipsalis cereuscula	Coral Cactus
Sansevieria intermedia	Pygmy Bowstring
Sansevieria trifasciata	Snake Plant
Sansevieria trifasciata 'Golden Hahnii'	Golden Birdsnest Sansevieria
Sansevieria trifasciata hahnii	Birdsnest Sansevieria
Sansevieria trifasciata laurentii	Snake Plant
Schlumbergera bridgesii	Christmas Cactus
Schlumbergera truncata	Thanksgiving Cactus
Sedum morganianum	Burro Tail
Stapelia gigantea Giant	Toad Plant
Stephanotis floribunda	Stephanotis
Strelitzia reginae	Bird of Paradise
Yucca elephantipes	False Agave
Zamia furfuraceae	Cardboard Palm

Wind : The wind causes damage to ornamental plants several ways. High velocities winds blow away the fruits & break the branches. A situation which is exposed to wind causes. Water evaporation of soil moisture & thereby necessitates more frequent irrigation. Hot winds at the time of blossoming may cause failure of pollination due to drying of stigmatic fluid & reduced activity of the pollinating insects. The damage caused by winds can be reduced by planting wind breaks.

Hail/Frost : Fruit crops are greatly damaged by hail. Spring frost is more fatal. Because certain temperate fruit crops blossom during spring like apple, cherry, plum, peach. Flowers of these fruit crops are greatly damaged by frost.

Light : Light is a significant factor in plant growth and development as the permeability of protoplasm, intake and loss of water, enzymatic activity, reparation, photosynthesis, flower initiation, ripening of the fruits are all influenced by light directly or indirectly. Three aspects of light intensity (quality), duration (day length) and quality (spectral distribution) are important.

1. Intensity

Depending on growth performance under varying light intensities, plants are classified as helophyte (sun-loving) and sciophyte (shade-loving). Light intensity influence crop growth through photosynthesis. Production of photosynthesis depends on compensation point and saturation light intensity. Compensation point is the minimum light intensity at which respiration rate equals photosynthetic rate. Higher the temperature higher will be the compensation point. It can vary from 3-400 ft. candles. At normal air temperature, heliophytes, in general, reach this point at 50 ft. candle. Saturation light intensity is the maximum intensity at which the rate of photosynthesis reaches a maximum value. Most fruit crops reach light saturation point at $1/5^{th}$ of full sunlight, white sciophytic species reach it at about 1,000 ft. candles. Many woody species are light saturated at $1/9^{th}$ of the full sunlight with maximum rate of photosynthesis of 5-10 mg CO_2 $dm^{-2}hr^{-1}$.

A. Plants For High-Light Situations

These plants will tolerate or excel in situations with an interior light level of 1000 Foot Candles or more. Remember that temperature levels are not a consideration for this list.

Table 3.4 : Plants for high light situation

Scientific Name	Common Name
Abutilon hybridum 'Fireball'	Chinese Lantern
Abutilon pictum 'Thompsonii'	Flowering Maple
Acalypha hispida	Chenille Plant or Foxtail
Acalypha wikesiana macafeana	Copper Leaf
Aeonium arboreum 'Atropurpreum'	Tree Aeonium
Aeonium hawarthii	Pinwheel
Agave americana	Century Plant
Agave angustifolia marginata	Variegated Caribbean Agave
Agave attenuata	Dragon Tree Agave
Agave victoriae-reginae	
Aloe aristida	Lace Aloe
Aloe barbadensis	Aloe Vera
Aloe variegata	Tiger Aloe
Alternanthera versicolor	Joseph's Coat
Aphelandra squarrosa	Zebra Plant
Araucaria bidwilli	Monkey Puzzle
Araucaria heterophylla	Norfolk Island Pine
Ardisia crispa	Coral Berry
Asparagus meyeri	Plume Asparagus Fern
Asparagus plumosus	Plumosa Fern
Asparagus sprengeri	Sprenger Asparagus
Aspidistra elatior	Cast Iron Plant
Aspidistra elatior 'Variegata'	Variegated Cast Iron Plant
Beaucarnea recurvata	Ponytail or Elephant-Foot Tree
Beloperone guttata	Shrimp Plant
Bougainvillea glabra	Bougainvillea
Brassaia actinophylla	Schefflera or Umbrella Tree
Brassaia arboricola	Hawaiian Schefflera
Buxus microphylla japonica	Japanese Little Leaf Boxwood
Caryota obtusa	
Cephatocereus senilis	Oldman's Cactus
Chlorophytum comosum	Spider Plant
Chlorophytum comosum 'Variegatum'	Inside-Out Spider Plant
Chlorophytum comosum 'Vittatum'	Variegated Spider Plant
Chrysal idocarpus lulescens	
Cissus antarctica	Kangaroo Vine
Cissus arabica	Arabian Wax Cissus
Cissus rhombifolia	Grape Ivy
Cissus rhombifolia	Oak-Leaf Grape Ivy

Citrus mitis	Calamondin
Clusia rosea	Fat Pork Tree
Coffea arabica	Coffea Plant
Coldiaeum variegatum pictum	'Bravo' Croton
Cordyline terminalis	Baby Doll Dracaena
Crassula arborescens	Silver Dollar
Crassula argentea	Jade Plant
Crassula argentea 'Variegata'	Variegated Jade Plant
Crassula falcata	Propeller Plant
Crassula tetragona	Miniature Pine Tree
Dieffenbachia picta 'Rudolph Roehrs'	Gold Dieffenbachia
Dipteracanthus portellae	Monkey Plant
Dizygotheca elegantissima	False Aralia
Dracaena craigii 'Compacta'	Compact Dracaena
Dracaena deremensis 'Bausei'	Stripped Dracaena
Dracaena deremensis 'Janet Craig'	Janet Craig Dracaena
Dracaena deremensis 'Warneckei'	Warneckei Dracaena
Dracaena fragrans massangeana	Corn Plant
Dracaena godseffiana 'Florida Beauty'	Gold Dust Dracaena
Dracaena marginata	Dragon Tree of Madagascar
Dracaena marginata 'Tricolor'	Tricolor Dragon Tree
Dracaena sanderiana	Ribbon Plant
Dracaena thalioides	Lance Dracaena
Echeveria 'Doris Taylor'	Wooly Rose
Echeveria gilva	Green Mexican Rose
Echeveria pulvinata	Plush Plant
Epidendrum spp.	
Euphorbia mammilaris	Indian Corncob
Euphorbia pseudocactus	
Euphorbia pulcherrima	Poinsettia
Euphorbia splendens prostrate	Crown of Thorns
Euphorbia tirucalli	Pencil Tree
Fatshedera lizei	Tree Ivy
Fatsia japonica	Japanese Aralia
Ficus benjamina	Weeping Fig
Ficus benjamina 'Exotica'	Weeping Fig
Ficus deltoides	Mistletoe Fig
Ficus elastica 'Decora'	Rubber Plant
Ficus elastica 'Honduras'	Variegated Rubber Plant
Ficus elastica variegata	Variegated Rubber Tree
Ficus lyrata	Fiddle Leaf Fig
Ficus petifolia	

Ficus petiolaris var. *petiolaris*	Redvein Mexican Blue Fig
Ficus petiolaris var. *palmeri*	Mexican Blue Fig
Ficus pumila	Creeping Fig
Ficus retusa nitida	Indian Laurel
Ficus rubignosa	Rusty Fig
Ficus rubignosa 'Variegata'	Variegated Rusty Fig
Gardenia jasminoides	Gardenia
Graptopetalum paraguayense	Ghost Plant or Mother-of-Pearl
Gynura aurantiaca	Velvet Plant
Gynura sarmentosa	Trailing Velvet Plant
Haworthia cuspidata	Star Window Plant
Haworthia fasciata	Fairy Washboard
Hawthoria margaritifera	Pearl Plant
Haworthia reinwardtii	
Hedera canariensis variegata	Algerian Ivy
Hedera helix	English Ivy
Hedera helix 'Golddust'	Golddust Ivy
Hedera helix 'Needlepoint'	Needlepoint Ivy
Hedera helix 'Scutifolia'	Sweetheart Ivy
Hemigraphis 'Exotica'	Purple Waffle Plant
Hibiscus rosa-sinensis	
Huerina spp.	Dragon Flower
Iresine lindenii	Bloodleaf or Achyranthus
Iresine lindenii formosa	Bloodleaf or Achyranthus
Ixora spp.	Jungle Geranium
Jasminum sambac	Arabian Jasmine
Kalanchoe beharensis	Velvet Leaf
Kalanchoe daigremontiana	Maternity Plant
Kalanchoe marmorata	Pen Wiper
Kalanchoe tomentosa	Panda Bear Plant
Kalanchoe tubiflora	Chandelier Plant
Musa acuminata	Dwarf Banana
Oxalis regnellii 'Rubra Alba'	Oxalis
Oxalis rubra	Pink Oxalis
Pachyphytum oviferum	Moonstones
Pedilanthus tithymaloides 'Variegatus'	Redbird Cactus or Devil's Backbone
Pisonia umbellifera 'Variegata'	Bird Catcher Tree
Pittosporum tobira	Mock Orange
Pittosporum tobira 'Variegatum'	Variegated Mock Orange
Pleomele angustifolia honoriae	Narrow-Leaved Pleomele
Polyscias balfouriana	Balfour Aralia
Polyscias balfouriana marginata	Variegated Balfour Aralia

Polyscias fruticosa	Ming Aralia
Polyscias quilfoylei 'Victoriae'	Lace Aralia
Portulacaria afra	Elephant Bush
Portulacaria afra 'Variegata'	Variegated Elephant Bush
Sansevieria intermedia	Pygmy Bowstring
Sansevieria trifasciata	Snake Plant
Sansevieria trifasciata 'Golden Hahnii'	Golden Birdsnest Sansevieria
Sansevieria trifasciata hahnii	Birdsnest Sansevieria
Sansevieria trifasciata laurentii	Snake Plant
Sedum morganianum	Burro Tail
Senecio macroglossus variegatus	Variegated Wax Ivy
Senecio mikaioides	Parlor Ivy or German Ivy
Senecio rowleyanus	String of Pearls
Senecio serpens	Blue Chalksticks
Strelitzia reginae	Bird of Paradise
Syngonium podophyllum	Arrowhead or Nephthytis
Tupidanthus calyptratus	Tupidanthus
Vanda spp.	Vanda Orchids
Yucca elephantipes	False Agave

B. Plants For Low-Light Situations

These plants will tolerate light levels of less than 300 Foot Candles. Usually in these situations, plants will grow very slowlyor not at all, and care should be taken not to apply too much water or fertilizer.

Table 3.5 : Plants for low light situation

Aechmea rhodocyanea	Coralberry
Aglaonema commutatum 'Psuedo bracteum'	Golden Chinese Evergreen
Aglaonema commutatum 'Treubii'	
Aglaonema commutatum macultatum	Silver Chinese Evergreen
Aglaonema crispum	
Aglaonema modestum	Chinese Evergreen
Aspidistra elatior	Cast Iron Plant
Aspidistra elatior 'Variegata'	Variegated Cast Iron Plant
Aucuba japonica variegate	Gold Dust Tres
Chlorophytum comosum	Spider Plant
Clivia miniata	
Epipremnum aureus	Golden Pothos
Epipremnum aureus 'Marble Queen'	Marble Queen Pothos
Ficus elastica 'Decora'	Rubber Plant
Maranta leuconeura kerchoveana	Prayer Plant
Monstera deliciosa	Split-Leaved Philodendron

Peperomia obtusifolia	Baby Rubber Plant
Philodendron cannifolium	Flask Philodendron
Philodendron oxycardium	Heart-Leaf Philodendron
Rhoeo spathaceae	Moses on a Raft
Spathiphyllum clevelandii	Peace Lily or White Flag

2. Duration

Duration of day light period (photoperiod) influence time of lowering of many fruit crops. Response of plants to day light period (light) is known as photoperiodism.

Based on flowering behaviour to the photoperiod, ornamental crops are grouped into 4 broad Categories. Long-day plants flower when theday length exceeds a certain critical length (and the night is less than acritical length). Plants need day lengths of 14-16 hours (night lengths less than eight to 10 hours) to flower. Short-day plants flower when the day length is less than certain critical length (and the night length exceeds some critical length). Generally, crops need photoperiods of 11 hours or less (night length greater than 13 hours) to flower. Day-neutral plants do not flower in response to day length. Because it is the length of the night that a plant measures, long day plants can be induced to flower by turning lights on for four or more hours in the middle of the night (known as night-interruption or mum lighting), traditionally from 10 p.m. to 2 a.m. Plants need to be exposed to approximately 2 micromoles per square meter per second (about 10 foot-candles) for night interruption lighting to be effective. Within the long-day and short-day plant classification, crops can be further categorized as having an obligate or facultative photoperiod response. Plants with an obligate photoperiod response must be grown under that photoperiod forflowering to occur. Plants with a facultative photoperiod response will eventually flower under any photoperiod, but will flower much earlier if produced under the appropriate photoperiod.

Classification of ornamental plants on the basis of day length requirements

1. Obligate long-day plants

Ammi, asperula, callistephus, catananche, centaurea, dill, fuchsia, gazania, ipomopsis, lathyrus, lavatera, legousia, leptosiphon, limnanthes, linum, lobelia, mimulus, nigella, nierembergia, oenothera, petunia, platystemon, strawflower.

2. Facultative long-day plants

Ageratum, basil, calendula, collinsia, dianthus, dimorphothica, linaria, pansy, petunia (grandiflora), phacelia, reseda, salvia, snapdragon, statice, sunflower, tithonia.

3. Obligate short-day plants

Mina vine, hyacinth bean, African marigold.

4. Facultative short-day plants

Celosia, cosmos, gomphrena, ipomea, morning glory, sanvitalia, signet marigold, zinnia.

5. Day-neutral plants

Amaranthus, carpanthea, centranthus,cleome, cobea, nemophila, oxypetalum,stock, verbascum.

3. Quality

The process of photosynthesis, assimilation of nutrients dry matter distribution in horticultural plants depends on quality of light. The wavelength in the range of 400-700 nm is considered to be only useful part. It is also known as photo-synthetically active variation (PAR).

4. Soil

Soil is another most important determining factor for horticultural corps production. While examining soil, it is necessary to pay attention on both its physical & chemical properties. Suppose for establishing of an avenue plants the physical condition of soil should be:

i. Porous and well aerated
ii. 2 m to 2.5 m deep
iii. Having uniform texture upto this depth
iv. Free form heard pan
v. Well drained
vi. Water table should be about 4 m
vii. Extreme conditions of top and sub-soils, very heavy (clayey) and very light (highly sandy) should be avoided.

Soil Texture : Soil texture is an expression of the distribution of the various particle size present in the soil. On the basis of soil particles; soil are broadly grouped into sandy, clayey and loamy soils.

Soil reaction : Growing medium, pH determines whether or not nutrients in the medium are available for plant uptake. Nutrients must be in solution to be taken up by the roots; different nutrients are only soluble over a particular pH range. For most of the ornamental plants pH between 6-7 is most suitable. But some crops can even be grown below or above this pH range.

Table 3.6 : List of landscape Plants suitable for different climate

Common Name	Latin Name	Flower Colour	Height (cm)	Preferred Environment	Life Zone	Comments
Herb						
Aspen daisy (showy daisy)	*Erigeron speciosus*	Blue to violet	30-60	Part sun, Dry/Moist	F, M	Abundant
Black-eyed susan	*Rudbeckia hirta*	Orange-yellow	30-75	Part sun, Dry/Moist	F, M	Frequent in dry mountain meadows
Blanket flower	*Gaillardia aristata*	Yellow to orange yellow	30-60	Sun, Dry	P, F, M	Verify source of commercial varieties, many are hybrids;common mid summer in sagebrush lower montane
Blue flax	*Adenolinum (Linum) lewisii*	Blue	15-30	Sun, Dry	P, F	Watch for native, inter breeds freely
Blue vervain	*Verbena hastata*	Dark blue/purple	90	Sun, Mois	P	Usually many spikes, available by seed
Broom snakeweed	*Gutierrezia sarothrae*	Yellow	15-45	Sun, Dry	P, F	Flowers late summer
Bush sunflower	*Helianthus pumilus*	Yellow	15-75	Sun, Dry	P, F	
Fringed sage, small	*Artemisia frigida*	Yellow	10-20	Dry	P, F, M	Woody perennial with fragrant, silver gray foliage
Golden banner	*Thermopsis divaricarpa*	Yellow	30-60	Part sun, Moist	F, M, S	Perennial, rhizomes
Goldenrod	*Solidago spp.*	Yellow	Variable	Sun, Dry/Moist	P	Many and varied species
Lupine (silvery)	*Lupinus argenteus*	Bluish lavender to purple	30-75	Sun, Dry/Moist	5 P, F, M,	Abundant
Nelson's larkspur	*Delphinium nelsonii*	Purple, purplishblue	30	Sun, Dry	F, M	
Prairie sage (Sagewort)	*Artemisia ludoviciana*	Yellow, small	30-75	Dry	P	Perennial, fragrant silvery white/ greenish foliage
Prickly pear cactus	*Opuntia polyacantha*	Yellow, orange, pink	10-20	Sun, Dry	P	Very spiny
Pussytoes	*Antennaria parvifolia*	White, pink, rose	5-15	Dry	P, F, M, S	Perennial, forms large mats
Rocky Mountain beeplant	*Cleome serrulata*	Pink to lavender	3-8	Sun, Dry	P, F	Abundant, mid summer along roadsides

Rocky Mountain penstemon	*Penstemon strictus*	Blue	3-6	Part sun, Dry/Moist	F, M	Common west of Continental Divide
Rocky Mountain wild iris	*Iris missouriensis*	Blue, purplish, white	2-4	Part sun, moist	F,M,S	
Western wallflower	*Erysimum asperum*	Yellow/oran ge	2-6	Sun, Dry	P, F	Watch for native species; biennial or perennial, flowers late spring to early summer
Grasses						
Alkali sacaton	*Sporobolus airoides*	Ornamental	4-7.5	Moist	P, F	Damp, alkaline
Arizona fescue	*Festuca arizonica*	Turf	2.5-7.5	Dry	F, M	Pine forests
Big bluestem	*Andropogon gerardii*	Ornamental	7.5-20	Dry	P, F	Showy summer and fall
Blue grama	*Chondrosum gracile (Bouteloua gracilis)*	Ornamental /Turf	1-3	Dry	P, F, M	Prefers sandy or gravelly soils; State grass
Buffalo grass	*Buchloe dactyloides*	Turf	10-15	Dry	P	Prefers full sun, tolerates clay
Indian grass	*Sorghastrum avenaceum (nutans)*	Ornamental	7.5-20	Dry-moist	P, F	Prairies, bottomlands, open woods, meadows
Indian rice grass	*Achnatherum (Orzyopsis) hymenoides*	Ornamental	2.5-5	Dry	P, F, M	Sandy plains, mesas
June grass	*Koeleria macrantha*	Ornamental /Turf	2.5-5	Dry	P, F, M	Prairies, open woods
Little bluestem	*Schizachyriu m scoparium*	Ornamental	4-13	Dry	P, F, M	Prairies, open woods, dry hills; Plants turn copper-red in fall
Mountain muhly	*Muhlenbergia montana*	Ornamental /Turf	2.5-5	Dry	F, M	Open woodlands, hillsides; May only be available by seed
Prairie drop seed	*Sporobolus heterolepis*		2.5-5	Dry	F	Pine forests
Tufted hair grass	*Deschampsia cespitosa*	Ornamental /Turf	75-100	Moist	F, M	Open wet meadows, stream banks
Western wheat grass	*Pascopyrum (Agropyron) smithii*	Turf	2.5-5	Dry-moist	P, F, M	Adaptable to variety of habitats

TREES & SHRUBS						
SMALL/MEDIUM SHRUBS						
Antelope bitterbush	*Purshia tridentata*	Light yellow	150	South-facing slopes, dry well drained soils	P, F	Attractive, birds eat seeds
Boulder raspberry	*Oreobatus (Rubus) deliciosus*	Showy white flowers	150	Rocky ground, foothills and canyons	P, F, M	Very attractive; wildlife eats the fruits
Common juniper	*Juniperus communis* ssp. *alpina*	Not Applicable	90	Coniferous forest understory	P, F	Attractive, broad evergreen shrub
Golden currant	*Ribes aureum*	Bright yellow flowers, black fruits	180	Plains to foothills, along roadsides and streams	P, F	Flowers sometimes have a clove scent
Kinnikinnick	*Arctostaphylo s uva-ursi*	Small white to pinkish; red fruits	30	Semi-dry areas; needs acidic, welldrained soil	P, F	Attractive, waxy leaves and red berries; ground cover. Browsed by a wide variety of wildlife.
Mountain ninebark	*Physocarpus monogynus*	White to rose colored, showy	120	Rocky canyonsides, outer foothills	P, F, M	Very attractive, showy flowers; birds eat seeds
Oregon-grape, Holly-grape	*Mahonia repens*	Small yellow in fragrant clusters; berries blue to purple	30	Dry slopes	F, M	Holly-like leaves; attractive ground cover. Birds eat the fruits.
Red-berried elder	*Sambucus microbotrys (racemosa)*	White to rosecolored, showy	240	Along streams and moist slopes	F, M, S	Attractive dark green compound leaves and red berries
Red osier dogwood	*Swida sericea (Cornus stolonifera)*	Tiny white clusters	240	Streamsides and canyon bottoms	P, F, M	Attractive glossy leaves and red bark
Sand cherry	*Cerasus (Prunus) pumila* ssp. *besseyi*	White	30-90	Sandy or gravelly prairie hillsides	P, F	Purplish-black fruit are sweet and edible. Leaves turn reddish in fall.

Shrubby cinquefoil	*Pentaphylloides floribunda (Potentilla fruticosa)*	Yellow	90	Variety of habitats	F, M, S	Often cultivated as an ornamental; attractive
Snowberry	*Symphoricarpos albus*	Small white to pinkish flowers	90	Forest foothill canyons; dry rocky soil and gravelly banks	P, F	Attractive red stems
Twinberry honeysuckle	*Distegia (Lonicera) involucrata*	Showy reddishpurple leaf-like bracts with twin yellow flowers	210	Moist soils; forests, stream sides	F, M, S	Fruits eaten by birds; hummingbirds attracted to the tubular flowers
Waxflower	*Jamesia americana*	Waxy white	180	Cliffs and cliff bases	P, F, M	Aromatic
Winterfat	*Krascheninnikovia (Ceratoides) lanata*	Inconspicuous	30-90	Sandy alkaline soils	P, F	Flower clusters become fluffy resembling lambs' tails
LARGE SHRUBS/SMALL TREES						
Boxelder	*Negundo aceroides (Acer negundo)*	Inconspicuous	600	Stream banks	P, F	Leaves yellow in fall
Chokecherry	*Padus (Prunus) virginiana ssp. melanocarpa*	white Fragrant,	350	Hillsides, gulches, canyons & stream sides	P, F, M	Red to purple fruits a favorite of bears
Hackberry	*Celtis reticulata*	Inconspicuous yellowish-green	350	Dry rocky hillsides and ravine banks	P, F	Small reddish-brown fruits eaten by birds
Hawthorn	*Crataegus erythropoda*	White	300	Dry hillsides and stream banks	P, F	Fruits purplish-red, thorns are up to 2" long, shiny red
Mountain mahogany	*Cercocarpus montanus*	Small yellow flower	450	Open rocky woods and stony soils	P, F	Sometimes evergreen
Rocky Mountain maple	*Acer glabrum*	non-conspicuous	450	Moist sites, streams, and canyons	F, M, S	Attractive red stems. Birds eat seeds, buds, and flowers.

Silver buffaloberry	*Shepherdia argentea*	Tiny yellowish white	350	Streamsides and river bottoms	F, M	Leaves silvery on both sides; branches spiny. Native cousin of Russian Olive.
mooth sumac	*Rhus glabra*	Greenish-yellow	300	Slopes and canyonsides	P, F	Similar to staghorn sumac (*R.typhina*), but grows shorter and without velvety branches. Leaves turn crimson in autumn.
Thinleaf alder	*Alnus incana* ssp. *tenuifolia*	Catkins; reddish brown	550	Swampy ground or sandy soil; montane stream banks & pond borders	P, F, M	Female catkins are cone-like and persist through winter
Wild plum	*Prunus americana*	White	360	Gulches, outwash mesas; moist soil along stream banks	P, F, M	Often forms thickets; 1" yellow to red fruits are edible.
LARGE TREES						
Lanceleaf cottonwood 40 - 60	*Populus x. acuminata*	Not Applicable (male)	1200-1800	Streambanks and valleys	P, F	First generation sterile hybrid between *P. deltoides* and *P.angustifolia*; leaf shape is indicitive of this cross.
Narrowleaf cottonwood 30 - 60	*Populus angustifolia*	Not Applicable (male)	900-1800	Streambanks and valleys	P, F	Long narrow willow-like leaves turn yellow in fall. Available as a graft onto *P.xacuminata* rootstock which should reduce suckering
Plains cotton wood *a* 60 - 80+	*Populus deltoides* ssp. *monilifer*	Not Applicable (male)	1800-2400+	Moist soils; floodplains, riparian areas and valley bottoms	P, F	Nursery stock should be male and therefore "cottonless"; broad leaves turn yellow-gold in fall.
Ponderosa pine 60 - 80+	*Pinus ponderosa* ssp. *scopulorum*	Not Applicable	1800-2400+	Variety of habitats - adaptable	P, F, M	Tall evergreen tree grows up to 150'. Seeds eaten by birds and small mammals.

P - Plains 1200–1800 M

F - Foothills 1800M–2400M

M - Montane 2400M–3000M

S - Sub-alpine 3000M–3500M (or timberline)

A - Alpine above 3500M or timberline

Electrical conductivity (EC) : To maintain different nutrition levels between plant groupings, we can either have multiple fertilizer injectors set up to deliver different feed rates or alternate between fertilizer and clear water for crops with lower nutrition requirements. Check your fertilizer injector regularly. Many EC problems are simply due to clogged or malfunctioning fertilizer injectors.

Plants suitable for Light electrical conductivity (1:2 dilution of 0.26 to 0.75 mS/cm) (SME of 0.76 to 2.0 mS/cm)(Pour Thru of 1.0 to 2.6 mS/cm) Aconitum, African violet, ageratum, anemone, anigozanthus, asclepias, aster, astilbe, balsam, begonia (fibrous, hiemalis, rex and tuberous), bracteantha, caladium, calceolaria, calla lily,celosia, cineraria, coleus, cosmos, cyclamen, freesia, geranium (seed), gerbera, gloxinia, impatiens, marigold, New Guinea impatiens, orchids, pansy, plugs, primula, salvia, streptocarpus, snapdragon, zinnia.

Plants suitable for medium electrical conductivity (1:2 dilution of 0.76 to 1.25mS/cm) (SME of 1.5 to 3.0mS/cm)(PourThru of 2.0 to 3.5 mS/cm):Angelonia, alstroemeria, alyssum,arygeranthemum, bidens, brachyscome, bougainvillea, calendula, calibrachoa, carnation, centaurea, cleome, clerodendrum, coleus, crossandra, dahlia, delphinium, dianthus, diascia, dusty Miller, exacum, fuchsia, geranium (vegetative), helichrysum, hibiscus, hydrangea, ipomoea, Jerusalem cherry, kalanchoe, lantana, lily, lobelia, morning glory, nemesia, ornamental kale, ornamental pepper, osteospermum, oxalis, petunia, phlox, platycodon,ranunculus, rose, sanvitalia, scaevola, schizanthus, sunflower, sutera (bacopa), torenia, verbena, vinca vine.

Plants for Small Spaces

These plants will keep a low bushy shape in most situations.

Table 3.7 : Plants for small spaces

Scientififc Name	Common name
Adiatum cuneatum	Delta Maidenhair Fern
Adiatum microphyllum	Maidenhair Fern
Aechmea rhodocyanea	Coralberry
Aglaonema commutatum 'Pseudo-bracteum'	Golden Chinese Evergreen
Aglaonema commutatum 'Treubii'	
Aglaonema commutatum macultatum	Silver Chinese Evergreen
Aglaonema crispum	
Aglaonema modestum	Chinese Evergreen
Asparagus meyeri	Plume Asparagus Fern
Aspidistra elatior	Cast Iron Plant

Aspidistra elatior 'Variegata'	Variegated Cast Iron Plant
Asplenium nidus	Birdsnest Fern
Aucuba japonica variegata	Gold Dust Tres
Buxus microphylla japonica	Japanese Little Leaf Boxwood
Calthea insignis	Rattlesnake Plant
Calthea makoyana	Peacock Plant
Chlorophytum comosum	Spider Plant
Chlorophytum comosum 'Variegatum'	Inside-Out Spider Plant
Chlorophytum comosum 'Vittatum'	Variegated Spider Plant
Cissus rhombifolia	Grape Ivy
Cissus rhombifolia 'Danica'	Oak-Leaf Grape Ivy
Clivia miniata	Cafir Lily
Cordyline terminalis	Baby Doll Dracaena
Crassula argentea	Jade Plant
Crassula argentea 'Variegata'	Variegated Jade Plant
Cycas revoluta	Sago Palm
Dieffenbachia 'Exotica'	Dumbcane
Dracaena craigii 'Compacta'	Compact Dracaena
Dracaena deremensis 'Warneckei'	Warneckei Dracaena
Echinocactus grusonii	Golden Barrel
Fatsia japonica	Japanese Aralia
Ficus pumila	Creeping Fig
Kalanchoe tomentosa	Panda Bear Plant
Maranta leuconeura 'Erythroneura'	Red Veined Prayer Plant
Maranta leuconeura kerchoveana	Prayer Plant
Nephrolepis exaltata bostoniensis	Boston Fern
Peperomia astrid	
Peperomia caperata 'Emerald Ripple'	Emerald Ripple Peperomia
Peperomia griseoargentea	Silver Leaf Peperomia
Peperomia griseoargentea 'Blackie'	Dark Silver Leaf Peperomia
Peperomia incana	Felted Peperomia
Peperomia obtusifolia	Baby Rubber Plant
Peperomia obtusifolia variegata	Variegated Baby Rubber Plant
Peperomia rubella	Pepe Peperomia
Peperomia sandersii	Watermelon Peperomia
Peperomia scandens	
Peperomia scandens 'Variegata'	
Peperomia viridis	
Philodendron cannifolium	Flask Philodendron
Rhoeo spathaceae	Moses on a Raft
Sansevieria trifasciata	Snake Plant

Sansevieria trifasciata laurentti	Snake Plant
Schlumbergera bridgesii	Christmas Cactus
Schlumbergera truncata	Thanks giving Cactus
Scindapsis aurerus	Golden Pothos
Scindapsis aureus 'Marble Queen'	Marble Queen Pathos
Spathiphyllum clevelandii	Peace Lily or White Flag

Plants for use in Situations Requiring Height

These plants, in general, can be easily purchased in sizes that allow for a five foot plant or taller.

Table 3.8 : Plants for higher altitude

Araucaria bidwilli	Monkey Puzzle
Araucaria heterophylla	Norfolk Island Pine
Bougainvillea glabra	Bougainvillea
Brassaia actinophylla	Schefflera or Umbrella Tree
Brassaia arboricola	Hawaiian Schefflera
Caryota obtusa	Fishtail Palm
Chamaedorea seifrizii	Bamboo Palm
Chrysalidocarpus lulescens	Areca Palm or Butterfly Palm
Coffea arabica	Coffea Plant
Coldiaeum variegatum pictum 'Bravo'	Croton
Dizygotheca elegantissima	False Aralia
Dracaena fragrans massangeana	Corn Plant
Dracaena marginata	Tricolor Dragon Tree
Ficus benjamina	Weeping Fig
Ficus benjamina 'Exotica'	Weeping Fig
Ficus elastica 'Decora'	Rubber Plant
Ficus elastica 'Honduras'	Variegated Rubber Plant
Ficust elastica variegate	Variegated Rubber Plant
Ficus lyrata	Fiddle Leaf Fig
Ficus retusa nitida	Indian Laurel
Gardenia jasminoides	Gardenia
Hibiscus rosa-sinensis	
Howea belmoreana	Pigmy Date Palm or Dwarf DatePalm
Leea coccinea	West Indian Holly
Musa acuminata	Dwarf Banana
Pandanus veitchii Veitch	Screw Pine
Pisonia umbellifera 'Variegata'	Bird Catcher Tree
Pleomele angustifolia honoriae	Narrow-Leaved Pleomele
Podocarpus macrophylla	Buddhist Pine

Podocarpus macrophylla maki	Japanese Yew Pine
Polyscias balfouriana	Balfour Aralia
Polyscias balfouriana marginata	Variegated Balfour Aralia
Polyscias fruticosa	Ming Aralia
Rhapis excelsa	Lady Palm
Strelitzia reginae	Bird of Paradise
Tupidanthus calyptratus	Tupidanthus
Yucca elephantipes	False Agave

CHAPTER 4

Cultivation and Management of Annual Flowers

Annual flowers are a group of herbaceous plants which grow from seeds, produce flowers, set seeds and complete their life-cycle within one year or one season. They provide a beautiful display of colours in the garden. Whether it is a small home garden or a big public garden, it is incomplete without beds of annual flowers. They enhance the decorative value of a garden within a short span of time. At their blooming time, one feels elated when a rain of beauty drizzles in the garden.

Classification of Annuals

Based on season, the annual flowers are divided into 3 groups—winter season, summer season and rainy season.

Winter season annuals

This group includes a large number of seasonal flowers. The important ones are acroclinium, agrostemma, amaranthus, anchusa, annual carnation, annual chrysanthemum, antirrhinum, arctotis, bells-of-Ireland, brachycome, calceolaria, calendula, campanula, candytuft, celosia, China aster, cineraria, clarkia, cornflower, daisy, delphinium, dianthus, dimorphotheca, echium, eschscholzia, gazania, godetia, helichrysum, helipterum, hollyhock, gypsophila, limonium, linaria, linum, lupin, marigold, matricaria, mesembryanthemum, mignonette, mimulus, myosotis, nasturtium, nemesia, nicotiana, nemophila, nigella, pansy, petunia, phlox, pimpinella, poppy, rudbeckia, salvia, saponaria, scabiosa, schizanthus, senecio, stock, sweet alyssum, sweet pea, sweet sultan, sweet william, venidium, verbena, viola and wall flower.

Summer season annuals

The common summer season annuals are coreopsis, cosmos, gaillardia, kochia, marigold, portulaca, sunflower, tithonia, zinnia etc.

Rainy season annuals

This group includes all summer season annuals and others like amaranthus, balsam, celosia, gomphrena and torenia.

Annuals for cut Flowers

China aster, gypsophila, statice, gaillardia, annual carnation, annual chrysanthemum, cornflower, sweet sultan, bells-of-Ireland, piminella and larkspur are grown commercially for cut flowers for interior decoration.

Dwarf Annuals

Acroclinium, Ageratum, Sweet Alyssum, Arctotis, Daisy, Brachycome, Cacalia, Calceolaria, Campanula, Wallflower, Cuphea, Cynoglossum, Dianthus, Echium, Eschscholzia, Felicia, Gaillardia, Gazania, Godetia, Candytuft, Linaria, Linum, Lobelia, Matricaria, Stock, Mesembryanthemum, Mimulus, Nemesia, Nemophila, Nierembergia, Pansy, Petunia, Phacelia, Phlox, Portulaca, Mignonette, Salpiglossis, Scabiosa, French Marigold, Torenia, Nasturtium, Venidium, Verbena and Viola.

Medium-Tall Annuals

Agrostemma, amaranthus, anchusa, antirrhinum, calendula, celosia, sweet sultan, cornflower, cineraria, clarkia, coreopsis, cosmos, larkspur, carnation, digitalis, dimorphotheca, gomphrena, gypsophila, balsam, kochia, limonium, lupin, bells-of-ireland, nicotiana, nigella, garden poppy, rudbeckia, salvia, schizanthus and african marigold.

Tall Annuals

Hollyhock, Chrysanthemum, Sunflower, Helichrysum, Heliotrope, Pimpinella And Zinnia.

Annuals for Shade

Phlox, Viola (Violet), Impatiens, Begonia, Nicotiana, Torenia.

Fragrant Annuals

Sweet Sultan , Sweet Peas, Lemon Bergamot, Jasmine Tobacco, Sweet William, Nasturtium, Moonflower Vine *(Ipomoea Alba)*

Annuals with a Short Season of Bloom : Several Sowings should be made for Succession of Bloom. *Coreopsistinctoria* (Coreopsis), *Iberis umbellate* (Candytuft), *Dimorphotheca aurantiaca* (Cape-marigold), *Centaurea cyanus* (Cornflower) *Myosotis* spp. (Forget-me- not), *Gypsophila elegans* (Gypsophila), *Nigella damascena* (Love-in-a-mist), *Reseda odorata* (Mignonette), *Phlox drummondii* (Phlox), *Papaver spp.* (Poppy), *Alyssurn maritimum* (Sweet Alyssum).

Annuals forvery poor soil

Mentzelia aurea (Bartonia), *Eschscholtzia californica* (California poppy), *Coreopsis tinctoria, Papaver rhceas, Celosia plumose* (Cockscomb), *Mirabilis jalapa* (Four-o'clock), *Gaillardia lorenziana* (Gaillardia), *Impatiens balsamina* (Garden Balsam), *Godetia grandiflora* (Godetia), *Amaranthus caudatus* (Love-lies-bleeding), *Tropeolum majus* (Nasturtium), *Petunia hybrida* (Petunia), *Portulaca grandiflora* (Rose Moss), *Cleome spinosa* (Spiderflower), *Alyssum maritimum* (Sweet Alyssum), *Centaurea moschata* (Sweet-sultan).

Annuals for a neutral or alkaline soil

Impatiens, lberis, Papaver, Phlox, Reseda odorata (Mignonette), *Tropeolum majus, Zinnia elegans.*

Annuals which benefit from pinching

Ageratum houstonianum, Calendula officinalis, Chrysanthemum, Petunia, Phacelia Phlox, Dianthus chinensis, Salpiglossis sinuate, Schizanthus. Piñatas, Antirrhinum, Verbena, Zinnia

Annuals with White Flowers

Ageratum houstonianurn (Mexican Ageratum), *Alyssum maritimurn* (Sweet Alyssum) *Ammobiurn alaturn* (Winged Everlasting), *Antirrhinum majus* (Snapdragon), *Arctotis grandis* (Arctotis), *Argemone grandiflora* (Pricklepoppy), *Brachycome iberidifolia* (Swan River Daisy), *Browallia demissa, Campanula* (Annual Canterbury Bells), *Centaurea* (Royal Sweet-sultan, Cornflower, etc.), *Centranthus macrosiphon* (Spur-valerian) *Chrysanthemum, Clarkia , Cleome spinosa* (Spiderflower), *Clintonia pulchella , Collinsia bicolor, Crepis barbata* rubra (Hawkweed), *Datura fastuosa, Dimorphotheca aurantiaca* (Cape-marigold), *Echiurn plantagineurn* hybrids, *Gilia tricolor , Godetia amcena* and *G. grandiflora , Gypsophila elegans, Helichrysurn bracteatem* (Strawflower), *Iberis amara* (Candytuft), *Impatiens* (Balsam), *Lathyrus odoratus* (Sweet Pea), *Lavatera trimestris* (Tree mallow), *Limoniurn sinuaturn* (Statice or Sea-lavender), *Lobelia erinus* and *L.tenuior Lupinus mutabilis* (Lupine), *Malcomia maritima* (Virginian-stock), *Mathiola incana* (Stock), *Nemesia strumosa, Nemophila maculata, Nicotiana* (Flowering Tobacco), *Cenothera americana* (Evening primrose), *Papaver somniferurn* (Opium Poppy), *Petunia hybrida, Phlox drummondii, Portulaca grandiflora, Saponaria vaccaria* (Soapwort), *Scabiosa atropurpurea, Schizanthus pinnatus* (Butterfly Flower), *Senecio elegans* (Purple Groundsel), *Verbena erinoides* (Moss Vervain), *Vinca rosea* (Periwinkle), *Viola tricolor* (Pansy), *Xeranthemurn annuum* (Immortelle), *Zinnia elegans.*

Annuals with red, rose and pink Flowers

Abronia umbellata (Sand-Verbena), *Alonsoa acutifolia* (Maskflower), *A. warscewiczii Amaranthus caudatus* (Love-lies bleeding), *Antirrhinurn majus* (Snapdragon), *Calandrinia grandiflora, C. speciosa, Callistephus chinensis* (China Aster), *Campanula* (Annual Canterbury Bells), *Celosia* (Cockscomb), *Centaurea cyanus, C. imperialis, C. moschata Centranthus macrosiphon, Clarkia, Collinsia bicolor* (Chinese-houses), *Collomia coccinea, Cosmos bipinnatus, Crepis barbata rubra, Cuphea ignea* (Fiery Cuphea) *Diascia barberm* (Twinspur), *Echium creticurn, E. plantagineurn* hybrids, *Emilia flammea* and *E.sagittata, Eschscholtzia californica* (California Poppy), *Gilia, Godetia Gypsophila muralis, Helichrysurn bracteaturn, Helipterurn manglesii* and *H.roseurn lberis umbellata* (Candytuft), *Impatiens* (Balsam), *Lathyrus odoratus* (Sweet Pea). *Lavatera trimestris* and *L. rosea* (Treemallow), *Leptosiphon hybrida, Limoniurn suworowi* (Statice or Sea-lavender), *Linaria bipartita* (Toadflax), *Linurn grandiflorurn* (Flowering Flax), *Lupinus hartwegii* (Hartweg Lupine), *Lychnis* (Rose-of-heaven) *Malcomia maritima* (Virginian stock), *Malope trifida grandiflora* (Mallow wort) *Mathiola* (Stock), *Mimulus* (Monkeyflower), *Myosotis dissitiflora* (Forget-me-not) *Nemesia strumosa, Nicotianasanderc* and *N.*sylvestris , *Papaverrhocas* (Shirley or Field Poppy), *P. somniferum* (Opium Poppy), *Petunia hybrida, Phlox drummondii, Portulaca grandiflora* (Portulaca), *Rehmannia angulata, Salvia* splendens, *Silene armeria* (Sweet-William), *Saponaria calabrica* (Calabrian Campion), *Tropeolum* (Nasturtium), *Scabiosa atropurpurea, Schizanthuspinnatus* (Butterfly-Periwinkle), *Senecio elegans* (Purple Groundsel).

Flowering annuals are grown in beds and pots. They are used for various purposes in the garden. In beds, they are grown individually with or without perennial plants in borders. They are grown in hanging baskets, window boxes or rock gardens, for training on walls and trellises and for planting in the form of edges, borders or ground covers. Dried flowers of some annuals are also used for interior decoration. Some annuals provide sweet fragrance.

Climate and soil

The annual flowers are grown successfully in all parts of the country. In north Indian plains, winter season annuals are grown during winter, while summer and rainy ones during summer and rainy seasons respectively. In northern hills, all annual flowers are grown during summer season. Under warmer climate of south India, where there is no distinct winter, common winter annuals, i.e. carnation, sweet pea, antirrhinum, pansy, viola etc. do not grow well. In the milder climate, particularly in Bangalore and Pune, almost all types of annuals can be grown.

Annuals thrive best in well-drained, sandy loam soil rich in organic matter. The soil should be neither too acidic nor too alkaline. The soil pH should be between 6.0 and 7.5. In acidic soils, lime should be added for making it neutral. In clay soil, organic matter is added to make it porous. In sandy soils, add sufficient organic matter for improving its texture.

Propagation

Almost all annuals can be propagated by seeds. Marigolds can be propagated by cuttings also. Their sowing time depends upon climate. It varies in different parts of India. In the northern plains, winter season annuals are sown during September–October. However, late-flowering types – China aster, carnation and cineraria are sown during August–September. Summer season annuals are sown during January–February, while rainy season annuals during May–June. In south India, September–October is ideal time for sowing annuals. In northern hills, March–April is most appropriate time for sowing annual flower seeds, whereas March–May in southern hills.

The seeds of annual flowers are sown in nursery beds, earthen pots, seed pans or wooden seed trays. The seeds of a few annuals like sweet pea, morning glory, lupin nasturtium and hollyhock which have bold seeds can be sown directly at permanent places. Mixing of fine sand in very small seeds is advisable for even sowing. The seeds of echium, lobelia and flowering tobacco do not germinate unless first exposed to sunlight, while seeds of nigella and cineraria germinate only in dark. The seeds recently harvested from the plant, although given required conditions, fail to germinate. This may be due to physical condition or chemical reaction of seed coat of seeds. Seeds of clianthus need stratification or scarification for germination. Some seeds require after ripening period for germination.

Location and Layout of seed bed

For selecting an area for establishing seed bed it is worthwhile to consider the following points:

- Seed bed should be prepared in such place where no water stagnation is experienced, and have good drainage system.
- Land for seed bed should be well drained and located at on a high level.
- The soil for seed bed should be sandy loam and normal in pH (around 6.5-7.0).
- The plot for seed bed should be selected near to a water source.
- Seed bed plots should be chosen near the farm building so that frequent supervision can be made easily.
- Seed bed should be away from the shady places.
- Seed bed plots should be selected at one side of the field to isolate the other fields for doing cultural practices easily.
- Site should be safe from stray animals and excessive diseases and pest attacks.

Methods

Soil Preparation for seed Bed

Seed bed preparation is an important step in crop management because it largely affects crop stand and its performance at field level. Therefore, soil should be worked to a fine tilth by repeated ploughing and spading. Dead plant parts which are seem to be dwellers of pathogens and pests should be collected, removed and burnt. Well decomposed organic manure @ 40-50 kg/10sqm, should be mixed thoroughly in the soil. The seed bed compost should consist of one part each of garden soil, coarse sand, farmyard manure and leaf-mould. For preparing the nursery beds, the soil should be dug up thoroughly and sufficient farmyard manure should be mixed in soil. Raised nursery beds of convenient size (normally 60 cm wide and 15 cm high) should be prepared. If soil is heavy, some quantity of sand may be added. It is better if the soil of nursery bed or earthen pots is sterilized with 2% formalin. For this, soil is drenched with formalin solution and is covered with polythene sheet for 45hr. afterwards, the polythene is removed and soil is dried before sowing the seeds. Before sowing, the seeds should be treated with Cerason (0.2%) and Captaf (0.2%) to prevent the seedlings from damping off disease. The seeds should be sown thinly and evenly as thick sowing causes damping off of seedlings.

Soil treatment

Soil treatment is an essential step in a successful Seed bed management because it the base for seedlings stand, source of nutrition and pathogens. In humid region damping-off caused by soil borne fungi like *Pythium, Rhizoctonia,*

Phytophthora etc, is a common disease in the nursery beds.Besides, pests like snails, cutworms and termite and mites are also damages young seedlings.There are various measures for soil treatment like soil solarization, chemical treatment, biocontrol treatment etc. Soil solarization can be done with transparent polythene of 25-100 mm thickness during the hot and dry periods. For this soil should be moist before mulching because it increases latent heat and thermal sensitivity for resting of soil borne pathogens, harmful pests and weeds which can be reduced to a sustainable level. Chemical treatment of nursery beds can be done by Formalin solution (1:100:: Farmaline: water) @ 5 lt/sqm. The treated area should be covered with polythene sheets for 7-8 days and after that it should be remained open for next 7-10 days for facilitation of formalin emission. Besides, the seed beds and seeds can be treated with some fungicides like Ceresan or Bavistin @ 2g/litre before sowing. The insect-pests can be controlled by treating soil with Chlorpyriphos or Sewin dust powder (20-25g/sqm) before sowing and or at the time of nursery preparation. Presently a large number of bioagents like *Trichoderma*, *Pseudomonas* and *Bacillus formulations* are available for soil as well as seed treatment which performs as antagonist to harmful microorganisms. Some of botanicals like NSK, Neem cake, Karanj cake etc. also can be used for nursery bed treatment.

Seed bed preparation

In nursery beds, the seeds are sown in rows spaced 6 cm apart. Then, they are covered with finely sieved leaf-mould. Watering is to be done with a watering can having a fine rose both in beds and pots. In beds, when germination is over, water is given for proper moisture. Thereafter, the beds should be kept weed-free. A working area of 60 cm between the beds is necessary. This facilitates ease in sowing of seeds, weeding, watering, spraying and lifting of seedlings. Irrigation channels are to be laid out conveniently.

Before sowing seeds the beds should be levelled and pressed gently to make it firm. Nearly 15-20 cm raised beds width are always preferred for raising nursery. Beds of 1-meter width of any convenient length are to be made. However, its length should be made according to the requirements or size of plots but should not exceed 5-6 m. In between beds, drains of about 30-45 cm width are prepared and connected to the main drain for removal of excess water during heavy pour. This space facilitates easy movement during intercultural operations and acts as physical barrier for inoculums spread. The drains are flooded during dry period to modify microclimate of nursery beds in favour of seedlings. In recent years various advancements have been made in Seed bed management for bed preparation to avoid possibilities of pathogen spread like use of soil less media, plug tray technique, perforated poly trays etc.

Cultivation

Planting

Seedlings are transplanted 25 days after sowing at 4-leaf stage. Before transplanting, seedlings are hardened off by withholding water for 1 or 2 days or by exposing them gradually to sunlight. Transplanting is, generally, done either on a cloudy day or in the evening. Transplanting in evening is good as the night cool temperature is beneficial for the establishment. Light watering every day in early morning or late in the afternoon is required for about a week for proper establishment of the seedlings.

Dwarf annuals should be planted at a distance of 30 cm × 30 cm, whereas medium-tall annuals should be planted at a spacing of 45 cm × 45 cm. The plants of tall annuals should be planted 60 cm apart.

Manuring and fertilization

The farmyard manure or compost @ 3kg/m^2 is mixed in the soil. Chemical fertilizers @ 20g urea, 60–120g superphosphate and 30–60g muriate of potash/m^2 should also be added. Half quantity of urea and full of superphosphate and muriate of potash should be applied at the time of bed preparation. The remaining quantity of urea must be applied one month after transplantation. Spraying plants with 2% urea twice or thrice is beneficial for good growth and flowering. Fertilizers should never come in the direct contact of the foliage since they cause scorching. Fertilizers should never be applied in the pot-grown annual flowers. However, some readymade pot-mixtures can be used. The pot-mixture should consist of 2 parts of garden soil and one part each of coarse sand and farmyard manure. Instead of fertilizers, it is better if pot-grown plants are given liquid feeding. The liquid manure is prepared by fermenting 1–2kg each of fresh cowdung and oil cake in 10 litres of water in a drum for one week. It is diluted to tea colour and sieved with the help of a muslin cloth. It is applied @ 500–1000 ml/pot at 7–10 days intervals.

Growth and flowering

Environmental factors and various cultural conditions affect growth and flowering of many annuals. The *Cosmos bipinnatus, Nicotiana* sp.and *Callistephus chinensis* are typical short-day plants for both vegetative growth and flowering, while *Dianthus barbatus* and *Nigella damascena* are typical long-day plants for vegetative growth as well as flowering. On the other hand, some annuals require long days for flower initiation followed by short days for flower development. An imbalance of N either hastens or delays flowering. Clarkia, candytuft and salvia flower prematurely if the soil is deficient in N. If N level is high in soil, sunflower, lupin, flowering tobacco and African marigold flower prematurely.

Growth ratarding or promoting substances play a major role in getting dwarf plants or higher flower yield. Growth retarding chemicals—CCC (1, 000–2, 000ppm), B-Nine (2, 000–5, 000ppm) and SADH (1, 000–3, 000ppm) —are used to control growth and flowering in hollyhock, arctotis, sweet sultan, coreopsis, cosmos, phlox and pansy. These chemicals retard plant height, produce more number of leaves and branches and improve flowering, Similarly, GA_3 (100–400ppm) gives beneficial results in respect of growth and flowering in African marigold, China aster and antirrhinum. Application of these growth substances is more effective at vegetative stage.

Aftercare

After transplanting, beds are weeded, hoed and watered regularly. As soon as seedlings are established in beds, pinching is done for making the plants bushy. Pinching is not practiced in antirrhinum, larkspur, lupin, stock and hollyhock. Sweet pea, carnation, morning glory and nasturtium, have weak, slender or straggling stems. They need support when they are 15–20cm tall. The stakes prepared from split bamboos are painted green so that they can be matched with the foliage colour of the plants. Sometimes, seedlings of carnation, marigold, China aster, cosmos and zinnia produce flower buds at an early stage. These buds should be removed as soon as they appear. The number of buds/ stem is reduced by disbudding the axillary buds, if large blooms are desired.

Irrigation

Little water is needed everyday up to 7–10 days after transplantation. When the seedlings start new growth, profuse watering once or twice a week is required in beds. Later, frequency and quantity of watering depend upon soil and season. In lighter soils, more frequent irrigation is needed than that in heavy soils. The season of planting also determines the frequency of irrigation. During summer season, irrigation should be done at weekly intervals in beds, while at 10–12 days intervals in winters. Irrigation during rainy season depends upon prevailing weather conditions. Potted plants need daily watering during summer, whereas on alternate days in winter.

Harvesting and Postharvest management

Most of the annual flowers are grown for garden display purpose in various ways. However, marigold, China aster, antirrhinum, gypsophila, statice, gaillardia, annual carnation, annual chrysanthemum, cornflower and bells-of-Ireland are grown commercially for cut flower or loose flower purpose. Flowers are harvested when they are fully open and are sold in the local markets. Antirrhinums are cut when one-third of the florets are open. China aster flowers are cut along with their stems when they develop their original colour. Marigold flowers are harvested when they are fully open. Gypsophila flowers are cut when these are open, but not over mature. The flowers, in general, are cut either late in the afternoon or very early in the morning. After harvesting, cut flowers should be put in a bucket of water filled up to one-fourth of the volume as it helps in their recovery from the shock of being cut away from the plant. As far as possible, the freshly opened flowers should be cut as freshness enhances their shelf-life.

In African marigolds, yield of 20 to 22 tonnes of fresh flowers is obtained from one hectare crop, whereas in French marigolds 10–12 tones/ha of fresh flowers is obtained. China aster gives the yield of 10–12 tonnes/ha of fresh flowers.

Some annuals are used as cut flowers. Therefore, proper postharvest management is necessary for prolonging their vase-life. The flowers are graded according to stem length, flower size, flower shape, flower colour and freshness. The cut flowers/loose flowers of most of the annuals are marketed in local markets. However, cut flowers of gypsophila, bells-of-Ireland and Limonium are traded in the international market.

Woody-stemmed flowers should have their stems split or crushed to enlarge the surface in contact with water. Remove all foliage from stems which are under water otherwise they decay and clog up the xylem vassels of stem. It is better to cut stems diagonally because it exposes more surfaces to water.

If flowers are not sold the day they are harvested, to store them in a cold storage is imperative. Antirrhinum flowers can be stored for short duration at a temperature -0.6°C to 1.7°C. For storing stock cut flowers, 4°C temperature is optimum. Gypsophila flowers can be stored for 1–2 days at 4°C. Statice cut flowers can be stored for 2–3 weeks at 2°C. A temperature of 0°–2°C is suitable for the storage of cut carnations.

Planning and Making A Herbaceous Border or Bed

Herbaceous border or herbaceous bed is simply an area planted up with annuals/ perennials which die down in the winter and re-grow the following spring to give a display of flowering colour. Many herbaceous borders or beds have other types of perennials in them, besides the herbaceous perennials. This is not a problem, for a mixed border or will give us a bit more scope for year round colour and interest.

We do not actually need a specific herbaceous border or bed. We can simply plant a few herbaceous plants in between shrubs or in specific places where it is not easy to get plants to grow. A light sunny aspect is best, but dappled shaded areas can also be used with the choice of the right herbaceous plants.Herbaceous perennials are a huge range of plants, and we can find one for any spot in the garden dry, damp; sunny, shaded; sheltered, exposed.

Modern herbaceous perennials come in all sizes, right from the newer (small) sedums and saxifrages, up through to stunning heleniums and more.Colour in the garden right through from January to December provided we have the space for plants.For a start, we can choose any ten plants and they will all grow to different widths. If for instance, we plant a group of 5 *Euphatorium maculatum* at the back of a border or bed, we will need a huge area to be able to plant other things. Herbaceous beds and borders tend to be better in informal shapes rather than straight lines. This is mainly because of the wide range of shapes, sizes and flowering periods of herbaceous plants are more suited to informal shapes rather than formal shapes. However, if our garden is already laid out in formal shapes and we prefer rectangles to kidney shape beds or straight border edges rather than curved edges, no problem go for what you want just bear in mind that the plants within the beds will take on an informal appearance.

General advice on planning a herbaceous border, usually advises the use of a bit of graph paper and pencil. Forget that unless you have a very good knowledge of herbaceous plants and you can also visualize what your border will be like from a two dimensional sheet of paper.By far the best way is to get out in your garden with a few canes of different lengths to coincide with the height of the plants you have bought, or are going to buy. Now, with the help of your canes, you can create a three-dimensional 'plan' taking into account the surrounding trees, shrubs, fences and other aspects of the garden that will have a visual impact upon your herbaceous plot to be. Place your canes in the places best suited for the height of your plants, and you will soon start experimenting on what to put where. It is surprising how different this will end up, compared with an inanimate sheet of graph paper. At the same time if you have not already done so, you can 'lay out' the edge shape with the use of a hose and adjust shape to suit from all angles. Bear in mind that the plants will be in the same position for a long time may be years so prepare the soil well and add rotted manure, peat or compost as you dig. Dig down a good spade or fork depth and then tread the dug over soil firm. Remove every bit of weed growth you can find.

Choice of Plants

Herbaceous plants can be used for more or less any position in the garden. The choice of perennials has never been greater, with new varieties coming onto the garden centre shelves each year. If you are planning on designing a herbaceous bed or border, you have two basic choices. You can either trawl through catalogues and web sites trying to find the plants that will suit your plan, or you can go out on a local flower nursery and buy what you like the look of and fit that into your proposed bed or herbaceous border. The second option might seem a bit hit and miss, but there are thousands of herbaceous perennials out there to choose from and a visit to a handful of garden centers or nurseries will probably furnish you with a few gems that would have been overlooked by going down the plan it and then find it route. Don't try to buy all your plants in one visit. Take your time. Spread your visits over a few weeks and you will find different herbaceous plants each time. Above all remember that herbaceous perennial plants don't always do what they are supposed to do in regards size.

CHAPTER 5

Cultivation and Management of Climbers

Climbers can be used to break the bareness of long walls, soften posts, frame doors and windows and beautify window boxes. With a fence or trellis, they can be used for screens, making a background for flower garden or quickly turning an ordinary porch or patio into a private, shady living centre. Some slower growing climbers are useful as container plants for patios. On the down side, most vine species have the potential to become invasive, spreading to areas where they may not be desirable or becoming too large for the intended area.

Types

Climbers are characterized as three types based on how they cling to a support. The kind of support needed depends on whether the Climbers cling by twining stems, orby specialized structures such as tendrils, adhesive disks and aerial roots. Climbers with twining stems wind around a support as they grow (Fig. 5.2). Examples of plant with twining stem sinclude bitter sweet, wisteria and trumpet creeper.Tendrils resemble thread like stems that wrap around any structure they contact (Fig. 5.1). Examples of plants with tendrils include grape, pea, and cucumbers. Some Climbers have specialized structures for clinging to a support with a disk-like suction cup (Fig. 5.3). Examples include Boston ivy and Virginia creeper. Other Climbers cling by means of small, aerial roots that attach to rough surfaces such as mortar, wood, and tree trunks (Fig. 5.4). Examples include English ivy, trumpet creeper and winter creeper euonymus.

Fig. 5.1

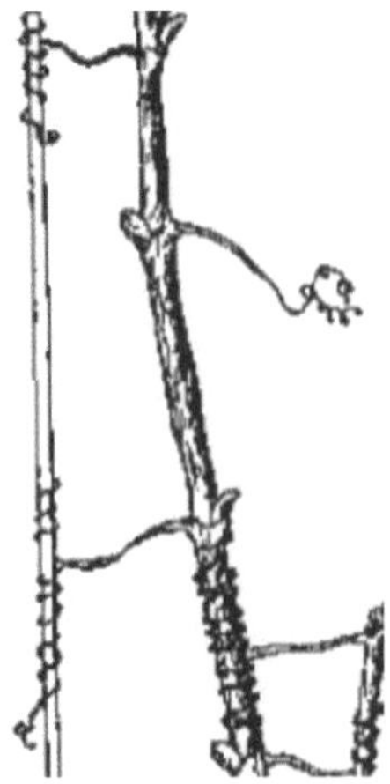

Fig. 5.2

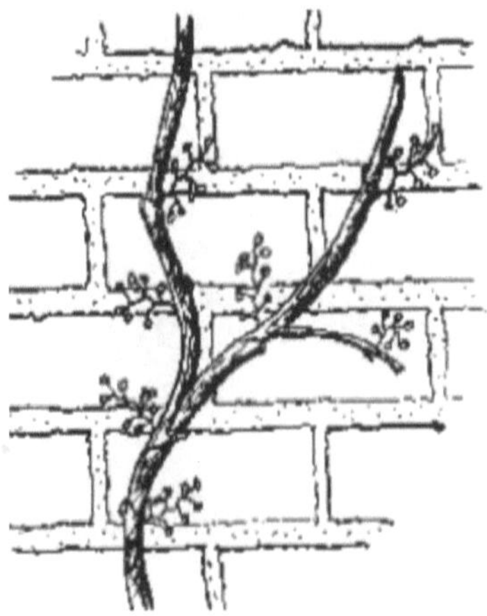

Fig. 5.3

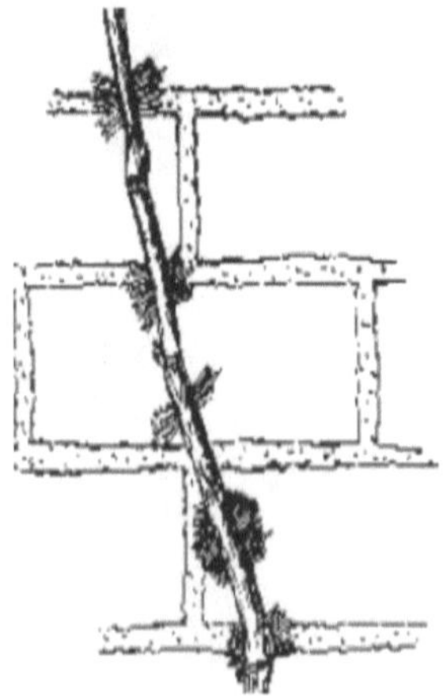

Fig. 5.4

Supports

Climbers that climb by means of twining stems or tendrils will need to be supported by wires, trellises, or arbors. Climbers that cling either by adhesive disks or aerial roots can be grown on either brick or stone walls but should not be used on wood framed structures due to the risk of structural damage. Clinging Climbers tend to collect moisture leading to wood decay. Provide a trellis or similar support far enough away from a wood structure to allow proper ventilation. Support structures for vines should be made from durable materials such as metal or wood. When using metal wire or tubing, copper or aluminium is preferred over other metals to avoid rusting. For wood structures, naturally rot-resistant woods such as redwood, cedar, or cypress are the most durable. Treated lumber and synthetic lumber made from recycled plastic provide additional durable options.

Annual Vines

Annual Climbers are those which must be sown from seed every year. Relatively inexpensive, most types bloom orhave attractive foliage all summer. We can change the planting each year to vary the colour and foliage characteristics.Most annual vines do not transplant easily, so the seed should be sown directly into the pot. Some of the important annual climbers are as under:

Balloon Vine (*Cardiospermum halicacabum*): The small white blooms are rather inconspicuous. Fruits (seed pods) look like an inflated green balloon. The seed pods are highly decorative on the vine, or they can be cutand used for flower arrangements. The vine produces a feathery and delicate growth. It is best used on a small support or as a container-grown plant. It climbs by tendrils.

Black-eyed Susan Vine, Clockvine (*Thunbergia alata*): Small, bright orange blooms with jet black centers are produced in great profusion on this dainty vine. Vines can be grown on a small, erect support or in a container.Tendrils may hang downward, so provide some type of support.

Cardinal Climber (*Ipomoea x multifida, I. sloteri*): Produces red blooms with yellow throats. The foliage is feathery, giving the vine a delicate appearance. The plantis ideal for patio containers. Ipomoea quamoclit, or cypress vine, are similar but produces small, phlox shaped, red blooms.

Cup and Saucer Vine (*Cobaea scandens*): Also called Cathedral Bells, produces cup-shaped blooms of pure white or lilac on a heavy-growing vine. It climbs by means of twining stems.

Gloxinia, Climbing (*Asarina erubescens,Lophospermum erubescens*): The flowers resemble miniature gloxinia blooms in colors of white, velvety purple and rose. Ideal for a container plant with a support needed, it also grows well in hanging baskets.

Hyacinth Bean (*Dolichos lablab, Lablab purpureus*): The pea-like blooms of either white or purple are simultaneously borne on one plant. The purplish, decorative seed pods follow. The foliage is also pea-like and the vines make an excellent screen. Stems twine on supports.

Moonflower Vine (*Ipomoea alba*): Similar to a morning glory, this vine produces extra-large and pure white blooms which usually appear only in late afternoon or early evening. An interesting effect can be created by interplanting with common morning glory.

Morning Glory (*Ipomoea tricolor, I. purpurea, and hybrids*): Vines are available in single or double flowered types in red, blue, white, rose, purple, or two-toned colors. Vines grow rapidly in warm soil and sunlight and need a sturdy support. They can also be used for non traffic ground cover areas.

Scarlet Runner Bean (*Phaseolus coccineus*): Vine produces typical bean foliage and flowers and the fruits are green pods. It grows fast and the twining vines need ample support. It is ideal for screening.

Sweet Pea (*Lathyrus odoratus*): Known for their delicate, fragrant flowers in a rainbow of colors, sweet peas needfull sun for best growth.

Sweet Potato, Ornamental (*Ipomoea batatas*): Ornamental sweet potatoes grown for their foliage are relatively new to the horticultural market. Sweet potatoes are warm season plants, heat tolerant and adapt well to growing on trellises or in hanging baskets and other garden containers. Currently available cultivars offer foliage in dark purple, bright chartreuse green, and atricolor blend of green, pink and white. They are usually grown as annuals in Indiana, though the tuberous roots could be stored over winter and replanted in spring.

Perennial Climbers

Perennial Climbers usually are best started similarly to other woody ornamental trees and shrubs. Set 20 to 30 cm from the desired support. Add soil and set the plant at the original soil level. Use a planting mixture of 2 parts soil to 1 part peat moss. Settle by watering and mulch with a 5 cm layer of peat moss or bark chips. Fertilize perennial vines sparingly in the late fall or early spring. Use 1/4 cup of a general analysis fertilizer suchas 12-12-12 or 16-8-8 per plant. Be sure to water it in thoroughly.

Akebia, Fiveleaf (*Akebia quinata*): This dense, fine textured twining vine grows rapidly. Small purple flowers are produced in clusters during May. The fleshy pod-like fruits are interesting and conspicuous, but seldom produced. This is one of the neatest and easiest foliage vines to train around a wire or waterspout.

Bittersweet, American (*Celastrus scandens*): This is one of the most ornamental twining vines. Staminate (male) and pistillate (female) flowers are produced on separate plants, so plant at least one male plant for adequate pollination. The yellow capsules open to reveal scarlet red seeds and remain attached all winter.

Bittersweet, Oriental (*Celastrus orbiculatus*): This is the Oriental counter part of our native bitter sweet. The leaves are more rounded and the fruits are produced in small lateral clusters, not terminal clusters, like the native species. Staminate (male) and pistillate (female) flowers are produced on separate plants. The fruit is a yellow capsule which splits and exposes the red, fleshy seed.

Boston ivy or Japanese Creeper (*Parthenocissus tricuspidata*): This clinging vine climbs by tendrils with adhesive discs. It climbs rapidly and densely, even under trying city conditions. It produces a scarlet fall color.Small-leaved varieties are available.

Clematis (*Clematis sp.*): There are many species and cultivars available, with new hybrids being introduced yearly. All climb by twisting the petioles and all prefer a light loamy soil. Ample soil moisture and a cool soil are important, thus mulching is beneficial. Deep preparation of the planting site is beneficial. Vigorous-growing species need little pruning, although cold winters often kill the above-ground portions of some hybrids. A few popular species include:

Jackman Clematis (*Clematis x jackmannii*): Large violet-purple flowers in mid-July.

Big Petal Clematis (*Clematis macropetala*): Azure blue, large flowers in the spring.

Sweet Autumn Clematis (*Clematis terniflora*): White, very fragrant flowers in late August.

Scarlet Clematis (*Clematis texensis*): Bright scarlet, bell-shaped flowers in July.

Virgins Bower (*Clematis virginiana*): Small, white flower in late August. Use for wild or wooded gardens.

Dutchmans Pipe (*Aristolochia macrophylla*): This vigorous twining vine produces large leaves, sometimes 30 cm long, giving it a coarse texture. The inconspicuous flowers are almost hidden by the leaves, but do resemble a small, curved Meerschaum pipe. It is ideal where a large-leaved, vigorous vine is desirable.

English Ivy (*Hedera helix*): This dark green vine climbs by aerial rootlets. Mature types will produce a blackberry. Many named types are available, but all need some type of winter protection as they usually brown off quite badly.

Gloryvine (*Vitis coignetiae*): This rapid grower with redfruit and red fall colour is one of the parents of most of the American hybrid grapes. Choose hybrids according to fruit desired.

Grapes (*Vitis sp.*): Climbing by tendrils, grapes are dual purpose vines for both screening and fruit production. Having large leaves, they are rather coarse in appearance. Choose types which producc both attractive fruit and foliage.

Hydrangea, Climbing (*Hydrangea anomala subsp.petiolaris*): This outstanding climbing deciduous vine climbs by aerial rootlets. The white flowers are produced in clusters in mid-June. The reddish shredding bark on older stems is of interest in winter time. Use on brick walls, old walls, or stone piles.

Kiwi, Hardy (*Actinidia arguta*): An ideal screening twining vine useful for its dense foliage. The glossy, green, heart shaped leaf blades contrast with the colorful red petioles. Flowers and fruits are inconspicuous.

Passion Flower (*Passiflora sp.*): Many variants are available in a wide selection of flower types. Plant it in sunny but wind-protected spots. Provide a sturdy support. Summer mulches keep the roots cool and moist. In some areas, certain varieties will be killed to the ground but will send up new vines each year, thus behaving as an herbaceous perennial.

Porcelain Ampelopsis, Porcelain Ivy (*Ampelopsis brevipedunculata*): Climbing by tendrils, this vigorous vine has deeply lobed leaves and colorful berries about 0.75 cm in diameter. Produced in clusters, the berries are colored green, yellow, white and porcelain blue, all in the same cluster.

Rose, Rambler, Climbing (*Rosa sp.*): A large number of hybrid roses are available. Most need support. Regular pruning and other cultural care, similar to garden roses, is necessary. Select for flower color, growing habits, and hardiness.

Silver Fleecevine (*Polygonum aubertii, Fallopia aubertii*): This vine produces a dense and bright green foliage and profuse, small, greenish-white flower clusters in August. It is a vigorous grower (often 20-30 feet peryear).

Trumpet Creeper (*Campsis radicans*): This shrubby vine climbs poorly by aerial rootlets, but may need additional support since it becomes quite heavy. Its brilliant orange-scarlet mid summer flowers are its chief attraction.

Winter creeper Euonymus (*Euonymus fortunei*): This vine produces aerial footlets to become a climbing vine on low walls and fences. It can also be used as a ground cover.

Wisteria, Chinese (*Wisteria sinensis*): This strong growing, twining vine produces blue-violet, slightly fragrant flower clusters 18 to 30 cm long in late May. The flower clusters are dense and compact, giving a beautiful flower show. This species characteristically has 7 to 13 leaflets compared to 13 to 19 for the Japanese Wisteria.

Wisteria, Kentucky (*Wisteria frutescens, var.macrostachys*): The lilac-purple flowers are produced in flower clusters up to 10 inches long. It blooms later than the other species, in early June.

Wisteria, Japanese (*Wisteria floribunda*): This vine produces flower clusters 12 to 36 inches long of fragrant violet to violet-blue flowers in late May. Flowers open progressively. Many varieties have been carefully selected for color and are asexually propagated. Root pruning and applications of super phosphate seem to encourage bloom. This is reputedly hardier than Chinese Wisteria (*W. sinensis*).

CHAPTER 6

Cultivation and Management of Shrubs

Shrubs are usually woody or semi-woody perennial plants; their branches arise from the base of the plants and grow up to the height of 3 to 4 meter. Many shrubs, which are more or less prostrate, make excellence ground flower. Availability of a large number of species & cultivars, a wide range of variation in shape & size of the plants, diversified types of leaves, growth habit, remarkable ranges of size, shape, colour of flowers & the variety of fragrance have made the shrubs as important garden plants. Several species of shrubs have unusually attractive colours.

Most of the shrubs are hardy & grow in all types of soil with little attention. Although some plants flower almost throughout the year, in most cases they are seasonal. A selection can be made in such a way that some plants in the shrubbery will be in bloom in any season of the year. Most of these plants are fairly quick growing and when vegetatively propagated, start flowering in the first year. Plants like *Bauhinia acuminata* (Kanchnar), *Cassia alata* (Candle Bush), *Vinca rosea*,which are usually grown from seeds also produced flowers in the first year.Shrubs can be classified into three groups according to the requirement of the sunlight for growth & flowering:

- Show normal growth & flowering in full sun.
- Grow only in partial shade.
- Grow well in both in sun & semi-shady condition.

Suitable in Full Sun

Most of the tropical flowering shrubs are sun loving, growth & flowering are adversely affected in shade like *Coleus, Duranta plumerii variegata, Hibiscus rosa sinensis, Tabernaemontana coronaria.*

For Shady Areas

Plants like *Magnolia pumilla, Nandina domestica, Calliandra brevipes, Mussaenda erythrophylla* grow well in partial shade in ful sun these plants show signs of scorching during the hot weather & suffer in dry heat. Foliage shrubs like *Codiaeum* (Croton), *Polyscias* (Aralia), *Eranthemum* grow well in shade.

Suitable Both For Shady & Sunny Areas

The following species grow well in full sun and also in semi shade *Jatropha panduraefolia, Ravenia spectabilis, Brunfelsia undulata, Crossandra* (Firecracker Plant), *Turnera ulmifolia,* etc.

In addition to their utility as flowering or foliage plants, shrubs are also used as hedges & for screening. Hardy dwarf shrubs like *Gardenia jasminoides, Lantana camara Depressa, Malpighia coccigera, Plumbago capensis, Russelia juncea* are also grown in rockery to add color.

Description of Shrubs

Size of the plant and location of planting should carefully think before planting shrubs. Shrubs should be placed in such a place where they could not obstruct other flower beds, plants etc. though you can prune them but carefully think about the size when they will be fully-grown so that they should not towared over other plants.

1. *Abelia* (*A.grandiflora*):There are about 30 different deciduous to evergreen species in this genera. It is native to eastern Asia to Mexico, the United States and many European countries. It will do well in full sun to partial shade and like well-drained soil. The flowers are trumpet shaped and come in white and mauve colours.
2. *Abutilon* - Chinese lantern (*A.striatum*): There are more than 100 species and most are evergreen varieties. This plant does well in warmer climates. The Chinese lantern loves well drained soil and can take full sun or partial shade.
3. *Acacia* (*A. farnesiana*): This plant genus contains over 1200 species of different trees and shrubs. *Acacia* is native in tropical and sub-tropical Africa. This genus is considered warm climate specimens. It love full sun and its fragrant, globular flowers range in colour from a rich yellow tone to a creamy off white.

4. *Allamanda* (*A.cathartic, A.neriifolia, A.violecea*): This is a wonderful evergreen climbing shrub. It does well in a sunny and sheltered location. It loves rich soil and thrives on fences or decorative wall. The Allamanda shows its beauty with trumpet shaped yellow flowers. In cooler climates, this genus need to be in a container and protected from the outside elements.
5. *Aphelandra (A.tetragona)*: A common variety is known as the Zebra plant. The plant is noted for its beautiful foliage of vibrant green and yellow colour tones. The Aphelandra can not tolerate cold temperatures below 15°C. This is a fantastic evergreen in the tropics and will do well in part shade and rich moist soil. A wonderful specimen for a tropical ground or container garden.
6. *Artemisia* (Worm wood): A large genus of evergreen and deciduous shrubs that do well in the northern hemisphere. The Worm wood is used in gardens and landscapes for its soft foliage and many varieties are fragrant. They prefer a sunny location and well-drained soil. This is a great genus to create a border in a garden.
7. *Aurinia*: This genus has seven species of biennials and evergreens. The Aurinina is a small, spreading and mound forming plant. In the spring time it produces clumps of yellow or white flowers. Aurinia prefers full sun and light well-drained soil. The species are frost hardy and look sensational around dry stone walls and planted throughout decorative rocks.
8. *Barleria* (Philippine Violet): A group of 250 species of evergreen shrubs those are native to tropical Asia, the Americas and Africa. All of these species are frost sensitive, but can be an indoor plant in cooler climates. They need ample light to survive indoors. In warmer climates the Barleria loves full sun and humid conditions. The leaves are simple in nature and the flowers range from a beautiful mauve, purple, pink and white. The Philippines variety is actually not a native of the Philippines, but grows naturally in eastern India. This variety prefers a sheltered location and partial shade. The Barleria looks stunning in mass plantings and accents well as a shrub border.
9. *Barberry*: There are over 400 species of the Berberis genus. They are either hardy evergreens or deciduous. They are native to East Asia, some from South America and Europe. They can be propagated from seeds or cuttings and will do well in full sun and adapts to a variety of soil conditions. The Barberry is a beautiful specimen throughout the growing seasons. The Berberis genus is known for its sharp leaves and also brilliant fall colours. Some varieties have yellow, cream, and reddish-orange tone flowers.

10. *Boronia*: Around 100 species of compact evergreen shrubs that are native to Australia. This genus of shrubs has aromatic foliage and dominant pink four petal flowers. A few varieties have an off white and red tone flower. The Boronia does well in a sheltered area, full sun to partial shade and well-drained soil. It can tolerate a light frost.

11. *Brachyglottis*: Around thirty species makes up this attractive shrub. The Brachyglottis is native to New Zealand. This plant genus does well in a cool and mild environment that provides ample rainfall. It will grow in a well drained soil with full sun. The beautiful yellow flowers reveal this plants beauty and can be a focal point in the right garden setting. This plant has attractive foliage.

12. *Brunfelsia* (Yesterday-today and tomorrow): This genus is native to south and central America. They are evergreen shrubs or small trees with fragrant flowers. The unique aspect of this genus is the flowers will change colours throughout the growing season. They are slow growers and love full sun, but some afternoon shade. This is a species that will do well in containers but needs watering during the hot and dry months. The flowers produce a bluish-purple tone and create a wonderful look in a mass planting setting. The Brunfelsia can mature up to eight feet tall.

13. *Buddleia* (Butterfly Bush): This genus is native to China, Africa, Madagascar, southern Asia and South America. The Butterfly bush is one of the most popular genera in many garden settings and especially a butterfly garden. It love full sun and good drainage, but will adapt to a variety of soil conditions. It can propagate from cuttings in the summer and a pruning should be done in fall or early spring. It can trim back the butterfly bush to create a full look for the next year. Flower colours range from pinks, mauves, reddish purples, orange and yellow. This is a great specimen for accents throughout a landscape. Butterflies are attracted to this family, as well as bees and some varieties of birds.

14. *Buxus* (Boxwood): The Boxwood is a traditional plant in many places throughout the world. The Boxwood is common as a hedge row, along a foundation, or intermingling with other evergreen shrubs. It is known for its different green tone foliage throughout the year.

15. *Calceolaria* (Pocket book Flower): This genus contains over 300 species. It is native to North and South America, as well as Mexico and points further south. One of the most popular varieties of the Calceolaria genus is the slipper flower. This is a great cut flower used by many florists today. The genus loves a cool and shady environment, with well drained and moist soil. The shrub can be propagated from seeds or cuttings in

the late spring or summer time. The flowers form in clusters of yellow, orange and many have red or purple markings on a part each cluster. A favourite species that is native to Chile is *Calceolaria integrifolia.* This spreading bush can mature six feet tall, and produces vibrant yellow flowers from late spring into the fall season.

16. *Callandra* (Powder puff): There are over 200 species native to North and South America. A few varieties are also native in India, Africa and Madagascar. The Powder puff is a gorgeous hardy shrub that loves full sun and well drained soil. The red flowers resemble that of Mimosa, but are soft in nature with their long and slender petals. The unique leaves create a tropical look in many style gardens. This shrub will do great in a unique part of your garden. In the summer, these shrubs need to be on a good watering schedule to make sure the root systems don't dry out in the heat and humidity.
17. *Calicarpa* (Beauty Berry): There are over 140 species of this genus native to East Asia and North America. It can either be deciduous or evergreen shrubs, or small trees. One of the highlights of some varieties is the tiny cluster of pink or purple flowers that bloom in the summer. The cooler temperature in the fall allows these species to produce shiny red to purple colour berries. The Beauty Berry loves full sun and many good nutrients in the soil. Propagation can be done from cutting a part of the tips.
18. *Callistemon* (Bottle brush): Even though there are only twenty five species in this genus, the Bottle brush is a stunning looking shrub and native to Australia. The dominant colour of the spike flowers is red. Other varieties have flowers in pink and white. The Bottle brush can tolerate some frost conditions and loves full sun. The Bottle brush will also adapt to poor soil conditions. This is a great accent shrub and can be propagated by seeds or tip cuttings.
19. *Camellias*: The Camellias are native to North America, Europe, China, Korea and Japan. It is one of the most popular fall and winter species with many different varieties. The beautiful array coloured flowers peak once the temperatures cool down in the autumn. There are over 300 delightful species and look wonderful in natural settings, as well as throughout many different gardens.
20. *Caragana* (Pea Shrub): There are about 80 species and native to part of eastern Asia, Europe and North America. It is a unique looking shrub that has tiny leaves and tiny yellow and orange tone flowers. The Pea Shrubs thrive in cold winters and hot and dry summers, but due to the nature of their root systems, this shrub will adapt to a variety of locations. Their trunks are long and very narrow and make an interesting specimen or group of specimens in a natural setting.

21. *Carya* (Hickory): There are twenty species of the Hickory family and native to Asia and North America. They are considered medium to large trees and are noted for their edible nuts as well as the wood used for hard wood floors and furniture. Male and female flowers appear on different parts of the tree and blossom in the late spring. The Hickory has large green leaves that arrive in the warming of spring and in the fall, the leaves transforms into magnificent colours of yellow, orange and a rich gold tone. They love well drained and fertile soil and can take cold winters. A beautiful specimen in a front or backyard.

22. *Daboecia* (Irish Heath): There are only a couple of species that make up this genus. They are small leafed evergreen shrubs that are native to Western Europe and the Azores Islands. These shrubs are low growing and spread wonderfully in rock gardens and along retaining walls. This genus can be frost hardy and loves a sunny location. Moist and acidic soil works best for these evergreens. The highlight of this family is in the summer throughout the fall, when urn shaped striking purple tone flowers appear. These look stunning also in a mass planting. Once the flowers have fallen, the shrubs should be trimmed back for bountiful blossoms the next year. A few delightful species are *cantabrica* which is native along the Atlantic coast of France. The *alba* variety has white flowers and the *atropurpurea* has deep purple flowers. This specimen is a showcase and a great accent in many landscapes.

23. *Dais* (Pompon Bush): This is a small group of delightful species that are native to southern and eastern Africa. In late spring clusters of showy tubular pink flowers appear in a series of five lobes. This specie is considered an evergreen in warmer climates and deciduous in cooler climates. They require abundant sunshine and do like some shelter. The Dais can be propagated in the spring by seed or cuttings.

24. *Daphine*: The Daphine consists of more than 50 species and is native to Europe, North Africa and temperate Asia. The flowers are very fragrant and this species is a wonderful winter time bloom.

25. *Desfontainea*:This is a delightful evergreen species that is native from Andes to western South America. This bushy shrub has attractive dark green foliage and long tube shaped flowers that appear in mid summer throughout the fall season. The colours are a light yellow-orange tone, to a bright orange colour. Hummingbirds love to pollinate on this plant family. The Desfontainea matures to around five feet tall. A great sub-tropical to tropical plant specimen.

26. *Dodonaea* (Hopbush): There are around 60 species that make up this genus. Most of the specimens are native to Australia. The Hopbush is

native to not only Australia, also New Zealand, tropical regions of America, Africa and Asia. The Dodonaea genus loves full sun to partial shade. This genus is frost tender but can take the salty coastal winds as well as drought tolerant. It can mature up to five feet. The Dodonaea genus is considered an upright shrub and some of the larger shrubs produce large fruit. The flowers aren't really noticeable.

27. *Dracaena*: There are approximately forty species in this genus. They are native to Africa and Asia. The variety of delightful plants from this genus is grown indoors or outdoors. The outdoor specimens can take full sun or partial shade.
28. *Echium*: This genus consists of over forty specimens and is native to the Mediterranean, Canary Islands and Madeira in the western part of Europe. This genus is fascinating as it takes on the appearance of the eloquent Butterfly bush. Bright blue and purple large flowers bloom in late spring through the summer months. The *Echium* is considered either an annual or perennial shrub. This specimen thrives in a dry climate and full sun. This is a great shrub to mix with evergreens or plant in a mass setting. The flowers when in full bloom will mature to twenty four inches.
29. *Elaeagnus*: This species has around forty five different specimens and is native to Europe, Asia, and North America. Elaegnus does well in full sun to partial shade. The fragrant white flowers put this evergreen in the class of the Gardenia. After the flowers have died off, silver tone fruit appears and will last from late fall into the winter months. The maturity of the Elaegnus will reach ten to fifteen feet tall and wide.
30. *Ephedra*: There are around forty species of this genus. The Ephedra is native to southern Europe, North Africa, temperate Asia and the Americas. This evergreen shrub produces cone like flowers, and will bear fruit after the flowers die out. Species of this family do well in sandy soil or stone environment. The Ephedra can be propagated from seeds.
31. *Eriogonum*: There are around one hundred and fifty species that make up this genus. The Eriogonum species are annual or perennial shrubs that show off their silvery tone leaves, and produce pink tones and yellow flowers. It loves sunny and well drained environment, and does well in dessert.
32. *Euonymus*: About one hundred and fifty species make up this eloquent genus. This genus is native to the northern hemisphere as well as East Asia.It grows best in a sheltered area, a good amount of sunlight and well drained soil. The variegated Golden Euonymus is a favourite hill of India.

33. *Euphorbia* (Milkweed): This is a huge genus made up of over two thousand species. The species are native to Madeira and Madagascar. Many of the species will also grow well in a sub-tropical and tropical setting. They all flourish in well drained soil and can take sun to partial shade. All of the Euphorbia contain a milky sap and bear red, pink and off white flowers. Many of these evergreens are seen growing wild in a lot of locations.

34. *Euryops*: There are about one hundred species in this genus. Native to southern Africa and grown for their eloquent and vibrant colour flowers. They will thrive in full sun and well drained soil. The flowers resemble a daisy and their foliage is similar in nature to a fern leaf. These evergreen shrubs will do well if kept pruned. If they are allowed to grow without pruning, these plants will become leggy. A great specimen to accent with other solid evergreens.

35. *Felicia* (Blue Daisy): There are about eighty species that make up this genus. The Blue Daisy is native to Africa and Arabia Felicia species can be annuals, perennials and evergreens. In warm tropical climates, the fabulous flowers will appear throughout the entire year. Purple and blue tones create a wonderful setting for mass planting. The foliage is highly fragrant and wonderful near a bench or gazebo. The Blue Daisy thrives in full sun and needs well drained and rich humus in the soil. Dead heading will give a longer flowering season.

36. *Forsythia*: This fabulous genus consists of seven different species and has a long history. They are grown today throughout many different climate regions. The Forsythia comes alive in the spring time with massive yellow and gold tone flowers. Different varieties range from three to ten feet high and five to twelve feet wide. It is a deciduous shrub and loves full sun. The Forsythia can be drought tolerant, but does need occasional watering. The flowers aren't as abundant in the warmer climates. A great specimen for spring showcase.

37. *Fortunella* (Kumquat): There are five evergreen small trees or shrubs in this family. The Kumquat or Cumquat is native to China and Europe. In spring the plants emerge with white flowers, followed by the delicious edible mall orange-yellow tone fruit. This genus loves full sun and well drained soil.

38. *Fuchsia*: This genus has some of the most fabulous colour flowers and consists of around one hundred species. It is native to South and Central America. Fuchsia needs sun to partial shade and moist, well drained soil. This genus does need protection from the hot summer winds and afternoon sun. Pink, red and purple tones are in mass quantity on these

plants. Absolutely a wonderful specimen for a ground or container garden.

39. *Gardenia* : The sub-tropical and tropical climates are the ideal place to have one of the most fragrant evergreen shrubs. There are over two hundred species of the Gardenia genera. This genus is native to tropical Asia and South Africa. Most specimens will grow four to six feet tall and wide. The creamy off white flower appears in late spring and continues throughout the summer months. This great shrub is ideal species to plant near a deck or patio. The sweet aroma is one of the most magnificent smelling flowers in their climate zones.

40. *Gaultheria* : There are around one hundred and seventy species in this genus. These evergreen shrubs are native to moist climates all over the world. Gaultheria can also be seen in the mountainous areas of the tropics. This plant family has dark glossy leaves and bell shaped pink or white flowers. Gaultheria prefers well drained soil with an acid base. The plants should be some what sheltered and prefers partial shade. After flowering fragrant berries appear on the shrub and make this species a great season to season plant for many style gardens.

41. *Halesia* (Silverbell - Snowdrop): There are five species that make up this genus. They are native to the eastern United States and China. The Silverbell and Snowdrop are seen in moist woodlands and along natural streams. They are known for their bell shaped white or pink flowers that make their appearance in late spring. This genera can survive in a cool climate and do well if in a sheltered area. Partial shade to full sun is the ideal condition for these species.

42. *Hamamelis* (Witch Hazel): There are around five species of the genus and are native to North America and East Asia. They are deciduous shrubs or small trees. Red and yellow fragrant flowers dance on the branches in the springtime. These species are wonderful for a cool climate and in full sun. Great shrub to plant in an open area. In the autumn, the foliage changes to a yellow with red and orange tones in each leaf.

43. *Jasminum* (Jasmine): This genera consists over two hundred members. They are considered deciduous, semi-evergreen and evergreen, depending on the climate zone. The Jasmine is native to Asia and Africa but also in parts of North America. Some of these plants can be seen along the edges of wooded areas. Most are frost tender, but a few species can take some colder temperatures.

44. *Juniperus* (Juniper): A very popular genus made up of about fifty species. They are native to different areas of the Northern Hemisphere. Many of

the species are great trees and shrubs for both landscape and specific gardens. The Junipers love well drained soil and a sunny location. Many of these specimens are used for banks or on slopes.

45. *Kolkwitzia* (Beauty Bush): This genus is made up of one member and is native to China, but planted in many landscapes and garden throughout the world. It loves well drained soil, full sun or light shade. In the spring mass of pale pink clusters of flowers appear on this upright shrub. This species is deciduous but a wonderful accent plant in front of an evergreen backdrop. After the flowers die off, small fruit cover the shrub. A very showy specimen for the spring and summer months.
46. *Lagerstroemia* (Crepe Myrtle): This genus consists of around fifty members. The genus produces some of the most vibrant blooms in the late spring and throughout the summer months. The Crepe Myrtle is native to southern and eastern Asia and also Australia. The *indica* species is the pride of India.
47. *Magnolia*: This magnificent specie has around one hundred family members. The species was named after a French botanist named Pierre Magnol. These species are deciduous or evergreen trees or shrubs. Magnolias require well drained and fertile soil. This species can take full sun or partial shade but should be protected from strong or salty winds. The magnificent aspects of these plants are their incredible fragrant flowers.
48. *Mahonia*: This genus is made up of around seventy species. Native to East Asia, North and Central America. These species can take full sun in cooler climates, but require partial to full shade in warmer regions. It will thrive in well drained soil and regular watering. The plants don't need much pruning but they do produce basal suckers that can be cut to keep in control. Tiny yellow flowers appear in spring before bearing fruit. Other species will produce the yellow flowers in summer and purple-blue tone berries will emerge in the autumn and throughout the winter months. The leaves are pointed and can be sharp to touch.
49. *Melianthus* (Honey Flower or Touch-Me-Not): This magnificent family is made up of six members. All of these shrubs are all native to South Africa. The Honey Flower or Touch-Me-Not species will mature anywhere from six to around ten feet tall. The Melianthus loves full sun, but can take some partial shade. Well drained to moist soil conditions will suit this shrub. Usually after the growing season, these shrubs are cut back as they can become leggy. If the leaves are bruised, they can produce and unpleasant order. Flowers on this species are a dark red tone and are followed by seed pods. This species is grown for its foliage, but when blossoming, the flowers take centre stage.

50. *Mimulus* (Monkey Flower - Musk): There are about one hundred and eighty specimens that make up this genus. They love full sun or partial shade. These species can be annuals, perennials, or evergreen shrubs. Flowers range in colour from crimson, orange, cream and yellow. The Monkey or Musk plant will grow to around three feet high and wide.

51. *Nandina* (Heavenly Bamboo): There is only one specimen in this genus (*N.domestica*). The Nandina is native to Japan and China. In the summertime, white flowers appear and are followed by vibrant red berries throughout the fall and winter months. This shrub will come to life in the winter as its leaves turn into a deep red tone, and also become a showcase in many garden settings.

52. *Nerium* (Oleander): This genus has only about two species. The Pacific Islands are native to these evergreen shrubs. The flowers are funnel shape and vibrant colours of red, pink and white show off in many different landscapes in warmer climates. Oleander thrives in full sun and well drained soil.

53. *Neviusia* (Alabama Snow Wreath): There are only two specimens of that make up this genus. It is a deciduous shrub that can mature to around five feet tall. The white flowers make their appearance in the spring and this shrub will spread due to the nature of its root system. After flowering, the shrub should be trimmed back. The Neviusia needs an open space to maintain its health.

54. *Notospartium*:This genus is made up three species. It is primary native to New Zealand but has been popular in gardens and landscapes in other mild winter climates. In spring and the summer months, this deciduous shrub becomes full of pea shape pink and lilac tone flowers. The specimens love full sun and may have to be staked. The shrubs will flower for many weeks if protected from any strong winds. A beautiful spreading shrub that creates a nice look along a fence or natural stone wall.

55. *Paeonia*: There are around thirty five species that make up this genus. Paeonia's grow in many different climates and are native to China. The Paeonia is a gorgeous perennial shrub. This species loves cooler winters. The shrub will thrive in full or filtered sunlight. This entire genera is deciduous and will need proper pruning after flowering takes place. The flowers themselves are precious and colours range from yellow, white, and pink. It is a beautiful specimen for a variety of gardens and landscapes.

56. *Paliurus*: This is a small genus of around eight species. They are deciduous or evergreen spindly shrubs and trees. The Paliurus is native to the

Mediterranean region. These shrubs and trees both contain thorns and make a good specimen for hedging or as an accent plant away from children. In the summertime, yellow and green tone star shape flowers appear. This species loves full sun and well drained soil. After the flowers have died out, yellow coloured berries arrive.

57. *Pandanus*: This is a large family of about six hundred species. Pandanus is native to East Africa, Malaysia, Australia, and the Pacific. These species love a warm climate. They love full or partial sun and a well drained soil. This palm like shrubs have tiny white flower and fruit that resembles a pineapple. The male flower has a rich scent to it. One of the unique aspects of these plants when they are young, they can make precious indoor plants. The Pandanus are stunning in a landscape or garden near the ocean.

58. *Penstemon*: The Penstemon is large genus of about two hundred and fifty members. It is also native to Central America. The family is somewhat frost hardy. The plants love full sun and well drained soil. In late spring and throughout the summer months, eloquent tones of white, blue, red and variegated flowers appear on this woody shrub. There are tall and shorter varieties from hybrids.

59. *Phebalium*: This genus is made up of about forty evergreen shrubs and small trees. These species are native to Australia. These plants are fragrant and also considered members of the citrus family. The Phebalium can tolerate light frosts and love full sun or partial shade. They need well drained soil. The Satinwood species will mature to around twenty feet. This is a great privacy shrub or tree. The flowers are simple and are yellow and white. The leaves can be all green or variegated green and yellow tones. This family is known for its foliage.

60. *Philadelphus* (Mock Orange): This is an extremely popular genus having about sixty specimens. Southern Europe, Asia, and many northern hemisphere countries are native to these deciduous shrubs. These precious shrubs all grow to around ten feet in height and can reach ten feet in width. The clusters of white to cream shade flowers appear in the late spring and summertime months.

61. *Rauvolfia* (Quinine shrub): There are over fifty species that make up this genus. This genera is native to both the northern and southern hemisphere tropical areas, and especially southern Africa.Rauvolfia specimens are fast growers and can mature up to forty feet. The fragrant flowers have green and white tones and followed by shiny black colour berries.

62. *Rhabdothamus*:This is a rare one specie family and native to New Zealand. The Rhabdothamus is actually a member of the African violet family.

The flowers are tubular in stunning shades of orange and red tones. This plant is an evergreen shrub that likes partial shade and humus rich soil. It can be a frost hardy shrub and should be pruned often to produce a fuller plant. The Rhabdothamus is a wonderful specimen that grows about six feet high and wide.

63. *Rhamus* (Buckhorn): This genus is made up of deciduous and evergreen shrubs. The Buckhorn is native to a variety of different climates. There are about one hundred and twenty five species in this family group. The shrub loves well drained soil and will do well in full sun, but partial shade is best for hot areas. Some of the species have thorns and the leaves are jagged. The flowers aren't noticeable, but the berries are popular with birds.

64. *Rhalphiolepis*: There are nine species that make up this genus. These species are native to sub-tropical and tropical Southeast Asia. These shrubs love full sun or partial shade and well drained soil. The Rhalphiolepis is known in spring for its tiny clusters of white and pink flowers. The coastal area is an ideal area to grow these specimens.

65. *Rhododendron*: This is a very large genus of over eight hundred species. The Rhododendron can be found in so many different geographic locations. Many of are native to southern China, Himalayan region, Japan, Southeast Asia, Australia and North America. Also known as the Rhodi this specimen grows wild in many mountainous areas. This genus is known for its brilliant colour flowers that appear in the late spring throughout the summer months. There are many different coloured flowers and a wonderful focal point in gardens and landscapes.

66. *Ribes* (Currant): This genus is made up of around one hundred and fifty species. This genus is a group of wonderful deciduous and evergreen ornamental and fruit bearing shrubs and trees. In late winter and early spring, clusters of yellow, pink, or red flowers appear. The berries are in colours of white, green, purple or black and many are edible. These plants can be planted in masses to produce different coloured berries. They will thrive in rich, well drained soil and can take full sun or partial shade. Another great specimen for gardens.

67. *Richea*: This is a rare family of twelve evergreen shrubs. They are native to Australia and are also large numbers in Tasmania. The leaves are a rich green color, and long and slender. The clusters of pink, white, yellow or cream tone flowers come to life in the summer months. The popular Richea will do well in ground gardens and also popular in container gardens. Richea needs a rich base humus and acid base soil and needs protection from full sun and strong winds. These shrubs are generally

small in height but a couple of the varieties can mature up to fifty feet. A very unique genus.

68. *Salix* (Willow): This genus consists of around three hundred species. They are deciduous trees, shrubs and dwarf shrubs. The Willow generally thrives in cooler climates. Salix is native to most of the northern hemisphere including Europe, northern Africa, northern and central Asia. The Willow prefers a moist soil with partial shade conditions. Their height varies with the different species in this family, but some will mature up to eighty feet tall. Others will max out around fifteen feet.

69. *Sambucus* (Elderberry): This genus consists of around twenty five members and is native to North America, Europe, northern Africa and western Asia. The Elderberry is a sensational deciduous shrub and small tree. Sambucus is a fairly easy specimen to grow and can take sun or shade, as long as it has a moist and well drained soil. In late spring and early summer clusters of small white flowers emerge. Some of the varieties have edible flowers and berries. A great plant to put as an accent around evergreen shrubs and trees.

70. *Santolina* (Cotton Lavender): There are about eighteen species that make up this genus. The Cotton Lavender is an evergreen shrub that is mainly native in the Mediterranean region. Santolina requires full sun and well drained soil. This is a low growing evergreen herb, and a sensational fragrant shrub for an herb garden. Cotton Lavender is low spreading and mature to around two feet tall and can spread three feet wide. Planting Santolina in front of Rosemary creates a wonderful look in any landscape.

71. *Senna* (Cassia): This is a large genera consisting of around two hundred and fifty species. They are native to Americas, Australia, and Africa. In the summertime, clusters of rich yellow cup shaped flowers appear and will blossom until winter. They should be pruned after bloom time. The Cassia loves full sun but needs protection against strong winds. These species are considered evergreen and perennials.

72. *Taxus* (Yew): This genus has a small number of members and even though they are slows growing. Taxus is native to Asia, North Africa, and Europe. The Yew is very adaptable to many soil and climate conditions, but does need some protection in warm and dry climates during the summer months.

73. *Thrptomene*: This is a genus of about forty members and native to Australia. One of the highlights of this specimen is when the leaves are crushed; they create a tremendous fragrance to enjoy. The white and

pink flowers arrive in winter and throughout the spring. Thrptomene will mature up to five feet tall and can also spread up to five feet wide. They prefer full sun to partial shade and a lime free soil.

74. *Thuja* (Aborvitae): Great genus with around five family members. The Arborvitaes are evergreen shrubs and trees that are native to North America and north-eastern Asia.

75. *Thunbergia*: This is a attractive family of around one hundred species. They are evergreen and mostly climbers with soft purple-blue tone flowers. The blossoms appear during the summer months and will die out around the winter time. Thunbergia specimens are native to Africa, Japan and other parts of Asia. They can grow well in sub-tropical to tropical climates. These shrubs prefer full sun but do need some protection during the hot summer months. They love rich soil and well drained soil and look tremendous in with a variety of natural outdoor features.

76. *Ugni* : This is a small family of evergreen shrubs. Ugni is native to South and Central America. Ugni average around ten feet tall and are highlighted by their edible and splendid fruit. This species needs well drained and sandy soil conditions. Ugni can take full sun but also partial shade. These shrubs produce colourful pink tone flowers and the wonderful berries are red and purple. Ugni is a great landscape shrub and a nice back drop.

77. *Ulex*: This genus is made up of about twenty or so species. Ulex is native to Western Europe, and North Africa. One of the unique aspects of this genera, is the shrub does not bare many leaves. The fragrant and colourful rich yellow tone flowers emerge in springtime. Ulex can take poor soil conditions, and can mature up to six feet tall. This species is frost hardy, and can take full sun. In sub-tropical and tropical climates, these shrubs should be protected from the summer sun and heat. This is also a great specimen in coastal areas. The Ulex species are benefited from pruning after flowering.

78. *Viburnum*: There are about one hundred and fifty species that make up this genus. This family is native to Asia, with some varieties in North America, Europe and Northern Africa. This genus of evergreen, semi-evergreen and deciduous shrubs are known for their fragrant and colourful flowers and berries. Viburnum love well drained soil and sun or partial shade. They can be pruned after flowering, but if pruned, will produce fewer berries.

79. *Vitex* (Chaste Tree): This genus is made up of about two hundred and fifty species. The Vitex species are usually found in sub-tropical to tropical

locations. It is native to many parts of Asia and New Zealand. This genera consists of evergreen and some deciduous trees and shrubs. Vitex adjusts to a variety of soil conditions and needs some shelter from the hot summer months. It will thrive in moist conditions. The flowers are stalks of rich red, pink and lavender tones. This is another great specimen for landscapes and a variety of garden settings.

80. *Weigela*: This genera is made up of around twelve species. They are all considered deciduous and are native to Japan. The Weigela species loves full sun but can take partial shade. Moist and well drained soil will allow these species to flourish. In late spring to early summer, crimson, pink and white tone flowers appear and can accent any landscape or garden. The flowers are shaped like a trumpet and go well with evergreen shrubs or trees planted around them. In fall, the foliage turns to vibrant yellow, red and purple tones. This is great species to incorporate in a garden setting.

81. *Westringia* (Coast Rosemary): This genus is made up of around twenty five evergreen species. Westringia is native to Australia, and also known as Australian Rosemary. This species of Rosemary will thrive in warmer climate regions, especially in coastal regions. Westringia loves an open area that gets full sun and will adapt to most soil conditions. In summer the species should be kept on a good watering schedule. In spring light purple tone flowers make their appearance and will last for months. This is an attractive evergreen shrub and also creates a long focal point in landscapes.

82. *Xanthoceras* (Yellow Thorn): This genus is native to China. This species can tolerate cold temperatures, but should be protected against late season frost. The Yellow Thorn will thrive in hot summer conditions; the warmth of the summer allows the plant and flowers to achieve its growth and beauty. The Yellow Thorn is considered a deciduous shrub or small tree and it can mature up to fifteen feet tall. The white flowers emerge in late spring and throughout the summer months.

83. *Yucca*: There are about forty evergreen perennial shrubs and small trees that make up this genus. Yucca is noted for its long slender solid and variegated leaves that have a sharp point at the end. This specie should be put out in an open area, where there is low humidity and full sun. Yucca does better in a sandy soil environment, but can survive in some clay base soil. The roots of this plant are deep and also invasive. Another great look for a sub-tropical to tropical setting.

84. *Zelkova*: A cool climate genera that is native to China, Japan, and western Asia. Zelkova is grown for its foliage, and is a deciduous specimen. It is frost tolerant but needs shelter from extreme cold conditions. Full sun

will help this specimen develop a healthy growth pattern. Great specimen in front or backyards.

85. *Zizipus*: There are eighty or so members of this genus of deciduous and evergreen shrubs and trees. They are native to both North America and South America.

Low-Maintenance Shrub and Hedge Plants for Landscape

It is important to note that even the most drought tolerant plants require watering during the establishment period. The term "low-maintenance" refers to a plant that does not require frequent maintenance such as regular watering, pruning, or spraying to remain healthy and to maintain an acceptable aesthetic quality. A low-maintenance plant has low fertilizer requirements and few pest and disease problems. In addition, low-maintenance plants suitable for India must also be adapted to or at least tolerate our poor, alkaline, sand or limestone-based soils. Both native and non-native species are included herein. Also, a variety of plants that are not commonly available in Indian nurseries or garden centres were intentionally included with the hope that increased demand will generate increased supply. Putting the right plant in the right place is of foremost importance in creating a healthy and successful low-maintenance landscape. However, this principle is dependent upon one's ability to accurately select the right plant species for a given location. Common names are often misleading, and sometimes more than one plant species may be referred to by the same or similar common name.

Table 6.1 : Low-maintenance shrubs and Hedge Plants

Common name/ *Scientific name*	Size (cm)	Light preference	Salt tolerance	Remarks
Aglaonema (*Aglaonema commutatum)*	45	Partial to full shade	Low	Small, herbaceous perennial with fleshy stems, branching from the base and Slender-stalked leaves up to about 30 cm long. Flowering stems have a narrow, pale green spathe enclosing a small white spadix. Other Aglaonema are also drought tolerant.
Scarlet milkweed (*Asclepias curassavica)*	90-120	Full sun to partial shade	Low	Erect, evergreen perennial with narrow, elliptical leaves and terminal clusters of scarlet and orange flowers spring to fall. Attracts butterflies spring to Fall. Can tolerate wet soil.

Butterfly weed *(Asclepias tuberosa)*	45-90	Full sun to partial shade	Low	Perennial with orange, summertime flowers. Essential component of a butterfly garden. Attracts butterflies and humming birds.
Cast iron plant *(Aspidistra elatior)*	90	Partial to full shade	Moderate	Grown for its deep green foliage. Named for its ability to tolerate harsh, adverse conditions.Spreads to form clumps. Excellent for shady areas.
Star begonia, winter begonia *(Begonia heracleifolia)*	60-90	Partial to full shade	None	Rhizomatous (spreads by rhizomes) begonia with large, hairy, deeply lobed, toothed, bronzy green leaves and fragrant, white to pink flowers on long, reddish stalks.
White begonia *(Begonia popenoei)*	60-90	Partial to full shade	None	Rhizomatous begonia with huge, round, dark glossy green leaves to two feet wide and large white, fragrant flowers held in large panicles well above foliage.
Castor bean begonia *(Begonia ricinifolia)*	60-90	Partial to full shade	None	Rhizomatous begonia with apple green leavess haped like stars and with rings of red hairs around leaf petioles. Winter blooms are deep pink.
Strawflower *(Brachteantha bracteatum)*	90	Full sun	Low	Annual or short-lived perennial with erect habit. Weak, hollow stems. Thin green leaves bear golden yellow blooms up to two inches in diameter at branch tips from summer to early fall. Good ground cover or small shrub.
Spiral ginger *(Costus scaber)*	170	Partial shade	Low	Tall spiral ginger with attractive roundish jade green leaves with fuzzy underside, and long, hard, waxy 20 to 30 cm red bracts

				with small,yellow flowers that peek out as bracts open from bottom up. Long lasting flower; excellent cut flower.
Crossandra (*Crossandra infundibuliformis)*	30-60	Partial to full shade	Low	Simple leaves arranged in opposite pairs and bearing erect spikes of showy yellow to red flowers, with petals opened flat into a hand-like shape.
African iris *(Dietes iridioides)*	60-90	Full sun	Low	Drought tolerant perennial, though extra water needed when in bloom. Easy to grow. Attractive white flowers with blue and brown shading appear in spring. Forms spreading clumps. Also known as *D. vegeta; D. bicolor*has yellow flowers.
Twinflower *(Dyschoriste angusta)*	15-45	Full sun	Low	Native perennial with small, delicate, purple flowers. Grows best in dry sand. Plant close together for best cover.
Dwarfcrown-of-thorns *(Euphorbia milii)*	30-90	Full sun	High	Small Thai dwarf varieties used as bedding plants and ground covers
Indian blanket *(Gaillardia pulchella)*	30-60	Full sun	High	Colorful annual or perennial. As easily grown as it is beautiful. Considerable variation in flower color; typical variety bears red flowers with yellow-tipped petals. Plant in open site with good drainage. Good for beach-front plantings.
Whirling butterflies *(Gaura lindheimeri)*	60-120	Full sun to partial shade	Low	Herbaceous perennial that grows in loose,bushy clump. Stems slender, wiry and covered with tiny hairs. Leaves spoon-shaped with toothed margins. Flowers above foliage on long spikes;

				open white at dawn, fading torose-pink by day's end. Requires well-drained soil.
Moss verbena *(Glandularia puchella)*	30	Full sun to partial shade	Unknown	Freely branching, sprawling perennial with3-lobed, finely dissected, aromatic leaves and clusters of pink, lilac, purple or white flowers. Requires well-drained soil. Attracts butterflies. Good bedding plant.
Tampa verbena *(Glandularia tampensis)*	45-60	Full sun	Low	Sprawling, short-lived perennial. Sometimes treated as annual. Flowers purplish-pink. Also known as *Verbena tampensis.*
Rain lilies (*Habranthus* spp.)	60-120	Full sun	Moderate	Flowers yellow, pink, red or white produced spring/ summer after rain. Goes dormant in cool season, requiring very little water. Flowers best when crowded.
Beach sunflower (*Helianthus debilis)*	90-120	Full sun	High	Erect or prostrate, spreading plant with sand paper-like leaves and 2.5 - three-inch yellow flowers. Suitable for beach-front plantings. Good for attracting butterflies.
Daylily *(Hemerocallis spp.)*	30-60	Full sun or partial shade	Moderate	Clump-forming perennial with grass-like leaves and lily-like flowers from late spring to fall, depending on variety. Available in yellow, orange, red, pink, purple, and near-white.
Amaryllis *(Hippeastrum* spp.)	60	Full sun to partial shade	Moderate	Spectacular trumpet-shaped, tropical lilies. Many species and varieties. Moderately drought tolerant.
Beach elder *(Iva imbricate)*	60-90	Full sun	High	Scraggly, coarse, shrubby perennial. Good forbeach-front plantings.

Shrimp plant *(Justicia brandegeana)*	90-150	Full sun to full shade	None	Perennial or shrub with stems topped with 15 cm spikes of red, rusty brown, or green brackts containing small, white, shrimp-like flowers. A related species bears yellow brachts. Unique, adaptable, easily grown plant. Also known as *Beloperone guttata.*
Kalanchoe *(Kalanchoe blossfeldiana)*	30	Partial shade	Moderate	Small, shrubby African species. Multiple, upstretched branches covered with round to rectangular, deep green leaves with red margins and notched tips. Thick racemes of small, deep red, cylindrical flowers appear from winter to early summer; may also be pink, yellow, or orange.
Statice *(Limonium sinatum)*	45	Full sun	Low	Bushy, upright perennial. Dense rosettes of oblong, deeply waved, dark green leaves and masses of tiny, papery flowers on winged stems summer/ early fall. Fairly slow growing. Available in golden, lemon-yellow, white, cream, salmon-pink, purple or blue spikelets.
Pennyroyal *(Piloblephis rigida)*	15-60	Full sun	Low	Low-growing perennial. Not picky about soil, but likes it dry. Small, shrimp-like flowers emerge from green, scale-like brackts. Attractive to butterflies. Also known as *Pycnothymus rigidus* and *Satureja rigida.*
Plumbago *(Plumbago auriculata)*	90-180	Full sun	Low	Attractive light blue flowers in terminal spikes. Drought tolerant once well established. Prone to root and stem rots. Freely draining soil and good air circulation.

Florida mountainmint *(Pycnanthemum bloridanum)*	120	Full sun to partial shade	None	Erect, loosely-branched, shrub like perennial. Leaves smell like spearmint or camphor. Tiny, lavender flowers in dense terminal clusters.
Firecracker plant *(Russelia equisetiformis)*	60-120	Full sun	High	Grown for weeping habit, masses of small, red tubular flowers, and scale-like leaves. Often used in planters. Requires freely draining soil.*Russelia sarmentosa* is similar, although with actual leaves and smaller red flowers that are highly attractive to butterflies.
Stonecrop *(Sedum spectabile)*	60	Full sun to partial shade	Moderate	Hardy plant, well adapted for rockeries, withshowy fall flowers. Many cultivars (white, pinkor red); at least one with variegated leaves.
Blue porterweed *(Stachytarpheta jamaicensis)*	30-90	Full sun to partial shade	Moderate	Small, shrubby perennial. Flower shighly attractive to butterflies—must for any butterfly garden. Name derived from traditional foamy, porter-like beverage brewed from the plant.
Pink porterweed *(Stachytarpheta mutabilis)*	150-240	Full sun to partial shade	Moderate	Small, shrubby perennial with somewhat weedy habit. Flowers highly attractive to butterflies and humming-birds. Variety *violacea*is violet-flowered form.
Mexican sunflower *(Tithonia deversifolia)*	150-180	Full sun	Low	Warm-season annual or perennial. Leaves coarse, three-lobed, and covered in soft,downy fuzz. Flowers orange to red-orange.
Society garlic *(Tulbaghia violacea)*	60-90	Full sun to partial shade	None	Clump-forming, herbaceous perennial with narrow, grayish leaves and large clusters of lavender flowers. Related to garlic.

Purpletop verbena *(Verbena bonariensis)*	90-180	Full sun to light shade	Low	Erect, clump-forming perennial with stiff, widely branched stems. Small, purple flowers in clusters held high above foliage. Best when densely planted.
Verbena (*Verbena* x *hybrid)*	90-180	Full sun	Low	Erect clump-forming perennial or annual with stiff, widely branching stems. Flowers purple and borne in rounded clusters. Treat as annual. Good bedding plant.
Rain lily *(Zephyranthes spp.)*	15-40	Full sun to partial shade	Moderate	Charming, clump-forming, flowering bulbs. Most species have open flowers like small Asiatic lilies and are deciduous. Can be white, yellow or various shades of warm pink. All bear one flower to a stem and most close their petals at night.
Orach, French spinach *(Atriplex hortensis)*	60-180	Full sun	Unknown	Usually grown for its tender, spinach-like leaves. Leaves can be green, yellowish-green, red or purple. Sometimes grown as bedding plant.
Cosmos*(Cosmos bipinnatus)*	150-180	Full sun	Low	Feathery-leaved annual. Showy, daisy-like flower heads in summer and fall, in shades of pink, red, purple or white. May require staking. May not survive rainy season.
Gazania *(Gazania rigens)*	15-60	Full sun	Moderate	Mat-forming plant with crowded rosettes of mostly unlobed leaves that are green aboveand whitish beneath. Orange flower heads with a black eye spot at petal bases.
Globe amaranth *(Gomphrena globosa)*	30-60	Full sun	Low	Small, branching annual with hairy leaves and bright purple, pink, or white, clover-like flowers. Benefits from mulching.

Butter daisy *Melampodium divaricatum*	15-60	Full sun	Low	Small, summer annual with small, yellow flowers and bright green foliage. Prolific self-seeder. Great bedding and border plant.Also known as *M. paludosum*.
Moss rose, rose Moss (*Portulaca grandiflora)*	10-20	Full sun	High	Small, annual, low-growing succulent with small, lance-shaped, fleshy, bright green leaves on reddish stems. Yellow, pink, red or orange flowers in summer; close at night andon cloudy days. Suitable as ground cover, in rockery, or as border.
Black-eyed Susan *(Rudbeckia hirta)*	90	Full sun to light shade	Low	Typical daisy-like perennial or annual (depending on variety). Large, yellow-orangeto reddish-orange flowers with darker centers. Attracts butterflies. Does not tolerate prolonged, wet, humid weather.
Chenille plant *(Acalypha hispida)*	180-360	Full sun to partial shade	Moderate	Medium to large shrub with large, oval, evergreen leaves. Separate male and female plants. Females have long, velvety clusters ofpurple, bright red or crimson flowers resembling fluffy cattail. Can wilt in full sun during drought.
Copperleaf (*Acalypha wilkesiana)*	240-360	Full sun to partial shade	Moderate	Purple/red, green and pink foliage. Excellent for hedge or border, but can be over-powering.
Desert rose *(Adenium obesum)*	90-210	Full sun	Moderate	Very showy flowers year-round, particularly during hot, dry weather. Must have perfect drainage since this plant is very prone to rootand stem rots. Will lose leaves during cool weather. Excellent for rock gardens. Scale soccasional pest.

Century plant *(Agave Americana)*	180-240	Full sun	High	Dramatic foliage and form. Evergreen, silver/gray to blue-green foliage. Showy, green-brown fruit. Armed with spines. Gritty, free-draining soil required. Blooms anywhere from 12-20 years of age then die. Excellent for rock gardens. Many other agaves available; some are variegated; all are drought tolerant.
Purple allamanda *(Allamanda blanchetii)*	300	Full sun	None	Evergreen, vining shrub with purple, tubular flowers. Attractive to butterflies and hummingbirds. Also known as *Allamanda violacea.*
Bush allamanda *(Allamanda neriifolia)*	90-150	Full sun to partial shade	Moderate	Colorful, bright yellow, trumpet flowers. Glossy leaves in whorls of 3-6 on smooth stems, which bleed milky sap if cut.
Shell ginger *(Alpinia zerumbet)*	180-240	Full sun to partial shade	Moderate	Tall stems topped with drooping inflorescences bearing white, pink, red and yellow shell-shaped flowers. Forms dense clumps. Green and variegated forms. Best growth in moist soil, but can tolerate drought.
Marlberry *(Ardisia escallonioides)*	150-600	Partial shade	High	Native plant with attractive foliage and berries.Can be used as small specimen tree or barrier shrub. Does poorly in full sun.
West Indian sea lavenderv *(Argusia gnaphalodes)*	120-180	Full sun	High	Rounded shrub with fine, grayish-green leaves and small white flowers. Also known as *Mallotonia gnaphalodes.*
Sea oxeye *(Borrichia arborescens)*	60-120	Full sun	High	Yellow, daisy-like flowers. Widely used on banks and slopes. Excellent for beach-front plantings in sand. Elsewhere, use freely

				draining soil to prevent root rot. *Borrichia frutescens* is a native silver oxeye, and has similar requirements.
Bougainvillea (*Bougainvillea* spp.)	150-900	Full sun	None	Very showy, thorny vine-like shrubs with bestbloom during dry season. Numerous cultivars available in a striking array of colors, magenta being the most common. Some with variegated foliage. Best effect as a vine butcan be trained as a shrub. Requires freely draining soil.
Yesterday-today tomorrow (*Brunfelsia grandiflora)*	90-240	Full sun to partial shade	Moderate	Aptly named shrub with pansy-like flowers that open purple on the first day, then turn pale lavender on the second day, and finally white on the third day. Some leaf drop may occur during persistent drought.
Butterfly bush (*Buddleia officinalis)*	180-360	Full sun to partial shade	Moderate	Fast-growing shrub with attractive pink, purple, or white flowers and gray-green leaves. Highly attractive to butterflies-hence the name. *Buddleia madagascariensis* is also drought tolerant.
Dwarf poinciana (*Caesalpinia pulcherrima)*	150-600	Full sun	Moderate	Large shrub to small tree. Showy, orange/yellow flowers during warmer months.Needs good drainage. Leaves and seeds poisonous.
Beautyberry (*Callicarpa americana)*	180-240	Full sun to partial shade	Moderate	Flowers insignificant, though long-lasting purple berries are quite attractive and provide a good food source for birds.
Giant milkweed (*Calotropis gigantean)*	180	Full sun	Moderate	Large, thick, grayish-green leaves and purpleto white, crown-like flowers. Tolerates

				poor soil, but requires excellent drainage. Host plant for monarch butterfly larvae.
Spicewood *(Calyptranthes pallens)*	150-450	Partial Shade	Moderate	Small, shrubby, native tree with burgundy-tinged new growth and insignificant white flowers. Can be sheared and grown as ahedge. Prefers moist soil.
Jamaica caper *(Capparis cynophallophora)*	540	Full sun to partial shade	High	Slow-growing, undemanding plant with attractive, white, fragrant, spider lily-like flowers. With time, can be shaped into attractive hedge.
Natal plum *(Carissa macrocarpa)*	90-360	Full sun to light shade	High	Large shrub to small tree with attractive, thick, glossy foliage, fragrant, jasmine-like flowers and edible fruit. The spiny leaves and stem smake this a good barrier plant once established.
Bamboo palm *(Chamaedorea erumpens)*	180-300	Partial to full shade	None	Heavily clumping palm with tall, slender stems resembling bamboo canes. Can be used as a shrub or a hedge. Also known as *C. seifrizii*.
Cocoplum *(Chrysobalanus icaco)*	90-450	Full sun to partial shade	High	Shrub with insignificant flowers, but new foliage is quite attractive. Often used as a hedge, but will thin if planted in too much shade. Spreading and erect forms available.
Snowberry *(Chiococca alba)*	60-90	Full sun	Low	Vining shrub with glossy green, elliptical to lanceolate leaves and tiny white flowers. *Chiococca parviflora*, the pineland snowberry,is a similar, but less common, native shrub.
Fiddlewood *(Citharexylem spinosum)*	360-900	Full sun to partial shade	High	Large, densely leafy shrub to small tree with glossy, ellipitical leaves and small,

				white, fragrant flowers followed by round, orange-brown berries on female plants.Attractive to birds and other wildlife. Also known as *C. fruticosum*.
Pigeon plum *Coccoloba diversifolia*	150-900	Full sun to partial shade	High	Evergreen large shrub to small tree withdense, narrowly rounded crown and attractive, peeling bark. Small, dark purple berries on female trees attractive to birds.
Sea grape *(Coccoloba uvifera)*	300-1500	Full sun	High	Large shrub to medium tree with large, thick, saucer-like leaves bearing attractive venation. Also has edible fruit. On occasion,can be susceptible to a number of pests.
Croton (*Codiaeum variegatum)*	150-300	Full sun	Moderate	Grown for the colorful, variegated foliage in greens, yellows, reds and pinks. Scales can be a problem. Can be leggy, especially ifgrown in shade.
Coffee colubrina *(Colubrina arborescens)*	600	Full sun to light shade	High	Large, course-leaved shrub or small tree with open crown of large, tri-lobed leaves with inconspicuous green flowers and 3-seeded capsules.
Buttonwood *(Conocarpus erectus)*	150-1500	Full sun to partial shade	High	Shrub or tree. Both green and silver leaved forms available; the latter is more attractive. Insignificant flowers followed by small, button-like seed pods. Can be used as a hedge but bottom is prone to thin out.
Ti plant *(Cordyline terminalis)*	90-300	Full sun	None	Slender palm-like plant with unique long,narrow leaves in red, pink, white, or purple.Also known as "red sister."

Christmas berry, ground holly. *(Crossopetalum ilicifolium)*	30	Full sun to partial shade	Low	Prostrate, evergreen shrub with small, holly-like, spiny leaves and attractive redberries.
Desert spoon *(Dasylirion wheeleri)*	90-900	Full sun	High	Linear, grey-green leaves with toothed edges. Spectacular inflorescence bearing many creamy-white flowers may reach 30 inches after which the plant dies. Takes extreme drought, but must have excellent drainage and air circulation.
Varnish leaf *(Dodonaea viscosa)*	540	Full sun to partial shade	High	Grown for its attractive, stiff, shiny green leaves. Showy, yellowish, 3-winged capsules produced in terminal clusters; brown, pink or purple at maturity. Dense and fast growing. Used as free-standing specimen or hedge.Excellent for beach-front.
Dracaena *(Dracaena marginata)*	450	Partial to full shade	None	Narrow, sword-like leaves with red margins."Tricolor" cultivar has a cream stripe and red edge.
Golden dewdrop *(Duranta erecta)*	150-540	Full sun to partial shade	Moderate	Sprawling, sometimes vine-like, evergreen shrub or small tree. Beautiful purple flowers and yellow, grape-like, poisonous fruit. Also known as *D. repens*.
Lingaro *(Elaegnus philippensis)*	90-450	Full sun	High	Shrubby, evergreen vine with attractive light green foliage, the underside a reddish brown.Small, fragrant, white flowers followed by edible fruit. Grows well on poor/ calcareous soils.
Beach creeper, golden creeper *(Ernodea littoralis)*	30-90	Full sun	High	In tolerant of over watering, but excellent as a ground cover for beach-front. Flowers insignificant.

Coral bean tree *(Erythrina herbacea)*	90-450	Full sun	Low	Prennial shrub to small tree. Deciduous, 3-lobed compound leaves. Showy scarlet blossoms on tall stalks in spring,followed by large beans that split to reveal bright red seeds. May grow into small tree in India. All parts of this plant are poisonous.
White stopper *(Eugenia axillaris)*	150-600	Full sun to dense shade	High	Large shrub or small tree with pale, whitish bark and aromatic foliage that can be overpowering. Small, white flowers in mid summer followed by small, purplish berries.
Redberry stopper *(Eugenia confuse)*	180-540	Full sun to partial shade	High	Slow-growing native plant with attractive glossy leaves and red berries. Used as a specimen plant or hedge. Upright growth suitable for a restricted site.
Scarlet plume *(Euphorbia fulgens)*	150	Full sun to partial shade	High	Evergreen, arching shrub. Very showy red flowers all year. Needs excellent drainage. Ideal for a rock garden.
Crown-of-thorns *(Euphorbia milii)*	30-90	Full sun	High	Slow-growing, thorny, semi-succulent shrub with bright green, obovate leaves and small to medium yellow, orange, or red flowers. Excellent in rock gardens. Often used as low hedge in coastal areas.
Green aloe *(Furcraea foetida)*	90-240	Full sun	High	Large rosette of thin narrow leaves up to 240 cmlong. Requires freely draining soil. Inflorescence borne on 15-30" stem; flowers pungent. Rosette dies after flowering. New plants arise from bulbils. Variegated form available.

Thryallis *(Galphimia glauca)*	90-150	Full sun to partial shade	Low	Versatile, evergreen, tropical shrub. Non-stop yellow flowers bloom year-round. Excellent forlow hedges. Wood brittle. Requires sheltered position.
Star flower *(Grewia occidentalis)*	180-360	Full sun	Moderate	Adaptable evergreen shrub with deep green,oval, to othed, leathery leaves. Bears star-shaped, mauve-pink flowers about 1.5"across during spring and summer, followed by brownish, 4-lobed berries. Excellent plant for espaliers.
Lignum vitae *(Guaiacum sanctum)*	180-750	Full sun	High	Attractive, slow-growing large shrub to small tree, with blue flowers year-round but mostly in spring. Flowers followed by yellow seed pods which pop open to reveal shiny, bright red seeds. Extremely dense, prized wood. *Guaiacum officinale* is a larger relative with blue flowers that turn white, and light green leaves with yellow petioles.
Firebush *(Hamelia patens)*	150-450	Partial shade to shade	Low	Produces attractive orange/ red flowers at any time. In different to soil, as long as drainage is good.Less shrubby and more tree-like in shade. Can be used as hedge, but pruning interferes with flowering.
Heliconia *(Heliconia episcopalis)*	90-210	Full sun to partial shade	Moderate	Smaller heliconia with tight, arrow head-shaped inflorescence bearing dense but colorful bracts which begin red then fade to orange then yellow toward the tips. Year-round bloomer.
Heliconia *(Heliconia latispatha)*	180-450	Full sun to partial shade	Moderate	Tall heliconia with long-stalked, yellow-green, banana-like leaves with a

				red margin and erect inflorescence with widely spaced, triangular-bracts that are yellow or orange at base, changing to scarlet at tips.
Lobster claw heliconia *(Heliconia rostrata)*	90-600	Full sun to partial shade	Moderate	Tall heliconia with long-stalked, green to bluish-green, 5' long banana-like leaves, and pendant inflorescence with zig-zagging bracts of scarlet and yellow.
Chinese hat plant *(Holmskioldia sanguinea)*	90-180	Full sun to partial shade	Moderate	Unique flowers are main interest of this scrambling shrub/vine: each is a narrow, orange-scarlet tube backed by a circular calyx, appearing in dense terminal clusters fall-spring. Leaves oval and slightly serrated. Rampant growth can be contained by pruning after flowering. Old canes can be removed. Loses leaves during drought.
Joewood *(Jacquinia keyensis)*	300	Full sun to partial shade	High	Very slow-growing, native shrub with attractive foliage and year-round, showy, fragrant blooms. Good resistance to wind and salt spray. Some shade from hottest sun beneficial.
Primrose jasmine *(Jasmimum mesnyi)*	150-300	Full sun to partial shade	Low	Rambling, open, evergreen shrub with long,arching stems that will climb if given support.Without support, grows as fountain-like mound. Fragrant, yellow, trumpet-shaped flowers.
Downy jasmine *(Jasmimum multiflorum)*	150-300	Full sun to partial shade	Low	Evergreen, branching vine that can be trained as a shrub. Stems and leaves covered with downy pubescence, giving the plant a grayish-

				green appearance. Small, white, star-shaped flowers in clusters year-round.
Peregrina *(Jatropha integerrima)*	450	Full sun to partial shade	Moderate	Deeply-lobed leaves and showy display of small red flowers throughout the year. Needs good drainage.
Coral plant, physic nut *(Jatropha multifida)*	180-600	Full sun to partial shade	High	Shrub to small tree with loose, spreading crown. Large, distinctive leaves with 7-11 narrow lobes, each divided into narrow, pointed segments. Flowers coral red, in flat-topped clusters on long stalks above foliage. All parts are poisonous.
Chinese juniper *(Juniperus chinensis)*	60-300	Full sun	Low-Moderate	Evergreen ground cover, shrub, or tree,depending on variety. Both adult and juvenile foliage found on adult trees. Berries fleshy and glaucous white.
Crepe myrtle *(Lagerstroemia indica)*	240-750	Full sun	None	Grown for peeling bark and outstanding latespring to summer blossom in various shades of pink, red purple and white. Miniatures used for edging; standards used as specimen plants.
Barbados cherry *(Malpighia emarginata)*	150-360	Full sun to partial shade	Moderate	Attractive shrub. Produces clusters of small, pink flowers, followed by red, cherry-sized, edible fruit that is tart in flavor and rich in vitamin C. Makes good hedge. Also known as *M. glabra.*
Monstera *(Monstera deliciosa)*	150-180	Full sun to partial shade	Low	Slow-growing vine or shrub with huge, broad,glossy, perforated and deeply cut leaves and woody stems with aerial roots. Mature plants bear thick, cream spathes followed by sweet-smelling, cone-like, edible fruit.

Simpson's stopper *(Myrcianthes fragrans)*	180-600	Full sun to partial shade	High	One of the most beautiful and ornamental native woody plants. Densely branched,densely leafy large shrub to small tree. Pure white, puffy flowers followed by large, bright orange berries that contrast with dark foliage.
Wax myrtle *(Myrica cerifera)*	300-750	Full sun to partial shade	Low	Clumping, fast-growing shrub to small tree with small, evergreen leaves, inconspicuous flowers, and waxy gray fruit attached to twigs. Leaves and fruit smell like bayberry. Tolerates periodic flooding. Natural insect repellant.
Mazari palm *(Nannorrhops ritchiana)*	180-600	Full sun to partial shade	None	Medium-sized, shrubby palm with branching trunks and whitish-blue, fan-shaped leaves. Each stem flowers only once and dies back. Exceptionally cold-tolerant.
Firespike *(Odontonema strictum)*	180	Full sun to partial shade	Low	Small, showy, evergreen shrub with sparse, thick, upright branches, shiny, dark green leaves with wavy margins, and 9-12" upright panicles of brilliant red, tubular flowers in late summer to early winter. Tolerates all but most severe droughts.
Prickly-pear cactus *(Opuntia* spp.*)*	15-900	Full sun	High	Many species, from tiny plants with tuberous roots to over 30' tall trees. Branches constricted, forming broad/ flattened or cylindrical "joints." Most have sharp spines and small bristles. Flowers generally yellow or red, followed by edible fruits (prickly pears).

Jacob's ladder *(Pedilanthus tithymaloides)*	180	Partial shade	High	Flowers insignificant; surrounded by showy, slipper-like red bracts. Zig-zag stems, with prominently keeled leaves. Well adapted to poor, dry soil. Variegated cultivars. Poisonous.
Selloum, tree philodendron (*Philodendron selloum)*	150-450	Full sun to partial shade	Low	Prostrate to upright trunks. Huge, deeply divided, dark green leaves. Called tree philodendron because it does not climb. Also known as *P. bipinnatifidum*.
Pittosporum (*Pittosporum tobira)*	300-450	Full sun to partial shade	Moderate	Broad-leaved, evergreen shrub with bright green, glossy leaves (can be variegated) and clusters of small, white flowers that smell like orange blossoms. Moderate to fast grower.
Podocarpus (*Podocarpus macrophylla)*	150	Full sun to partial shade	None	Evergreen shrub or tree with needle-like leaves. Makes an excellent hedge. Bears small, bluish berries.
Aralias (*Polyscias* spp.)	60-750	Full sun to partial shade	Moderate	Many species and cultivars used as shrubs,hedges and ground covers. Foliage differs in form (usually compound), and can be variegated.
Elephant Bush *(Portulacaria afra)*	90	Full sun to partial shade	High	Grown for jade colored leaves and red stems. Rarely flowers in cultivation. Grow in gritty, freely draining soil. Provide good air circulation.Variegated form exists. Excellent for rock gardens.
Bahama wild coffee (*Psychotria ligustifolia)*	180-270	Partial to full shade	Low	Small, under story shrub with dull green leaves. Related to gardenia and similar in overall appearance. Flowers abundant over much ofthe year, followed by decorative red berries.

Wild coffee *(Psychotria nervosa)*	180-270	Partial to full shade	Low	Small under story shrub similar to *P.ligustifolia*, but with distinctive, glossy, dark green leaves with deeply impressed side veins.
White indigoberry *(Randia aculeate)*	45-300	Full sun to partial shade	High	Shrub with no outstanding features. Fragrant flowers and white berries on female plants. Main attribute is ability to grow under adverse conditions.
Buckthorn *(Rhamnus* spp.)	450	Full sun	High	Deciduous plant with glossy, dark green, oval leaves up to 3" long. Insignificant green flowers followed by small fruit that change from red to black as they ripen. No outstanding landscape value, but well adapted to dry,infertile sites. Numerous spines make these plants useful as security screens.
Lady palm *(Rhapis excels)*	450-300	Partial to full shade	None	Small, densely clumping fan palm with numerous slender, dark brown trunks bearing distinctive fibers and leaf scar patterns. Small, deeply segmented, glossy, dark green leaves on thin, delicate petioles. Often used as shrub or potted plant.
Rosemary *(Rosemarinus officinalis)*	90	Full sun	High	Evergreen, woody shrub with aromatic, needle-like leaves and gray, scaly bark. Easy to propagate from cuttings.
Scrub palmetto *(Sabal etonia)*	90-150	Full sun	Moderate	Small, trunkless, solitary, shrub-like fan palm. Bears large clusters of small, purplish black fruit. Grows best in sandy soils.
Dwarf blue palmetto *(Sabal minor)*	60-120	Partial shade	Moderate	Small, trunkless, shrub-like palm resembling *S. etonia*, but smaller and with bluish-

				green leaves. Can tolerate wet soils.
American elderberry *(Sambucus canadensis)*	300-450	Full sun to partial shade	Low	bushy, multi-stemmed, wide-spreading shrub with deciduous, compound leaves. Tiny, star-shaped, white flowers followed by shiny, blue-black fruit. Provides colorful autumn display of yellows, oranges, and reds.Branches brittle. Forms dense thickets by suckering from roots.
Inkberry *(Scaevola plumieri)*	60-120	Full sun to partial shade	High	Succulent leaves, insignificant, small, pink/ white flowers. Spreads by under ground stems. Well-suited to sandy soils at beach-front. *Scaevola sericea* (=*S. taccada v. sericea, S. frutescens*) is in invasive category.
Dwarf schefflera *(Schefflera arboricola)*	180-450	Full sun to partial shade	Moderate	Many cultivars, some with variegated foliage. Used as specimen or informal hedge. Takes pruning well. Mealy bugs and scales can be a problem.
Desert senna *(Senna polyphylla)*	360	Full sun	Low	Small tree with a beautiful cascading habit.Prefers well-drained soil. Small, yellow flowers mostly in dry season.
Saw palmetto *(Serenoa repens)*	90-240	Full sun to partial shade	High	Clumping fan palm with subterranean, prostrate, or upright trunks. One of the most abundant native palms in Florida. Green and silver forms available. Spreading tendency can be a problem when left uncontrolled. Petioles are heavily armed.
Necklace-pod *(Sophora tomentosa)*	180	Full sun	High	Large, densely branched, native shrub with natural rounded shape. Bears clusters of yellow flowers at

				tips of branches. Fast-growing and easily cultivated. Seeds poisonous.
White bird-of-paradise *Strelitzia nicolae*	240-600	Full sun to partial shade	Low	Named for the appearance of the spectacular flowers. Foliage dark green, banana-like. Drought tolerant once established.
Orange bird-of-paradise *(Strelitzia reginae)*	90-150	Full sun to partial shade	Low	Named for spectacular flowers. Foliage gray-green, fan-like. Drought tolerant once established.
Marmalade plant *(Streptosolen jamesonii)*	180	Full sun	Low	Fast-growing, evergreen shrub grown for showy yellow to dark orange flowers produced winter to spring. Needs open, sunny site with excellent drainage. Sprawling growth benefits from support. Prune hard after flowering is complete.
Bay cedar *(Suriana maritime)*	180-360	Full sun	High	Small-leaved, evergreen shrub or small tree. Clusters of small, leathery leaves and attractive, peeling bark. Will grow in sand or on bare rock. Good choice for ocean and beach-front sites.
African milkbush *(Synadenium grantii)*	500	Full sun	High	Large shrub to small tree. Fleshy leaves and stems, becoming woody with age. Showy clusters of red flowers. Milky latex is poisonous.
Cape honeysuckle *(Tecomaria capensis)*	60-300	Full sun to light shade	High	Sprawling, rampant, evergreen, sometimes vine-like shrub. Compound leaves with 5-7 toothed, diamond-shaped leaflets. Clusters of brilliant red-orange to scarlet, tubular flowers. Can be pruned to shrub, trained to espalier, or grown as vine or ground-cover. Requires well-drained soil.

Tetrazygia, West Indian lilac *(Tetrazygia bicolor)*	180-350	Full sun to light shade	Low	Extremely ornamental native shrub or small tree. Beautiful white and yellow flowers followed by purple-black berries, highly attractive to birds. Elegant, glossy, dark green foliage. Can not be used as hedge because branches die when pruned.
Limeberry *(Triphasia trifolia)*	90-300	Full sun	Moderate	Small, glossy, trifoliate leaves. Fragrant, diminutive white flowers. Edible red berries. Well adapted to rocky, calcareous soil.
Walter's viburnum, black haw *(Viburnum obovatum)*	360-900	Full sun to partial shade	Low	Large shrub to small tree with attractive, pure white blossoms in spring, before leaves fully developed. Flowers followed by berries that turn from green to pink to red to purplish-black in the fall. Ideal as a hedge.
Spanish bayonet *(Yucca aloifolia)*	150-600	Full sun or partial shade	High	Trunk-forming yucca with dangerously pointed, strap-like leaves. Trunks often topped with large, upright clusters of creamy white flowers.
Bear grass, Adam's needle *(Yucca filamentosa)*	90-180	Full sun	High	Leaves resembling long green spears edged with white threads form basal rosette. White, bell-shaped flowers bloom in terminal spikes from mid to late summer.
Spineless yucca *(Yucca elephantipes)*	360-450	Full sun	High	Attractive form. Leaves with soft spines. Showy, white, fragrant blooms. Requires freely draining soil.
Cycad *(Zamia amblyphyllidea)*	90-120	Partial shade	High	Mounding, clumping cycad with elegant leavesto 4' and bearing wide, paddle-shaped leaflets. Requires well-drained soil.

Cardboard "palm" *(Zamia maritime)*	90-180	Full sun to partial shade	High	Medium sized cycad with broad, stiff, carboard-like leaflets. Name recently changed from *Z.furfuracea*. Can be weedy if both male and female plants are in the same vicinity. Requires well-drained soil.

Hedge and shrubbery border

Hedge is an important adjunct to every garden. It is necessary to demarcate the garden from public road and adjacent gardens particularly in flat gardens. Besides its utility as screens, hedge is necessary between flower and vegetable garden, to provide a background for annuals if shrubbery does not exist and to enclose a rose garden or a children's corner in a park or public garden. A wind screen is also wanted when strong breezes, hot or cold blow from one direction during certain seasons of the year and frosts. It is useful to screen the manure pit, potting area, servants' quarters and other unsightly views.

Growing the hedges

The majority of hedges are propagated by cuttings but where seed can be obtained the cost of raising a large number of plants in their delicate stage is minimize. Seed rates per 100 metres of some important hedge plants are as under; *Duranta* 1 kg, Inga dulcis 1kg, Jaint (*Sesbania*) 150 gms, Mendi (*Lawsonia*) 250 gms, Dodonea (*Sanatta*) 250 gms. Hard seed such as that of *Acacia modesta* should be soaked in warm water for 24 hours.

Selection of plants

1. Shrubs

- A hedge of *Breynia rhamnioides* is often seen in gardens and is very ornamental when in berry.
- *Bauhinia acuminata* in spite of its large leaves makes an effective display of white flowers.
- *Caesalpinia pulcherrima* ages fast when severely pruned but if trimmed slightly it provides tall hedge.
- *Clerodendron inerme* forms neat hedge or edge of varying height for m20 cm to 2 metres. It forms a compact hedge of 20-120 cm.
- *C.inerme* is not touched by cattle and white ants leave it alone.
- *Duranta* in its many varieties holds first place in many places, *D.variegata* being very ornamental.

- *Lawsonia* (Mendhi) is better in dry localities.
- *Dodonea viscosa* (Sanatta) is an up country hedge but it should not be cut back at latter stage as it is liable to die off in three or four years.
- *Nerium* and *Tecoma* are two others that might be used.
- *Dracaena fragrans* and *D.sticta* are good hedges for a shady spot.
- *Hibiscus* form excellent hedges particulary *H.liliflorus* and *H.schizopetalus.*
- *Ehretia buxifolia* , *Serissa foetida* and *Malpighia coccigera* are small leaved evergreens hedge for shady situations.
- *Polyscias* (Aralia) of many kinds may be used as hedge.
- *Codiaeum* (Croton) *Eranthemum* and *Graptophyllum* are ornamental shade loving hedge plants. Crotons of the narrow leaved and small foliage varieties are better hedges than those with large foliage.
- *Plumbago capensis* and *P.zeylancia, Meyenia* (*Thumbergia*) *erecta* and *M.affinis* are dwarf types of flowering shrubs that can be used for a hedge.
- *Plumbago capensis* may be cut down to 15 cm for hedge.
- *Caryopteris, Mussaenda, Rhapiolepis* and *Rondeletia* may also be used for hedges.
- *Euphorbia pulcherrima,* if backed or having a supporting hedge in front can provide a fine mass of colour in the cold months. The last pruning should be done in August –September.
- *Jacquinia ruscifolia* is an evergreen hedge with orange red flowers and needled–tipped leaves and ideal hedge but difficult to propagate.
- *Justicia gendarussa* can be clipped to15 cm and allowed to grow upto a metre.
- *Lagerstroemia indica* in many colour variations, though bare in the cold weather, is a fine flowering hedge in April-June and if two colours are planted alternately provides a pretty effect when in bloom.
- *Ligustrum neilgherriensis,* one of the privets, is not to be despised and the *Lantana,* though so frightful a weed, can be retained as a dwarf plant with frequent clipping though the non-seeding types are recommended. *Malyaviscus* can be cut back to form a 1 metre hedge or allowed to grow 2-3 metres high. When in bloom it is very effective.
- *Murraya* (Kamini) and *Tropis aspera* (Sheora) are slow growing but clip well.

- *Jasminum pubescens, Coffea bengalensis, Barleria* (Jhati), *Ixora (I.ragoosula,I.coccinea I.strica), Daedalacanthus* and *Strobilanthes* are flowering hedges that are well known and easily grown.
- Roses form excellent hedges and many of the cacti such as *Cereus, Opuntia* as well as *Euphorbia antiquorum, E.trigona and E.tirucalli* are used in exposed positions for hedge work.
- *Pedilanthus tithymaloides* will be found as hedges surrounding village huts.
- *Sesbania aegyptiaca* is a fast-growing hedge reaching a height of 2 metres in a couple of months; it must be grown from seed and is a great favourite in the drier parts of India for a tall temporary boundary hedge. After two years it commences to become rugged and should be replaced.
- Another quick hedge is *Cajanus indicus* (Arhar).
- *Thuja orientalis* is an excellent evergreen hedge, 2-3 metres in height.
- At an elevation of over 1000 metres a different type of plant is cultivated for hedges. Here are some varieties: *Spirea, Kerria japonica, Berberis* (Barberry), *Habrothamnus elegans, Fuchsia, Heliotrope, Euonymus, Ligustrum* (Privet), *Buxus* (Box)*Hydrangea, Philadelphus, Hibiscus syriacus, Cupressus* and *Ancuba.*

2. Trees

For very tall hedge several trees lend themselves to pruning:-

- *Polyalthia longifolia* is one of the best tall hedge.
- *Putranjiva roxburghii, P.acuminata, Anogeissus, Diospyros embryopteris Grevillea robusta* and *Pithecolobium* are others that can be recommended .
- *Pongamia glabra* and *Inga dullics* are also used.
- *Thevetia nerifolia, Erythrina* of many kinds, *Haematoxylon campechianum, Schinus terebinthifolius, Cryptomeria, Cupressus* and other conifers are considered good when cut back to a height of 3-4 metres.
- *Casuarina equistifola* if pruned before large trunk forms, makes a very neat hedge.
- *Acacia farnesiana* and *Parkinsonia aculeata* are both useful but do better in dry situations.
- *Pisonia alba* (tree lettuce) which forms hedges of pale coloured foliage, is grown in South India. Commence cutting back when the seedlings are 20 cm high and repeat when the shoots are 40 cm a quick thick hedge from the very beginning is then ensured.

- *Thevetia nerifolia* is poisonous to cattle and goats and make a neat-clipped hedge 3-4 metres high.

3. Climbers

On a wire fencing *Jacquemontia violacea* is ideal, *Ipomea palmata* (Railway Creeper) and *I.sincrata* and *Antigonon* (Sandwich Island Creeper), *Clerodendron splendens, Ficus repens, Passiflora pruinosa. Tristellatia austrulis, Thunbergia fragrans* and *T. alata* and *Vallaris hynei* are good for screening. For a very shady situation *Rhycospermum jasminoides, Asparagus plumosus* and *Lygodium scandens* are suggested.

4. Palms

Ptychosperma mcarthuri will grow 4-6 metres and not spread more than 1.5 to 2.0 m in width. *Rhapis flabelliformis,* on the other hand, while growing form 2-3 m in height, sends out runners, which must be constantly removed. *Areca lutescens* grows as tall as the *Kentia* but bushes out more.

Planting Distance

This depends on the size of the plant. Trees should be planted 2-3 m apart, shrubs 15-50 cm apart and *Alternanthera,* etc., (edging) 2 cm apart if planted in a single row but when staggered double the distances. *Duranta, Lawsonia* etc. can be thickened when the hedge has grown up, by having the pruned cuttings, planted in between or on either side of the growing bushes. For hedge planting calculate 9 plants per mete row of planted singly 10 cm apart and 12 in a double row if planted 15 cm apart staggered.

Care of hedges

The first requirement for a hedge is impenetrability and the next neatness. Any prickly plant, such as *Inga dulcis* or the thorny type of *Duranta,* will not ensure an absolute cattle or thief proof hedge in its earlier stages, so run a barbed wire through the bushes form the ground or plant a row of pineapple, *Agave,* prickly cactus or thorny *Euphorbia* at the base and the case is altered. Weeds should be removed from the roots of a hedge. In the dry season soil should be forked up and flooded.During Mendhi (*Lawsonia*) or *Duranta* are pruning, in the wet weather, cuttings should be planted to fill gaps.

Fertilizers

Usually hedges receive little attention till they become thin. The plants should be pruned heavily in the rains.Manure should be added at the close of the Monsoon. Hedges should be fed, especially those that are clipped regularly by giving superphosphate of lime and bone meal 50 gms each, 12 gms sulphate of potash and 50 gms sulphate of ammonia per two metres row, every six months. Once in a year soil should be dug to add cow manure as the rains break. This might seem superfluous but the results will amply repay the attention given.

Pruning of hedges

Some varieties have to be left to grow as in nature, i.e. *Pandanus, Euphorbia,* etc but kept within bounds by the removal of side growth or the central head. In *Euphorbia* and *Opuntia,* the taller portions broken off can be replanted in the gaps to thicken the hedge. For low hedges, 50 cm to 1.0 meter high, clip frequently during the rains but once a year cut back the old wood below the maximum height to restart a new top do this in May or June. A sharp pair of pruning shears should be used for small twigs but Secateurs would be used for larger stems. It is best to stick to a simple shape instead of attempting battlements, turrets, balls, columns, etc. arches can be grown on an iron rod bent to shape, the shrub kept in position by being loosely tied with strong wire to the rod.

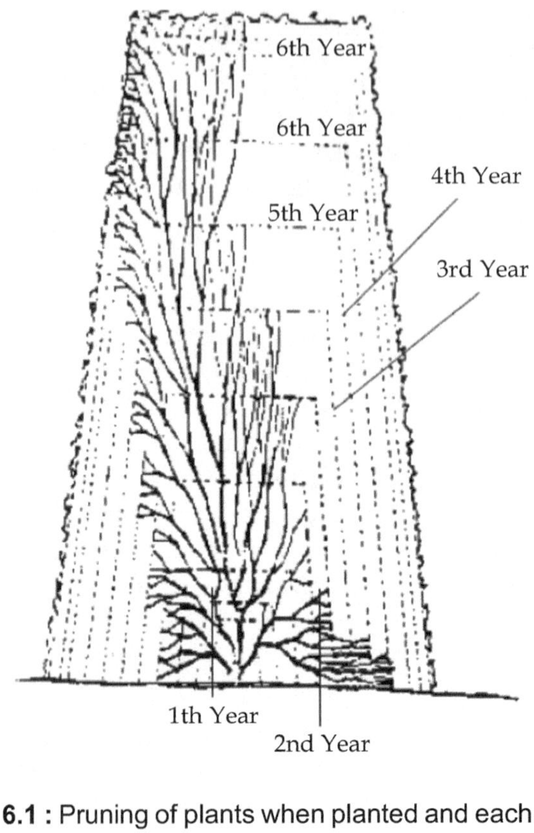

Figure 6.1 : Pruning of plants when planted and each year after

Shapes of hedges

A hedge is apt to become wedge-shaped the top widening out, if not carefully clipped square. A rounded top is difficult to keep unless the type of hedge is slow, compact growing and small leaved. Hedges are also cut on a slope but here too the Mali often makes a mess of the shape. It is best to have wooden or iron stakes painted green, sunk in the hedge at intervals on which string can be tied and used as a guide. Stick to a square cut hedge.

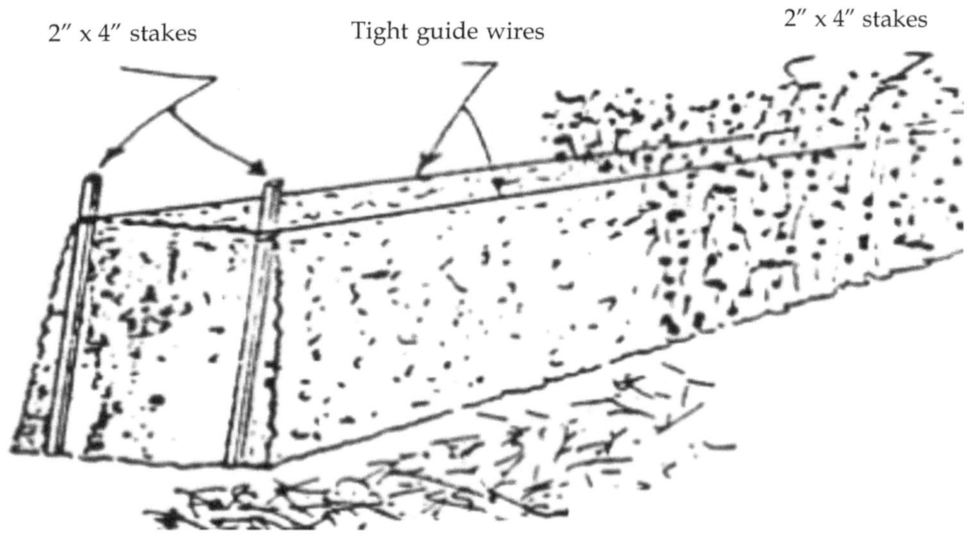

Figure 6.2 : Stakes and tight guide wires or ropes for even surface

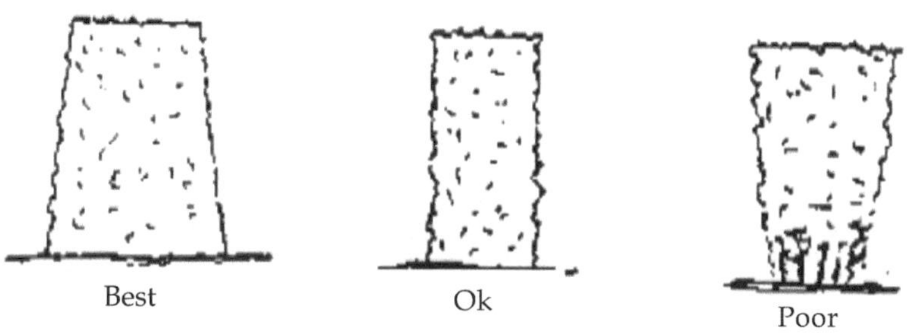

Figure 6.3 : Overhanging top edges shade sides

Topiary

Topiary is practised only in few traditional gardens. The plants that have small foliage and are of slow growth are good for using in topiary. The results required takes years to attain and one careless clipping will spoil the entire design. Stating with a ready-grown plant, stems are bent and held in position with wire if they do not naturally grow at the required angle. A rough clipping to get the general outline follows and as new shoots develop these are either trained to fit in with the design or removed. Simple shapes, a ball, spiral, table, cube etc., are not so difficult to obtain, it is when birds or beasts or even man is to be moulded that the topiarist tackles a tough proposition. Topiary work gives an old world appearance to a garden but these clipped bushes should not be scattered aimlessly all over the garden. A formal garden is best suited to topiary work. For simple shapes the following can be used *Hibiscus, Thuja* and *Cupressus, Putranjiva* and *Polyalthia* but for the more intricate *Murraya exotica, Duranta repens, Ehretia buxifolia* and *Tropis aspera* are used.When using wire to model the bush or tree to the rough outline care should be taken that this does not cut too deeply into the bark. Stoppage of sap will mean poor growth or death of a stem.

CHAPTER 7

Cultivation and Management of Bulbous Plants

Bulbs

Floriculturists use the term "bulbous" to include a wide array of perennial plants that grow from underground organs, including bulbs, corms, tubers and rhizomes. Any plant described as bulbous will normally have a swollen root or stem or even leaves underground (Some orchids and other plants have swollen stems - above ground). Bulbs, corms, tubers and rhizomes are perennial plants organ that are part of a plant type called bulbous plants in which part of the plant has evolved into a below-ground storage unit where food created one year is used to nourish the plant in the next. They are valuable in the garden for many reasons. Spring bulbs appear early, before most perennials have started to grow again after the winter. They can also add color to containers when other plants are not ready to face the danger of frost. In summer, bulbous plants such as Alliums and lilies can provide stately interest, while autumn crocuses and tiny hardy cyclamen can cheer up the garden toward the end of the season. The purpose of the bulbous part of the plant is to store food for the plant parts that live above ground to enable the plant to survive adverse conditions. The leaf bases may resemble scales, or they may overlap and surround the centre of the bulb. A modified stem forms the base of the bulb and plant growth occurs from this basal plate. Roots emerge from the underside of the base, new stems and leaves from the upper side. Most true bulbs have their scales closely packed together such as *Hyacinthus* but some types, such as the Lily *(Lilium)*, they are loose and swollen. Most popular bulbs have a papery skin, the tunic, whose function is to protect the tissues within. Some true bulbs such as the Lily do not have a tunic and are therefore, easily damaged by rough handling. Cardio crinums dies after flowering but other bulbs are perennial. Vegetative reproduction in true bulbs is by means of offsets called bulblets.

Bulbous plants in the garden

Adaptation to tough environments makes bulbous plants invaluable in a garden setting. Many are suitable for the kind of thin, dry and nutrient deficient soil found at the foot of a hot and sunny wall. Others can cope with dry shade and are therefore, ideal for planting under deciduous or evergreen trees, where nutrients and soil moisture are deficient.

Spring-flowering, summer-flowering and autumn-flowering bulbs provide the opportunity to add extra color to any garden. They can be planted under shrubs or between perennials in the flower border or can be naturalized informally in lawns and orchards, or formally in seasonal bedding display. For splashes of color all around the garden, bulbous plants can also be planted in pots. Bulbous plants ring the seasonal changes throughout the year with glorious flower display. Some possess handsome foliage, others are valued for their flagrance, but above all, their blooms are the most essential. They offer a wide variety of color and form, from bright, primary shades to delicate, pastel hues. Traditionally, bulbous plants are cultivated in parterres and herbaceous borders. The choice of the species depends on a number of factors, among them, the site in the garden (shady or sunny) and the type of soil, the color and the effect that is pursued in the garden and the part of the year when the flowering is wanted. Some examples of bulbous plant genera and their flowering season are given below.

- Spring flowering species belong to the genera *Allium, Arum, Asphodelus, Camassia, Convallaria, Crocus, Cyclamen, Eranthis, Freesia, Fritillaria, Galanthus, Hyacinthus, Hippeastrum, Iris, Ixia, Leucojum, Muscari, Narcissus, Ornithogalum, Ranunculus, Scilla, TrilliumyTulipa* and *Zephyranthes*.
- Summer flowering species are included in the genera *Achimenes, Agapanthus, Allium, Alstroemeria, Amaryllis, Anemone, Begonia, Calochortus, Canna, Crinum, Crocosmia, Dahlia, Dierama, Eucomis, Galtonia, Gladiolus, Gloriosa, Haemanthus, Hymenocallis, Lilium, Oxalis, Pancratium, Paradisea, Polianthes, Sprekelia, Tritonia, Watsonia, Zantesdechia.*

- Genera that usually flower in the fall: *Crocus, Colchicum, Cyclamen, Nerine, Sternbergia.*
- Winter flowering genera are *Galanthus, Crocus, Cyclamen Eranthis.*

Bulbs for different sites

Bulbous plants for shade

Many species of bulbous plants have woodland or shady places as their natural habitats. In the garden they are especially valuable, ornamenting places which may otherwise be difficult to plant. Some species for shade are *Allium ursinum, Anemone blanda, Anemone nemorosa, Arum italicum, Convallaria majalis, Corydalis flexuosa, Cyclamen purpurascens, Disporum flavescens Erythronium, Fritillaria pallidiflora, Galanthus, Hyacinthoides non-scripta, Iris douglasiana, Leucojum vernum, Lilium martagon, Ranunculus ficaria, Sanguinaria canadensis, Smilacina racemosa, Trillium* and *Uvularia grandiflora.*

Rock garden

A rock garden, also known as a rockery or an alpine garden, is a type of garden that features extensive use of rocks or stones, along with plants native to rocky or alpine environments. It is possible to cultivate a great number of species in it. The only limitation is the size of the plants, they must be small. Some of the genera of bulbous plants with species well suited for the rock garden are:*Allium, Anemone, Anthericum, Bulbocodium, Chionodoxa, Cyclamen, Eranthis, Erythronium, Galanthus, Ipheion, Muscari, Ornithogalum, Oxalis, Romulea, Rhodohypoxis* and *Scilla.*

Bulb Maintenance

Total removal of the Winterizing mulch, where applicable, should not be done until about mid-April. Mulch prevents frost heaving, premature growth and frozen growth tips. Most bulbs are quite accustomed to coming up through mulch and snow. Some bulbous plants actually produce their own heat to help them push up through the ice and snow. In native environments, snow, leaves, mulch and grasses protect them through the winter and keeps them healthy in the spring.

Planting Time

The optimum time to plant bulbs is usually from October to about December. Planting them too early may result in premature above ground autumn growth and reduced spring bloom. Storage in a warm garage or a damp basement corner may kill the bulbs. Longer refrigeration may abort the internal flower buds. Refrigerator can be used for storing bulbs.

Planting Depth

A. *Bulbs*:Planting should be done at 2.5 Cm depth for example *Anemone* (Wind Flower), *Anemonella* (False Rue Anemone), *Anemonella* (False Rue Anemone), *Claytonia* (Spring Beauty), *Corydalis* (Fumeworts), *Dentaria* (Toothwort), *Eranthis*(Winter Aconite).

B. *Bulbs*:Planting should be done at 5-7.5 Cm depth for example *Brodiaea, Ipheion, Cyclamen* (Cyclamen) ½ of bulb in ground, remainder a top with mulch light.

C. *Minor Bulbs* : Planting should be done at 7- 10 cm deep for example *Allium* (Onion small species), *Arisaema* (Jack-in-the-Pulpit) *Bellevallia* (Bellevallia), *Brimeura* (Wild Hyacinth), *Bulbocodium* (Spring Saffron), *Chinodoxa* (Glory-of-the-Snow), *Crocus* (Spring & Autumn Crocus), *Cypripedium* (Lady Slipper Orchids), *Dicentra* (Dutchman's Breeches), *Endymion* (Woodland Hyacinths), *Erythronium* (Trout Lily), *Fritillaria* (Checker Lily species), *Galanthus* (Snow Drops), *Hycinthoides* (Woodland Hyacinths) *Irisdanfordiana* (Danford Iris), *Irisreticulate* (Reticulated Iris), *Muscari* (Grape Hyacinth), *Oxalis* (Sorrel), *Ornithogalum* (Star-of-Bethlehem), *Puschkinia* (Squill) *Scilla* (Squill) *Trillium* (Wake Robin)

D. *Intermediate Bulbs* : Planting should be done at 10-12 cm deep for example *Colchicum*(Meadow Saffron), *Iris bucharica* (Buchar Iris), *Eremerus* (Foxtails), *Leucojum* (Spring Snowflakes), *Liatris* (Gayfeather), *Narcissus, Tulipa* (Tulip species)

E. *Major Bulbs* : Planting should be done at 15-20 cm deep for example *Allium* (Onion species), *Camassia* (Camas), *Fritillaria* (Crown Imperial & Lg. spp.), *Hyacinthus* (Hyacinth), *Iris hybridum* (Dutch Iris), *Iris x iphioides* (English Iris), *Lilium* (Lily), *Lycoris* Magic Lily), *Narcissus, Tulipa* (Tulip flowering cultivars).

Generally plant bulbs deeper in light sandy soils and shallower in heavy clay soils. If in doubt, cover a bulb to a depth of three times their maximum diameter. Plant bulbs with their growth tips facing any direction except directly down. Most bulbs are tear or drop shaped. The point is the top. On other bulbs, check for dried roots, plant with roots downward. *Anemone, Eranthis* and *Fritillaria* are almost impossible to tell which way is up, just plant them and they will find their way.*Liatris*is slightly concave on the top.*Trillium* is planted horizontally as it is a rhizome.*Anemone* and *Eranthis* should be soaked in 21°C water overnight. This plumps up the bulbs and promotes more rapid development before the onset of winter.

Watering

If spring or the location is dry, bulbs appreciate additional water to keep the soil moist. Most bulbs enjoy copious amounts of water in the spring but very little during dormancy. If there is snow, keep it piled on them as long as possible. Watering is generally not necessary when planting but if the soil is dry, irrigate two days before planting for ease in planting.

During a bulb's dormancy, excessive watering reduces its vitality or potentially kills many of your favorites. Most species and cultivars of *Bulbocodium, Eremerus, Hyacinthus, Oxalis* and*Tulipa* enjoy hot dry conditions during the summer months. Planting these bulbs with heavy water requiring annuals or perennials, may not allow the bulbs to properly mature. Insufficient maturation often results in improper flowering in subsequent years or terminal rotting of your beloved bulbs.

Allium, Anemone, Brimeura, Camassia, Chinodoxa, Crocus, Endymion, Erythronium, Galanthus, Leucojum, Lilium (certain types), *Lycoris, Muscari, Narcissus, Scilla* and *Trillium* tolerate moist conditions and are better suited to the moist flower border. There are always exceptions to the above, as many species and cultivars are involved, being native to many different environmental conditions.

Mulching

Anemone, Camassia, Autumn Crocus, Colchicum, Corydalis, Cyclamen, Eremurus, Iris and Oxalis should be mulched in November or December with a 2.5-5 cm deep light and airy shredded bark, evergreen boughs, marsh straw or leaves. Generally, mulching for the first winter is important. Other bulbs may be mulched simultaneously but it is not as critical for their survival. A year round shredded bark mulch of 5 cm prevents erosion, maintains moderate soil temperatures and moisture, and provides for clean blooms. Ground covers, (e.g., *Dianthus deltoides, Ajuga reptans* and *Vinca minor)* are excellent permanent mulches and may be planted with many bulbs to mask dying foliage.

Foliage

When the bulb's foliage begins to turn yellow, it may be cut off do not tear or pull it off as the bulbs may be damaged. Damaging the bulbs allows for pest and disease opportunities. Wait six to eight weeks from the time of flower before cutting off the foliage on *Narcissus* and *Tulipa*. This may be done even if it has not completely turned yellow. Certain species of bulbs do not bloom at the same time as the foliage. *Lycoris* produces foliage in the Spring with the foliage dying in early Summer. It then blooms in August without any leaves. *Muscari* begins its growth in August, surprising many gardeners into believing they have weeds or wondering about Autumn blooming. *Colchicum* and Autumn *Crocus* follow *Lycoris*pattern.Occasionally *Anemone x fulgens* also comes up in the Autumn.

Maintanance

Bulbs require little care other than the above. If desired, remove spent flower pods and flower stalks - pedicels, on hybrid or non-species Tulips, Daffodils and Lilies. Known as dead-heading, this practice is not required in the home garden. If desired, on *Narcissus* the pedicel may be cut to within 5 cm of the ground. On *Tulipa* hybrids, you may remove the pedicel just above the second highest leaf on the stem. On *Lilium*, cut off the stem just below the lowest flower's pedicel. On minor bulbs: *Allium, Anemone, Arisaema, Eranthis, Erythronium, Galanthus, Iris, Muscari, Scilla, Trillium, Tulipa* species, etc., allow the seeds to ripen. The seeds are ripe when the pods begin to open. Let the seeds drop to the ground or collect them and plant them by broadcasting in a desired location.Most bulbs are monocots with the sprouts looking like grasses. Bulb seedlings feel rubbery when gently pinched, grasses do not. Bulb seeds generally produce flowering size bulbs in 3-10 years.

Maturation

Anemone, Anemonella, Arisaema, Bulbocodium, Camassia, Claytonia, Corydalis, Cypripedium, Dentaria, Dicentra, Endymion, Eranthis, Eremerus, Erythronium, Frittilaria, Leucojum, Liatris, Ornithogalum, Oxalis, Trillium, and certain species of *Tulipa* may not bloom the first year. Often these bulbs require some time for environmental adaptation. Prolonged dry storage commonly renders the bulb useless.Occasionally, the foliage on many of the above bulbs may not even show the first spring but it often flowers quite nicely the following season. This may be due to planting depth, time of planting, weather, age of bulb, storage before and after you received the bulbs and a host of other environmental conditions.*Erythronium* bulbs expand and contract together with soil temperature and moisture. Gradually, over a period of years, the bulb finds its proper soil depth and only then commences blooming. *Erythronium* from seed to proper planting depth requires 8-10 years.

Spring Growth

Newly planted bulbs commonly make their spring arrival some what later than bulbs that have had a year or two in the ground. Some bulbs such as, *Anemone, Anemonella, Arisaema, Claytonia, Dentaria, Dicentra cucularia, Erythronium, Leucojum, Lycoris, Lilium martagon, Narcissus poeticus,* and *Trillium* may require a full year before emerging from the soil.

Hardiness

When using any list of bulbs, it is imperative that we interpret the information correctly. Soil composition, moisture, exposure, acidity, alkalinity, quality and size of bulb, genetic strain, nutrition, abnormal seasonal variations all play their part in whether bulbs make it or not. When trying new types of bulbs for the first time, don't plant them all in one location try many locations.

Bulb Design

A bulb design may be included in almost any landscape and provides many years of enjoyment with minimal maintenance. Naturalization is the ability of a bulb to mature and multiply in a given environment without any additional care. Naturalization occurs in hundreds of different species and cultivars of bulbs. Many hardy bulbs such as *Anemonella, Arisaema, Dentaria, Erythronium, Liatris, Lilium* and *Trillium* provide countless years of enjoyment with minimal, if any maintenance.

CHAPTER 8

Nutrition Management in Ornamental Plants

The term essential element was proposed by Arnon and Stout (1939).They concluded three criteria must be met for an element to be considered essential. These criteria are:

1. A plant must be unable to complete its life cycle in the absence of the mineral element.
2. The function of the element must not be replaceable by another mineral element.
3. The element must be directly involved in plant metabolism.

The number of elements considered essential for the growth of higher plants now varies from 16 to 20 or more, depending upon the definition of essentiality. Other scientists such as Nicholas (1961) believed that an element should be considered essential if its addition enhances plant growth even though it merely substitutes for one of the 16 elements that Arnon declares to be essential. For example, sodium can substitute in plant nutrition for some potassium and vanadium for some molybdenum, Nicholas would consider both sodium and vanadium as essential, but Arnon would not. On the basis of the criteria used, Arnon specifies 16 elements and Nicholas 20 elements as being essential for the growth of higher plants. Three other debated nutrients are nickel (urea transformations), cobalt (N_2 fixation) and silicon. These criteria are important guidelines for plant nutrition but exclude beneficial mineral elements. Beneficial elements are those that can compensate for toxic effects of other elements or may replace mineral nutrients in some other less specific function such as the maintenance of osmotic pressure. The omission of beneficial nutrients in commercial production could mean that plants are not being grown to their optimum genetic potential but are merely produced at a subsistence level.

Since some are needed in relatively large quantities and others in very small amounts, the former are referred to as "major" elements and the latter as "minor" or "trace" elements, or as micro nutrients. The terms "major" and "minor" do not refer to the relative importance of the functions of the elements in plant growth and for this reason the term "trace" element is preferable for the latter class. Carbon ($C0_3^{-2}$, HCO_3, CO_2), hydrogen (H^+, OH^-) and oxygen (O_2, OH^-) are supplied by air and water. The six macronutrients, nitrogen (NO_3^- , NH_4^+), phosphorus (HPO_4^{-2}, H_2PO_4), potassium (K^+), calcium (Ca^{+2}), magnesium (Mg^{+2}) and sulfur (SO_4^{-2}) are required by plants in large amounts. The rest of the elements are required in trace amounts (micronutrients). These trace elements include boron (H_3BO_3, $B_4O_7^{-2}$), chlorine (Cl^-),copper (Cu^{+2},Cu^+), iron (Fe^{+2} , Fe^{+3}), manganese (Mn^{+2} , MnO_4), sodium (Na), zinc (Zn^{+2}), molybdenum ($HMoO_4^-$, MoO_4^{-2}) and nickel (Ni).Beneficial mineral elements include silicon (Si) and cobalt (Co). The beneficial elements have not been deemed essential for all plants but may be essential for some.

Nitrogen

Nitrogen is a major constituent of several important substances, which occur in plants. N is biologically combined with C, H, O and S to create amino acids, which are the building blocks of proteins. Amino acids are used in forming protoplasm and nucleus, the site for cell division and thus for plant growth and development. It is of outstanding importance among the essential elements in that nitrogen compounds comprise from 40 to 50 per cent of the dry matter of protoplasm. Since all plant enzymes are made of proteins, N is essential for all of the enzymatic reactions in a plant. It is a major part of the chlorophyll molecule and is therefore, necessary for photosynthesis. It is a necessary component of several vitamins. For this reason nitrogen is required in relatively large quantities. Proteins are compounds of nitrogen whilst chlorophyll, the green colouring matter of the leaves, also contains the element. Chlorophyll is the green pigment that plays an important part in photosynthesis. If nitrogen is limiting, chlorophyll molecules can not be synthesized. The plant loses its green color. Nitrogen is a necessary element for leaf growth and blossom formation. As a result, the N-deficient plant does not produce required carbohydrates. An adequate supply of nitrogen encourages the formation of the growth hormones cytokinin, auxin and gibberellic acid but a reduction in its availability affects the concentration and balance of these hormones and encourages the formation of the senescence hormone, abscisic acid.With a nitrogen deficiency the growth of plants is retarded as fewer structural and enzyme-functioning proteins are produced. Certain compounds of nitrogen are very mobile in plants and this enables them readily to mobilize supplies of the element at vital growing points and to transfer stored supplies to points where they are most required. Such transference is common from old tissues to young growing points when supplies of the element are short. This mobility and re-utilization of nitrogen explains why deficiency symptoms of the element always appear first in the older parts of plants and why growing points are the last to be affected.The mobility in soil is dependent on the chemical form of the element used.

Deficiency symptoms

- Stunted growth may occur because of reduction in cell division.
- Pale green to light yellow colour (chlorosis) appearing first on older leaves, usually starting at the tips. Depending on the severity of deficiency, the chlorosis could result in the death and/or dropping of the older leaves. This is caused by the translocation of N from the older to the younger tissues.
- Reduced N lowers the protein content of seeds and vegetative parts. In severe cases, flowering is greatly reduced.
- N deficiency causes early maturity in some crops, which results in a significant reduction in yield and quality.

Nitrogen is applied as fertilizer in organic forms as animal manures, composts and green manure crops. Inorganic N fertilizers are available in liquid, granular and slow-release formulations. Ammonium-N fertilizers delay leaching because NH_4^+ is held on the soil exchange complex. When NH_4^+ is converted to NO_3^-, N is more subject to leaching loss. Urea fertilizer leaches readily until it is converted in the soil to $NH_4^+$by the naturally occurring enzyme urease, usually within a few days of application. Urea is a cheap form of nitrogen often used in fertiliser. Solutions of urea in water have low conductivity. Urea in its better, low biuret grades makes a good nitrogen source for plants grown at temperatures that are above 16°C. It is not a good source of nitrogen for cool growing crops, particularly during the winter months because bacteria populations, needed for nitrifying urea, are in low population. Urea requires bacterial action before its nitrogen is available to plants. Organic N is not available to plants until it has been converted to an inorganic form by soil bacteria. Uptake inhibited by high phosphorus levels. N/K ratio is important: high N/low K favours vegetative growth; low N/high K promotes flowering and fruiting. Nitrogen is generally made available to plants in fertiliser solution as ammonium ions, nitrate ions or urea. Most of the nitrate ions that enter the roots from the surrounding solution are carried upwards through the sap-conducting vessels (the xylem) into the cytoplasm of the leaves. Excess nitrate is stored in the vacuoles. For their main function as a nitrogen nutrient, nitrate ions are first converted to nitrite ions in the cytoplasm with the aid of an enzyme (nitrate reductase) containing molybdenum and iron. These nitrite ions are then transported to the chloroplasts where an iron-containing enzyme (nitrite reductase) catalyses their conversion to ammonium ions. Because of their toxicity to plants, even in quite low concentrations, the ammonium ions are rapidly converted to several nitrogen compounds, one of which (glutamine) undergoes further enzyme-controlled reactions to produce many different amino-acids. Finally in ribosomes (mainly in the cytoplasm) twenty of these amino-acids are combined together in the genetically coded sequences defined by DNA and guided by RNA to form the wide variety of proteins required for their structural and enzymatic roles in the plant.If ammonium ions are present in fertilisers, their toxic effect on the plants is avoided because bacterial action on the root surface converts the ions to amino acids that are then translocated from the roots to the leaves for protein synthesis. Many simpler nitrogen compounds are synthesised in plants from nitrate and ammonium ions. These include compounds (guanine, adenine, cytosine, thymine and uracil) required for the formation of DNA.

Excess N may cause K deficiencies. With excess nitrogen the leaves become darker green, wider and longer but reduced in thickness. This sometimes causes the plant to become droopy. The roots are generally smaller and more bunched than normal.

Phosphorus

This element is closely concerned with the vital growth processes in plants as it is a constituent of nucleic acid and nuclei in which this occurs are essential parts of all living cells. Hence a deficiency of this element will also be expected to result in greatly restricted growth. Phosphorus is also of importance in seeds and in connection with the metabolism of fats. Compounds of phosphorus are also concerned with the processes of respiration and with the efficient functioning and utilization of nitrogen. This relationship to nitrogen probably accounts for the fact that several of the symptoms of phosphorus deficiency are identical or similar to those which result from a deficiency of nitrogen. It is of special importance in the processes concerned in root development and the ripening of seeds and fruits.DNA (Deoxyribose Nucleic Acid) and RNA (Ribonucleic Acid), which are responsible for replication of a plant in the same form through generations, are phosphorus compounds composed of long sequences of four different building blocks called nucleotides. Each nucleotide in DNA contains a phosphate group, a sugar (deoxyribose) and one of four nitrogen compounds (adenine, guanine, cytosine and thymine) while those in RNA are similar but sugar is ribose and uracil is present instead of thymine. Nucleotides also serve another equally important function in plants by acting as chemical agents for the storage and transfer of energy between substances during reactions. The most important, adenosine triphosphate (ATP), containing three phosphate groups, participates in a great number of reactions during photosynthesis, respiration and other plant processes. Many other complex phosphorus compounds are involved in the 50 or more reactions which give rise to the final products of both photosynthesis and respiration. For example in utilising the light energy stored during photosynthesis carbon dioxide must first react with a suitable compound to begin the process, function is performed by ribulose bis phosphate. When reserves of protein are being deposited during the formation of seeds (bulbs and tubers) most of the phosphate is converted to phytic acid, a compound with six phosphate groups.Phytic acid becomes associated with calcium, magnesium, potassium, zinc and iron as phytates. These substances regulate the deposition of starch during the development of the above-mentioned storage organs of the plant. When growth begins again, the proteins release amino-acids that are then translocated with other mineral nutrients (phosphate, calcium and magnesium) to the developing roots and shoots. Seeds have the highest concentration of P in a mature plant and are required in large quantities in young cells, such as shoots and root tips, where metabolism is high and cell division is rapid.P aids in root development, flower initiation and seed and fruit development.It has been shown to reduce disease incidence in some plants and has been found to improve the quality of certain crops.

Because P is needed in large quantities during the early stages of cell division, the initial overall symptom is slow, weak and stunted growth. It is relatively mobile in plants and can be transferred to sites of new growth, causing symptoms of dark to blue-green coloration to appear on older leaves of some plants.Older leaves turn a darker green followed by a purplish tint starting from the leaf margins and leaf tips dry off. The plants are often darker green than normal and a reddish colouration is occasionally seen due to the pigment, anthocyanin.In ornamental plants flowering and fruiting may be affected.Lack of P can cause delayed maturity and poor seed and fruit development. Phosphorus-deficient plants are retarded in growth because fewer phosphorus-containing compounds essential for photosynthesis and respiration are formed. A deficiency also affects the plant hormone balance thereby delaying flower initiation and decreasing the flower count. Deficiency symptoms are more prevalent during cold, wet conditions.

Excess phosphorus hastens the maturity of a plant. Root growth is more extensive than normal. Second to N, P is often the limiting element in soils.Phosphorus is supplied to plants as water-soluble compounds such as calcium dihydrogen-phosphate. It enters the roots as phosphate ions.To correct the deficiency apply solid or liquid phosphate fertilisers.Phosphorus is relatively immobile in soil and does not leach. It is stored in seeds and fruit. It is most readily available to plants between a pH of 6 and 7.5 (unavailable in very acid or alkaline soils). Phosphorus absorption is reduced at low soil temperatures.

Potassium

Unlike all the other major elements, potassium does not enter into the composition of any of the important plant constituents, such as proteins, chlorophyll, fats and carbohydrates, concerned in plant metabolism. The element is present in all parts of plants in large or fairly large proportions. It seems to be of special importance in leaves and at growing points, as these are especially rich in potassium. Probably the whole of the potassium in plants is present in soluble form and most of it seems to be contained in the cell sap and cytoplasm. Potassium ions play an essential role, not by being part of the structure of plants, but by exerting a regulatory function on reactions and processes. Potassium ions activate over 50 enzymes in plants. This activation has been likened to a "lock-and-key" effect. For example the enzyme, starch synthetase, assists in the rapid conversion of glucose to starch in amyloplasts but only if potassium ions are present to modify the shape of the enzyme (activate). Similarly these ions activate the enzyme, nitrate reductase, as it converts the nitrate ion to nitrite ion in a step towards synthesis of proteins.It is important in many enzymes that are essential for photosynthesis Potassium ions play an important role in the steps involved in the conversion of amino-acids to proteins. These ions also protect plants from disease by promoting the growth of thicker outer walls of leaves. In addition they regulate the level and balance of growth and flower initiation (senescence) hormones in plants. The most important function of potassium ions in plants is in the control of the movement of water into and out of, cells by a process called osmosis. The vacuoles of cells contain a high concentration of dissolved substances of which potassium ions are the most prominent positively charged ion. This results in water moving through the cell-wall membrane into the concentrated solution of the vacuole thereby building up a pressure within the cell. This pressure can be maintained because of the rigidity of the cell wall. The resultant turgidity of the cells plays a major part in producing the rigidity of the plant as a whole. This same mechanism is observed in two other important plant processes:

- The outer surface of leaves contains holes (stomatal pores) whose size can be increased by the swelling of adjacent sausage-shaped guard cells. When light falls on a leaf, potassium ions move from cells surrounding the stomata into the two guard cells. Water then enters the more concentrated solution and the resultant pressure build-up causes guard cells to bow outwards. This leaves an opening through which carbon dioxide from the air can enter for photosynthesis. The oxygen which is formed moves out through this opening. At night the potassium ions move away from the guard cells and the stomate closes. The non-growth hormone, abssicisic acid assists in bringing about this closure.
- Cell extension occurs because potassium ions accumulate in the central vacuole of the cell with other substances and water moves in by osmosis. As the vacuole swells to occupy 80-90 per cent of the whole cell volume, the cell wall extends uni-directionally under the influence of the auxin.

It promotes vigor and disease resistance, helps development of root system, improves plant quality and increases winter hardiness due to carbohydrate storage in roots. Increases protein production and is essential to starch, sugar and oil formation and transfer and in water relations. Potassium is also required for flower and fruit formation and thickening of cell walls.

The great mobility of potassium in plants,its special importance for and its reutilization by young tissues and its apparent functions as a regulator of plant processes on a large scale are in harmony with the observations that, when potassium is moderately deficient, the effects are seen first in the older tissues and progress from these towards the growing points, but, when the deficiency is acute, growing points are severely affected and die-back and general collapse of the plants commonly occur. Plants lacking potassium are more susceptible to fungal diseases, probably due to an accumulation in plant tissue of unused sugars and amino-acids which provide the food for the attacking organism. Under these conditions the roots are readily affected by rotting. Some potassium-deficient plants are more susceptible to frost damage. Lack of potassium causes leaf margin chlorosis, followed by necrosis from the outside to the mid vein. Mature leaves show a browning and drying of the upper surface and puckering on the margins. Darkening appears between the leaf veins. Bronzing and dying of leaf margin. Chlorotic spotting with brown spots between veins of leaf occurs. Stunted internodes and roots. The stalks are thin and shortened. The fruits may fail to develop full colour.

Potassium (potash - K_2O) is generally immobile in soil.To correct the deficiency apply sulphate of potash. Inorganic fertilizer K is sold in various formulations including liquid, granular, and slow-release. Relatively small amounts of K become available from decomposition of manures and crop residues. It has been shown in many instances that the potassium content of plants is frequently much higher than is necessary for healthy growth and it is generally considered that luxury (i.e. unnecessary) absorption of potassium often takes place.An excess of potassium can affect the uptake and availability of magnesium and calcium to a plant.

Calcium

Calcium occurs in plants chiefly in the leaves and the amounts present in seeds and fruits are relatively low. One of its main functions is as a constituent of the cell wall, the middle lamella of which consists largely of calcium pectate. This function appears to be of fundamental importance since, if calcium is replaced by any other of the essential elements, such as magnesium or potassium, the organic materials and mineral salts in the cells are readily leached through the walls. It provides a base for the neutralization of organic acids; it is concerned with activities of growing points (meristems), especially with root tips; it may be of importance in nitrogen absorption. Although a large proportion of the calcium contained in the plant may be soluble in water it does not appear to move freely from the older to the younger parts of plants, and hence, young tissues contain lower proportions of calcium than older ones. This may explain why calcium deficiency effects begin at the tips of shoots.It has a major role in the formation of the cell wall membrane and its plasticity, affecting normal cell division by maintaining cell integrity and membrane permeability. It is an activator of several enzyme systems in proteinsynthesis and carbohydrate transfer. It combines with anions including organic acids, sulphates and phosphates. It acts as a detoxifying agent by neutralizing organic acids in plants.Calcium is necessary for cell elongation and division, protein synthesis, root and leaf development, and plant vigor.It influences intake of other nutrients and increases calcium content of plants. Its deficiency is not often observed in plants because secondary effects of high acidity resulting from soil calcium deficiency usually limit growth, precluding expressions of Ca deficiency symptoms.

When calcium is deficient, terminal bud dies, young leaves are hooked, because Ca^{++} is not easily translocated inside the plant. Dying back occurs at tips and margins, foliage may become distorted. Stalk dies off at the terminal bud. Root systems may be damaged by the root tip death. The growing tips of roots and leaves turn brown and die.Without adequate Ca, which in the form of calcium pectate is needed to form rigid cell walls, newly emerging leaves may stick together at the margins,which causes tearing as the leaves expand and unfurl. This may also cause the stem structure to be weakened. Buds and blossoms fall prematurely in some crops. Symptoms are more prevalent on the fruits than on the leaves.

Calcium is immobile in plants and relatively immobile in soil.Calcium is often limited in acidic soils that receive abundant rainfall. Sometimes it is difficult to differentiate between calcium deficiency and magnesium toxicity. Excess Ca may cause boron or magnesium deficiencies.

Magnesium

Magnesium is essential for photosynthesis, proteins and chlorophyll formation.It is the central atom in the green pigment chlorophyll found in the chloroplasts of leaf cells. In the photosynthesis, certain wavelengths of light are absorbed by the chlorophyll to begin the many reactions which ultimately replace carbon dioxide from the air with oxygen while forming the high energy sugar, glucose. Magnesium ions activate some of the enzymes involved in these reactions especially those in which ATP are used as a source of energy. Magnesium ions play a major role in activating those enzyme-controlled reactions using ATP that occur during plant respiration. In this process glucose and oxygen react to release energy:

- Mostly as heat to maintain plant life throughout day and night;
- While the remainder constitutes to be entrapped in still more ATP (Adenosine Tri Phosphate) and other high-energy chemicals which ultimately lead to the formation of many important plant substances including chlorophyll, amino-acids and nucleotides.
- Magnesium also plays a part in the formation of DNA and RNA; in the activation of the enzymes that assist in the conversion of ammonium ions to amino-acids and, together with calcium, as phytate during the storage of starch in seeds and bulbs.
- Magnesium is also regarded as a carrier of phosphorus in the plant.

It moves freely within the plant and is taken from the older leaves to supplement new growth. Lower leaves are first affected,yellowing from the tip downwards. Mottling and yellowing between veins of older leaves while veins remain green. Yellow areas may turn brown and die. Yellowing may also occur on older leaves. Leaves may turn reddish purple due to low P metabolism. In severe cases, symptoms may appear on younger leaves and cause premature leaf drop. Plant stems are stiff, thin and woody. Symptoms may be similar to N deficiency and are most often found in sandy soils that are low inorganic matter and receive moderate to heavy rainfall. Symptoms occur most frequently in acid soils and soils receiving high amounts of K fertilizer or Ca.

Deficiencies occur on sandy acid soils.To correct the deficiency apply magnesium sulphate (Epsom salts) or dolomite (magnesium calcium carbonate $MgCO_3$, $CaCO_3$).

Deficiencies most likely on leached sandy soils and where high levels of N and K have been applied.Magnesium deficiency is rarely a problem in most soils. Excessive magnesium, on the other hand, can induce potassium deficiency due to interference with K uptake and utilization.

Zinc

It is required in the synthesis of tryptophan, which in turn is necessary for the formation of indole acetic acid in plants. It is an essential component of several metallo-enzymes in plants (dehydorgenases) and therefore, is necessary for several different functions in plant metabolism. The enzyme carbonic anhydrase is specifically activated by Zn. It has a role in RNA and protein synthesis. Interveinal chlorosis occurs on younger leaves, similar to Fe deficiency. However, Zn deficiency is more defined, appearing as banding at the basal part of the leaf, whereas, Fe deficiency results in interveinal chlorosis along the entire length of the leaf.

It is not a very mobile nutrient so a foliar spray of zinc and manganeseis recommended for fruit set. Zinc is mobile in plants; mobility in soil decreases as pH increases. Availability is reduced by high pH, low levels of organic matter in mineral soils, soil compaction, excessive rates of P, and low temperature and wet soil. Organic matter can both inhibit and stimulate zinc uptake. Deficiency is more common in sandy soils. Spotted older leaves, usually yellow to white between veins (interveinal chlorosis). Early twigs loss of plant. Distorted or puckered leaf margins. Reduction in size of leaves (little leaf) and internodes of plants occurs.

Iron

Many fertiliser mixtures contain this micronutrient in the form of pale green iron sulphate crystals. However, unless the pH of the nutrient solution is below 5, very little iron is absorbed through the roots because iron ions, react with water producing an almost insoluble substance. Iron chelates are far more effective. They dissolve in, but are unaffected by water and on contact of the nutrient solution with the root surface they are reduced to Iron ions which are immediately bound to certain chemicals in the sap (e.g. citric acid) and carried throughout the plant. In plant growth iron mainly functions as an electron-transfer agent by alternating between the Iron (II) and Iron (III) states - ferrous and ferric states. Several iron-sulphur proteins are present in plant cells, the most important being ferredoxin. In photosynthesis this compound plays a key role in the presence of chlorophyll, in releasing oxygen from water. In the reaction leading to protein synthesis ferredoxin transfers electrons to reactants in enzyme-controlled reactions which convert nitrite ions to the amino-acid, glutamine. It also takes part in the conversion of sulphate ions to sulphide ions in the series of reactions that ultimately form the two sulphur-containing amino-acids, cysteine and methionine.Other iron-proteins also act as enzymes for reactions during plant growth.One such hemeprotein (peroxidase), catalyses the formation of lignin in cell walls. Other hemeproteins (cytochromes) provide electrons for the final steps of respiration which Oxygen forms water in a reduction reaction. Yet another protein of this type (catalase) is involved in the synthesis of chlorophyll and also plays a part in a reaction releasing oxygen during photosynthesis.Iron is closely concerned with chlorophyll formation but is not a constituent of it. Its role appears in this connection to be that of a catalyst. As a result of this function of iron, chlorosis is invariably an outstanding symptom when the element is deficient. Iron may also act as a catalyst, in the role of an oxygen carrier, in respiration. A point of great importance in connection with iron is its relative immobility in plant tissues. Its mobility seems to be affected by several factors, such as the presence of manganese, potassium deficiency and high light intensity. There is evidence that the amount of chlorophyll is related to active (readily soluble) iron in plants. It will thus be seen that so-called iron deficiency in the plant may in fact usually mean iron immobility. Lack of mobility may also account for the fact that iron deficiency is first shown in the younger tissues. Interveinal chlorosis in younger leaves.The youngest leaves maybe white, because Fe, like Mg, is involvedin chlorophyll production. Young leaves are yellow with contrasting narrow green veins; older basal leaves remain darker green. Exposed leaves are bleached and will eventually exhibit apical or marginal scorch. Leaves may be small, symptoms will be severe in cold, wet springs. Shoot length is usually normal, but diameter will be small; twig die back and defoliation will occur when the deficiency is severe. Usually it observed in alkaline or over-limed soils. Iron is immobile in plants and mobility decreases in soil with increasing pH. Deficiencies can be corrected by lowering pH. Excessive P may induce deficiency.

Manganese

Manganese is usually supplied to plants as Manganese sulphate. It is absorbed through roots from the nutrient solution as manganese ions, which are then transported in the sap of the xylem to the growing leaves. The functions of manganese are regarded as being closely associated with those of iron and as being concerned with chlorophyll formation.Mn is necessary in Photosystem II, where it participates in photolysis. It fulfils its main function during the first stage of photosynthesis. In the presence of chlorophyll in the chloroplasts of leaf cells a Manganese-chloride protein brings about the light-induced reaction in which water is split with the release of Oxygen. Hence, when manganese is deficient, chlorosis is a common symptom. It may decrease the solubility of iron by oxidation and hence, an abundance of manganese within the plant may lead to iron deficiency and chlorosis. It is a constituent of pyruvate carboxylase. Mn activates indole acetic acid oxidase, which then oxidizes indole acetic acid in plants.

A Manganese deficiency restricts the formation of chlorophyll.Symptoms first appear as chlorosis in young tissues.Manganese deficiency causes yellowing between the veins of new foliage. Greenish-grey specks appear at the lower base of younger leaves. The specks may eventually become yellowish to yellow-orange. From its role in photosynthesis, an inadequate supply of Manganese reduces the amount of sugars and indirectly affects the formation of amino-acids and therefore, protein synthesis.An excess of Manganese can reduce the uptake of Magnesium and Calcium.

Manganese is immobile in plants; mobility in soil decreases with increasing pH. Soils very high in organic matter or poorly drained are deficient at pH 5.8 to 6.5. For other soils, deficiency usually occurs between pH 6.5 and 8.0, especially where soil has been heavily limed. It is toxic in very acid soil. Excessive water, poor aeration and excess heavy metals influence Mn uptake.

Manganese increases availability of Ca, Mg and P and is necessary for chlorophyll synthesis and photosynthesis.To correct the deficiency apply manganese sulphate.

Sulphur

Sulphur is generally received by plants as sulphate ions. These enter the roots from the surrounding solution and most are conveyed unchanged in the sap-conducting vessels (xylem) to the leaves.Its functions in connection with proteins and chlorophyll are doubtless account for the similarity of its deficiency effects to those due to deficiency of nitrogen. In the chloroplasts sulphate ions are converted to sulphide ions, using energy supplied by ATP with the assistance of an iron-sulphur enzyme, ferredoxin. A similar reaction occurs in the roots. Most of these sulphide ions are rapidly converted into the two sulphide-containing amino-acids, cysteine and methionine that are required with other essential amino-acids for protein synthesis. Traces of cysteine are used to form an important sulphur compound (coenzyme A) that is involved in the step during respiration in which carbon dioxide is formed. The sulphur-containing vitamin B_1 (thiamine) coupled with phosphate takes part in this reaction. Small amounts of methionine are converted to a compound (S-adenosylmethio-nine) from which can be formed lignins to strengthen cell walls, brightly coloured compounds (anthocyanins) and chlorophyll. Two iron-sulphur proteins and ferredoxin (in which they are similar) participate in the initial reactions of photosynthesis in which absorption of light by chlorophyll releases oxygen. Deficiencies result in yellow leaves and stunted growth.Sulphur can be used to lower the pH of alkaline soils.A sulphur deficiency in plants, although not very common, affects protein synthesis and, to a lesser extent, photosynthesis. In most plants sulphur is not easily translocated. As a result, the deficiency symptoms are a light green to yellow chlorosis over the whole of new leaves including the veins. The leaves may not grow to their normal size. Growth rate is retarded and maturity is delayed. Plant stems are stiff, thin, and woody. Symptoms may be similar to N deficiency and are most often found in sandy soils that are low in organic matter and receive moderate to heavy rainfall.

Organic sulfur is converted into available sulfate sulfur by soil bacteria. Deficiency, when it occurs, is most likely on sandy, low-organic matter soils. Because enough sulphate is present in most soils, sulfur deficiency is fairly uncommon. Inorganic (mined) gypsum (17 per cent sulfur), elemental sulfur (30-99 per cent S), ammonium sulfate (24 per cent S), potassium sulfate (17 per cent S), super phosphate (12 per cent S) may be used for contol the sulphur deficiency.

Boron

Boron is an important constituent in cell-walls and plasma membranes of growing tissue including the roots.It is necessary in the synthesis of one of the bases for RNA formation and in cellular activities. It has been shown to promote root growth.It is essential for pollen germination and growth of the pollen tube. It has been associated with lignin synthesis, activities of certain enzymes, seed and cell wall formation, and sugar transport. It can delay the onset of calcium deficiency effects but cannot replace calcium; it tends to keep calcium soluble; it may act as a regulator of potassium/calcium ratios, and of the absorption of nitrogen; it may be concerned with the oxidation-reduction equilibrium in cells.Boron is extremely immobile in plants and is not translocated to new growth, but moves readily in soil.Boron is necessary to increase the yield and quality of Orchids and is associated with calcium utilization. Generally, B deficiency causes stunted growth, first showing symptoms on the growing point and younger leaves. The leaves tend to be thickened and may curl and become brittle.Deficiency is more apparent during drought stress. It causes internal breakdown and external necrosis of stems and roots. It reduces flowering and failure to set seed.

Boron is present in fertiliser mixtures either as sodium borate ($Na_2B4O_7)^3BO_3$). Plants can tolerate only a narrow range in concentration (about ten fold) between deficiency and toxicity limits - about 0.5 to 6 ppm of boron in the nutrient solution given to the plants.Deficiencies often occur when pH is between 6.0 and 7.5, on deep sandy soils and when high rates of N, K, and Ca are used. Excess of boron causes blackening or death of tissue between veins.

Molybdenum

Very small amounts of ammonium molybdate are necessary in fertiliser mixtures. The dry tissue analysis of healthy plants shows only about one part of molybdenum in ten million is present, much lower than for any other essential micronutrients. Molybdenum enters the plant as molybdenum ion. It is a structural component of the enzyme that reduces nitrates to ammonia. Without it, the synthesis of proteins is blocked and plant growth ceases. Seeds may not form completely and nitrogen deficiency may occur if plants are lacking molybdenum. It is a necessary component of two major enzymes in plants, nitrate reductase and nitrogenase, which are required for normal assimilation of N. It is required by some soil microorganisms for nitrogen fixation in soils. Deficiency symptoms resemble those of N because the function of Mo is to assimilate N in the plant. Older and middle leaves become chlorotic, and the leaf margins roll inwards. In contrast to N deficiency, necrotic spots appear at the leaf margins because of nitrate accumulation. Deficient plants are stunted and flower formation may be restricted. Deficiency signs are pale green leaves with rolled or cupped margins. Molybdenum is mobile in plants and in soil. It is less available at a lower pH. Generally no deficiency over pH 6.0 was observed.

Copper

Copper is usually present in fertiliser mixtures as blue crystals of copper sulphate and enters the roots as either copper ions, or in a chelated form (with amino-acids). About fifty per cent of plant copper is present in chloroplasts where it performs most of its functions, generally as an electron-carrier between the copper (II) and copper (I) states. During photosynthesis a copper-protein (plastocyanin) together with another enzyme (plastoquinone) transfer electrons to reactants linking two series of reactions that take place in chloroplast membranes. Although plastoquinone contains no copper it requires a copper enzyme (laccase) for its formation.In most plants a copper-zinc protein (copper-zinc dismutase), rather than the corresponding manganese compound, assists in the removal of the highly reactive oxygen species, O^{2-},formed during photosynthesis as previously mentioned. In the membrane of chloroplasts another copper enzyme (phenolase) plays a part in synthesising both lignin and the brown substances which exude when plant tissue is damaged thereby preventing fungal growth.

It is relatively immobile in soil and plants. Its deficient is more in organic than mineral soils and often in sandy than heavy soils. It is deficient at high pH. A copper-deficiency results in younger leaves being stunted in growth due to reduced photosynthesis. Reduced growth, distortion of the younger leaves, and possible necrosis of the apical meristem .Leaves distorted because of insufficient lignin deposited in the walls of the sap-conducting cells. Multiple sprouts occur at growing points, resulting in a bushy appearance. Young leaves become bleached, and eventually there is defoliation and dieback of twigs. A copper-deficiency results in the non-viability of pollen. Excess copper may affect the uptake of iron by plants.

Chlorine

Chloride ions assist in achieving a charge balance with positive ions in plant solutions andby entry into vacuoles, help to maintain the turgidity of cells.It has been shown to increase the water content of the tissues and affect carbohydrate metabolism, leading to an accumulation of starch in the leaves. The element is present in plants as chloride and is wholly soluble. Chloride ions activate as enzyme (asparagine synthetases) for one of the series of reactions involved in the conversion of ammonium ions to amino acids. Cl is essential in photosynthesis, where it is involved in the evolution of oxygen. Deficiency symptoms first appear on younger leaves as chlorosis, wilting, stubby roots and bronzing. Odours in some plants may be decreased due to its deficiency. Most plants absorb up to 100 times very small amount of chloride that is essential for optimal plant growth. Deficiency seldom occurs because Cl is found in the atmosphere and rainwater.

Sodium

Sodium is not strictly an essential element it cannot be expected to have a specific role in the metabolic activities of plants. It is regarded as a conserver of potassium and as being able partly to replace that element in its role. In no instance, however, has it been shown that sodium can wholly replace potassium where the latter is acutely deficient. Sodium seems to affect the water relations of plants and often enables crops to withstand drought conditions which would otherwise produce severe adverse effects.It involved in osmotic (water movement) and ionic balance in plants.

Nickel

It is required for the enzyme urease to break down urea to liberate the nitrogen into a usable form for plants. Nickel is required for iron absorption. Seeds need nickel in order to germinate. Plants grown without additional nickel will gradually reach a deficient level at about the time they mature and begin reproductive growth. If nickel is deficient plants may fail to produce viable seeds.

Cobalt

The demand for cobalt is much higher for nitrogen fixation than for ammonium nutrition. Deficient levels could result in nitrogen deficiency symptoms.

Silicon

It is found as a component of cell walls. Plants with supplies of soluble silicon produce stronger, tougher cell walls making them a mechanical barrier to piercing and sucking insects. This significantly enhances plant heat and drought tolerance. Foliar sprays of silicon have also shown benefits reducing populations of aphids on field crops. Improved leaf erectness, stem strength and prevention or depression of iron and manganese toxicity has all been noted as effects from silicon. Silicon has not been determined essential for all plants but may be beneficial for many.

Factors affecting Availability of Nutrients

Each type of plant is unique and has an optimum nutrient range as well as a minimum requirement level. Excessive nutrient uptake causes poor growth because of toxicity. Therefore, the proper amount of application and the placement of nutrients are important.If only one nutrient is missing from the soil solution the plant will not grow and develop normally. This notion was postulated by Justus von Liebig in his Law of the Minimum. The Law of Minimum maintains that yield is proportional to the amount of the most limiting growth resource. The interrelationship between a nutrient's availability and soil pH, air and soil temperature, available moisture, excesses of other nutrients, soil organic matter and soil mineral content are complex. Soil testing is the best way to accurately determine whether primary nutrient levels are optimum for plant survival and growth. There are several factors which can inhibit proper nutrition of orchid plant:-

pH : One factor which can introduce nutrient deficiencies in orchid plants is the substrate pH. The general pH range for Orchid plants is 5.4 to 6.8, but maintaining the pH from 5.6 to 6.2 is recommended. Poor uptake of nutrients, particularly boron, copper, iron, manganese, and zinc can occur if the pH of a substrate is above 6.5. Certain macronutrients like calcium and magnesium can become less available at pH below 5.4.

Cation Exchange Capacity

The Cation exchange capacity refers to the ability of the growing medium to hold exchangeable mineral elements within its structure. These cations include ammonium nitrogen, potassium, calcium, magnesium, iron, manganese, zinc and copper. Peat moss and mixes containing bark, sawdust and other organic materials all have some level of cation exchange capacity.

Soil Structure : Soil particle size and chemistry affect the way in which water and nutrients are retained and taken up by the plant. Soils high in loam or clay have a higher water and nutrient retention compared to sandy soils where nutrients are easily leached out.

Roots : undeveloped or insect damaged roots are ineffective in taking up nutrients.

Working equipment : Improperly working fertilizer equipment can cause nutrients to be less than optimum in the substrate. The significant problem with equipment failure is the introduction of multiple nutrient deficiencies.

Water Balance : Under or overwatering the plants may lead to nutrient deficiencies. Constant saturation of the substrate can lead to macro and micronutrient deficiencies. As oxygen levels are inhibited by overwatering, root growth can be limited and slow water uptake by plant. Elements such as calcium are transported via water flow and deficiency symptoms can develop rapidly on new growth. Also the inactivity of root systems due to saturated conditions can lead to inefficient uptake of iron or phosphorus.

Soluble salts : Soluble salts refer to the total dissolved salts in the root substrate at any given time and are measured in terms of electrical conductivity (EC). When the EC of the root zone is too low, plant growth is stunted and mineral deficiencies are observed. Low salts are usually due to too many clear water irrigations. Deficiencies appear among plants like lower leaf yellowing (nitrogen), lower leaf purpling (phosphorus), and lower leaf interveinal chlorosis (magnesium) are common when values are below 0.75 mS/cm.

Mineral antagonisms : When certain elements are provided in excess to plants, uptake of other nutrients may be hindered. One example of a mineral antagonism is the nitrogen-potassium interaction. Other types of antagonisms are the potassium-calcium-magnesium interaction. Any one of these elements in excess can cause a decrease in the uptake of the others. Excess phosphorus can cause a decrease in uptake of zinc, iron, and copper.

Environmental Conditions : Extremes of temperature will inhibit nutrient uptake. Temperature can also play a role in the introduction of nutrient deficiencies. Purpling of the lower foliageis the common symptom.

Disease : Organisms like *Pythium* feed on the nutrients in roots causing an inefficient uptake of minerals. Iron deficiency (upper foliage interveinal chlorosis) can occur if root rot pathogens infect the root system. Foliar diseases, particularly fungal diseases can cause chlorosis of leaf tissue, a direct reflection of harvesting nitrogen from plant cells.

Mycorrhizae and nutrient availability

Mycorrhizae are symbiotic associations of roots and fungi that enhance plant nutrition.Mycorrhizae (fungus roots) are modified roots, consisting of mutualistic associations of fungi and roots. The fungus benefits from a hospitable environment and a steady supply of sugar donated by the host plant. Fungus provides several potential benefits to the host plant. First the fungi increase the surface area for water uptake and selectively absorb phosphate and other minerals in the soil and supply them to the plant. The fungi also secrete growth factors that stimulate roots to grow and branch. The fungi produce antibiotics that may help protect the plant from pathogenic bacteria and fungi in the soil. Almost all plant species produce mycorrhizae. This plant-fungus symbiosis may have been one of the evolutionary adaptations that made it possible for plants to colonize land in the first place. Fossilized roots from some of the earliest land plants include mycorrhizae. Mycorrhizal fungi are more efficient at absorbing minerals than roots, which may have helped to nourish pioneering plants, especially in the nutrient-poor soils present when terrestrial ecosystems were young. Mycorrhizae take two major forms: ectomycorrhizae and endomycorrhizae. In ectomycorrhizae, the mycelium forms a dense sheath over the surface of the root.Some hyphae grow into the cortex in extracellular spaces between root cells. Hyphae do not penetrate root cells but form a network in the extracellular spaces to facilitate nutrient exchange. The mycelium of ectomycorrhizae extends from the mantle surrounding the root into the soil, greatly increasing the surface area for water and mineral absorption. Compared with "uninfected" roots, ectomycorrhizae are generally thicker, shorter, more branched, and lack root hairs.Ten percent of plant families have species that form ectomycorrhizae. Endomycorrhizae have fine fungal hyphae that extend from the root into the soil.Hyphae also extend inward by digesting small patches of the root cell walls, forming tubes by invagination of the root cell's membrane.Some fungal hyphae within these invaginations may form dense knot like structures called arbuscles that are important sites of nutrient transfer. Roots with endomycorrhizae look like "normal" roots with root hairs, but the microscopic symbiotic connections are very important. Endomycorrhizae are found in more than 85 per cent of plant species, including important crop plants such as corn, wheat, and legumes.Roots can be transformed into mycorrhizae only if they are exposed to the appropriate fungal species.In most natural systems, these fungi are present in the soil and seedlings develop mycorrhizae. However, seeds planted in foreign soil may develop into plants that show signs of malnutrition because of the absence of the plant's mycorrhizal partners. An epiphyte is an autotrophic plant that nourishes itself but grows on the surface of another plant, usually on the branches or trunks of trees. Epiphytes absorb water and minerals from rain, mostly through their leaves. Examples of epiphytes are staghorn ferns, some mosses, Spanish moss and many species of bromeliads and orchids. There is growing evidence that the molecular biology of root nodule formation is closely related to mechanisms that first evolved in mycorrhizae. The nodul in genes activated in the plant during the early stages of root nodule formation are the very same genes activated during the early development of endomycorrhizae. Mutations in these genes block development of both root nodules and mycorrhizae. The signal transduction pathways of both systems share some components. Experimental application of plant hormones called cytokinins to root cells of legumes activates expression of early nodulin genes even in the absence of bacterial or fungal symbionts. "Infection" by either symbiont causes cytokinin concentrations to increase naturally. Cytokinins may be one of the links between the "I'm here" announcement of the microbes and the changes in gene expression in the plant that lead to structural modifications of the roots.Further evidence comes from the similarities in the chemical cues used by both microbes.The Nod factors secreted by *Rhizobium* bacteria are related to chitins, the same compounds that make up the cell walls of fungi.A reasonable hypothesis is that root cells have a family of closely related receptors that detect their particular bacterial and fungal symbionts.Mycorrhizae evolved very early, probably over 400 million years ago, in the earliest vascular plants. In contrast, the root nodules in legumes originated only 65-150 million years ago, during the early evolution of angiosperms. The common molecular mechanism in the root's two major symbiotic relationships suggests that root nodule development was at least partly adapted from a signaling pathway that was already in place in mycorrhizae. On sandy soils small amounts of fertiliser should be used frequently, to ensure a steady supply of nutrients to the plant. On heavier soils apply fertiliser in greater quantities, but less often. Plants can also suffer from nutrient toxicity if too much of a particular nutrient is applied. Make sure that fertiliser is watered in after application, to avoid burning of the plants.

In nature, orchids obtain their supply of inorganic nutrients like calcium, magnesium, iron, potassium, nitrogen and traces of manganese, boron, copper, zinc etc. from the tree on which they are growing and also from atmosphere and decaying vegetables and dropping of birds. However, under controlled conditions they have to be supplied with all major and minor nutrients. Taking into consideration the special need of different orchids, a large number of fertilizer mixtures, both solid and liquid, are available in market. Liquid fertilizers are much more quickly absorbed and can be applied more frequently. As the orchids are slow growing, slow release fertilizers like osmocote can be used to get very good result. Usage of fertilizers should also depend on stage of growth. During vegetative growth, large quantities of nitrogen are required while during flowering, nitrogen should be reduced and amount of phosphate increased. It have been getting excellent results by using slow release fertilizer mixtures (NPK 20:20:20) with trace elements and coconut water (20-25 %) applied every week for three successive weeks followed by a 10:20:30 (NPK). Orchids are relatively light feeders compared to other plants.

Some orchids, like members of the genus *Disa*, are exceptionally sensitive to feed and are injured at modest feed strengths. *Odontoglossums* are particularly sensitive to moderate feed strengths. The important factor is not to exceed a certain total value of dissolved salts. The concentration of salts in a nutrient solution, held by the mix, affects a parameter known as the "salt index" or osmolarity. As the osmolarity of the substrate solution increases less water transport occurs from the root. Typically, orchids are watered and allowed to partially dry, between each watering, thus a further increase in osmolarity occurs as water evaporates from the substrate. With excessive substrate osmolarity a plants water loss from transpiration will exceed the water replaced by the roots. This results in several problems. A common problems is leaf tip die-back, a condition where the leaf tips dies because of lack of water. Warm days, a poor root system and excessive drying out of the substrate in the presence of high feed rates will exasperate leaf tip die-back.

Bio-fertilizer and growth or ornamental plants

Fertilizer for crops may be classified as chemical fertilizer, bio-fertilizer, green manure and compost. Bio-fertilizer contains microbes or enzymes and is available in the forms of powder, granule or liquid. When bio-fertilizers are applied on the seeds, seedlings or soils, not only the effectiveness of nutrient uptake can be improved, but also the amount of beneficial microbes can be replenished to maintain soil in good eco-environment. In general, the advantages may be grouped into 3 categories: (1) nitrogen fixation, to fix atmospheric nitrogen into ammonia and convert into nitrate which can be used by crops, (2) solubilization of soil minerals such as phosphorus, calcium and iron to be used by plants, (3) promotion of the nutrient uptake by roots and plant growth.The bio-fertilizer composition comprises of a symbiotic organism, which is selected from the group consisting of *Rhizoctonia* sp. (BCRC930076 and BCRC930077), a growth substance and a medium, which is used to mix and evenly disperse the symbiotic organisms and the growth substance. The bio-fertilizer can be applied to promote seed germination in test tubes or be applied in seedling growth after deflasking to promotegrowth, preventing disease, and increasing the beneficial components toward human health.

Fertilizer Concentration Calculations

A. Units Used

ppm = parts per million

mM = milli molar

meq/l = milliequivalent per liter

B. Conversion Factors (metric vs. British system)

1 ounce = 28.35 g

1 pounce = 0.45 kg

1 gallon = 3.78 liters

1 g = .035 ounce

1 kg = 2.205 pounce

1 acre = 43,560 ft^2

1 liter = l kg

C. Fertilizer Concentrations

a. *Parts per million* (ppm)

The term, parts per million, is an expression of concentration used often to describe very dilute solutions. The term states how many parts of solute there are in a million parts of the whole solution. Parts per million almost always expresses concentrations on a mass basis.

For example, a 10 ppm solution is one in which every million grams of solution contains 10 grams of solute. The ppm designation is most often applied to dilute solutions in water. For example,1 kilogram (1000 gram) of water contains 1 million milligrams of water; thus

1 kg = 1 kg x 1000 g/kg x 1000 mg/g = 1,000,000 mg

At normal temperatures, 1 liter of a dilute water solution has a mass of approximately 1 kilogram. If we have 10 mg of solute in 1 liter of solution, it will be 10 ppm.

10 mg solute = 10 mg solute = 10 ppm

1 liter solution 1,000,000 mg solution

Thus when we say that the concentration of nitrogen in water is 200 ppm, we mean that 1 liter of the solution contains 200 milligrams of nitrogen. The important thing to remember is:

1 kg = 1,000,000 mg

1 liter water = 1 kg therefore, 1 liter water = 1,000,000 mg

b. *Milli-molar* (mM)

One millimolar (mM) concentration refers to a solution containing one-thousandth of molecular weight (g) of the solute per liter of water. One molar (M) concentration equals 1000 millimolar (mM) concentration.

Table 8.1 : Chemicals and their molecular weight

S.No	Chemical	Molecular or atomic weight
1	N	14.01
2	K^+	39.10
3	NH_4NO_3	80.05
4	$NH4^+$	18.01
5	NO_3^-	62.01
6	Ca^{++}	40.08
7	Mg^{++}	24.32
8	S	32.07
9	SO_4^{-2}	96.07
10	$MgSO_47H_2O$	246.50
11	NO_3^-	62.01

1 mM NH_4NO^3 = 80.05 mg per liter (mg/l)

1 mM NO_3^- = 62.01 mg/l

1 mM SO_4^{-2} = 96.07 mg/l

1 mM $MgSO_47H_2O$ = 246.5 mg/l

c. *Milliequivalent per liter (meq/l)*

Milliequivalent per liter (meq/l) concentrations are often used to show the strength of fertilizer ions (anion or cation) in a solution. Since one equivalent weight is the molecular weight divided by valency, one meq/l refers to the ionic concentration of a solution that contains one millimole/valency per liter of water.

$NH_4NO^3 \rightarrow NH_4^+ + NO_3^-$ (Monovalent ions)

(80) (18) (62)

$MgSO_47H_2O \rightarrow Mg^{++} + SO_4^{-2}$ (Divalent ions)

(246.5) (24.3) (96.1)

1 meq/l $NH4NO_3$ = 80 mg/l

1 meq/l NH_4^+ = 18 mg/l

1 meq/l NO_3^- = 62 mg/l

1 meq/l $MgSO_47H_2O$ = (246.5 mg/2)/l = 123.3 mg/l

1 meq/l Mg^{++} = (23.3 mg/2)/l = 11.6 mg/l

1 meq/l SO_4^{-2} = (96.1 mg/2)/l = 48.0 mg/l

D. Fertilizer Analysis

a. Commercial Analysis

Commercial analysis is given by the percentages of nitrogen (N), phosphorus (P_2O_5) and potassium (K_2O) in that order. For example, Peters 20-16-20 fertilizer contains 20% N, 16% P_2O_5, and 20% K_2O by weight.

b. Elemental Analysis

Elemental analysis is used for more technical and scientific purposes. It is expressed as Percent weights of elemental nitrogen (N), phosphorus (P) and potassium (K) in that order.

c. Conversion of Commercial Analysis to Elemental Analysis

By using the ratios of elemental to oxides for phosphorus and potassium, the commercial analysis can be converted to elemental analysis.

Nitrogen - always expressed as elemental N

Phosphorus - $P_2 / P_2O_5 = 0.44$, or $P_2O_5/P = 2.99$

Potassium - $K_2 / K_2O = 0.83$, or $K_2O/K = 1.20$

Thus, Peters 20-16-20 commercial analysis fertilizer can be labelled as a 20-7.04-16.6 elemental analysis fertilizer.

20% N - 16% P_2O_5 - 20% K_2O = 20% N - 7.04% P - 16.6% K

For example, if you want to apply 200 ppm nitrogen to your plants and were going to mix up 1 liter of solution you then would have to put 1000 mg or 1 gram of fertilizer into the liter of water.

1 liter water = 1,000,000 mg, therefore, 200 mg of N are needed.

However, the fertilizer is only 20% N. So:

200 = x x = 1000 mg = 1 g. 20 1.00

Problems

1) You wish to prepare 5 litres of a 100 ppm nitrogen (N) fertilizer. How much 15-10-5 commercial analysis fertilizer will you need to add to 5 litres of water to get the desired concentration?

2) You are mixing 5 litres of concentrate fertilizer to apply with a hose-on (1:15 proportion), andyou want the final concentration to be 200 ppm nitrogen (N). What amount of fertilizer, if youare using 20-20-20 commercial analysis fertilizer, do you need to add to 5 litres of water?

Trouble Symptoms for Foliage Plants and Possible Causes

1. Lower leaves turn yellow and drop off easily
   a. Over-watering
   b. Insufficient light

2. Burned margins or brown tips on leaves
   a. Accumulated salts in soil
   b. Drought or low humidity

3. Pale leaf color, long internodes, loss of vigor, dropping lower leaves
   a. Lack of sufficient light

4. Growing tips chlorotic or growth slow
   a. Accumulated salts in soil
   b. Too high a soil pH

5. Brown spots on leaves no pathogen present
   a. Excessive light
   b. Water spotting

6. Interveinal chlorosis
   a. Iron (Fe) deficiency
   b. High pH

7. Poor flowering
   a. Insufficient light intensity
   b. Vegetative growth encouraged (N fertilizers, too large pot, improper photoperiod)

8. Lower leaf drop, yellowing and/or dieback, roots brown and rotting (lower stems may be soft)
   a. Excessive watering
   b. Poor drainage
   c. Accumulated salts (over fertilization)
   d. Root rot caused by pathogens: *Pythium, Rhizoctonia, Phytophthora*

CHAPTER 9

Weed Management in Ornamental Crops

Weed is a plant that is growing in the wrong place. Weeds have many mechanisms and characteristics that enable them to compete, having thick waxy cuticle leaves that can be resistant to some chemicals, fast reproduction methods, the ability to reseed in 6 week cycles and deep tap roots enabling the weed to survive in compacted dry ground conditions.

There are many reasons to control weeds in annuals/biennial and perennial ornamentals. Weeds compete with the plants by absorbing essential nutrients and water, or by blocking light or air circulation. Weeds dramatic reduce the growth of the plants. The greatest damage is often done when the weeds and ornamental plants are small. Weeds also interfere with harvesting by physically hindering. They lower flower quality further by harbouring insects, such as whiteflies, thrips mites, slugs, snails and/or diseases.

Growing under optimal conditions for the flower will reduce the ability of weeds to compete against the crop. Controlling humidity/ventilation, water, light, pH, nutrients and temperature all figure prominently in providing flower optimal conditions. A plant health strategy to greenhouse weed control contains three principal elements, prevention/sanitation, evaluation of water quality and using all controls available physical, cultural and chemical. Prevention and sanitation can include the use of mulches in orchidhouse weed control. Rice hulls, perlite, pumice, corn gluten meal bark, Penn Mulch - pelletized recycled paper and human hair have all been used for this purpose. Prevention also includes, eliminating air borne weed seeds from entering the greenhouse by screening vents and other openings and preventing perennial weeds from growing under foundations. A 1-3 meter weed free zone should be maintained immediately adjacent to the greenhouse.

Weeds Flora Associated with Ornamental crops

Description of different weeds associated with nursery and pot ornamental plantsis as under:

Cardamine hirsuta (Hairy Bitter cress, hairy cress, popping cress)

Hairy bittercress is variable in size and leaf shape. Waved bittercress, *C. flexuosa*, closely resemles hairy bittercress and is also variable in habit. It is usually annual or biennial but occasionally perennial. A related introduced weed, New Zealand bittercress, *C. corymbosa*, it is similar in appearance to hairy bittercress and has the explosive seedpods but is generally smaller. Hairy bittercress is found in flower all through the year but mainly from March to August. Seed is shed in May and June and sometimes into the autumn.

Fig. 9.1 : *Cardamine hirsute*

Plants can be found in fruit for 8 months of the year. The seed after-ripens at high temperatures. Germination is increased by a period of dry storage. Hairy bittercress seed germinates from April to December. There are peak flushes of seedling emergence from July to August and November to December but this varies in different years. Hairy bittercress can complete its lifecycle in 5-6 weeks. Seedlings can survive the severest frost.Hairy bittercress forms a relatively persistent soil seed bank.When the seedpods are ripe the seeds are dispersed explosively for up to 1 m if the plants are shaken by the wind or by weeding operations. The seeds become sticky when wet and can be spread on tools and clothing. Seedlings should be killed by early cultivations and regular hoeing to prevent hairy bittercress plants flowering and setting seed. Stem fragments are capable of re-rooting following cultivation in moist conditions. Hairy bittercress seedlings should be removed from container plants before the weed can set and shed seeds.

The standing areas must also be kept weed-free. Improved drainage may discourage this moisture-loving weed.

Fig. 9.2 : Chenopodium album

Chenopodium album (Pig Weed Fat-hen goose foot, lamb, Bathua)

Chenopodium is a very variable plant and is sometimes treated as an aggregate species. It shows morphological plasticity in response to soil fertility and plant density. It is known to hybridise with related species but the hybrids are difficult to identify due to the variability of the main species. Some populations of fat-hen have developed resistance to some widely used herbicides. It is one of the most troublesome annual weeds. It may act as a host to insect pests. It has long day flowers, usually from July to September. It is wind pollinated and the flowers may be cross or self-pollinated. Seeds are produced in abundance but mature relatively late in the season. It produces several different types of seed on the same plant. Most seed is black and hard coated with either a rough or smooth surface. The brown seeds germinate much more readily while the black seeds are able to persist longer in soil. Damage to the seed coat of the black seeds will encourage germination if conditions are favourable. Immature seeds are capable of germination and may do so more readily than ripe seeds due to a thinner seed coat. Seedlings with 2-6 leaves are killed by flame weeding. Fat-hen seed is susceptible to soil solarisation and can be killed by direct heating of soil. The seedlings are moderately susceptible to ultraviolet-B radiation. Leaf area and biomass decline, shoot height is reduced and root biomass is also affected. Covering soil with a 3 cm layer of compost reduced seedling emergence. Leachate from green-waste compost has been shown to decrease seed germination. In the field, a mulch of cover-crop residue over the soil surface reduces seedling emergence.

Plantago major (Broadleaved Plantain, ripple grass, waybread, slan-lus, waybroad, snakeweed, cuckoo's bread, englishman's foot, white man's foot, buckhorn plantain, dog's ribs, hock cockle, lance-leaved plantain, rub grass)

This is a broad leaved evergreen perennial weed. It spreads by seed which cling to feet. Persistently remain for many seasons, forming a rosette close to the ground and so it survives mowing. Plant height varies from 15-20 cm, flowers in summer, usually remains squat. The weed is controlled by manual removal from the pot at the early stage.

Gnaphalium uliginosum, C.sylvaticum (Marsh Cudweed)

This is a plant of damp habitats growing in marshy places, along banks of rivers and lakes, in damp meadows, drained bogs, on soils of various mechanical structures. The plant is 5-20 cm tall, with short roots (5-18 cm). It is covered with tufted white tomentum, especially above, at anthodia. Stalk more or less branchy from the base. Leaves are linear-oblong or lanceolate, pointed, narrowed at base, petiolate, frequently crescent, entire. Anthodia 3-4 mm long and 3-3.5 mm wide, gathered at ends of branches in glumes in groups of 2-4, less often single, located in axilla of convergent upper leaves exceeding anthodia; external leaflets of involucre are ovoid, blunt, somewhat woolly at base; internal ones are pointed, bare, green, later brownish or brown, shining. Marginal flowers are pistillate, thread like; median flowers are bisexual, tubular; receptacle bare, cellular. Pappus is formed of falling breakable hairs.

Hemicarp is greenish-gray or light brown, oblong, 0.5 mm long and 0.15-0.2 mm wide, slightly flattened, with narrow short apical rostrum having appreciable ring torulus. Hemicarps sprout well in damp ground during summer. Germinating capacity of seeds is maintained for about 5 years. Blossoming occurs during the second half of June until August. The weed spreads by seed. Sometimes another species of *Gnaphalium* (*Gnaphalium sylvaticum*) were also recorded in some places. Manual removal by hand pulling before flowering is usually practiced for controlling the weed.

Fig. 9.3 : Gnaphalium

Wood Sorrel *(Oxalis corniculata, O. stricta, O. acetosella, O. latifolia)*

Oxalis, commonly known as Wood Sorrel, is one of the ugliest of weeds. It has two properties that make it particularly problematical. One is the vigorous network of bulbs, rendering ineffective, hand or mechanical weeding. Secondly, while most perennial weeds are active during the summer, the primary growing season for Oxalis is the winter. Creeping red sorrel (*O. corniculata*) or Yellow wood sorrel (*O. stricta*) are very common weed species associated with ornamental crops and its hybrids in the Uttrakhand Himalayan region. They may be observed singly or in association with other species of *Oxalis*. There are so far four species of oxalis were recorded in the *Cymbidium* house in this region. These are *Oxalis corniculata*, *Oxalis stricta*,

Oxalis acetosella and *Oxalis latifolia*. *O. corniculata* has purple green foliage and grows in a more prostrate habit. O. *stricta* has green foliage and grows in a clump shape habit. *O. corniculata* foliage is more pubescent, particularly along leaf margins. O. *stricta* foliage is pubescent, but to a lesser extent than *O. corniculata*. The plant grows as small rounded clump, 10-20 cm tall and wide. *O. Corniculata* grows low and more prostrate, while *O. stricta* is taller and more mounded. Leaves are palmately compound with 3 heart shaped leaflets. First true leaves are typical trifoliate. Flowers are yellow with 5 petals and occur in cluster. Seedpods are long, thin, angular and pubescent (more so for *O.corniculata*). The weed species are perennial. They spread by seed, rhizome or stolons along with the potting media.

Fig. 9.4 : Oxalis

Sonchus arvensis (Field milk thistle, Perennial sowthistle)

All plants contain white sap not palatable to cattle. It is regarded as a good bee plant. The main requirement is weed eradication no later than rosette stage, when newly formed root system is not able to vegetatively propagate. It a deep rooted perennial herbaceous weed plant. Roots are long, well developed, with adventitious buds. Root system is shallow. Main root does not reach a depth of more than 50 cm. Long horizontal roots branching off this root elongate more than one meter in length, although they lie no deeper than 6-12 cm. Root system is very fragile, and even small root fragments (up to 3 cm in length) are able to produce sprouts. Propagation of this weed in crops is due to the formation of root offsets. Stem is upright, in upper part it is adenotrichous or glabrous. Leaves are lance-pinnate with triangular lobes; near their base there are rounded auricles. Upper leaves are entire. Inflorescences are rather large flower heads, up to 3 cm across when open. It flowers from July till September. Seed production is highly variable, but a typical head produces many seeds, which are generally dispersed by wind. However, the hairs on the end of the seeds have hooked cells that allow them to stick to clothes, fur, vehicles and farm implements.

Fig. 9.5 : Sonchus arvensis

Solanum nigrum (Black nightshade)

Solanum nigrum is commonly known as black nightshade. Morphologically, it is a brittle stemmed weed that grows up to height of one metre.

It has ovate leaves, tiny white star shaped flowers and small round black berries in small hanging clusters. The unripe fruits are green but turned into black when ripe. Stems green to purple. Leaves 3–8 (rarely to 13) cm long, 2–5 (rarely to 7) cm wide, dark green, ovate-lanceolate, margins undulate, entire or shallowly-lobed; leaf stalk 1–3 (rarely to 7) cm long, narrowly winged in the upper section. Seeds numerous (20–35 seeds per fruit), flattened, minutely pitted. Distinguished by absence of prickles; almost hairless to covered with long shaggy hairs which are simple and may be non glandular or glandular; flowers in groups of 4–12, petals white; berries globeshaped, green at first but maturing dull black to purple-black, 6–8 (rarely to 11) mm wide; seeds pale yellow, 1.8–2.2 mm long.

Fig. 9.6 : *Solanum*

Crassocephalum crepidiodes (Redflower and Ragleaf)

Redflower Ragleaf (*Crassocephalum crepidiodes*) is a dicot weed of the Asteraceae family. The plant is a strong herbaceous weed with an erect little branched to 1 m tall.

Plant stem is glabrous or finely puberulous. Leaves are pinnate, coarsely serrate, petioles 6-32 mm long, blades 3.8 15.9 cm long, 1-7 cm wide, upper leaves more or less sessile. Inflorescences appear terminately with axes of corymbose. Plumed seeds are distributed by wind.Thick head is an annual herb which grows between 30 and 100 cm tall with ribbed stems, soft alternated leaves which are separate and leaf blades which are 7 to 15 cm long and 2 to 7cm broad. Flower heads are on slender branched stalks in open, dropping clusters at the top of the plant. Heads are cylindrical with narrow greenish bracts which enclose slender orange tipped flowers mixed with long hairs. These spread and release numerous small, brown ribbed seeds which have a parachute of fine hairs at the top.Thick head is most commonly found in all areas in our district.

Ferns : Different types of ferns are found associated with Orchid. Ferns are found to grow very closely with the orchid plants. When the old media are removed from the pots and the plants are replanted in the same pot with new media, the perennial stem bits of the ferns are also come to the new pots and regenerate with the emergence of new branches and leaves. They are perennial and spread by spores and stem bits along with the growing media.

Ferns such as the Hay Scented Fern (*Dennstaedtia punctilobula*), have proven to be difficult to control with herbicides.

Fig. 9.7 : *Crassocephalum crepidiodes*

Hydrocotyle sp: (Dollarweed, Pennywort)

Pennywort is a perennial broadleaf plant with creeping underground stems. It usually forms dense low-growing mats on wet soil near water or in shallow water. Because of its creeping habit, it can be problematic in irrigation and drainage ditches. In both floating pennywort and lawn pennywort leaves are hairless, stems root at joints (nodes) and leaf stalks have papery structures (stipules) at their base. Leaves are round or kidney shaped and are alternate to one another along the stem. Floating pennywort has fleshy leaves, mostly 2/5 to 3 inches (1-8 cm) wide, which are deeply lobed or have smooth to scalloped edges and sometimes have a reddish spot at the point where the leaf stalk attaches to the stem. Lawn pennywort leaves are usually 2/5 to 4/ 5 of an inch (1-2 cm) wide and are shallowly lobed with finely scalloped margins.Small white or greenish flowers with five petals cluster on a flowering stalk called an umbel stalk, forming an umbrella like flower head. Floating pennywort blooms from March through August; five to ten flowers each with their own stalks attach to the umbel stalk, which is 2/5 to 2-2/5 inches (1-6 cm) long, forming a dense flower head.The fruit are flattened and separate into halves at maturity. Each half contains one seed. Floating pennywort seeds are on stalks, football shaped to round, and 1/25 to 1/8 of an inch (1-3 mm) long with inconspicuous ribs. Lawn pennywort fruit do not have stalks, are round, about 1/25 to 1/21 of an inch (1-1.2 mm) in diameter, and have conspicuous ribs.

Fig. 9.8 : *Hydrocotyle* sp.

Erigeron canadensis, syn *Conyza canadensis* (Canada fleabane, horseweed and mare's-tail)

Winter or summer annual, reproducing by seeds, short taproot with laterals, seedlings form a basal rosette of dark green, sparsely hairy leaves, rarely more than 1 cm wide, with distinct petioles and coarsely toothed margins. The basal rosette deteriorates after the stem begins to elongate; stems are erect, 10–180 cm high, hairy, with many small flowering branches in the upper portions; stem leaves alternate, numerous and crowded on the stem, often appearing opposite or even whorled, lanceolate to linear, with nearly entire margins; upper stem leaves only 5 mm wide; flower heads (capitula) small, 3–5 mm in diameter, and numerous on short branches near the top of the main stem; pistillate ray florets are white, ligulate and very short, often concealed by involucre bracts around each head; perfect disk florets are

Fig. 9.9 : *Erigeron canadensis*, (Conyza canadensis)

yellowish-green and very fluffy at maturity;seeds (achenes), averaging 60-70 per capitulum, are small, 1-2 mm long, transparent and have an attached pappus 3-5 mm in length. It serves as a wild host of the tarnished plant bug, and of aster yellows, a mycoplasma disease transmitted by the aster leaf hopper. Large numbers of small, wind-dispersed seeds, ranging to over 200 000 seeds per plant, are produced in late summer.

Sagina procumbens (Pearlwort, Beads, Birdseye, Procumbent)

It is a perennial weed, spreading by seed and rooting of stems. It forms a dense mat close to the ground. Hand weed, regrowth from the roots is not a problem, but it can be difficult to remove so with a large area it may be easier to slice off the surface. Small fragments can regrow. Remove at seedling stage if possible.

Fig. 9.10 : *Sagina procumbens*

Integrated Weed Management

Integrated weed management consists of the combination of different control methods, in order to ensure good control of weeds. Combination of methods does not mean that they are simultaneously applied. There are preventive and curative methods and the combination of different practices will depend on the weed socio-economic environment and on his technical level of expertise. For technical and economic reasons, integrated management should be used rather than any single method on its own. Or in other word Integrated Weed Management (IWM) is a coordinated, interdisciplinary, multistrategy approach to managing weeds that combines the most appropriate weed control methods in a way that minimizes economic, health and environmental risk to meet weed management plan objectives. IWM is based on eliminating noxious weeds and preventing their reproduction while protecting and establishing desired plant species and communities.

The optimum method or methods for weed management are varying depending on a number of site-specific variables. Factors to be considered should include soil type and stability, grade, local climatic conditions, native vegetation, presence or absence of the weeds natural enemies, existing and proposed land use, proximity to wetland or riparian areas, availability of irrigation water, weed type and stage of growth and severity of infestation. The management method selected should be the least environmentally damaging, yet practical and reasonable in achieving the desired results.IWM methods include:

1. Prevention and detection

Prevention is the most effective noxious weed control method available for non-infested lands. Protecting weed-free plant communities is the most economical and efficient weed management practice and therefore, the highest priority. Prevention is the most important but often least used control strategy. Wind, water, wild animals, livestock and man are the major agents of weed dispersal. When weeds are spread by natural agents, control is very difficult or impossible. When man is the weed dispersal agent, carelessness or a lack of understanding of his actions is responsible. Following is the preventive measure for weed management :

- Weeds should be control in feed and bedding grounds. Many weed seeds pass through the animal's digestive tract intact and viable.
- Well rotted manure should be used.
- Total farm weed management should be practiced. Clean fencelines, irrigation ditches, stockyards, farm roads, or any annual weed seed nursery to prevent spread by natural or other means.

2. Cultural Control Methods

The use of cultural control methods primarily refers the prevention of noxious weed infestation through the modification or elimination of land use practices by humans which may indirectly cause or aid in the spread of noxious weeds. Cultural control uses plant competition or cropping practices to suppress weeds, either through use of smother or competitive crops and crop rotation.

2.1. Plant Competition

The use of plant competition is one of the cheapest and most useful general weed control practices available to all farmers. Weeds are strong competitors by nature. If not, they would fail nature's survival tests. Certain weeds that can best compete under a certain set of circumstances always tend to dominate. Usually, early weed competition reduces crop yield far greater than late season weed growth. It naturally follows then, that early weed control is exceptionally important. Late season weed growth may not seriously reduce yields, but it can make harvesting difficult, lower crop quality, and add to the reservoir of weed seeds in the soil.

2.2. Crop Establishment

As a general rule, the first plants to germinate and emerge in an area tend to exclude all others. Therefore, it is critical when considering plant competition as a weed management tool to establish a vigorous dense crop. Important factors that affect germination and emergence include the viability of the seed, soil temperature, availability of moisture and physical resistance to seeding emergence by the soil. These factors are influenced by the soil type, physical condition of the soil, depth of planting, the firmness of soil around the seed, the degree of soil compaction above the seed and the formation of surface crusts after planting. Seed germination, if planting has been correctly done, is generally rapid and predictable. If pre-seeding weed growth has been killed, at least a temporary advantage has been gained for the crop. This initial advantage can be lost however, if effective post-seeding herbicide application or tillage is not undertaken to control late emerging weeds.

3. Physical control

The use of physical control is a control strategy commonly used on target species and often serves as the foundation of the integrated weed management effort. Physical control usually falls under three categories: 1) manual control, which can be as simple as hand pulling of the weed species to eliminate individuals and reduce the seed source for usually very small infestations, to using hand tools like a hoe, loppers or a machete; 2) mechanical control, which involves the use of power tools (chainsaws/ clearing saws) and heavy equipment (tractors); 3) Control via fire, which is normally achieved through the use of prescribed burns.

3.1. Tillage

Soil tillage enables to attack many weed survival mechanisms. For annual weeds, the tillage objective is to prevent seed production and deplete current seed reserves in the soil. This is accomplished by encouraging weed seeds to germinate, then subsequently killing them. With perennials, destruction of the underground parts is sought, as well as the prevention of seed production and reduction of seed reserves. Tillage kills weeds by:

- Burial of the entire plant. If complete burial is not accomplished and a small portion of the plant is exposed, life may continue. Burial is usually only effective with seedlings.
- Depleting food reserves. This is accomplished by repeatedly removing top growth whenever it reaches sufficient size. Food manufacture can be halted by cutting off the plant tops or burning them. Tillage that breaks the underground parts into pieces is also very effective. This creates more growing points to use up food, and hastens food depletion.
- Exposure of underground parts to frost. The roots of most plants are killed when left on or near the soil surface during freezing temperatures.
- Exposure of root systems to drying. If many roots remain under moist soil, growth will continue, therefore, tillage must be thorough, with the plants fully exposed on the soil surface.
- Encouraging rotting of underground parts.

3.2. Mowing

When weeds are too numerous to hand pull, too large to effectively destroy by cultivation, or in an area where cultivation is impractical or impossible; they can be destroyed by mowing. This should be done before they produce seed and as close to the ground as possible. Perennial weeds usually require several cuttings before the food reserves in the roots are exhausted. If only a single cutting can be made, the best time is just prior to blooming because (1) the reserve food supply in the roots is at its lowest level, and (2) viable seed is often produced just after blooming.

3.3. Burning

In situations when seed production has already occurred, some of the seeds can be destroyed by burning. The effectiveness of burning depends on the duration and intensity of heat produced, plus the maturity and location of the seeds. Mature, dry seeds are more heat resistant than green seeds, which have high moisture content. Although intense heat will destroy most seeds remaining in plant heads, only a relatively small number of seeds on or below the soil surface can be destroyed by burning surface trash. Burning weeds over an extended area destroys valuable surface trash that would normally be returned to the soil through decay or cultivation.In most farming operations, the most appropriate use of burning would be to selectively burn patches of weeds that have headed out by using a large propane torch. The flame can be directed at the mature heads of the weeds and the undesirable side effects of burning can be largely avoided.

3.4. Mulching

The principle of mulching is to exclude light from the tops of the weeds until the reserve food supply in the roots is depleted and the weeds starve. Mulches include clean straw, hay or manure, tar paper, sawdust and black plastic. Black plastic is very effective. When the vegetation under the mulch has been destroyed, the resultant bare patch must be reseeded with competitive vegetation to prevent new weed introductions.

4. Biological control methods

Biological weed control involves the release of organisms that attack plants to control weeds. The aim of biological control is to shift the balance of competition between the weed and the crop in favour of the crop and against the weed. The biological control agent, normally a fungus or insect, may not necessarily kill the target weed but should, at the least, reduce its vigour and competitive ability. From a practical point of view the organism or agent should prevent the weed setting seed or producing other reproductive parts. There is considerable potential for encouraging the use of native biological control agents against weeds and substantial research effort has been put into biological control in general. In practice, there are three basic types of biological control:

4.1. Allelopathy

Allelopathy refers to the direct or indirect chemical effects of one plant on the germination, growth, or development of neighbouring plants and can be legitimately regarded as a component of biological control whatever the type. The effect is exerted through the release of allelochemicals, either by growing plants or their residues.

The assessment of the potential risks involved in introducing biological control agents remains a difficult and contentious issue as any predictions of biological control may affect the interaction between species, and influence the life cycle of non-target species is extremely complicated. Even if there were no risk to non-target species, there could still be a conflict of interests because some may perceive a particular plant as a weed while others see it as a desirable wild flower, or even a potential crop. For this reason it seems difficult to imagine that of the shelf biological control of weeds is a realistic prospect in the short to medium term.

Factors Affecting Herbicide Effectiveness

The Weed Leaf

For foliar active herbicides, the chemical must be retained on and penetrate through the leaf surface. Wax cuticle, hairs on the leaf and the angle of the leaf all determine retention and adsorption. Surfactants or surface active ingredients can be added to herbicide formulations to increase their wetting ability and penetration into the leaf.

Climate

Under dry or cool conditions, plants grow slowly and translocation of systemic herbicides is slow. Cloudy conditions interfere with food production in the plant thereby slowing translocation and reducing effect of systemic herbicides. High temperatures and low humidity increase evaporation of the herbicide from the weed leaf. The weed leaf surface is also difficult to penetrate under these conditions as the plants tend to develop a heavy wax cuticle. High humidity often damages the leaf cuticle allowing for a high absorption of herbicide into the weed. Warm temperatures and high humidity are generally favoured for best control with systemic herbicides. Spraying of weedicide should be avoided in wind as herbicide will be lost to drift. Rainfall during or after application can wash the herbicide off the leaf, hence, reducing effectiveness; but rain or irrigation is required for many soil applied herbicides such as Dual and Primextra.

Time of Day

This factor relates both to temperature and light. Generally carbohydrate levels in the leaves are lowest in the morning and highest in the late afternoon. Spraying early in the morning is more effective.

Age of Weed

As a rule, weeds in the seedling stage are usually most susceptible to herbicides. Young weeds are growing actively and there is a great deal of movement of food and water throughout the plant. Systemic herbicides, once inside the plant, will be moving with this material to the growing points.With established perennial weeds, such as Canada thistle, treatment at the bud or early flower stage is usually more effective. This is because the root food reserves are expended and newly produced food along with herbicide is being moved back down to the root system to again build up reserves. With Canada thistle and other perennials, it is the root we must attack for control. It is critical to follow herbicide label directions regarding stage of weed growth to obtain optimal control.

Nutrition

Herbicide injury to weeds is greatest when nutrient levels are high and the weeds are growing vigourously. Adequate fertility also improves weed control by increasing competition by the crop.

Cultivation

Chemical weed control is more effective in fields that are cultivated because more weed seeds are brought to the surface to germinate. With perennial weeds, cultivation lowers the root reserves making them more susceptible to herbicides.

Soil Organic Matter

Soils with high O.M. such as peats or mucks require higher rates of chemical than do sandy soils. This is because more of the chemical is bound to the soil colloids in high O.M. soils.

Soil Texture

Higher herbicide rates are needed on silts and clays because they provide more surface area for chemical binding. Sandy soils require the lowest rates.

Soil Moisture

For soil applied herbicides, even mechanically incorporated products, some moisture is required to place them in the soil solution for weed uptake. Soil herbicides generally work best in a warm, moist soil.

Chemical Compatibility

Some chemicals, when present in combination with other chemicals, are more toxic to plants and animals than when present alone. If the field has been sprayed with a herbicide previously, spraying with another chemical may cause great crop damage. Read the label to determine compatibility with other pesticide products.

Some Problems Associated With Herbicide Use

Injury to the crop

Faulty applications and environmental factors can lead to crop damage from herbicides.

Residues in the crop

Herbicide residues in crops cannot exceed the safe and acceptable levels.

Residues in soil

Some chemicals persist in the soil in phytotoxic concentrations. The land manager must be aware of this if susceptible crops are to be planted subsequently.

Compatibility with other pesticides

Two chemicals mixed together might cause crop injury where either one would not when applied separately.

Occupational hazard

Herbicides as a rule are quite safe to use, having a low toxicity to man and animals. All precautions must be taken, however, to protect the applicator and persons who may come in contact with the chemical.

Air pollution

a) Spray droplet drift.
b) Vapour drift (volatile gases).

Water Pollution

Herbicides may enter water courses and groundwater either by direct application through accident or indirectly via leaching and runoff from sprayed areas.

Disposal of used containers

Irresponsibility in disposal of used pesticide containers sometimes leads to poisonings of animals and man.

Resistant weeds

Weed populations contain a very small number of individual plants that are naturally able to withstand a particular herbicide (or herbicide group) which is registered to control the weed. This naturally occurring resistance is not due to weather conditions or application technique but is an inherent characteristic of the genetic makeup of the resistant plants. Because of their low frequency of occurrence, these resistant plants go undetected. The continual use of the same herbicide (or herbicide group) year after year allows them to set seed and multiply. Futhermore, by removing the susceptible weeds, there is more opportunity for the resistant strain to flourish. It is important therefore to know the herbicide groups so that herbicides with different mechanisms of action can be selected as part of the strategy in preventing a build up of resistant weeds.

CHAPTER 10

Pruning and Training of Ornamental Plants

Pruning is a regular part of plant maintenance involving the selective removal of specific plant parts. Although shoots and branches are the main targets for removal, roots, flower buds, fruits and seed pods may also be pruned. Pruning wounds plants, but plants respond differently to wounding than do animals. In plants, damaged areas are covered by callus tissue to close wounds. Another response to pruning occurs inside plants. Chemical boundaries form around wounded areas, walling off or compartmentalizing the wounds. Compartmentalization limits any decay that results from wounding, or from the natural death of branches. Use pruning techniques that minimize plant wounding and speed wound closure.

Why to Prune?

Anyone can prune, but not everyone prunes properly.Improper pruning, or pruning at the wrong time of the year, can result in misshapen plants, reduced flowering or plants that are more likely to be damaged by insects, diseases or winter cold. Ornamental plants are pruned for many reasons:

1. *To improve the aesthetic quality of the plant*: Each plant in the landscape has its own growth habit and different requirements for pruning. Pruning improves light penetration into the canopy and light is required for

flower-bud development, fruit set and growth, and red colour development.

2. *To improve the appearance or health of a plant*: Prompt removal of diseased, damaged, or dead plant parts speeds the formation of callus tissue and sometimes limits the spread of insects and disease. For trees, pruning a dense canopy permits better air circulation and sunlight penetration. To avoid future problems, remove crossing branches that rub or interfere with each other and those that form narrow crotches.
3. *To control the size of a plant*: Pruning reduces the size of a plant so that it remains in better proportion with your landscape. Pruning can also decrease shade, prevent interference with utility lines and allow better access for pest control.
4. *To prevent personal injury or property damage*: Remove dead or hazardously low limbs to make underlying areas safer. Corrective pruning also reduces wind resistance in trees. Prune shrubs with thorny branches back from walkways and other well-travelled areas. Have trained or certified arborists handle any pruning work in the crowns of large trees.
5. *To train young plants*: Train main scaffold branches (those that form the structure of the canopy) to produce stronger and more vigorous trees. You'll find it easier to shape branches with hand pruners when a plant is young than to prune larger branches later. Pruning often begins with young plants for bonsai, topiary, espalier, or other types of special plant training.
6. *To influence fruiting and flowering*: Proper pruning of flower buds encourages early vegetative growth. You can also use selective pruning to stimulate flowering in some species and to help produce larger (though fewer) fruits in others.
7. *To rejuvenate old trees and shrubs*: As trees and shrubs mature, their forms may become unattractive. Pruning can restore vigour and enhance the appearance of these plants.

Anatomy and Physiology of Pruning

Meristems are regions of cell division and there are two types of plant meristems. An apical meristem is located at the tip of every shoot and root. As cells divide in these apical meristems, the shoots and roots elongate as cells are piled one on another. Behind the region of cell division is a region of cell differentiation, where cells enlarge and differentiate into various tissues. In the axil of each leaf is a small apical meristem called an axillary meristem that forms an axillary bud, which usually remains dormant until subtending leaf is fully developed. An axillary bud may remain dormant or develop into a lateral branch or a flower.

There are two distinct layers of meristematic tissue within the stem or root responsible for secondary growth, the *vascular cambium* and the *cork cambium* (Fig. 1). The vascular cambium is a cylinder of specialized cells, usually five to ten cells thick, running the length of the plant, including the rootsand is responsible for the radial growth of plant parts. Phloem cells are produced to the outside of the cambium and xylem cells are produced to the inside of the cambium.

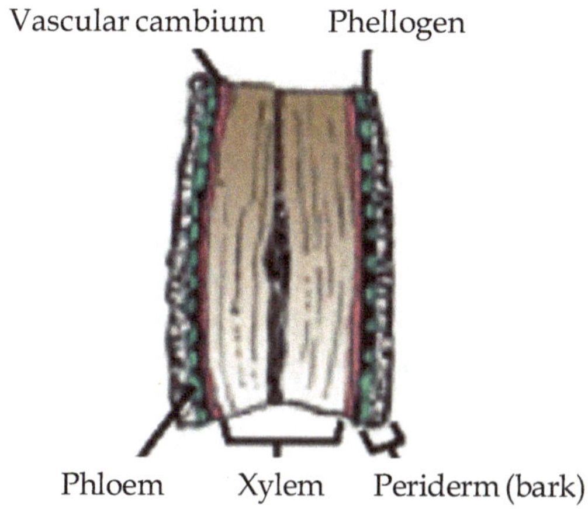

Fig. 10.1 : Longitudinal cross-section of a tree trunk.

Downward transport of sugars, nutrients and hormones from the top of the tree to the roots occurs in the phloem tissue. Xylem cells are tube shaped, become hollow and die to form a pipe-like system through which water, hormones and mineral nutrients move from the roots to the top of the tree. Most of the radial growth of woody plants is due to activity of the vascular cambium, but a small amount results from activity in another lateral meristem, the cork cambium, located outside the vascular cambium. The cork cambium (phellogen) together with the cork cells, constitute the periderm: a protective layer of suberized dead cork cells forming the bark. Suberization is the impregnation of cell walls of cork tissue with a fatty substance called suberin. Each season new layers of cells are produced and appear as growth rings when viewed in cross-section.

Buds

Buds are actually undeveloped shoots.Pruning primarily involves bud manipulation of trees to some extent. During late June and July some of the shoot apices will flatten out and develop into flower buds. Flower buds are actually modified shoots and the various flower tissues (petals, stigmas, anthers, etc.) are modified leaves. Although the process of switching from vegetative to reproductive buds is not fully understood, hormones that can be influenced by environmental factors, stresses and plant nutrition may play a vital role.

There are several things to influence whether or not a bud becomes a flower bud or remains vegetative. In general, factors that favour rapid growth, such as high nitrogen levels in the shoot tissues, inhibit the development of flower buds. Applying growth-promoting plant growth regulators such as gibberellins usually inhibits flower-bud induction, whereas, ethylene may promote flower-bud development. Mild stresses such as shoot bending and water stress may also promote flower-bud development.

Buds are dormant when they are not visibly growing. When shoots develop around large pruning cuts, they usually are sprouting from dormant buds. Adventitious buds form irregularly on older portions of a plant and not at the stem tips or in the leaf axils. They form on parts of the root or stem that have no connection to the apical meristems. They may originate from either deep or peripheral tissues. For example, shoots often arise from adventitious buds growing from callus tissue around wounds. Root suckers develop from adventitious buds on the roots.

Plant Hormones and Pruning

While pruning, it is useful to consider the activity of the general types of hormones, promoters (gibberellins and cytokinins) and inhibitors (auxins and abscisic acid). Promoters generally cause bud growth, cell division, elongation and stem growth. Inhibitors are usually associated with dormancy and inhibit shoot development from seeds and buds and may be involved in flower-bud induction. It is often the ratio of promoters and inhibitors, rather than their absolute concentrations, that determines how a plant will grow. The production of plant hormones is usually controlled by environmental conditions such as temperature or day length. Vegetative growth is usually associated with low ratios of inhibitors to promoters and dormancy is usually associated with high ratios of inhibitors to promoters.

Dormancy is a condition characterized by temporary growth cessation and suppressed metabolism. During the winter trees appear not to be growing, but the tissues are alive, there is metabolic activity, cells are slowly expanding and differentiating. By early October all the flower parts (petals, stigmas, anthers, etc.) can be seen in a flower bud and vegetative buds contain leaves. During the winter these various tissues continue to enlarge and differentiate. Given favourable growth conditions, some buds will develop into shoots or flowers, but others may remain dormant. By understanding the factors influencing bud dormancy we often can influence certain aspects of tree growth. Buds of deciduous trees go through several stages of dormancy. Plant physiologists currently describe dormancy in four stages.

Para-dormancy occurs in the mid to late summer when buds do not grow because inhibitors produced in the leaves and terminal buds inhibit bud growth. Para-dormancy can often be overcome by removing leaves (leaf stripping) along a section of a shoot so the axillary buds develop into shoots. Using heading cuts to remove the terminal portion of a shoot will allow several axillary buds just below the cut to develop into shoots. Sometimes an application of growth promoters (gibberellins and/or cytokinins) will induce bud growth.Sometimes axillary buds do not become dormant and develop into shoots within a few days of being formed. Such shoots are referred to as sylleptic shoots and are fairly common on vigorously growing peach trees, but are rarely produced on apple trees.

Ecto-dormancy occurs in the early fall before defoliation, when plants do not grow because the environmental conditions are not conducive for growth. Growth will resume if the plants are exposed to suitable temperatures and day lengths.

Endo-dormancy occurs during the winter because there are high levels of inhibitors (abscisic acid) within the buds. During this phase of dormancy the trees will not grow even under ideal growing conditions. The concentration of inhibitors declines as buds are exposed to chilling temperatures. Temperatures near 7^0C are ideal for chilling, but temperatures between 1°C and 12°C will provide some chilling. When the chilling requirement is satisfied, the level of inhibitors within the bud is low enough that growth may commence when environmental conditions are appropriate for growth.

Eco-dormancy occurs in the late winter, usually by mid January, after the chilling requirement has been satisfied. At this time the trees do not grow because conditions are unsuitable for growth. Growth will commence when trees are exposed to warm temperatures.

Apical dominance is a type of para-dormancy, where axiliary bud growth is inhibited in the apical meristematic zone. Axiliary buds on trees typically remain dormant for a prolonged period while the main shoot continues to grow. Growth of axiliary buds is inhibited by high concentrations of auxin produced by the terminal bud. Auxin moves down the shoot from cell to cell by gravity, so concentrations are highest near the shoot tip. Promoters are produced in the roots and are transported upward in the tree. Growth of axiliary buds may occur at the base of shoots where concentrations of inhibitors are relatively low and concentrations of promoters are relatively high.

We can overcome apical dominance by removing the shoot tip, which is the source of auxin . The three or four buds immediately below a heading cut usually develop into shoots. Pinching annual plants to induce branching is a form of heading. Another way to overcome apical dominance is to notch buds. Notching involves cutting through the bark to hard wood, with a knife or hacksaw blade at about bloom time, just above a bud. The cut interrupts the downward flow of inhibitors, but not the upward flow of promoters, and releases the bud from dormancy. On vigorous upright one-year-old shoots, notching often successfully overcomes dormancy in about 70 percent of the buds. Sometimes apical dominance can be overcome by spraying shoots with promoters (gibberellins and/or cytokinins) just before bloom time.

Shoots bend in response to an auxin gradient within the shoot. Everyone who has grown plants in the house has noticed that plants tend to grow towards the light. This phenomenon is known as photomorphism and is caused by varying concentrations of auxin in different sides of a stem or shoot. Auxin causes cells to elongate, but auxin is destroyed by light. Therefore, there is a higher concentration of auxin on the dark side of a shoot and the cells on the dark side elongate more than cells on the sunny side of the shoot, causing the shoot to bend towards the light. Horticulturists have noticed a similar phenomenon where the tips of growing branches tend to bend upward, even when the branch was physically oriented to the horizontal. This condition, known as gravimorphism, is also caused by an auxin gradient within the branch in response to gravity. Auxin flows by gravity to the lower side of a limb. The subsequent accumulation of auxin is responsible for increased cell elongation on the underside of the limb, and the growing tip bends upward.

Another consequence of gravimorphism is the development of watersprouts from the upper surface of horizontally oriented limbs. Watersprouts are vertically growing shoots that develop from the upper surfaces of branches or near pruning cuts. High auxin concentrations on the underside of the limb inhibits growth of buds on the underside of the limb, but the concentration of auxin on the upper side of the limb is inadequate to inhibit bud growth and many of these buds develop into watersprouts. Watersprouts are usually undesirable and their development can be suppressed by orienting limbs no more than 45 degrees from the vertical. Trees are sometimes trained as espalier (tree fence). There are several ways to espalier trees, but one method involves orienting limbs to a horizontal position. This system induces many waterspouts along the length of the branches. Watersprout development can be greatly suppressed by orienting limbs 45 to 60 degrees above horizontal. Reducing tree height by cutting into large diameter branches or trunks often results in the development of vigorous watersprouts around the cut. There are buds buried in the bark that normally remain dormant. However, a severe pruning cut will release these buds from dormancy.

Pruning Time

Because flowering ornamentals form their flower buds at different times of year, pruning times must be adjusted accordingly. Many spring-flowering plants such as azalea, dogwood, forsythia, redbud and rhododendron set flower buds in the fall, so pruning during the fall and winter months eliminates or decreases their spring flower display. Plants that typically flower during the summer form flower buds on new growth and can be pruned during the winter with no effect on their flowering. Examples of this type of plant are crape myrtle, hibiscus and abelia. As a general rule, plants that flower before June should be pruned after they bloom while those that flower after June are considered summer-flowering and can be pruned just prior to spring growth. One exception to this rule is the oakleaf hydrangea, a summer-flowering shrub that forms flower buds the previous season. Late-flowering azalea cultivars that bloom during June or even July are another exception. Both (oakleaf hydrangea and the azalea cultivars) should be pruned after they bloom. Ornamental plants that are not grown for their showy flowers should be pruned during the late winter spring or summer months. Pruning should be avoided during fall or early winter because, it encourages tender new growth that may not be sufficiently hardened to resist the winter cold.

Renewal pruning means cutting the plants back 15 to 30 cm above ground level. In this instance, timing is more important than technique. The best time to prune severely is when spring growth begins. Pruning in late fall or mid-winter may encourage new growth that can be injured by cold. Renewal pruning should result in abundant new growth by mid-summer. Once the new shoots are 15 to 30 cm long, prune the tips to encourage lateral branching and a more compact shrub. Most broadleaf shrubs (such as azaleas, camellias, ligustrum, abelia, nandina, cleyera and crape myrtle)respond well to renewal pruning. Boxwoods, however, recover slowly and may even die when severely pruned. Junipers, pines, cypress, cedar, arborvitae andother narrow-leaf evergreens do not respond well when pruned severely and may die when over half of their foliage is removed. Skirting, lifting the canopy by removing low growing branches and shaping a tree form, is an alternative to renewal pruning.

A. Prune after Flowering

Azalea, Japanese Pieris, Beautybush, Lilac, Bigleaf hydrangea, Mockorange, Bradford Pear Oakleaf hydrangea, Bridalwreath, Spirea, Pearlbush, Clematis, Pyracantha, Climbing roses Redbud, Crabapple Saucer,

Magnolia, Deutzia Star Magnolia, Dogwood Shrub, Honeysuckle Doublefile Vibernum , Thunberg Spirea, Flowering Almond,Vanhoutte Spirea, Flowering Cherry, Weigelia, Flowering Quince, Winter Daphne, Forsythia, Wisteria, Japanese Kerria Witchhazel.

B. Prune before Spring Growth Begins

Beautyberry, Goldenrain Tree, Camellia Japanese, Barberry, Chaste Tree (Vitex), Japanese Spirea, Cranberrybush, Viburnum, Mimosa, Crape myrtle, Nandina, Floribunda roses Rose-of-Sharon (*Althea*), Frangrant Tea, Olive Sourwood, Grandiflora roses, Anthony Waterer Spirea Glossy Abelia, Sweet shrub Hedges.

Some shade and flowering trees tend to bleed or excrete large amounts of sap from pruning wounds.Among these trees are maple, birch, dogwood, beech,elm, willow, flowering plum and flowering cherry. Sap excreted from the tree is not harmful, but it is unsightly.

To minimize bleeding, prune these trees after the leaves have matured. Leaves use plant sap when they expand, and the tree excretes less sap from the wound.

Pruning Technique

Heading Back or By Thinning

Heading back (Fig. 10.2) or stubbing trees is rarely warranted in landscape sites and often results in undesirable multiple leaders or trunks. If it is necessary, for instance, to prune beneath power lines or to clear a tree from interfering with a structure, it should always be pruned back to a fork where there is a live branch that is at least half the diameter of the limb being removed. This technique is called drop-crotching. Within several months, prune out all sprouts growing in response to the cut. Never hat-rack a landscape tree by cutting all of its branches back to an arbitrary length. This type of pruning has no place in horticulture. Thinning (cutting selected branches back to a lateral branch or main trunk) is usually preferred over heading back (Fig. 10.2). Trees can be thinned to increase light penetration and encourage turf grass growth beneath the tree. First, remove branches that are rubbing, crossed over each other, dead, diseased or dying. Removing upright branches creates a more spreading tree while removing horizontal branches results in an upright form. If further thinning is desired, remove branches back to major limbs to create an open

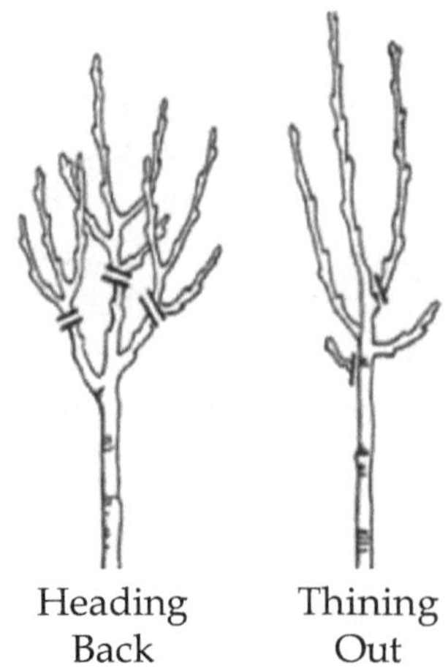

Fig. 10.2

crown. This is a specialized technique best performed by a professional arborist. Space remaining branches along the major limbs so that each one has room to develop. Trees with properly thinned crowns resist wind damage better than unpruned trees.

Making the Cut

When we cut a branch back to the main trunk, to a lateral branch or to a lateral bud, a higher concentration of hormones in these areas causes the wound to heal rapidly. When we leave a stub, the distance from the hormonal source increases and the wound heals slower, if it heals at all. Insects and diseases may enter the cut portion of a stub and cause it to die back.Therefore, regardless of whether pruning a small twig or a large branch, avoid leaving a stub by always cutting back to a bud, a lateral branch or the main trunk. When prune back to a bud, make the cut at a slight angle just above the bud. This allows moisture to flow readily off the wound. A hormonal stimulus from the nearby bud accelerates the healing process. However, avoid making the cut at a sharp angle because it will produce a larger wound.

Branches Selection for Pruning

In many landscapes, little or no attempt should be made to significantly change these characteristic growth habits. Instead, prune in such a way as to enhance and encourage the natural shape of the tree. First remove dead, diseased or broken twigs and branches. Now study the tree's form and select the best-spaced and positioned permanent branches,removing or shortening the others. To shorten, use thinning cuts. Permanent branches should be spaced 15 cm to 60 cm apart on the trunk, depending on the ultimate mature size of the tree. On smaller trees, a 15 cm spacing is adequate, whereas spaces of 45 to 50 cm are best for large maturing trees. Remove fast-growing suckers that sprout at the base of and along tree trunks or on large interior limbs.

To prune a young tree to a single leader (the stem that will become the trunk), locate the straightest and best leader to retain. In shaping the tree crown, remove lateral branches that are growing upright. They will compete with the leader and form a weak, multi-leader tree. Most trees can be grown with a single leader when they are young, but the growth habit of some species will change to a multiple leader spreading form at maturity. There should be no branches leaving the trunk at an acute angle or narrow forks either between branches or between a branch and the trunk. Branches that are less than two-thirds the diameters of the trunk are less likely to split off than larger branches.

Deciduous Shade and Flowering Trees Pruning

When training a young tree, prune back those branches below the lowest permanent branch 20 to 30 cm from the trunk; these are temporary branches. Remove any lower branches that are larger than a one cm in diameter. By keeping the smaller-diameter branches on the trunk, the tree will grow faster and develop a thicker trunk. The trunk will also be better protected from sunburn and vandalism or accidental damage. Removing the lower branches too soon will result in a poorer quality plant. When the tree trunk approaches 3 cm in diameter (measured 15 cm up from the ground), remove the temporary branches. Once the framework (trunk and main branches) of the tree is established, some annual maintenance pruning is required. Each tree is different in its growth habit, vigor and pruning requirements, but there are some general considerations that may help direct pruning decisions:

- A major limb growing at a narrow angle to the main trunk (less than a 45-degree angle) is likely to develop a weak crotch and may split during heavy winds and ice loads. Remove branches that have narrow crotch angles.
- Remove branches that grow inward or threaten to rub against nearby branches.
- Remove branches that grow downward from the main limbs, which may interfere with mowing and other maintenance practices.
- Prune branches damaged by insects, diseases, winter cold or storms below the damaged area. Prune branches of pear, pyracantha or loquat damaged by fireblight disease several cm below the infection.
- Trees such as Bradford pear, ornamental cherry, crabapple and ornamental plum form vigorous shoots or suckers at the base of the trunk and many upright succulent shoots (water spouts) along the main branches. These shoots starve the tree of valuable nutrients and detract from the tree's overall appearance. Remove them while they are young.
- Some trees develop upright shoots, which compete with the main trunk for dominance. Remove these shoots if want to maintain a conical or pyramidal growth habit.

Large Tree Pruning

Branches larger than 4 cm in diameter require three separate cuts to prevent trunk bark stripping (Fig. 10.3). The first cut is made on the underside of the branch about 40 cm away from the trunk and as far up through the branch as possible before the branch weight binds the saw. The second cut is made downward from the top of the branch about 45 cm from the main trunk to cause the limb to split cleanly between the two cuts without tearing the bark. The remaining stub can then be supported easily with one hand while it is cut from the tree. This final cut should begin on the outside of the branch bark ridge and end just outside of the branch collar swelling on the lower side of the branch. The branch bark ridge is usually rough, always darker than the surrounding bark and fairly obvious on most species. The collar is a swollen area at the base of a branch. This region between the branch and the trunk acts as a natural barrier to decay-causing organisms. Note that the cut is usually made angling down and outward from the tree. If the cut must be made straight down (parallel to the trunk), do not make it flush with the tree trunk. Painting wounds with tree wound dressing has become a controversial practice. Research has shown that wound dressings do not prevent decay. When exposed to the sun, the protective coating often cracks, allowing moisture to enter and accumulate in pockets between the wood and the wound covering. This situation may be more inviting to wood rotting organisms than one with no wound cover.

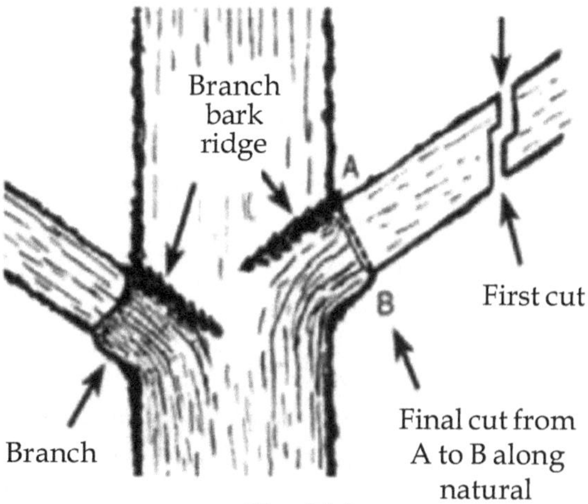

Fig. 10.3

Pruning of Broadleaf Evergreen Trees

Broadleaf evergreens, like magnolias and hollies, usually require little or no pruning. In fact, most broadleaf evergreens develop a naturally symmetric growth habit when left alone. We may want to prune some during the early life of the tree to balance the growth or to eliminate multiple trunks and/or multiple leader branches. Otherwise routine annual pruning is not recommended.

Pruning of Conifers (Needle-Type Evergreens)

Most upright-growing plants in this group such as spruce, pine, cedar and fir have branches spaced evenly around the main trunk. They develop a symmetrical growth habit and become quite large at maturity. If planted in open areas and given plenty of room to grow, they require minimal pruning.If we remove about one-half of the new shoots while new growth is in the "candle stage" (small immature needles packed around the stem resembling a candle), we can thicken the growth of pines and spruce. Avoid cutting back into the hardened older wood because new shoots will not grow and the form of the plant will be destroyed. Upright and broad-spreading junipers sometimes outgrow their sites and must be reduced in size. We can make thinning cuts within the canopy to reduce plant size without destroying the natural shape. We can also shear, but shearing is recommended only when desire formal shapes. Like pines and spruces, junipers do not generate new growth from old wood, so should never severely prune more than one-half of the foliage. We can reduce the length of individual branches by cutting them back to a lateral branch. This technique maintains a natural appearance while it decreases the size of the shrub.

Shrubs (Deciduous and Broadleaf Evergreen)

Some shrubs in the landscape have an inherent dwarf growth habit and require little or no pruning. Japanese boxwood, dwarf yaupon holly, helleri holly, rotunda holly and gumpo azaleas are a few examples. These shrubs can be sheared, but routine shearing will encourage a dense outer canopy and sparse foliage onthe inside. Do not shear too many shrubs in to uniform shapes the landscape may appear monotonous and unnatural. If you must prune dwarf shrubs, thin their growth instead of shearing. Pruning requirements of larger shrubs depend on a number of factors including age, growth habit, vigor, flowering time and location in the landscape. Begin pruning with the shrub is young to encourage vigorous, compact growth. Occasional thinning cuts control shrub size and decrease the need for severe pruning. Prune azaleas after they bloom. Because azaleas set their flower buds for the following season during mid to late summer, prune them after they bloom. Light pruning after bloom encourages branching and increases the number of flower buds the following year. Pruning after mid-July reduces spring flower display. Use hand pruners to cut back shoots that are dying from disease or winter cold or that have become leggy. Always cut dying shoots back to healthy green wood to prevent further dieback. Once new shoots are 15 to 20 cm long, they can be pinched or "tip pruned" to encourage branching, compact growth and more blooms. A young, vigorous azalea may require light pruning several times during the growing season to develop a dense, compact growth habit.

Plants are more likely to withstand shock if they are transplanted before flowering and just prior tothe spring growing season.

Camellia japonica varieties require little pruning.If a plant is misshapen or too large, remove undesirable limbs by cutting them back to a lateral branch or lateral bud inside the canopy. Eliminate suckers that rise from the base of the shrub. Some camellia growers also recommend that remove small interior foliage that can become a haven for scale insects.

Camellia sasanqua varieties may require considerable pruning. For a more natural appearance, thin branches instead of shearing the plant. Some camellias bloom during the fall, some during the winter and some during the spring. If pruning is needed, it's best to prune the fall and winter bloomers in early March and the spring bloomers just after they bloom.

In terms of pruning, crape myrtleis one of the most versatile plants in the landscape. It can be pruned to a tree form by selective thinning or to a shrub by pruning it to within a foot of ground level. It is a summer-flowering shrub, so prune it during late winter or early spring. We can encourage a second flush of bloom during late summer if we remove the seed clusters after the flowers fade and if we thin some of the new growth. Crape myrtle shaped as a tree form often sends up vigorous shoots or suckers from the base that can easily be removed by hand when they are young and succulent.

Nandina and Chinese mahonia(*Mahonia fortunei*) are pruned differently than most shrubs. They grow like bamboo, sending up numerous shoots from the root. As the plants age, thin out older canes by cutting them back to within 15-20 cm of ground level. Old, over grown plants can be severely pruned to ground level. Because the berries are often used for holiday decorations, prune nandina after the Christmas season.

Pyracantha blooms during the spring from buds formed the previous season. Therefore, it's best to wait until after blooming to prune this plant so don't remove the clusters of berries that provide fall colour. Pyracantha forms a lot of lanky growth, so light periodic pruning during the summer season may be necessary to encourage branching and more compact growth. It can also be trained to grow against a wall (called an *espalier*), but we will need to prune frequently to maintain its shape. Old, over grown plants can tolerate severe pruning, but expect the berry crop to be light for at least two seasons.

Pruning of Vines / Ground Covers

Certain vines in the landscape, such as honeysuckle, English ivy, clematis, winter creeper euonymus and trumpet creeper, climb trees or other supports and can grow rampant if they are not controlled. The amount of pruning these plants need each year depends on their vigor, growth habit and spread. Vines trained to an arbor may require only minor thinning or tip pruning to encourage branching, while those growing in trees or competing with other plants may need more severe pruning to control their size.Pruning time is important for flowering vines. Prune summer-blooming vines, such as star jasmine and trumpetcreeper,before new growth begins. Prune spring blooming vines, like wisteria, Japanese honeysuckle, winter jasmine and periwinkle, after they flower.

Ground covers are pruned primarily for three reasons:

- To thin their canopy when they grow thick and dense
- To keep them within bounds
- To rejuvenate their growth after a harsh winter has damaged their foliage.

Horizontal junipers, like Blue Rug, Bar Harborand Prince of Wales, tend to form new foliage on top of older foliage and become thick and dense once their canopies meet. Thinning cuts improve air circulation within the canopy and prevent insect and disease problems. Prune ground covers, such as English ivy, pachysandra, Algerian ivy, Japanese honeysuckle, St. John'swort, liriope and mondo grass, to rejuvenate them after a harsh winter has damaged their foliage. We can mow off the foliage of these ground covers with a lawn mower set at the highest possible cutting height. Beure not to injure the crown or main trunk of the plant when we mow. Prune ground covers before spring growth begins.

Pruning for Hedges Making

Hedges are used primarily as privacy screens; begin pruning early to encourage a compact growth habit. Head back newly planted hedge plants to within 30 cm of ground level, and prune new shoot tips during the growing season to encourage branching. To develop a dense, compact hedge that provides privacy, prune regularly while the plants mature. Once a hedge reaches the height, must decide whether an informal or formal pruning style. An informal style is best if a low maintenance landscape needed. Informally pruned hedges assume a natural growth habit. Prune only as needed to remove dead or diseased wood, to thin branches or to tip prune selected branches where more dense compact growth is needed.

Formally pruned hedges, on the other hand, require frequent shearing, sometimes three to five times during the growing season, to maintain their shape. If a formally pruned hedge best fits style of landscaping, prune in a pyramidal shape with the narrowest part of the hedge at the top tapering to a wider base. A pyramidal shape allows adequate light to reach the lower portion of the canopy.

Pruning for Special Types of Training (Espalier)

Espalier (pronounced "es-PAL-er") is a French term used to describe a plant trained to grow along a flat plane, such as a wall or fence. The Romans origin at the pruning technique and the Europeans refined it into an art form still popular today.

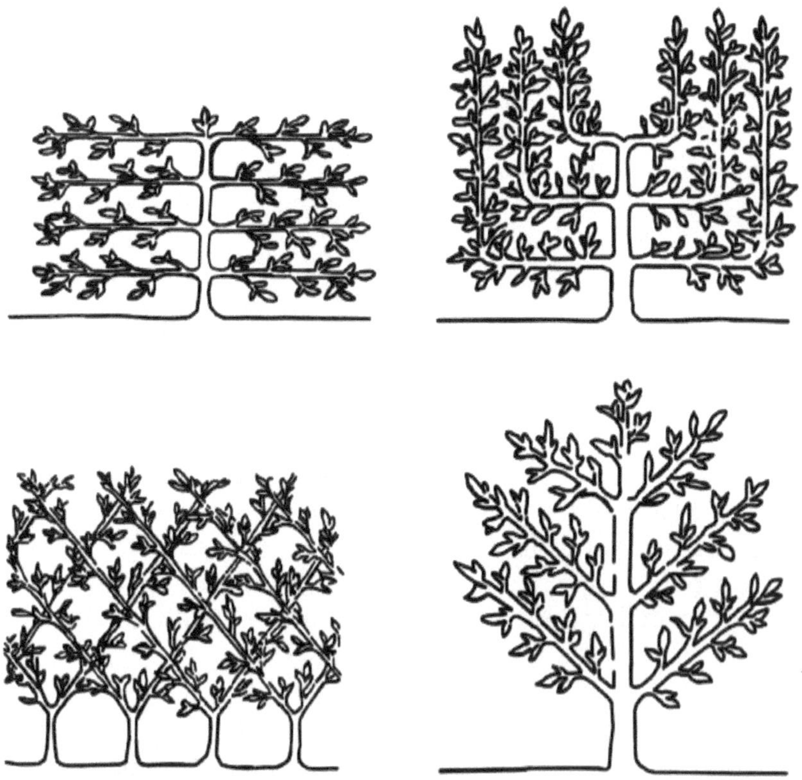

a b c d

Fig. 10.4 : Some common formal espalier patterns areTiered or Horizontal T (a), Palmette Verrier (b), BelgianFence (c), and Oblique Palmette (d).

In the typical landscape, espaliers are often used to add interest to a bare wall. Other times they accent a trellis or fence or add accent in confined areas where spreading trees and shrubs are not practical.Almost any plant can be espaliered. The type of pattern to achieve, formal or informal, will influence choice of plants. A slow-growing espalier may be best if do not like to clip and prune regularly. Pyracantha is the most commonly espaliered ornamental plant. Others include Bradford pear, southern magnolia, foster holly, ligustrum, crape myrtle, loquat, creeping fig and wisteria.

Formal espaliers can be simple or quite complex. Informal patterns are free-form and require less care than formal patterns. An informal espalier may be simply a vine climbing along a wall. Informal espaliers usually require no structural support. Framework for formal espaliers is the form on which the espalier is trained to grow. It may be constructed from rot-resistant wood such as cypress, cedar or redwood, or from wire or nylon cord. Avoid cotton rope because it decays rapidly. When attaching the framework to a wooden wall, use blocks of wood as spacers to keep the framework 15 to 20 cm from the wall. This will provide good air circulation around the plant and prevent the plants from staining the wall. Attach the framework to a masonry wall with special anchoring devices. Formal espaliers require regular pruning and training to maintain their shape. Prune branches that grow outward away from the intended pattern. Tie branches to the support to keep them in place. Be careful to prune flowering ornamentals according to the proper line.

Topiary

Topiary is another form of plant sculpture that originated with the Romans and became popular in European formal gardens. A topiary is a plant that has been pruned to an unnatural form, such as a geometric shape or a whimsical animal. Some topiaries, called"stuffed" or "mock" topiaries, are vines trained to grow on the outside of wire frames. Other topiaries, such as the life-like characters at Rashtrapati garden, are plants trained to grow within a wire frame. Plants often shaped into topiaries are boxwood, Japanese holly, yaupon holly, ligustrum, arborvitae, juniper, yew and podocarpus. Topiaries, like formal espaliers, require frequent pruning to maintain their shape.

How much to prune

In general, the "one-third" rule is a good guide. Try not to take out more than one-third of a plant at a time. This could be stressful to the plant, possibly causing it more harm than good. However, there are many broadleaf ornamentals that would benefit from a severe pruning, as every year they get bigger and bigger. The easy rule of thumb to remember is that the faster and larger a shrub grows, the more cut it back. Plants that respond to severe pruning are: Chinese and Japanese Holly cultivars, Photinia, Ligustrum, Osmanthus, Elaeagnus, Waxmyrtle, Rose of Sharon, Hydrangea, and Spiraea. Consider creating a multi-stem, small tree from the overgrown shrub. Instead of trimming the entire top off, remove many of the lower branches all the way to the ground. Leave 3-4 well-spaced branches and thin the top. If sprouts appear at the base, keep them trimmed off. Once created, the tree form plant will require much less attention over the years. Favorites for this category include Burford holly, photinia, camellia japonica and sasanqua, waxmyrtle and amelanchier. A variant of topping, called shearing, is actually quite useful with certain evergreen shrubs, especially those you want to train to form a hedge. To be successful, shearing requires that you clip tender new growth every year in early spring.

Training Trees

Most of the pruning required by young trees is referred to as thinning. Thinning is the selective removal of branches throughout the canopy to improve the structure of the tree. The first branches to evaluate for thinning are the scaffold branches, those branches 1.5 to 3 meter from the ground that will become the lower permanent branches in the canopy. These branches must be properly selected to maintain a good form. One of the most persistent canopy problems is the development of two main leaders rather than a single leader. This problem is most common with green ash, but it will also occur with maples, lindens and a number of other species. These double leaders, also referred to as codominant stems, are weakly attached and are prone to splitting or breaking. Avoid the formation of codominant branches by eliminating one of the two upright stems. This should be done as soon as the formation is noticed. Removing a codominant leader once it reaches a diameter of 7.5 cm or more may result in decay entering the trunk. Branches should be properly spaced along the trunk. As the tree matures and these branches increase in diameter, they can become too tightly spaced and begin to interfere with one another's development. Branches which are too close may also result in bark splits or cracking. The spacing for permanent branches for trees with a mature height of less than 15 meter is approximately 15 to 20 cm apart. Trees with an anticipated mature height greater than 15 meter should have a branch separation of approximately 30 to 45 cm. In addition to properly spacing the scaffold branches, there are several other types of branches that should be watched carefully. Many young trees will have sprouts coming up around the base of the tree. These sprouts come from the roots or root flares. These are referred to as suckers. Many ornamental trees are cultivars placed upon a seedling-grown rootstock. Thus, the shoots that come from the roots do not have the same ornamental qualities as the cultivar. Some tree species produce suckers annually throughout their life so removing suckers becomes part of the regular maintenance. The other branches to annually prune are watersprouts. These are rapidly growing upright shoots that form along the trunk or scaffold branches. The waterspouts are usually weakly attached and generally do not flower or fruit as profusely as the other branches.

CHAPTER 11

Lawn Management

Lawn

The lawn is a plot of closely mowed grasses. It is composed of thousands of individual plants crowded and forced into a very unnatural growth habit. It is an integral part of the garden. It is basically an area of land planted between gardens covered with grass and other low lying plants.

Lawn is one of the most important features of a garden. A well kept lawn makes a garden look extremely beautiful. It looks like a beautiful green carpet amidst colourful flowers in a garden. Lawn can be called as turf, pitch, field or green depending upon its plantation and usage. Selection of lawn grass depends upon it usage and its season of plantation. Though a well maintained lawn requires a little of patience and care, the final results are always very beautiful and soothing. Scientific reports say that lawns reduce energy consumption through it soothing effects and the green grass prevents soil erosion as well as cut down noise pollution in the urban areas.

Types of Lawn Grass

Though the lawn grass can be categorized in several categories; the most prominent categorization of lawn grasses is done according to their season of foliage. There are certain lawn grasses that prefer a warm and humid temperature however; others require cool and dry weather. Given here is a list of both kinds of grasses.

Table 11.1 Grasses suitable for lawn

Common Name	Botanical Name
Cool-Season Grasses	
Annual bluegrass	*Poa annua*
Annual ryegrass	*Lolium multiflorum*
Chewings fescue	*Festuca rubra* spp. *rubra*
Colonial bentgrass	*Agrostis tenuis*
Creeping bentgrass	*Agrostis palustris*
Creeping red fescue	*Festuca rubra* spp. *rubra*
Hard fescue	*Festuca ovina* spp. *duriuscula*
Kentucky bluegrass	*Poa pratensis*
Perennial ryegrass	*Lolium perenne*
Rough bluegrass	*Poa trivialis*
Tall fescue	*Festuca arundinacea*
Velvet bentgrass	*Agrostis canina*
Warm-Season Grasses	
Bahiagrass	*Paspalum notatum*
Bermudagrass	*Cynodon dactylon*
Buffalograss	*Buchloe dactyloides*
Carpetgrass	*Axonopus affinis*
Centipedegrass	*Eremochloa ophiuroides*
Kikuyugrass	*Pennisetum clandestinum*
Seashore paspalum	*Paspalum vaginatum*
St. Augustinegrass	*Stenotaphrum secundatum*
Japanese lawngrass	*Zoysia japonica*
Mascarenegrass	*Zoysia tenuifolia*

Management of Important Lawn Grass

A.Warm Season Lawn Grass

1.Bermuda grass (*Cynodon* spp.)

Doob grass is a major turf species for sports fields, lawns, parks, golf courses and general utility turfs in India, Australia, Africa South America and the Southern region of the United States. It is found in over 100 counties throughout the tropical and subtropical areas of the world. Common Bermuda grass, *C. dactylon,* naturalized throughout the warmer regions of the India.Bermudagrass has numerous turf uses.These grasses can easily be grown from their seeds. The genus *Cynodon* comprises nine species with *C. dactylon* being the most widespread. It is a tetraploid with broad genetic variability serves to explain its widespread distribution. Other *Cynodon* species have a more limited natural distribution and are often restricted to one particular habitat. *C. dactylon* is highly fertile, whereas the diploid species such as *C. transvaalensis* rarely produce viable seed.*C. dactylon* (L.) Pers.

Climatic Requirements

Bermuda grass is sun lover and grows well in warm seasons. They are easily maintainable by moderate lawn care, maintenance and mowing.Low winter temperature is the factor that limits the hills distribution of bermuda grass. Bermuda grass is a warm season perennial species adapted to tropical and subtropical climates. It grows best under extended periods of high temperatures, mild winters and moderate to high rainfall. Temperature is the main environmental factor that limits its adaptability to tropical and subtropical areas of the world. Research has demonstrated that bermuda grass will continue to grow with night temperatures as low as 1°C if day temperatures are near 21°C. However, when average temperatures drop below 10°C growth stops and the grass begins to discolour. At the onset of low temperatures in the fall and winter, bermuda grass begins to discolour, protein fractions change in composition and reserve carbohydrates increase in the stems and rhizomes. After the first killing frost, leaves and stems of bermuda grass remain dormant until average daily temperatures rise above 10°C for several days. The roots and rhizomes of bermuda grass continue to grow several weeks after the leaves and stems stop growth. In warm frost-free climates bermuda grass remains green throughout the year, but growth is significantly reduced at the onset of cool nights. The species makes the best growth where average daily temperatures are above 23°C. Optimum day time temperature for bermudagrass is between 35°C and 37°C.

Soil temperature, as influenced by air temperature, is also important to the growth and development of bermuda grass turf. Soil temperatures above 18°C are required for significant growth of rhizomes, roots and stolons. Optimum soil temperature for root growth is around 26°C. It has a high light requirement and does not grow well under low light (shaded) conditions. The duration of the light period (day length) also influences growth and development of this grass. Both increased light intensity and day length increase rhizome, stolon and leaf growth in bermuda grass. At low light intensities (less than 60% full sunlight) bermuda grass develops narrow, elongated leaves; thin upright stems; elongated internodes and weak rhizomes. Consequently, it develops a very sparse turf under moderately shaded conditions.

Bermuda grass is found in tropical and subtropical climates with 62 to 250 cm of annual rainfall, but it also survives in arid climates along waterways and in irrigated areas. Where annual rainfall is below 50 cm per year, bermuda grass requires irrigation to survive. Bermudagrass develops into a semidormant state during very dry conditions, but has the capability of surviving extreme droughts. Rhizomes of bermudagrass can lose 50% or more of their weight and still recover when favorable moisture develops. Generally, common bermuda grass, or tetraploids of *C. dactylon*, have the deepest root and rhizome penetration and better withstand prolonged drought periods.Common bermuda grass also has the characteristic of producing seedheads under stress conditions such as drought. Thus, the seeds provide another method by which the species can survive extreme drought. Some natural biotypes of *C. dactylon* produce numerous seeds.

Bermuda grass grows well on a wide variety of soils from heavy clays to deep sands, provided fertility is not limiting. It tolerates both acid and alkaline soil conditions and is highly tolerant to saline conditions. Bermuda grass survives some flooding but does best on well-drained sites. Although it may persist under low fertility, bermuda grass has a high nitrogen requirement for good quality turf.

A dense bermuda grass turf tolerates moderate wear and compaction and recovers rapidly from wear injury. Under moderate fertilization, frequent mowing and adequate moisture bermuda grass forms a dense, fine-textured turf. The only situation where bermuda grass cannot be used is in moderate to heavily shaded sites.

Selections of superior strains of common bermuda grass, natural hybrids between *C. dactylon* and *C. transvaalensis* and crosses resulting from grass breeding programs have been released. All of the hybrid Bermuda grasses are sterile and must be propagated by sprigs or sod. Some selections from *C. dactylon* produce viable seed. Nu-Mex Sahara and Princess bermudagrass are improved seeded varieties of bermudagrass.

U-3 (*C. dactylon*). Selected from Savannah Golf Club near Savannah, Georgia in 1936 from a series of fine-strain selections. Adapted for use on lawns, golf courses and athletic fields.

Sunturf (*C. magennissii*). Originated in S. Africa as a natural hybrid between *C. dactylon* and *C. transvaalensis*.Dark Green, fine-textured, low growing variety that forms a very dense turf.

Tiflawn (*Cynodon* sp.). A hybrid between two selections and it is a medium-textured, very fast spreading, wear resistant variety that forms a dense, weed-free turf.

Tifgreen (*C. dactylon* x *C. transvaalensis*). A hybrid between a fine- textured selection of *C. dactylon* from the Charlotte Country Club, Charlotte, North

Carolina and C. *transvaalensis*. It is a low growing, rapid spreading variety that develops a dense, weed resistant turf. Its density, fine texture and soft leave make it an excellent turf for golf greens. It tolerates over seeding with winter grasses better than most bermudagrass varieties. Tifgreen is highly susceptible to injury by ground pearls and is severely discolored by air pollution in areas where this is a problem. It is also highly susceptible to spring dead spot. It is recommended for golf greens and fairways, tennis courts, bowling greens and fine lawns with a high level of maintenance.

Texturf-10 (*C. dactylon*). Selected from common bermudagrass .It is medium texture, dark green colour, sparse seed heads and dense turf. It also has good wear tolerance and late fall color retention and makes earlier spring recovery than common bermudagrass. It was released by the Texas Agricultural Experiment Station in 1957. It is recommended for athletic fields, playgrounds and lawns. Texturf-10 is sensitive to chlorinated hydrocarbon insecticides, turning a straw color several days after an application of these materials.

Tifway (*C. dactylon* x *C. transvaalensis*). A chance hybrid that appeared in a lot of seed of *C. transvaalensis* from Johannesburg, South Africa, in 1954. It is very similar to Tifgreen except for its greater stiffness of leaf blades and darker green color. Recommended for golf course tees and fairways, home lawns, athletic fields and tennis courts.

Santa Ana (*C. dactylon* x *C. transvaalensis*). A selection from *C. dactylon* (Royal Cape) obtained from South Africa in 1954. The initial selection was made for its deep blue-green color, medium-fine texture and good fall color retention. Santa Ana was found to have good salt tolerance and a high degree of tolerance to smog which frequently discolored Tifway and Tifgreen varieties. It is recommended for golf courses, athletic fields, playgrounds and lawns.

Midway (*C. dactylon* x *C. transvaalensis*). A medium-textured lawn grass that produces relatively few seed heads. Released by the Kansas Agricultural Experiment Station in 1965 for its superior cold tolerance.

Bermudagrass seed should be planted at a rate of 450-500 g of seed per 1,000 sq. ft. Spring and summer plantings should utilize. Seeds should be hulled for faster germination. Late fall and winter plantings should be with unhulled bermudagrass seed to delay germination of a significant amount of the seed until more favorable conditions occur in the spring. Unhulled bermudagrass seed might be planted together with annual ryegrass in the fall to provide temporary cover. Annual ryegrass will delay the development of a bermudagrass turf, but it may be needed for cover and protection.When planting in the fall and winter on areas subject to severe erosion, wheat or rye can be drilled with unhulled bermudagrass seed. The wheat or rye will establish quickly and provide some cover during winter months. The small grains also provide less competition than ryegrass to seedling bermudagrass in late spring.

Bermudagrass sprigs or stolons for planting should be freshly harvested and protected from desiccation by wind and sun. They should not be subject to excessive heating which occurs when moist planting material is tightly packed or covered for several days. Sprigs are usually distinguished from stolons in that sprigs consist of stolons with roots and rhizomes; whereas stolons consist of above ground parts only. Sprigs are produced by shredding harvested sod or by sprig harvesters. Stolons are generally harvested with a vertical mower or a flail mower set close to the ground. Sprigs will tolerate slightly more environmental stress during planting and establishment because of the energy reserves in the roots and rhizomes. Sprigs or stolons should be planted at 5 to 15 bushels per 1,00 sq. meter depending on the rate of cover required. Higher planting rates up to 25 or more bushels per 1,00 sq. meter will provide a faster grass cover. A minimum planting rate should be 5 bushels per 1,00 sq. meter or 200 bushels per acre. Sprigs or stolons should be broadcast on a clean seedbed and pressed into moist soil with a roller or covered lightly with soil or mulch. Moist conditions must be maintained for 2 to 3 weeks after planting to obtain a good cover.

Seed or sprigs should not be planted before soil temperature is above 18°C. Planting too early may retard development of a turf and extend the critical establishment period several weeks. Soil temperatures of 20°C to 23°C are ideal for germination and rapid development of bermudagrass.

Fertilizer, as determined by a soil test, should be incorporated into the soil during seedbed preparation. Nitrogen fertilizer can be applied to the soil surface immediately prior to planting or at the time of planting at a rate of 500g per 1,00 sq. meter Nitrogen should be applied 3 to 4 week intervals until a cover is obtained.

Mowing should begin several weeks after planting to control weed growth and promote spreading of grass. If additional weed control is needed selective post emergence herbicides can be applied 3 to 4 weeks after planting. Pre-emergence herbicides should not be applied to bermudagrass turf during the first growing season. Weed control will greatly enhance bermudagrass growth and coverage.

Bermudagrass respond readily to irrigation. In general, water requirements of bermudagrass depends on turf use and climatic factors such as temperature, wind, humidity and light intensity. Water requirements increase with increasing levels of maintenance (golf green > sports field > lawn > roadside), higher temperatures, higher wind speed, lower humidity and greater light intensity. The longer the growing season the greater the water requirement for the year. Water use rates may range from less than 0.25 cm per day to 0.75 cm per day depending on these environmental conditions.The frequency of irrigation is dependent on water use rate and soil type. Clay soils, for example, hold more water than sandy soils and, consequently, require less frequent irrigations. The depth of the root zone also influences the frequency of irrigations. Grass roots can grow to a depth of 2 meter or more depending on soil profile characteristics. However, the majority of the root system, 80% or more, is found in the top 15 cm of soil. Where roots extend several meter into the soil, thorough and infrequent irrigation produces the most drought tolerant turf. Light, frequent irrigations such as practiced on golf greens produce shallow-rooted grass that shows drought stress very rapidly.

Bermudagrass does not tolerate poorly drained sites. On compacted sites and heavy clay soils, irrigation must be closely controlled to avoid waterlogged conditions. Hard, compacted sites can often be improved with respect to water penetration by core aeration and top dressing with sand or an aggregate material. The presence of a heavy thatch layer will also interfere with water penetration. Thatch removal by vertical mowing and core aeration also improves water penetration and reduces the frequency of irrigation required.

Mowing requirements for bermudagrass turf are dependent on variety, use and the level of maintenance. Common bermudagrass and other medium textured varieties produce dense, wear tolerant turf when mowed at heights between 1 and 2.5 cm. The lower heights being good for golf and sports turf and the tall heights for lawns. At mowing heights above 2.5 cm bermudagrass develops turf with an acceptable appearance but with poor wear tolerance. Fine-textured hybrid bermuda grasses such as Tifway should be mowed at a height of 2.5 cm or less. Taller mowing heights with these grasses produce puffy, stemmy turf that is easily scalped during mowing. As a general recommendation to maintain good turf density and color, no more than 40% of the leaf tissue should be removed at any mowing. Thus, the shorter the mowing height, the more frequent the turf must be mowed. Reel mowers produce the best cut on bermudagrass turf. However, the number of blades per cutting reel determines the smoothness of cut. Common bermuda grass mowed at 2.5 cm or higher can be cut with a reel with 5 or 6 blades. Common and hybrid bermuda grasses mowed at 1 to 2.5 cm should be cut with a reel containing 7 blades. At heights below 1 cm, 9 to 11 blades per reel are required for a smooth cut.

Bermudagrasses have a relatively high fertilizer requirement to maintain a high level of turf quality. The amount and frequency of fertilizer required depends on the desired appearance and growth rate of the turf, length of growing season, soil type, bermudagrass variety and the use of the turf. Where high quality is of critical importance and the turf is mowed frequently, 450 to 600 g of nitrogen per 1,00 sq. meter per month may be applied during the growing season. The lowest rate of nitrogen that can be applied and still maintain acceptable bermudagrass turf for sports fields and golf courses is about 225 g of N per 1,00 sq. meter per month.Soil types also influence fertilizer needs. Sandy soils require light but frequent applications of nitrogen because of low nitrogen retention. Sandy soils are also typically low in other nutrients such as phosphorus and potassium and these nutrients must also be provided through fertilization. Soil tests are required to determine phosphorus, potassium, calcium and other nutrient deficiencies. Potassium is particularly important because of its contribution to root growth, environmental stress tolerance (heat, cold and drought) and wear tolerance. Potassium has also been found to reduce susceptibility of bermudagrass to leaf spot diseases.Bermudagrass tolerates a wide range in soil reaction, but performs best between pH 6.5 and 8.0. At pH levels below 6.5 limestone should be added according to soil test recommendations.

Bermudagrass varieties also differ slightly in nitrogen requirements. Common bermudagrass and selections from common generally have a lower nitrogen requirement than the hybrid bermudagrasses. Tifgreen may have the highest requirement for nitrogen to maintain a dark green color and keep seed production to a minimum. Tifway bermudagrass, which has an inherent dark green color, requires less nitrogen than Tifgreen. Excessive nitrogen fertilization, beyond that required to maintain color and vigor, leads to increased mowing, irrigation, thatch control and pest problems all of which result in higher maintenance costs.

Turf use has a significant effect on the amount of fertilizer required. Golf greens, bowling greens and tennis courts have a very high nitrogen requirement; sports fields and golf course fairways, an intermediate requirement and lawns and other low maintenance areas, a low nitrogen requirement.

Hybrid bermudagrasses require regular cultivation practices - vertical mowing, aeration and topdressing - to maintain high quality turf. Bermudagrass golf greens may require weekly vertical mowing and monthly topdressing under heavy use conditions. Sports fields and golf fairways may need these cultural operations on an annual basis. Without cultivation bermudagrass turf tends to develop thatch, grain and spongy conditions that result in scalping and a nonuniform appearance.

2.Bahia grass (*Paspalum notatum*)

Paspalum notatum, is also known as Bahia Grass, Common Bahia, or Pensacola Bahia, is a tropical to subtropical perennial grass (family Poaceae).

Bahia grass is native to Mexico and South America. It prefers sandy soils and is tolerant of shade. Bahia grass is a warm season lawn and pasture grass and can survive period of drought. It involves moderate maintenance, mowing and is prone to less diseases and insect problems when compared to other warm lawn grass types. There are several varieties of Bahiagrass of which Argentine and Pensacola are more popular.It is a rough and tough kind of grasses for in lawn.

3. Carpet grass

Axonopus is a genus of grasses known generally as carpet grass. They are native to the tropical and subtropical regions of the Americas. They are sometimes rhizomatous and many are tolerant of periodic submersion. Selected species:

- *Axonopus affinis* - narrowleaf carpetgrass
- *Axonopus aureus* - golden carpetgrass
- *Axonopus compressus* - broadleaf carpetgrass
- *Axonopus fissifolius* - common carpetgrass
- *Axonopus furcatus* - big carpetgrass

Carpet grasses are warm season grasses that grow well on wet poor soil where other grass types do not grow. They are shallow rooted and hence not drought tolerant. They are cold tolerant and are ideal for shady, damp and moist areas.

4. Centipede grass (*Eremochloa ciliaris*)

Centipede is the lowest maintenance involving of all warm season grass types. They are uniform growing, thick sod forming and have a medium to light green coloured grass. It is a creeping perennial grass that adapts itself well to sandy and acidic soils consisting of low fertility.

5. Zoysia Grass

It is native to China, Japan and other parts of Southeast Asia. The species was named to commemorate an 18th century Austrian botanist, Karl von Zois. A highly versatile species, zoysia grasses make ideal lawn grasses in some situations and can be used on golf courses, parks and athletic fields. They can be grown in all kinds of soils ranging from sands to clays and both acid and alkaline in reaction. Zoysia grass is extremely drought tolerant. Although it does turn straw colored under severe drought conditions, it has the capacity to respond to subsequent irrigation or rainfall. Its water requirements are similar to those of bermudagrass. The leaf blades of zoysiagrass are among the first to roll under drought conditions, thus it tends to conserve moisture more effectively than other species. Zoysia grass also has a deep root system allowing it to more effectively extract water from greater soil depths. Zoysia grass is nearly as salt tolerant as bermudagrass. It is widely grown along sandy seashores where drainage is adequate. Zoysia grass does not tolerate poorly drained soils whether they are saline or otherwise. Zoysia grasses are among the most wear tolerant turfgrasses. However, their slow rate of growth gives them very poor recuperative potential. Therefore, they perform satisfactorily on lawns, golf course fairways and baseball fields.

There are three principal species of zoysia grass used for turf: *Zoysia japonica*, *Zoysia matrella*, and *Zoysia tenuifolia*. These species are differentiated by texture, cold tolerance and aggressiveness.*Zoysia japonica*, often called Korean or Japanese lawngrass is more cold tolerant than the other species, but is also the most coarse textured of the three species. *Zoysia japonica* is the only zoysia grass species that can be established fromseed. *Zoysia matrella* (Manilla grass). It is chiefly a tropical and subtropical grass, but can grow well in moderate shade and forms a thick mat in full sun. The leaf blades of *Zoysia matrella* are narrow, sharply pointed and wiry. In tropical climates the grass remains green year around. But, in cooler climates it turns brown after several hard frosts and remains brown until late spring. *Zoysia matrella* must be propagated from sprigs and is quite slow to become established. *Zoysia tenuifolia* is the finest textured, least winter hardy of the zoysia grasses. It has very fine, short, wiry leaf blades and forms a dense, fluffy turf. It is extremely slow to spread and is most often used as a ground cover.

Emerald zoysia grass is a hybrid between *Zoysia japonica* and *Zoysia tenuifolia*. Emerald combines the fine texture of *Zoysia tenuifolia* with the cold tolerance and faster rate of spread of *Zoysia japonica*. Emerald is similar to *Zoysia matrella* in appearance and habit. Zoysia grasses can be established from seed, sprigs or sod. *Zoysia japonica* is the only species that can be established from seed. Emerald zoysiagrass, Manila grass and *Zoysia tenuifolia* must be propagated vegetatively from sprigs, plugs or sod. The seedbed should be finely pulverized, smooth, firm and weed-free prior to planting. Zoysia grass sod may be shredded or torn apart to provide sprigs or it may be cut into 5 cm sod plugs for planting. A sprig should consist of a section of stem or rhizome with 2 or more nodes. Leaves do not need to be present. Sprigs should be planted no more than 5 cm apart in rows spaced 90 cm apart, or broadcast over an area at a rate of 10 bushels per 1,00 sq. meter If planted in rows it is important that the sprigs not be completely covered with soil. At least one node should be above soil level. If sprigs are broadcast over the surface, they should be rolled to insure good soil contact. Freshly sprigged zoysia grass must be kept moist for several weeks after planting. Special attention should be given to weed control since zoysia grass is much less aggressive than bermudagrass and some of the common turf weeds.

A newly planted zoysia grass turf should be fertilized with a 1-2-1 or similar fertilizer at a rate of 500 g of nitrogen per 1,00 sq. meter of area at the time of planting. Monthly applications of nitrogen at 500 g per 1,00 sq. meter will promote the spread of zoysia grass. Sprigging is the least expensive method of planting zoysia grass and usually gives a faster rate of cover than plugging. However, keeping the soil moist during the establishment period is most critical with sprigs. Small plantings of zoysia grass sprigs can be covered with a clear polyethylene tarp to maintain adequate moisture and increase soil temperature in the early spring. The cover can be left in place for several weeks, or until temperatures get too hot. The plastic cover can increase the rate of spread of zoysia grass and reduce the time required to obtain a complete cover. The best time for planting zoysia grass is late spring and early summer.

Zoysia grasses grow from early spring through late fall when moisture and nutrient requirements are met. Although zoysia grass is considered to be a drought tolerant species, it ceases growth and begins to discolor during extended dry periods. Water should be applied 2 to 3 times per week depending on temperatures and soil conditions. Sandy soils require more frequent irrigations than heavier clay soils; and as temperatures increase, irrigation frequency must increase. During prolonged droughts when it is impractical to water enough to maintain growth, weekly applications of as little as 0.5 inch of water are adequate to keep the grass alive. During dry winter months, zoysia grass requires occasional irrigation to prevent desiccation and serious loss of stand even through the grass may be dormant.

Zoysia grass lawns tend to build up a thatch layer, a layer of undecomposed organic residues just above the soil surface. Proper mowing is essential to prevent the accumulation of thatch in zoysia grass turf. Frequent mowing at recommended heights and clipping removal help prevent thatch accumulation. Avoiding excessive applications of nitrogen fertilizer also helps prevent thatch accumulation. Occasionally, thatch removal by mechanical means is required to prevent serious deterioration of zoysia grass turf. Vertical mowers or flail mowers may be used to remove excess thatch from zoysia grass turf. Thatch removal should be done well before fall to allow ample time for re-growth. Scalping the lawn in early spring to remove accumulated growth will also help prevent thatch accumulation.

Zoysia grasses are relatively free of serious pest problems. Brown patch, rust and leaf spot diseases can cause problems in zoysia grass turf, but the grass usually recovers when environmental conditions change. In intensively maintained lawns, fungicides may be needed to prevent these diseases. In the fall, applications of Banner, Daconil or Bayleton are required to prevent rust on zoysia lawns.

B. Cool Season Lawn Grass

1. Bent grass

Bentgrass (*Agrostis*) is a large genus with over 100 species. Throughout the New England States and the Pacific Northwest where climatic conditions are ideal for bentgrass, the species is used for lawns, athletic fields and golf courses. In Europe and parts of Asia, the grass is native and commonly found on lawns, pastures and sport fields.

It is considered the most beautiful of lawn grass owing to its texture, deep green colour, thick density and low growing habit. They thrive in cool climates and are called luxury grass that involves high maintenance. Colonial, creeping and velvet are its varieties that differ in growing habits and texture owing to climate and environments. Bent grass can be planted either by seed or sod. They are preferred in the gardens because of their deep green color and low growing habit. They are also called as velvet grasses. Bentgrass tolerates acid soil conditions and prefers full sun. Used primarily for golf courses, putting greens, baseball in fields, and very elegant lawns. It has the ability to grow in a wide range of soil types and can tolerate very acid soils.

A good seed bed is essential for establishing a good stand of bentgrass from seed. Till the soil 8 to 10 cm deep, remove all sticks, rocks, and other debris. Continue tilling until the soil is free of all clumps. Level and smooth the area to prepare a suitable seed bed.Using a over the shoulder or hand held spreader, spread 450 g or 900g of bentgrass seed per 100 sq. meter. Rake the seed into the seedbed very lightly then roll firm with a lawn roller. Care should be taken not to cover bentgrass seed more than 1 cm with soil.Water the area to moisten the top 5 cm of soil. In the absence of rain, water the seeded area lightly each day to keep the top 5 cm of soil moist. Continue water applications daily until the bentgrass seedlings are three weeks old. Reduce watering to three times per week for the next thirty days unless there is adequate rainfall.

2. Bluegrass (*Poa pratensis*)

Require medium amount of lawn care and are ideal for home as well as for sports areas. It is very beautiful with its dense, high quality deep bluish appearance or bright green appearance. It is deep bluish or bright blue green in appearance. It is a highly variable species, with cultivars that differ in color, texture, density, vigor, disease resistance and tolerance to close mowing. Bluegrass is best adapted to well-drained, moist, fertile soils with a pH between 6.0 and 7.0. It does best in full sun but will tolerate light shade. It will not perform well on shallow, compacted soils, or where the pH is excessively high or low. Under the right conditions and with proper management, Kentucky bluegrass can make a beautiful lawn.

3.Rye grass (*Lolium multiflorum)*

Rye grass grows well and fast from seed. It is not as aggressive as some other types of turf because it spreads by growing larger clumps, rather than sending out rhizomes or stolons. Therefore, the seeding rate is higher than for some other grasses.

General Lawn Care in summers

- In the summer season, it is always advisable to keep the fertilizer use to a minimum. This is because they contain nitrogen, which, when combined with the soaring temperatures of the summers, can lead to burning up of entire lawn. As far as possible, we should go for a slow-release fertilizer, which has lower nitrogen content than the usual.
- One of the most common mistakes that people tend to make, while taking care of their lawn in summers, comprises of inadequate watering. It needs to keep in mind that even though watering as usual, in summers, the evaporation is much more. So, it is advisable to keep on watering till the time moisture gets 15 cm into the soil.
- It has been seen that lawns become much more susceptible to fungal diseases in the summer season. At the same, even weeds have been observed to grow faster when the temperatures are high. Thus, it is advisable to take proper care of garden, keeping them free of fungi as well as weeds.
- In the summer season, we should always increase the height of the grass, mowing it to a height of 7.5-8.5 cm. This will help cool down the crowns of the grass and also discourage various diseases. Whenever mow lawn in the summer season, make sure never to remove more than one-third of the total leaf area, at any one time.

General Lawn Care in winters

- In the winter months, lawn will not grow much, remaining more or less dormant. Therefore, it is advisable not to use fertilizers therein. Rather, wait for later-winters, when the temperature is a bit higher, for adding fertilizers to lawn. Still, for fertilizing purposes, the best time would comprise of late fall or early summer.
- It is true that in winter season, lawn does not lose too much moisture. However, this does not mean that stop watering it altogether. Rather, water it at a gap of somewhere around 3-4 weeks. As and when the grass growing more rapidly, it is advisable to increase the frequency of watering.

- With the approach of the winter season, start lowering the height of mower i.e. mowing down the grass to a shorter length; say around 5-7.5 cm. Otherwise, lawn might become prey to winter diseases. However, make sure to be gradual in the process. Otherwise, might end up removing all the green leaf tissue and damaging the turf.
- Winter season present adverse condition for the growth of grass, but not weeds. In fact, weeds are known to grow faster in cold conditions. So, make sure to keep looking out for weeds and removing them before they get a chance to do any damage or else, grass will be in the danger of getting killed by the time spring comes.

General tips for lawns in shade

- Shady lawns require less fertilizer compared to lawns exposed to the sunlight. 450 to 900 g of actual nitrogen per 1,000 square feet per growing season at two intervals. The ideal time for fertilization is the first week of September.
- Shady lawns don't require ample watering but when watered do it thoroughly.
- Lawns in shade are not meant for rough use. As they are quite weak compared to sunny lawns so it is better not to walk or let the children play on it.
- Though moss and creepers like ivy grow largely on the shady lawns but it is not harmful. It only grows on the empty soil because the grass grows slowly.

Lawn Weeds Managements

Apart from interfering with the growth of garden plants weeds rob the soil of its rich nutrients. This makes it utmost necessary for to take due care and ensure that weeds do not get a chance to grow in lawn. In case they manage to sneak into lawn, even after the protective measures, we should get rid of them as soon as possible, trying to stick to the natural methods and avoiding the chemical-based products. Vigorous, healthy turf properly maintained provides the best means of weed control in bermudagrass turf. But, where turf thins due to environmental stress, pest problems or poor management, weeds rapidly invade grass. Some of the common method to manage the weeds of lawn are as under:

Mulching

Using mulch is one of the best and safest methods of killing the weeds in lawn. It has been seen that covering the garden soil with mulch, up to 5-7.5

cm, blocks weeds. For this purpose, we can make use of shredded bark, wood chips, straw, cocoa bean hulls, gravel and rocks. Apart from killing the weeds, mulch will also help the soil retain moisture, thus reducing the frequency of watering.

Pouring Boiling Water

Exposing the weeds to boiling water helps a great deal in killing them. Pour boiling water on the roots of the weeds. The only drawback of this method is that it may require several applications.

Washing with Soap

Using soap on the weeds cleans them thoroughly, in other words makes lawn free of them. For this purpose, we need to put one quart (4 cups) of water in a bowl and add 5 tablespoons of liquid soap (such as dishwashing liquid). Now, mix it thoroughly and pour into a spray bottle. Coasting the weeds with this solution will help get rid of them, especially on a hot day.

Salting the Weeds

In case we want to try another method of killing the weeds in ur lawn, rock salt will be the best bet. We can get large areas of weeds (and even plants) by putting rock salt on them. However, do make sure that there are no garden plants in close proximity of the weeds. This is because the salt spreads when it rains, harming the plants in the vicinity, if any.

Broadleaved weeds including clover, chickweed, dandelion, henbit, dichondra and others can be controlled with the hormone type herbicides such as 2,4-D, MCPP, dicamba. Grassy weeds including crabgrass and dallisgrass can be controlled with several applications of MSMA in spring or early summer. Annual grasses including crabgrass and annual bluegrass can be controlled with pre-emergence herbicides. However, all of these herbicides must be used together with good management to effectively reduce weed populations.

Pest Management in Lawn

Many insects attack our lawn grasses. Most of them are minor pests and their damage is usually overlooked. A few, however, can cause extensive serious damage, sometimes killing large areas of grass. Identifying features of various lawn pests are as under :

Table 11.2 : Some Lawn Pests, Appearance of Their Damage, and Cultural Control Methods

Pest (Scientific name)	Hosts	Damage appearance	Cultural control
Armyworms, cutworms *(Pseudaletia unipuncta, Peridroma saucia, Agrotis* spp.)	all grasses, dichondra	leaves and base of leaves chewed and cut beginning in small, irregular spots that can spread to patches extending many feet in width	reduce thatch; eliminate soggy areas; overseed lawn
Billbugs *(Sphenophorus* spp.)	all grasses	brown, thin, dying grass, beginning in small, irregular spots that can spread to patches extending many feet in width	irrigate and fertilize adequately; increase mowing height
Black turfgrass ataenius *(Ataenius spretulus)*	annual bluegrass, bentgrass, ryegrass, Kentucky bluegrass	brown, dying grass, few roots; lawn is easily peeled off soil	increase mowing height; aerate to improve root growth
Fiery skipper *(Hylephila phyleus)*	bentgrass, bermudagrass, St. Augustinegrass	1- to 2-inch-diameter spots of lawn turn brown; spots may join to form large, irregular dead patches; leaves chewed or missing	reduce thatch; overseed with grass species that are not preferred
Lawn moths, sod webworms *(Crambus sperryellus, Tehama bonifatella)*	all grasses, especially bentgrass, bluegrass, clovers	lawn brown; leaves chewed or missing	reduce thatch; irrigate and fertilize appropriately
Southern chinch bug *(Blissus insularis)*	primarily St. Augustine grass	irregular patches of lawn turn yellowish, then brown and begin dying during hot weather	reduce thatch; reduce nitrogen fertilization; irrigate adequately; plant resistant varieties such as Floralawn, Floratam, or FX-10 if growing St. Augustinegrass

White grubs–immatures of masked chafers (*Cyclocephala* spp.), May and June beetles (*Phyllophaga* spp.)	All grasses, especially bluegrass, ryegrass	brown dying grass; lawn can be rolled up if heavily infested	irrigate and fertilize appropriately; overseed lawn

Table 11.2: Lawn Pest Detection Methods, Treatment Targets and IPM-compatible Materials

Insect	Detection method	Treatment target	IPM-compatible materials
Armyworms, cutworms	drench test for fat, dull gray, green, or brownish larvae up to 2 inches long; inspect outdoor lights around dawn for 1-1/4 inch brownish to gray moths	crowns, leaves, thatch	A, Bt, P, Sc
Billbugs	dig around roots for whitish, C-shaped, legless grubs up to 3/8 inch longwith reddish heads; inspect outdoor lights around dawn for 1/3 inchbrownish to gray snout beetles	crown, roots	I, Sc
Black turfgrass ataenius (see also white grubs)	dig around roots for whitish, C-shaped grubs up to 1/3 inch long with 6 legs and reddish heads; inspect outdoor lights around dawn for shiny black adults 1/5 inch long	roots, thatch soil interface	Hb, I, Sc
chinch bug	drench test or inspect around grass bases for reddish, purple, black, or gray bugs up to 1/2 inch long	crowns, stems	P
Lawn moths (sod webworms)	drench test for slender, grayish larvae up to 3/4 inch long; whitish or brownish moths up to 3/4 inch long fly when grass is disturbed	crowns, leaves, thatch	A, Bt, P, Sc
Skipper, fiery	drench test for larvae up to 1 inch long with pink-green body and red and black head; orangish butterflies 1 inch wide with knobbed antennae feed at flowers; mere presence of this insect does not warrant control	leaves, stems	Bt, Hb, P

white grubs (the immatures of masked chafers, May and June beetles; see also black turfgrass ataenius)	dig around roots in late winter or summer for whitish to yellow, wrinkled, C-shaped grub up to 1-1/2 inches long with 6 legs and a reddish head; look for yellowish brown adults 1/2 inch long.	roots	Hb, I, Sg

IPM-compatible materials
A = azadirachtin or neem (Safer BioNeem)
Bt = *Bacillus thuringiensis* (BT WormKiller, Caterpillar Clobber)
I = imidacloprid (Bayer Advanced Lawn Grub Control, Grub Ex, etc.)
P = pyrethrin/potash soap (Safer Yard & Garden Insect Killer)

Predaceous nematodes
Hb = *Heterorhabditis bacteriophora*
Sc = *Steinernema carpocapsae*
Sg = *Steinernema glaseri*

CHAPTER 12

Plant Growth Regulator and Ornamental Plants

Any substance or mixture of substances intended through physiological action, for accelerating or retarding the rate of growth or maturation, or otherwise altering the behaviour of plants or the produce thereof are known as plant growth regulators. It does not include plant nutrients, trace elements, nutritional chemicals, plant inoculants and soil amendments.

Plant hormones are organic compounds other than nutrients, synthesize by plants and regulate plant growth and development. They are active in very low concentrations, produced in certain parts of the plants and are usually transported to other parts where they elicit specific biochemical, physiological and morphological responses. They are also active in tissues where they are produced. The term plant growth regulator is usually used to denote a synthetic plant hormone, but most of the synthetic compounds with structures similar to those of the natural hormones have also been called hormones. For instance, the synthetic cytokinin, kinetin is considered a hormone.

Classification of PGRs

Each plant growth regulators have many responses. The effects of different growth regulators overlap and may be stimulatory or inhibitory. Plant growth regulators may be classified into five groups:Auxin, Cytokinins (CKs), Gibberellins (GAs), Ethylene, and Abscisic Acid (ABA).The new plant hormones, Brassinosteroids (BAs) and Salicylates or Salicylic Acids (SAs) are further included into this category.

Growth hormones, Auxins and Cytokinins, stimulate metabolism and regulate growth. Stress hormones (GA's, Ethylene and BA's) are the complementary of growth hormones. The latter group is triggered by a deficit of nutrients, when there are not enough nutrients to allow growth. They allow a plant to survive during stress period and redistribute resources in order to resume growth. Shock hormones are the ABAs and SAs, allow a plant respond to rapidly developing stress situations (ABA) or return to normal conditions after the stress has passed. Flower initiation is controlled by hypothetical hormones called florigens, but these substances are still to be identified.

All plant hormones have a very broad spectrum with complex action. Each hormone performs its specific functions. However, nearly all measurable responses to heredity and environment are controlled by interaction between two or more hormones. Such interactions may occur at various levels, including the synthesis of hormones, hormone receptors and second messengers, as well as at the level of ultimate hormone action. Furthermore, hormonal interactions may be cooperative, antagonistic or in balance.

It have been indicated that phytohormones enter into cells and regulate intracellular processes. In several cases plant hormones mediate between an external signal and a physiological activity (a cell's response). Plant hormones act partially synergistic, partially antagonistic. Growth and development of plants is regulated by a combination of genetic factors and environment influences. Plants have receptors that sense and respond to a number of environmental cues including photoperiod, temperature, pressure and moisture changes.

Plant growth regulators mediate the effects of environmental cues. Genes code for the enzymes that catalyze the chemical reactions in plant growth and development. The transport velocity within the system of vascular bundles is dependent on the transpiration that is itself dependent on water supply, temperature and species-specific differences. Signal transduction pathways are important for chemical messaging in plants and both environmental cues and hormones serve as signal stimuli.

Plants growth regulators can function as signal molecules (ligands) that trigger signal transduction pathways in cells. Such pathways often result in the synthesis of transcription factors that in turn promote synthesis of enzymes that facilitate chemical reactions within the cell.

Fig. 12.1 : Production of Plant hormones and action centre

Auxins

This hormone was named auxin by Frits Went in 1926. Indole acetic acid (IAA) is the natural auxin, synthesized from indole or tryptophan. There are several synthetic auxins. It promotes growth in molar concentrations of 10^{-3} to 10^{-8}. They are synthesized in apical meristems of shoots, young leaves, seeds and fruits. Auxins always move down the stem parenchyma cells towards roots by polar (charged) transport (auxin becomes negatively charged) using proton pumps, an ATP requiring process. Auxin enters cells as IAAH passively, or as IAA- via active cotransport. IAAH dissociates within the cytosol and special auxin transport proteins in the basal end of the cell are needed to carry auxin through the plasma membrane to the top of the adjacent cell. Auxin destined for root tissue, however, moves through phloem sieve tubes.

Functions of Auxins

- Auxins are the common rooting hormones.
- Auxin produced in apical buds tends to inhibit the activation of buds lower on the stems. This is known as apical dominance. This effect lessens with distance from the shoot tip.
- It promotes elongation and cell enlargement. For cell elongation; proton pumps increase the H^+ concentration in cell walls, which stimulates expansions of proteins that disrupt hydrogen bonds and break cross linkages in cellulose. This facilitates wall expansion when cells take more water.
- Auxins are involved in tropic responses. They migrate away from light, which accounts for the uneven elongation of cells on the shaded side of a plant unevenly exposed to light.
- It stimulates the production of secondary growth by simulating cambium cells to divide and secondary xylem to differentiate. Wound tissues repair is initiated by auxin when portions of vascular bundles are damaged.
- Promotes other hormone production, especially ethylene when auxin concentration increases.
- Promote flower initiation.
- Loss of auxin initiates leaf abscission.
- Auxin also plays a role in fruit development after released from developing seeds.
- Synthetic auxin used as herbicides include 2-4-D and 2-4-5-T.

Cytokinins

Cytokinins are a group of phenyl urea derivatives of adenine, one of the molecules in DNA. The first cytokinin was chemically isolated in 1913. Cytokinins are found in actively dividing tissues of seeds, fruits, leaves and root tips and sites of wound tissue. Studies indicated that root tips are sources of cytokinin production and cytokinins are transported through xylem to rest of the plant.

Functions of Cytokinin

- It promotes cytokinesis. It is found primarily in root meristems embryos and fruits. It migrate from roots to the shoot systems of plants by xylem tissues.
- In tissue culture normally, cytokinins work in conjunction with auxins, which promote elongation and cell expansion. When auxin is added to a tissue culture medium, cells in the absence of cytokinins enlarge but division does not take place. When kinetin and IAA are both added to a tissue culture medium, cells divide rapidly producing many small cells. The ratio of cytokinins and auxin controls the tissue differentiation in the tissue culture.
- Cytokinins can promote the growth of axillary bud that over-come the apparent inhibiting effect of auxin. This is one of the ways in which plants balance root and shoot growth.
- It also retards leaf senescenc probably by stimulating RNA and protein synthesis and delaying degradation of chlorophyll. It inhibits dormancy.

Gibberellins

Ewiti Kurosawa is credited with the discovery of gibberellins when he determined that a fungus was responsible for abnormal rice seedling growth, called the "foolish seedling" disease. The fungus secreted a chemical that caused the rice plants to grow abnormally long, and then collapse from weakness. The fungus was *Gibberella fujikuroi*, hence the hormone named. Many seeds contain a variety of different gibberellins. Gibberellins are produced in roots and younger leaves, but have their highest concentration in seeds. Most effects of gibberellins are seen only in agreement with auxins.

Functions of Gibberellin

- Gibberellins work in combination with auxins to promote rapid elongation and division of stem tissue.
- It increases the growth of many plants and seedlings.

- It enhances bolting of biennials to produce flowers during the growing season.
- It reversed the genetic dwarfism.
- After imbibitions of the seed coat, gibberellins break the dormancy of seed. It signals to germination activities of seeds.
- Application of gibberellins to seeds will counter the normal environmental cues, such as exposure to low temperatures. They stimulate RNA to promote synthesis of enzymes that convert stored nutrients (starches) to sugars needed for rapid cell respiration during germination.
- It minimizes a plant's requirements of long day exposure or periods of low temperature in order to flower.
- Abscisic acid can encounter the effect of gibberllin to keep seeds dormant.

Abscisic Acid

Abscisic acid (ABA) is a hormone that functions by inhibiting growth activities in environmental stress condition. It often serves as an antagonist to the other growth promoting hormones.

Functions of Abscisic Acid

- ABA promotes activities of seed dormancy. ABA levels are high when seeds mature, promoting lowered metabolism and synthesis of proteins needed to withstand the dehydration associated with dormancy. Seeds germinate when ABA is degraded by some environmental action. Desert seeds must have the ABA washed out of the seed coat, as do plants of marshy areas; temperate area plants have ABA degraded by light-stimulated enzymes. In other cases breaking dormancy is relative to the ratio of ABA (which keeps seeds in dormancy) and gibberellins (which promote germination).
- Low levels of ABA in maturing seeds promote premature germination.
- It promotes winter bud scale formation on woody plants in preparation for winter dormancy. ABA derivatives, called dormins, are used in commercial nurseries to keep materials to be shipped in dormant conditions. The dormancy can be reversed with gibberellins.
- ABA promotes stomata closure during leaf water deficit conditions by activating K^+ ion transport out guard cells. This involves signal transduction pathways with secondary messengers of calcium. ABA in this case originates in roots and detects low water level in root tissues. ABA moves upward into leaves and activates stomata closure.

Ethylene

Ethylene is the single gaseous plant growth regulator. It is produced in many plant organs, though its effects are most studied in fruits. Ethylene is the signal molecule for signal transduction pathways that activate tissue growth and development activities throughout the plant life (fruit maturation, leaf abscission and senescence).It is synthesized from the methionine amino acid.Toxic substances such as air pollutants also trigger the step for ethylene synthesis. It is also produced in wounded or bruised tissues.

Functions of Ethylene

- Ethylene activates ripening in fruits by signaling chemical reactions that degrade the pectins of the middle lamella, softening fruit and promoting the conversion of stored starches and/or oils into sugars.
- Ripening of some fruit involves a significant increase in the rate of cell respiration with concurrent high O_2 uptake. This function of ethylene (fruit ripening) has extensive agricultural impact.
- Ethylene promotes female flower production, whereas, high gibberellins may promote formation of male flowers.
- It help to germinate seeds by overcoming mechanical stress. When the shoot tip cannot penetrate the mechanical obstacle, ethylene is synthesized, which slows cell wall expansion. The walls thicken (more resistant to the pressure) and the stem curves to bypass the obstacle via horizontal growth. The stem tip pushes upward periodically to test for the obstacle.If the obstacle is encountered, ethylene initiates another cycle of stem thickening and horizontal growth. If no obstacle is encountered, the shoot resumes negative geotropic growth.This growth response is known as the triple response (decrease in elongation, stem thickening and change in orientation of shoot growth to horizontal).
- Ethylene is responsible for initiating the programmed death (apoptosis) of sclerenchyma and xylem vessels and tracheids.
- Ethylene is the direct cause of leaf, flowers and fruit abscission. It promotes the degradation of the cell walls in the abscission zone. The decrease concentration of auxin in leaves signals the production of ethylene in the abscission zone during the end of summer.High concentrations of CO_2 inhibit ethylene production so CO_2 can be used to prevent fruit ripening in grocer warehouses.

Brassinolides (Brassinosteroids)

Brassinolides is a plant steroid discovered in pollen of members of the mustard family, and studied in Arabadopsis. Chemically the brassinosteroids are similar to animal steroid hormones and have similar genes that code for the steroid synthesis pathways.

The functions of brassinosteorids overlap those of other plant hormones and may be additive. Brassinosteoroids activate signal transduction pathways that promote cell elongation and cell division. Brassinosteroids also promote differentiation of xylem tissues. Brassinosteroids can also retard leaf abscission. Absence of brassinolides results in dwarf plants. Its effects overlap with those of auxin and gibberellins.

Salicylic Acid

Salicylic acid is known to activate defence genes for resistance against pathogens (hypersensitive response).

Salicylic acid is a phenolic extract obtained from willow bark and have been used as an analgesic. It is now prepared commercially and is the active ingredient of aspirin.

Oligosaccharins (Oligosaccharides)

Oligosaccharins are short chain of sugars present in cell walls that may have a role in defence against pathogens. When a pathogen attracted on the plant, the oligosaccharide activates a signal transduction pathway leading to plant defence responses. They may also help regulate growth, differentiation and flower development all by activating signal transduction pathways.

Systemin is a small peptide found in wound tissues. It may stimulate defence activities as the signal molecule (ligand).Systemin activates the signal transduction pathways that include the jasmonates.

Jasmonates

Jasmonates are a group of fatty acid derivatives. They play a role in seed germination, root growth and the storage of protein especially in seeds. Jasmonates synthesized in response to signal molecules produced in wound areas are involved in the signal transduction pathways that result in secondary metabolites (protease inhibitors) that poison the predator, or synthesis of volatile molecules that attract the predator's predators.

Plants Growth Regulators and Ornamental Plants

The key uses of plant growth regulators in ornamental horticulture and floriculture are height regulation, increased branching, propagation of cuttings, flowering control, enhancing stress tolerance and enhanced postharvest longevity during shipping, marketing, and display life of cut flowers and potted plants. In most cases, height reduction is desired in bedding, potted and woody ornamentals but in some cases height enhancement is also desired.Triazoles have greatly expanded the uses of PGR's especially in the production of woody crops and difficult to control crops. The use of auxins in plant propagation was one of the earliest adoptions of the use of plant growth regulators in the production of ornamentals. This use has led to shorter production times, decreased production costs and allowed mass production of many species that were difficult to root prior to the advent of synthetic auxins. Considerable success has been achieved in preventing injury induced by ethylene during storage, transport and retail display of ornamental plants and cut flowers. The use of silver thiosulphate and 1-methylcyclopropene to reduce ethylene sensitivity in flowers. Aminovinylglycine and aminooxyacetic acid to inhibit ethylene biosynthesis, has resulted in enhanced longevity and display life of several floriculture crops.

Height Control

1. A-Rest (ancymidol) reduces internodes elongation as the result of inhibition of gibberellins biosynthesis. When used properly, A-Rest produces no phototoxic effects. A-Rest is mostly applied using spray or drench applications and has been effectively used as a media spray. The label also allows for a broad range of chemigation applications including subirrigation. This chemical is effective on a wide range of plant varieties, and its growth regulating activity is greater than B-Nine and Cycocel, but less than Bonzi and Sumagic.
2. B-Nine (daminozide) reduces internodes length by blocking gibberellins biosynthesis. B-Nine generally does not produce any phototoxic effects; however, it is best to avoid applying B-Nine during periods of bright sunshine. This chemical is primarily used as a spray and recently has been used as a pre-plant dip. B-Nine is effective on a wide variety of plants, but its activity is less than most other growth regulators. This product works best when the spray is allowed to remain on the plant and dry slowly. Do not irrigate overhead for 24 hours after application, or the activity of B-Nine will be reduced.
3. Bonzi (paclobutrazol) also reduces internode elongation and is known to be a very active and persistent growth regulator. This chemical, when used properly, does not produce any phytotoxic effects, but over-application can lead to stunting. Bonzi is used primarily as either a spray or a drench but also has been used as a bulb soak. The Bonzi label recently has been expanded to include such applications as subirrigation and media sprays and has a broad spectrum label for use on most floricultural crops. Bonzi is absorbed within minutes of application, and overhead irrigation can occur 30 minutes after application without reducing its effectiveness.

4. Cycocel (chlormequat chloride) also reduces internode elongation within plants. It is a fairly active plant growth regulator, but has less persistence and activity than A-Rest, Bonzi and Sumagic. Phytotoxity with Cycocel can be a concern. Tank mixing Cycocel with B-Nine has been an effective strategy for reducing the phytotoxic effects of Cycocel and gaining more height control than the use of either chemical alone. Cycocel is labeled for both spray and drench applications, and like B-Nine, it is important to let it dry slowly and not irrigate overhead for at least six hours after application.

5. Florel (ethephon) is effective as a growth regulator for several ornamental plants. However, it is most often utilized for its ability to stimulate lateral branching and delay flowering. Florel is only effective as a foliar spray.

6. Sumagic (uniconazole) is the most active and persistent of all of the growth regulators. Internode elongation is reduced by the inhibition of gibberellin biosynthesis. There are no phytotoxic effects, but over-application can lead to stunting. Sumagic is most often used as a spray, drench or bulb dip. Recently, the label has been expanded to include such applications as media sprays and cutting dips. The weakness for Sumagic is the potential for over-application, but the strength is that Sumagic often shows activity on a wider array of plants when compared to the other plant growth regulators. Like Bonzi, Sumagic is absorbed into the plant within minutes of application and overhead watering can occur within 30 minutes of application without reducing its effectiveness.

Fruit Elimination

Fruit elimination is usually performed during the blossoming period. The tree must be sprayed during this period, before the fruit sets. Almost all trees that produce unwanted fruit can be treated. Read the label carefully to see if the product you plan to use lists your trees.Florel (Ethephon {(2-chloroethyl) phosphonic acid}is one of the more popular brands and will eliminate fruit on apple, carob, cottonwood, crabapple, elm, flowering pear, horse chestnut, maple, oak, olive, pine, sour orange, sweetgum, and sycamore. By eliminating the blosooms, the fruit will not develop. Sweetgum, however, has a more complicated flowering period and the timing must be more precise to control the "gumballs". Consulting an arborist may be necessary for precise timing.Florel can also be used to accelerate tomato ripening. When timed correctly and sprayed on late season green tomatos, the fruit will ripen faster so it can be harvested before the first frost.

Colours developments in flowers

There is evidence that plant growth regulators induce anthocyanin production during petal development. In emasculated petunia flowers or in isolated petunia petals bathed in sucrose solution, exogenously supplied gibberellic acid (GA_3) will induce anthocyanin production, which otherwise does not occur. Gibberellins are thought to be synthesized normally by the developing anthers and, in turn, to induce the expression of the anthocyanin biosynthetic genes in the petals to produce color. Exogenous GA_3 induces the transcription of EBGs such as CHS and CHI and LBGs such as DFR and AS. This type of mechanism may represent a method for coordination of anther development and color signal formation. Perhaps gibberellins induce the activity of transcriptional activators of the bHLH or my be type to induce color formation. Some tree leaves are noted for their spectacular fall color display. The changing fall colors are caused primarily by a photoperiodic response but also by a temperature response. As nights become longer in the fall, leaves stop producing chlorophyll. As the chlorophyll chemically degrades, it is not replaced. Other leaf pigments, the carotenoids, become visible and the green/orange splotches become more visible as the green chlorophyll turns orange. Carotenoids include the orange carotenes and the yellow xanthophylls. Anthocyanins produce the deep red and purplish-red colors in the fall display.

Producing a fuller tree, without lateral growth

Dikegulac-Sodium active ingredient (i.e., Atrimmec) temporarily promotes lateral branch elongation within the plant. This will help fill gaps in shrubs and other bushes or trees where a full look is desired without promoting outward growth.Where root damage from home construction, etc. is suspected, the use of canopy growth regulators may help. The reduced canopy growth will force the tree or shrub to redirect the energy toward root growth. This has saved a lot of trees that may have been lost.

Elimination of Tree Root Suckers

Root suckers are up-right growing stems that grow along roots and at the base of stumps or trees. Left alone and they will grow to become a congested mess. Others can grow to become unwanted trees. "Sucker Stopper" (Naphthalene acetic acid) is a plant hormone product that promotes the demise of suckers by inhibiting the elongation and growth of tree sprouts. First trim the sprout with a lawn mower and spray the product on the sprouts.Sucker Stopper has many uses in the nursery and landscape industry as well. Depending on the amount of the active ingredient, certain Sucker Stopper products are labeled for Christmas tree farms to slow growth and help determine the shape of the tree. Others can be used around pruned limbs to control unsightly sprouts around the cut.

Weed killers

Plant growth regulators may be used to control weeds thereby reducing competition of weeds in orchid house (2, 4-D and 2, 4, 5-T). Other compounds such as Dalapon are more effective against monocots than dicots, so they could be used in the opposite situation. In fact these weed killlers cause plants sensitive to grow too fast, damaging their cells so that their conducting tissue does not develop properly, or they use up their food reserves and die.

Use of PGRs for Turf Managements

With turf grass growth regulators, the original goal was to slow grass growth, for reducing the amount clippings and frequency of mowing. Reducing labour costs was a powerful motivator for the use and development of PGRs. They are now used along garden and cement edges to slow runner growth and reduce trimming time. They are used to improve grass appearance by preventing seedhead and stalk formations. They are even used for preventing energy depletion that can be used during stress periods. Some that are labelled for grass are also labelled for other landscape plants as well.

- *Slowing Grass Growth*: Plant Growth Regulators sprayed on the lawn will slow the growth of labelled turf species by up to 50 percent for up to four weeks. This can be important for those who are seeking a reduction in clipping collection or mowing frequency.
- *Improved Grass Health and Stress Relief*: Grass growth requires a substantial amount of energy, especially early in the season during the rapid growth phase of cool season varieties. As the grass enters the stresses of summer, energy reserves can be used up faster. Grass energy can be kept in reserve by slowing grass growth at least a month before the peak summer stress period begins. The energy stays in the roots instead of being used for growth, stalk and seed production. As a result, the grass will be able to handle stress and disease pressures better.
- *Reduced Trimming and Edging*: A plant growth regulator sprayed round the base of trees, along sidewalks and driveway edges, along planter edges, etc., will slow grass and runner growth. This can be especially helpful for warm season grasses that spread by the production of stolons. Stolons are the above ground runners that readily challenge any boundary or border. Although they will not prevent the production of stolons, they can greatly slow them down. Some plant growth regulators are also labelled for ornamental trees and shrubs for maintaining their shape longer.

- *Promoting the Growth of One Grass Species Over Another*: Example is spraying annual ryegrass that has been overseeded into a bermudagrass lawn. The annual ryegrass should be sprayed a week or more before the bermudagrass is expected to emerge from dormancy. This slows the growth of the ryegrass and allows the bermudagrass to recover will less competition.Certain plant growth regulators have also been used to promote the growth of fescue where bermudagrass was taking over. This is accomplished by using a product that is labelled for bermudagrass, but has no affect on fescue.

- *Reduce Mowing in Difficult Areas*: Hillsides, right-of-ways, and difficult to mow areas could benefit from plant growth regulators. This is especially necessary with species where grass stalks and unsightly seedheads may be the primary reason for mowing.

Other Uses of PGRs

In the past, plant growth regulators have, for the most part, been considered tools used by growers to control plant height. This will probably remain the primary purpose of these chemicals. However, there are some new areas where growers are using plant growth regulating chemicals.

1. EthylBloc (MCP) is an ethylene-action inhibitor, which inhibits the negative effects of ethylene. This product is being used to help prevent shattering of flowers, eliminate leaf yellowing, reduce flower browning, and stop flower and leaf drop. EthylBloc is a wettable powder and when mixed with a liquid buffer solution, releases a gas (MCP). This gas binds to the ethylene receptor in plant cells, inhibiting the negative effects of ethylene. EthylBloc is approved by the EPA and is manufactured by Floralife.

2. Fascination, a combination of GA_{4+7} and BA, is introduced and labeled for several new uses. The first labeled use of this product will be for the prevention of leaf yellowing in Easter, Oriental and LA Hybrid lilies. Foliar chlorosis is considered the primary factor contributing to low post-production longevity of potted Easter lilies. Many growers watch as their crop goes through a gradual leaf yellowing, which usually begins near or before the visible bud stage. Growers can also experience a post-harvest leaf yellowing of their crop, which may be dramatic, with rapid leaf yellowing occurring on finished plants prior to sale.

3. When used preventively, Fascination plant growth regulator can inhibit both types of leaf yellowing. This product must be used carefully, however. It is intended to be applied only to the lower foliage. If the immature leaves come into contact with the spray solution, an unwanted

stem elongation may result. Uniformity of coverage is extremely important because the spray is only effective where it comes into contact with the leaves.

4. The second labeled use of Fascination is to prolong the length of flowering in Easter, LA Hybrid and Oriental lilies. Current research on 'Nellie White' Easter lily has shown that one application of Fascination prolonged the length of flowering from five to eight days to 18 to 24 days.
5. ProGibb (GA3) is another plant growth regulator that is being developed for use in a number of different floricultural crops. Its uses include breaking flower bud dormancy, increasing flower number and size, promoting earlier flowering and increasing stem elongation for the production of tree forms of geraniums and fuchsias.
6. AVG (aminoethoxyvinyl glycine) is an amino acid that inhibits the onset of ethylene production in plants. The benefits of this product are very similar to EthylBloc, as it extends flower longevity and reduces flower senescence, flower bud abscission and leaf yellowing. AVG is applied as a spray, which could make it more predictable and easier to use than a gas such as MCP.

Application Techniques for Height Control

1. The most common method of applying growth regulators is spray applications. When using plant growth regulators as a foliar spray, it is important to achieve thorough, consistent and uniform coverage.
2. Drenching is the second most common method of applying plant growth regulators. Drench applications usually provide longer lasting, more uniform control of plant height than spray applications and typically use a larger, more diluted volume of solution than sprays. Drenches are primarily applied to the top of the media of a growing plant.
3. The sprench method is a hybrid of the spray and drench methods. When using this technique, a higher volume of spray is applied, achieving more of a drenching effect than that of a spray application. Typically, sprench volumes are two to four times that of a spray. The chemical concentration being applied is usually less than a spray but more than a drench.
4. Bulb dipping, or bulb soaking, is another innovative method that has been developed to help growers achieve height control for certain bulb crops. This method involves dipping or soaking bulbs in a growth regulator solution prior to planting. The concentration of the chemical and the length of time that the bulbs are submersed will vary depending

on the species. Another very similar method growers and researchers have been evaluating is a seed soak. However, at this time, no chemicals are labeled for this procedure.

5. Many growers are using a pre-plant dip of cuttings to obtain height control. This method can be used on either rooted or unrooted cuttings of some plant varieties. To achieve results, the cuttings are placed in a growth regulator solution just long enough to thoroughly wet all of the leaves and stems, then they are removed and planted.
6. Applying growth regulators as a media spray prior to, or shortly after, planting are another creative method that growers are employing to achieve height control. The purpose is to obtain effects similar to a drench.
7. Another innovative approach many growers are trying is called controlled residue. This application technique involves applying the growth regulator directly to the container prior to filling it with media. As the roots grow and come into contact with the container, they take up the chemical, and height control can be achieved.

Plant Growth Regulators and Enzymes

Genetic information directs the synthesis and development of enzymes which are critical in all metabolic process within the plant. Most enzymes are proteins in some form or another, are produced in very minute quantities and are produced on site—meaning they are not transported from one part of the organism to another. The major difference is that hormones are transported from one part of the plant to another as needed. Vitamins vital in the activation of enzymes and are produced in the cytoplasm and membranes of plant cells.

Plant movements

Plants appear immobile because they are usually rooted in one place. However, time lapse photography reveals that parts of plants frequently move. Most plants move too slowly for the passer by to notice. Plants move in response to several environmental stimuli such as: light, gravity and mechanical disturbances. These movements fall into two groups: tropisms and nastic movements.

Tropisms

A tropism is a plant movement that is determined by the direction of an environmental stimulus. Movement toward an environmental stimulus is called a positive tropism, and movement away from a stimulus is called a negative tropism. Each kind of tropism is named for its stimulus. For example, a plant movement in response to light coming from one particular direction is called a phototropism. The shoot tips of a plant that grow toward the light source are positively phototropic.

Phototropism

Phototropism is illustrated by the movement of sprouts in relation to light source direction. Light causes the hormone auxin to move to the shaded side of the shoot. The auxin causes the cells on the shaded side to elongate more than the cells on the illuminated side. As a result, the shoot bends toward the light and exhibits positive phototropism. In some plant stems, phototropism is not caused by auxin presence or movement. In these instances, light causes the production of a growth inhibitor on the illuminated side of the shoot. Negative phototropism is sometimes seen in vines that climb on flat walls where coiling tendrils have nothing to coil around. These vines have stem tips that grow away from the light, or better put, toward the wall. This brings adventitious roots or adhesive discs in contact with the wall on which they can cling and climb.Solar tracking is the motion of leaves or flowers as the follow the suns' movement across the sky. By continuously facing toward a light source, moving or not, the plant maximizes the light available for photosynthesis.

Thigmotropism

Thigmotropism is a plant growth response to touching a solid object. Tendrils and stems of vines, such as morning glories, coil when they touch an object. Thigmotropism allows some vines to climb other plants or objects, thus increasing its chance of intercepting light for photosynthesis. It is thought that an auxin and ethylene are involved in this response.

Thigmonastic movements are a type of nastic movements that occur in response to touching or shaking a plant. Many thigmonasties involve rapid plant movements, such as the closing of the leaf trap of a Venus flytrap plant or the folding of a plant's leaves in response to being touched. Some leaves of sensitive plants will fold within a few seconds after being touched. This movement is caused by the rapid loss of turgor pressure (water pressure) in certain cells, a process similar to that which occurs in guard cells in order to close stomata. Physical stimulation of the plant leaf causes potassium ions to be pumped out of the cells at the base of the leaflets and petioles. Water then moves out of the cells by osmosis. As the cells shrink, the plant leaves move. It is believed that the folding of a plant's leaves in response to touch is to discourage insect feeding.In addition, thigmonastic movements help prevent water loss in plants. When the wind blows across a plant, the rate of transpiration is increased. If the leaves of a plant fold in response to the "touch" of the wind, water loss is reduced.

Gravitropism

Gravitropism is a plant growth response to gravity. A root usually grows downward and a stem usually grows upward; that is, roots are positively gravitropic and stems are negatively gravitropic. Like phototropism, gravitropism appears to be regulated by auxins. One hypothesis proposes that when a seedling is placed horizontally, auxins accumulate along the lower sides of the root and the stem. This concentration of auxins stimulates cell elongation along the lower side of the stem, and the stem grows upward. A similar concentration of auxins inhibits cell elongation in the lower side of the root, and thus the root grows downward.

Chemotropism

Chemotropism is a plant growth response to a chemical. After a flower is pollinated, a pollen tube grows down through the stigma and style and enters the ovule through the micropyle. The growth of the pollen tube in response to chemicals produced by the ovule is an excellent example of chemotropism.

Nastic Movements

Plant movements that occur in response to environmental stimuli, but that are independent of the direction of the stimuli are called nastic movements. These movements are regulated by changes in water pressure in certain plant cells.Nyctinastic movements are plant movements in response to the daily cycle of light and dark. Nyctinastic movements involve the same type of osmotic mechanism as thigmonastic movements, but the changes in turgor pressure are more gradual.

Seasonal Responses

In nontropical areas, plant responses are strongly influenced by seasonal changes. For example, many trees shed their leaves in the fall, and most plants flower only at certain times of the year. Plants are able to sense seasonal changes. Although temperature changes are involved in some case and to certain degrees, plants mark the seasons primarily by sensing changes in night length.

Phytochrome Regulation in Plants

Plants monitor changes in day length with a bluish, light-sensitive protein pigment called phytochrome. Phytochrome exists in two forms, based on the wavelength of the light that it absorbs. It is generally produced in meristematic tissues in very minute amounts. The two stable forms can be converted to each other by absorbing light. P_{red} (P_r) which absorbs red light and $P_{far\text{-}red}$ (P_{fr}) which absorbs far-red light. In the daylight more P_r is converted to P_{fr} (the active form) than vice versa. P_{fr} will convert back to P_r over several hours in the dark where it would be stable indefinitely. The conversion in light is almost instantaneous. The phytochrome mechanism is what transforms the crook in the hypocotyls of the emerging seedling into a straight stalk. Stem elongation appears to be inhibited by $P_{fr.}$ However, if light levels are low, the shaded stems of a tree for example, more far-red light will reach them and cause the conversion to P_r which lowers inhibition and allows the stems to grow longer and out from under the shade. The interconversion abilities of phytochrome:

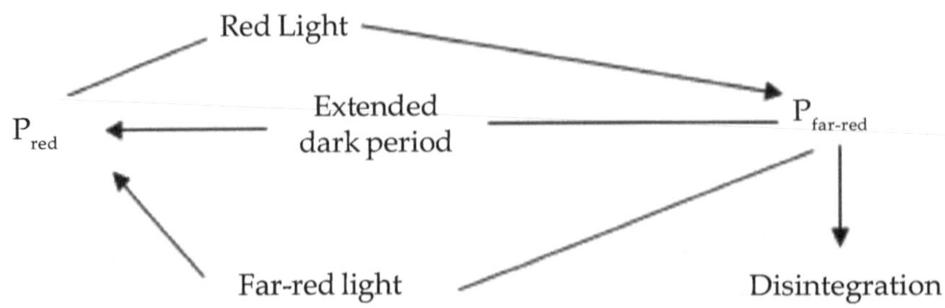

Long day (LD) exposure of rosette plants causes rapid stem/petiole elongation, a more vertical growth habit, and flowering; all changes are suggestive of a role for the gibberellin (GA) plant growth regulators.Vertical growth remains a critical problem in turfgrass management. Practical methods can be used to control the plant growth. For example, application of chemical regulators can inhibit gibberellin (GA) biosynthesis and reduce the stem growth.

Vernalization

Vernalization is the low-temperature stimulation of flowering. Exposure to cold weather causes the plants to flower in the early spring and an early crop is produced. Cold temperatures are not absolutely required for most crops, but they do expedite flowering. A biennial plant is a plant that lives for two years, usually producing flowers and seeds during the second year. In the spring their flowering stem elongates rapidly, a process called bolting. Most biennials must receive cold weather to vernalize before they flower during the second year. They will then die after flowering. Treating a biennial with gibberellin is sometimes a substitute for cold temperatures in vernalization, and will stimulate the plant to grow.

CHAPTER 13

Pot and Potting Management for Indoor Plants

Ornamental pot plants should be potted in small pots according to the size of the plants. Any kind of pot/container which can hold medium and provide aeration is suitable. The most common container is the plastic or clay pots. Plastic pots should be preferred because it retains moisture longer than mud pots. Plants should not be disturbed frequently and re-potting done only when absolutely necessary.

Potting media

Different types of growing media may be used for ornamental pot plant cultivation, but it is crucial that it must retain a certain amount of moisture and be well draining. Different plants require different substrate in which they would thrive. Conditions of the plant experiences in nature should think before choosing the media. Tropical plants require more water naturally compare to subtropical and temperate ones. For irrigating with sprinkling can or hose, it is better to use materials that drain well and decompose slowly while for drip irrigation heavy potting material should be used. The most important thing is to match watering habits to the potting material. Growing conditions play a most significant role in deciding on a medium. If there is no problem in providing a constantly humid environment due to climate or if growing in a greenhouse, the moisture retention on thepart of media is less of a concern. Ambient temperatures play a role here, as well, as warmer temperatures tend to increase the evaporation of water and along with the drying rate of the medium. The personal tendencies and how much time might be spending to water plants is another significant factor in choosing a medium. The combination of potting materials that will work best for garden plants mainly on Plant structure and root structures:

1. *Plant Structure:* Terrestrial orchids should keep slightly damp all the time, while epiphytes or lithophytes want to dry off thoroughly between watering. Orchid with pseudobulbs can generally store water for longer periods of time than those of without pseudo bulb. Those having thin, strap-like leaves, such as *Oncidium* and *Cymbidium*, will lose their stored water faster than those with shorter, thick leaves, such as *Cattleyas*. Slipper orchids lack pseudobulb for water storage, so they should not be allowed to dry out entirely. Plants that don't have pseudobulbs in some cases developed thick, fleshy leaves(*Phalaenopsis*) as their water storage scheme, but that's obviously not as effective as pseudobulbs. Others that have neither pseudobulbs nor thick leaves, such as *Phragmipediiums*, are needed a steady supply of moisture. Large plants usually do best in coarser potting materials and smaller plants do better in finer potting materials. Orchids which grow in the ground in a moderate climate will grow in regular soil (*Bletilla*), while terrestrial orchids from tropical areas (some species in the genus *Cymbidium* and *Paphiopedilum*) require very porous yet rich substrate. Young plants and seedlings need finer substrate than adult plants. Clearly, there is no universally applicable substrate which would suit all orchids. *Paphiopedilum* should not be grown in waterlogged soil because its roots will rot.

2. *Root Structure* : It is another important aspect that decides medium texture. In general, smaller roots grow better in finer water-retentive materials, while larger roots perform best in coarser materials. The factors to be considered are the extent of root growth(long versus short, branched versus single), the thickness of the roots, particularly that of the velamen layer. A very extensive root system suggests that the plant has to work hard for its water and nutrition, so wants to gather as much as it can when it is available. By contrast, a short root system suggests easy and/or frequent availability of them. Velamen is actually dead tissue that acts as a sponge to rapidly absorb water and nutrients, so they can be passed on to the plant. Thick velamen is generally found on plants that like to grab a lot of water, but prefer to have their roots dry rapidly. *Vandaceous* plants are good examples.It should not mean that roots with thin velamen like to stay moist. *Oncidiums*, for example, tend to have thin velamen on thin roots, but their root systems are quite extensive,

and they have pseudobulbs for storage, so they also like to dry fairly rapidly. *Phalaenopsis*, with their minimal storage capabilities, tend to have thick roots to add to the "reservoir," so do well in a damp, but not soppy environment. Some plants like *Arachnis* and *Renanthera* are planted directly in a well-drained bed or in pots with soil.

Media Ingredient

Commonly available media ingredients and their specific properties are as under.

1. *Fir bark*: Shredded bark is the basic ingredient contained in most substrates for ornamental plant growth. The most common bark found in these substrates is pine while some growers also use the bark of *Pseudotsuga menziesii*, known as Douglas fir. Shredded bark is sold in various fractions, from finely shredded (up to 25 mm in size), mid-sized (up to 40 mm) and roughly shredded (over 40 mm) particles. For larger orchids (some species in *Cattleyas*, *Angraecums* etc.) bark with fine shredded particles should be used. While for which demand good aeration of roots (*Vanda* and its hybrids), mid-sized and roughly shredded particle bark should be used which can be turn into smaller pieces if necessary. It is essential that bark is soaked in water for minutes and then washed at least once to get rid off tannins and fine dust particles, which are not good for tender roots of ornamental plants. Good quality bark is getting harder and harder to find, but the stuff that is available is typically steamed to extract the resins that could be toxic to plants. The removal of that resin increases the water-holding capacity of the bark, but also shortens the life of it. The irregular shape of the pieces of bark can lead to excess packing, which may reduce the airflow to the root system. Fine grade is often used for small seedlings, and the coarser grades for large plants. Bark should be moist by boiling water to facilitate wetting before use.
2. *Coconut husk chips (CHC):* It is made from the pithy covering outside of the spherical, hard coconut shell, is gaining popularity as a replacement for fir bark. It is readily available, relatively cheap and wets and rewets better than bark. It also tends to be more regular in shape - chunks or cubes as opposed to coarse flakes - so facilitates better air flow throughout the root mass. CHC holds a lot more water than bark, so keep that in mind when using it. It lasts considerably longer than bark as well, often going three years before decomposition becomes an issue, as opposed to a year for bark.
3. *Coconut husk fibre*:Itis the wiry result of stripping and shredding the interior of the coconut husk, rather than chipping it. It is springy and

flexible and is often matted to form liners to hanging baskets, but it can also be use straight as a medium for plants that really like to dry out fast. If fairly tightly compressed, water will stay in pockets in the mass for a longer time period. Coir, sometimes called "Coco-Peat," is the result of grinding the husk into coarse powder. It is often used as a substitute for peat moss in blends, but can be used alone for seedlings that like to stay damp.

4. *Osmunda fiber*: Roots of a number of ferns from the genus *Osmunda*, used to be a staple in the orchid growing community, but it is now difficult to find a high quality fibre. It is tough and springy and requires vertical alignment in the pot to ensure proper drainage. It almost totally ensures that the plant roots get all of the air they need, and as it decomposes, is an excellent source of nutrients, requiring little or no supplementation via fertilizers.
5. *Tree fern fibre*: The trunks of tropical tree ferns have replaced most osmunda in orchid culture. It is very stiff and airy, and can be obtained as shredded fibers, chunks or slabs. It has little moisture holding capacity, so aerates media blends very well. The slabs are often used for mounting plants directly.
6. *Redwood fibre*:It is sometimes known as redwood fibre, is a light, fluffy wood fibre that does not decompose, and is added as a moisture-retention aid and as a means of increasing the acidity of media blends.
7. *Sphagnum moss (Peat moss)*: Its characteristics are good water absorption and aeration. *Phragmipediums* and some other orchids are best grown in a substratum composed exclusively of sphagnum but, for most orchids, peat moss is mixed with tree bark which does not retain water very well. Peat moss can also be used in planting orchids on tree bark where the roots are wrapped in moss and then attached to large or small bark pieces. Generally, *sphagnum* becomes sour and in need of replacement in the 6 to 9 month range, although that can be extended by blending it with charcoal.
8. *Hardwood charcoal* : It is a very long-lasting medium ingredient having a moderate-to-low moisture holding capacity. Charcoal is added due to its absorption capabilities. It can extract a number of waste substances from the substrate and to a certain extent, also the nutrients contained in it, which is not desirable. Charcoal does not absorb much water and, similarly to shredded bark, it helps to aerate the substrate. *Vanda* orchid may be grower by using the coarse grade as the sole medium. It has been observed that the charcoal sweetens the medium by absorbing some of the foul chemicals produced during the decomposition of other media components or excess salts from fertilizers.

9. *Perlite*:It is often referred to as sponge rock for the coarser grades, is expanded volcanic glass and is a great aerator of blends, but still holds enough moisture to be a fairly good substrate for Semi-Hydroponic culture.
10. *Pumice*:It is similar in nature to perlite, but with a higher density and lower moisture-holding capacity.
11. *Lightweight Expanded Clay Aggregate*: It is a general term for any number of more-or-less spherical terra cotta pellets that were originally designed to be the aggregate in light weight concrete formulations, but later diverted to horticultural use. There are many brands available, including Aliflor, Hydroton, Prime Agra and others, and they vary greatly in their properties and applicability to orchid culture.
12. *Diatomite*: It is calcium diatomaceous earth mud. It is highly porous, and holds a tremendous amount of water.
13. *Styro foam*: These are the substrates contain fine particles of styro foam which improve aeration. These particles are completely inert (they do not decompose), are long-lasting and do not give out any nutrients as other ingredients in substrate do. Large pieces of styro foam can be placed at the bottom of the pot, which will provide excellent non-decomposing drainage material (small stones can also be used for this purpose).

Special Mixture (Substrate) For Orchids

In general a mixture of fine or middle size bark particles (depending on the orchid size) and sphagnum in a 4:1 ratio is recommended for most orchids. For *Cymbidiums, Paphiopedilums, Zygopetalums* and *Ludisia*, it is recommend to add an additional part of rough peat or coconut fibre. Charcoal may also be used by replacing one part of the bark. Large pieces of styro foam can be used for the purpose of drainage. Drainage material can also be provided by using large pieces of bark or small stones. *Masdevalias* and *Miltonias* require a 2:1 ratio of shredded bark and sphagnum. *Vandas, Ascocentrums, Arachnis* and similar natural and artificially bred types of orchids require only roughly shredded bark.Terrestrial orchids, like *Spathoglottis, Phaius* and *Calanthe*, should be grown in 20-25 cm size pots with 1:1:1 mixture of leaf mould, FYM and sand. For *Paphiopedilum*. A mixture of 2 parts leaf mould, 2 parts loam soil and 1 part each of brick pieces and charcoal is recommended. It is observed that orchid *Rhychostylis gigantea*,perform better on chunks of hard-wood charcoal alone as potting substance. Tree fern fibre also performed better than the other media while coconut husk and over-burnt brick as planting substance had adverse effects on growth and flowering of *R. gigen*tea plants. *Cymbidiums* like a slightly acid potting mix with good drainage facilities. For

Cymbidium planting mixes, fine bark mix with 20 - 25 per cent perlite, sand, coarse peat and redwood bark are often added to improve the waterretention. Larger grades of bark will require more frequent watering, but there is less risk of overwatering. Coir, coconut husk chips, coarse peat or leaf mold and sand may also be used. They can also be grown in mixes containing pumice or lava rock.

Pasteurization of potting media

Medium may contain harmful disease causing organisms, nematodes, insects and weed seeds, so it should be decontaminated by heat treatment or by treating with volatile chemicals like methyl bromide, chloropicrin etc.as mentioned in Table 13.1

Physical propagation facilities such as the propagation room, containers, flats, knives, working surface, benches etc. can be disinfected using one part of formalin in fifty parts of water or one part sodium hypochlorite in nine parts of water.An insecticide such as dichlorvos sprayed regularly will take care of the insects present if any.

Table 13.1 : Chemicals with recommended doses for pasteurization of media

Agent	Method	Recommendation
Heat	Steam	30 min at 85°C
Methyl bromide	10 ml/cu. ft. of medium	Cover with gas proof cover for 24-48 hr. Aerate for 24-28 hr before use.
Chloropicrin	(Tear gas) 3-5 ml/cu. ft. of medium	Cover for 1-3 days with gas proof cover after sprinkling with water. Aerate for 14 days or until no odour is detected before using.
Basamid	8.0 g/cu.ft. of medium	Cover for 7 days with gas proof cover and aerate for atleast a week before use.
Formalin	20 ml/l of water (37%)	Apply 2 l/cu.ft. cover for 14 to 36 hr and aerate for at least 14 days.

Care should be taken to disinfect the planting materials before they are moved into the new pots with a fungicide and insecticide combination. Disinfectant solution such as tri sodium phosphate or potassium permanganate placed at the entry of the orchid house would help to get rid off the pathogens from the personnel entering the orchid houses. Whatever potting material or mix choose, it must be wetted before use. Otherwise, it will never hold moisture properly and will always dry out. Here's how can do it:

1. Intend amount of potting material should be pour into a bucket that has about twice the volume of the mixture.
2. Bucket should be filled with hot water. Hot water penetrates the material better than cold water.
3. Let it soak overnight.
4. In the next day, pour out the mix into a colander or strainer.
5. Mix should rinse thoroughly to wash out the dust in it.

Now the mix is ready to use. The above media should have following physical and chemical properties which are distinct from field soils:-

1. An ideal medium should have a good balance between physical properties like water holding capacity and porosity.
2. The medium should be well drained.
3. Medium which is too compact creates problems of drainage and aeration which will lead to poor root growth and may harbour disease causing organisms.
4. Highly porous medium will have low water and nutrient holding capacity, affects the plant growth and development.
5. The media reaction (pH of 5.0 to 7.0 and the soluble salt (EC) level of 0.4 to 1.4 dS/m is optimum for most of the orchids).
6. A low media pH (<5.0) leads to toxicity of micronutrients such as iron, zinc, manganese and copper and deficiency of major and secondary nutrients while a high pH (>7.5) causes deficiency of micronutrients including boron.
7. A low pH of the growth media can be raised to a desired level by using amendments like lime (calcium carbonate) and dolomite (Ca-Mg carbonate) and basic, fertilizers like calcium nitrate, calcium cyanamide, sodium nitrate and potassium nitrate.
8. A high pH of the media can be reduced by amendments like sulphur, gypsum and Epsom salts, acidic fertilizers like urea, ammonium sulphate, ammonium nitrate, mono ammonium phosphate and aqua ammonia and acids like phosphoric and sulphuric acids.

Methods and timing of potting of Ornamentals

It is preferable always to use clean flowerpots. If old pots are reused, be sure that any soil adhering to the inside walls is entirely removed. Dirty containers make re-potting difficult, as the ball of dirt does not slip out readily. Then, too, tender feeding roots resent ancient, caked residue. New clay pots must be soaked in water for at least hours, and a whole week is better; then dry them off before using. Otherwise the thirsty clay insists upon absorbing moisture which is essential in starting newly potted plants into healthy growth. The condition and species of plant will determine the proper time for potting. The best time is just after a plant has started into growth. Then it is easier for the roots to withstand the shock of transplanting. When plants are resting (during the winter, usually), and root action is correspondingly dormant, the vitality of a plant is often too low to supply the energy necessary for roots to take hold of the soil and establish the plant in its new home, so it is apt to languish and die. Most house plants rest during the late autumn and early winter months, awakening into new growth in the spring. March, April, and May, then, are good potting or re-potting months for such plants. The end of a resting period may usually be determined by appearance of new green shoots and a general freshening in the whole appearance of the plant.

Plants potted in the spring should be plunged in the garden, where they enjoy a refreshing and invigorating season before taking up their duties in the house. Place some gravel under each pot and sink it to the brim. The additional drainage will dispose of overflow from exuberant summer showers, and it will also discourage earth worms from crawling into the pot through the bottom hole. The process of summering in the garden applies to most house plants, as well as those just recently potted. A few temperamental species, like the African Violet (*Saintpaulia)*, prefer to remain inside or on a protected porch. Cacti and succulents enjoy the sunniest place in the garden; other plants require light shade and ferns need the cool dimness of rather heavy shade not, however, where they will encounter drip from trees or bushes. It is necessary to leave house plants in their pots during the summer season in the garden. A common tendency, when potting plants, is to use too large a container. This fault is provocative of over watering, sour soil, and an early demise of the plant. Plants that continue to live under these circumstances just about hold their own; they seldom increase in size or show any signs of active growth. Also many species stub bornly refuse to produce flowers unless their roots are some what pot bound. This is commonly the case with Amaryllis, Hippeastrum, Clivia, Agapanthus, snake plant (*Sansetderia*), and jade plant (*Crassula portulacea*). There is nothing to be gained by giving a plant more soil than its roots need or can use. Many varieties, which you will eventually want to grow in house, may be seen in florists' windows or greenhouses. Look closely at them sometime when passing by. Note that four or five sizable cacti are thriving in a 3 inch 4 foot palms in 8 inch pots, and 6 foot rubber plants in 10-inch pots. Observation of the material in these windows will give an excellent idea of the proper size in pots for various plants. If any doubt exists while in the process of potting, under pot, rather than over pot; it is much safer and usually you will find that you really haven't used too small a container, after all.

Another planting must is firm soil about the roots. After placing the drainage layer in a pot, it should be filled about one quarter full with dampened soil and the roots spread out on top of the earth in a natural position not bent or cramped. Then add more potting mixture to within an inch of the container's rim. The whole is firmly compressed about the roots. With small pots, soil may be pressed down with the thumbs. When large pots are used, a section of broom handle makes a good tamping stick. Firm soil does not mean earth pounded down to the consistency of a brick. Such a condition prevents circulation of water and air, and invites curtailment of normal, healthy growth. There is a happy medium, which your own sense of judgment will determine.

Water well after planting; if the specimen is other than a cactus, succulent, or downy leafed plant like *Gloxinia* or *Saintpaulia*, it will also appreciate having its foliage sprayed.

When plants are first brought into the house, they should be kept in dim light no sun, for several days or a while they adjust themselves. This also applies to newly potted or re-potted plants. Water is most essential to plant life, of course, for it is through agency that food in the soil is made available to roots. Yet, without proper drainage, water becomes an enemy rather than a friend; it packs and sours soil, at the same time excluding all air, which is also a requisite for roots. Friable soil and healthy plants are boon companions. Although a drain age hole is incorporated in the bottom of most flower pots, this alone is insufficient. If nothing but soil is placed in the container, water soon washes it into the hole where it forms an impenetrable plug and all drainage stops. The hole is assisted in the performance of its duties by guardian pieces of broken pot, over which is placed a layer of pebbles or gravel. This drainage stratum varies in thickness according to the size the pot; it is always deeper in glazed containers which has no bottom holes.The use of saucers or supplementary vessels is universal. They protect table and shelf tops and serve to catch surplus water vented through drainage holes. It is sound procedure to fill them partly with pebbles. Then no harm can come to the plant if some overflow remains in the saucer, as it very often does, for the bottom of the pot is held above the water. In fact, under these circumstances, surplus water is a decided asset, as it adds to the moisture content of surrounding air. A few pieces of charcoal mixed in with the pebbles will keep saucer and contents clean. The simple steps to follow for potting are as under:

1. The medium is easier to handle if it has been sprinkled with water at the night before actual potting.
2. Remove the plants from the pot.
3. Remove the old, loose, rotted potting materials and any soft, damaged or dead roots.
4. If the roots are healthy, firm and filling the pot, put the plant in a pot just one size larger than the one removed it from, placing the older growth toward the back so the new lead or growth has plenty of room. Old clinging roots are severed from the pot with a sharp knife and the compost, containing the undisturbed plant, is forced from the pot with a screw driver. The roots are shaken free from decomposed material. Dead or long roots are trimmed slightly, since they would be injured in potting. If the roots are rotted and in poor condition, repot the plant in a container of the same or one size smaller than it was removed from. Growing material will stay too damp if placed in too large container, which will result in more of the roots rotting. It is better to add a coarse material like broken clay pots or Styrofoam in the bottom of the pots to improve drainage.
5. Place the plant in the pot so it's at the same depth as it was originally. The new shoot should be level with the pot rim. With the plant held firmly in one hand about one-half inch below the top of the pot, the back of the plant close to the edge of the pot, and the growing end toward the center, potting begins.The potting process for monopodial orchids is very similar to the *Cattleya* process, except that the orchid should be placed in the center of the container, rather than toward the back.
6. Press the fresh potting material into the pot and around the plants roots with thumbs and fore fingers. The plants should be secure in the pot so it doesn't wiggle otherwise, the new roots won't form properly.A piece of *Osmunda* is held at a right angle to the top of the pot with the grain also running at a right angle to the top; with a little flip of the potting tool, so that it never touches roots or plant, the chunk slips neatly in place within the pot, to rest vertically and against the plant.This procedure is repeated, from side to side, until all the gaps are filled, each chunk being made shorter as the edge of the pot is approached. If the plant leans to one side, can be straightened by adding potting mix to prop it up.Plants that are held weakly may wobble and roots may be broken. Firm potting also prevents plants from falling out of dry and shrunken compost and keeps water from settling in holes and rotting the roots. It is wise to spread the roots, packing potting mix among them for aeration.

7. Place a wooden or bamboo stake in the centre of the pot, and tie up the new and old leads with soft string or twist ties.
8. Bottom hole of pot should be enlarged to give the drainage and aeration needed by epiphytes.
9. Some growers use a single piece of crock in the bottom, while others fill as much as one-third of the pot with crockbits of brick, coke, hayditeor gravel may be used instead; in any case, the hole should not be tightly covered.
10. The potting material should be worked in without letting the potting tool touch the roots. Success depends largely on the root being disturbed as little as possible. A further and equally necessary precaution is to take care that the rhizome lies along the top of the potting material and particularly that the dormant eye is not buried.
11. When the pot is full, the stake is inserted at the back of the plant and the bulbs are tied up. They should be tied firmly enough to prevent a movement that will break off roots as they form, but not tight enough to choke the bulb or to force it so suddenly into an upright position that it will snap from the strain.
12. The label is placed toward the back of the pot and the pot is then set aside in a place especially designated for newly potted plants.

Re-potting

Repotting is needed when the plants have over grown the pots or when potting medium has deteriorated. When to re-pot is always a troublesome question to an inexperienced gardener. Often re-potting is resorted to when other conjectures as to what is a certain plant have proved erroneous. Under such conditions, the operation is seldom a success; the difficulty is more likely due to over watering or over-fertilizing. Generally speaking, house plants with exceptions which will be noted do not require larger containers oftener than every two or three years. All members of the amaryllis family and *Sansevieria*, for instance, should be left in the same pot until the roots actually break it; otherwise they will not bloom. Cacti are succulents with sparse root systems also prefer to remain in the same container for a long time; sometimes for years. It is a good plan, however, to renew the top inch of soil in the pot once a year.

Other plants, when plainly root bound, should be transplanted to another container one third larger. In common flower pots this means the next larger size. To find out if a plant is root bound, moisten the soil and gently knock out the earth ball. If the contents of the pot seem to be entirely a mass of tangled, tortuous roots, pushing against the clay walls in mute appeal for more room the plant is root bound. Un pot plants once a year and examine them, whether they need re potting or not. They will welcome having some of the old soil carefully scraped from around the roots and replaced with fresh earth. Also, the drainage layer may be examined at the same time and repairs made if necessary.

Seedlings, cuttings and bulbs of rapid growth like *Caladium* progress at a rate, which necessitates more numerous shifts. They are like children growing out of there shoes every three or four months. Young plants will have to be watched individually there is no rule to govern their transplanting requirements. Roots extending through the drainage hole advise that a larger container is necessary.

An overgrown orchid (*pot-bound*) will dry off much more quickly than one that has plenty of space in the pot. When pot space is limited, there is less potting material to hold onto the water, so the overgrown plant quickly uses it up. When repotting, the roots should be teased apart to remove old potting mix and then dead roots are to be removed. If required a plant may then be divided into two or more plants by twisting and tearing the bulbs apart. For best results each plant should keep as large as possible otherwise flowering would be retarded for 1 to 2 years.

The ideal time to repot most the orchids is when the plant starts new growth, usually right after it flowers. With certain orchids like the *Cattleyas*, swelling at the base of the plant might be seen, which is the beginning of the new lead or shoot that will form the next stem, leaf and flowers .It occur when orchids are putting out new roots. Pot-bound *Cymbidiums* will often be under watered which can reduce the vigor and blooming capability.When a *Cymbidium* fills the pot and starts to grow against or over the edge, it should be repotted even if the mix is not broken down. *Cymbidiums* generally need to be repotted about every three years under normal conditions, whereas, *Vandeceous* orchids and *Paphiopedilum* should not be disturbed unless very necessary. *Cymbidiums* should be repotted just after the old blooms have fallen off (spring season). If the old mix is broken down, the plant should be repotted as soon as possible to minimize root rot. Broken down mix tends to hold too much water and will reduce the air movement around the roots which often causes the older roots to die back. If the mix is very soft, the plant probably needs repotting. Finger may be inserted in the drainhole of a plastic pot to test the condition of the mix. If the mix is soft enough so a finger can enter the mix easily, it should repot. Vigorous plants will often outgrow the pot and may even break the pot before the mix starts to break down.How well a plant tolerates repotting, that can generally be judged by the plant's growth rate, slow growers being less tolerant of disturbance than others.

During repotting large plants may either be divided into muliple plants or kept together and potted up into larger pots for multispike specimens. Plant should take out of the pot and remove the old bark for repotting. Dead or compacted roots from the bottom of the root-ball should be removed. Between plants cuttings tools should be sterilized. The dormant bulbs (back bulbs) can be removed without destroying the strength of the division. These can be potted up to resprout and bloom in two to four years. A pot size which will allow the plant to grow unrestrained for three to four years should be selected. During placing the plant in the pot, bulbs should be positioned so they sit just a little into the surface of medium and so that the newest growth is in the center of the pot, with room to grow. Plants should be potted firmly by pouring the medium around the roots while tapping the sides of the pot and pushing down on the top with thumbs or a potting stick. Most large *Cymbidium* plants have leafless bulbs (backbulbs) toward the center of the plant. The green divisions of the plant will be connected to the backbulbs, and the location of the backbulbs usually marks the natural dividing points for the plant. Overgrown or potbound *Cymbidiums* are often difficult to remove from the pot. With clay or redwood pots, the root ball can be separated from the pot by working a long thin knife down to the bottom of the pot and cutting it away from the pot. Plastic pots will sometimes deform from the pressure of the roots inside and the only way to remove the plant is to cut the pot down one side. If the roots are a pale yellow or white color and firm, the root system is in good health. Brown and soft roots indicate the plant has either been in the pot too long or it has been kept too wet and the roots have rotted as a result. Some loss of roots is normal, but the loss of too many will slow down the growth of the plant.Once, the individual sections are separated, clean as much of the old mix and dead or broken roots from the rootball as is practical without damaging the good roots and the new roots. Be careful with new growths or spikes which may be emerging from the base of the psuedobulbs. Some root pruning may be necessary to fit the plant easily into the new pot. This does not hurt the plant, and will encourage new root growth. A mound of moist potting mix should be placed in the centre of the pot. Roots should be spread over the mound of mix and work mix in among the roots. Roots should not be balled in the centre of the pot. Pot should be filled with mix and should be shaked or pat the pot vigorously or tap the pot firmly on the potting bench to help the mix settle into the roots. Tamp the mix in around the edge of the pot firmly with the potting stick. The mix should tight in the pot, not loose. Root growth in the pot will usually fill some, if not all, of this gap in time. The last step in potting the division is to put a tag in the pot to identify the plant.

CHAPTER 14

Bonsai Management

Bonsai is a Japanese art form using miniature trees grown in containers. The first documented evidence of an oriental word for container planting emerged approximately 1600 years ago in China. Bonsai is a Japanese pronunciation of the earlier Chinese term pen *zai*. A bon is a tray-like pot typically used in bonsai culture. The purposes of bonsai are primarily contemplation (for the viewer) and the pleasant exercise of effort and ingenuity (for the grower). Bonsai is not intended for production of food, for medicine, or for creating yard-sized or park-sized gardens or landscapes. Instead, bonsai practice focuses on long-term cultivation and shaping of one or more small trees in a single container.

The practice of bonsai is sometimes confused with dwarfing, but dwarfing more accurately refers to research and creation of plant cultivars that are permanent, genetic miniatures of existing species. Bonsai does not require genetically dwarfed trees, but rather depends on growing small trees from regular stock and seeds. Bonsai uses cultivation techniques like pruning, root reduction, potting, defoliation and grafting to produce small trees that mimic the shape and style of mature, full-sized trees.

Bonsai can be created from nearly any perennial woody-stemmed tree or shrub species. Which produces true branches and can be cultivated to remain small through pot Some species are popular as bonsai material because they have characteristics, such as small leaves or needles that make them appropriate for the compact visual scope of bonsai. Some of the plants highly suited for making bonsai are as in Table 14.1.

Table 14.1 : Plants suitable for Bonsai

Botanical Name	Common Name
Abies	Fir
Acer buergerianum	Trident Maple
Acer campestre	Field Maple, Hedge Maple
Acer circinatum	Vine Maple
Acer ginnala	Amur Maple
Acer monspessulanum	Montpelier Maple
Acer palmatum	Japanese Maple
Acer rubrum	Red Maple
Adenium obesum	Desert rose
Alnus	Alder
Amelanchier canadensis	Amelanchier
Arbutus	Strawberry tree
Berberis	Barberry
Betula	Birch
Betula nigra	River Birch
Bougainvillea	Bougainvillea
Buxus	Box, Boxwood
Camellia	Camellia
Carmona microphylla	Fukien tea tree, *Ehretia microphylla*
Carpinus	Hornbeam
Cedrus	Cedar
Cedrus libani	Lebanon Cedar, Blue Atlas Cedar
Celtis sinensis	Chinese Hackberry
Chaenomeles	Flowering Quince
Chamaecyparis	Cypress
Chamaecyparis obtusa	Hinoki Cypress
Chamaecyparis pisifera	Sawara Cypress
Cissus antartica	Australian grape vine
Cornus	Dogwood
Cotinus coggygria	Smoke Tree
Cotoneaster	Cotoneaster
Crassula	Jades

Crataegus	Hawthorn
Cryptomeria	Sugi
Cupressus	Cupressus
Cydonia oblonga	Common Quince
Dasiphora fruticosa	Shrubby Cinquefoil
Elaeagnus	Elaeagnus
Enkianthus	Enkianthus
Eugenia	Brush cherries
Euphorbia balsamifera	
Euonymus	Spindle
Fagus	Beech
Ficus benjamina	Weeping Fig
Ficus microcarpa	Chinese Banyan Fig
Ficus neriifolia	Willow-leaved Fig
Ficus retusa	
Ficus rubiginosa	Port Jackson Fig
Fraxinus	Ash
Gardenia	Gardenia
Ginkgo biloba	Ginkgo
Hedera helix	Ivy
Gledista triacanthos	Honey Locust
Hibiscus syriacus	Hibiscus
Ilex	Holly
Jacaranda mimosifolia	Blue jacaranda
Jasminum nudiflorum	Winter Jasmine
Juniperus procumbens	Dwarf Japanese Garden Juniper
Juniperus chinensis	Chinese Juniper, Shimpaku
Juniperus squamata	Blue Juniper
Juniperus virginiana	Eastern Juniper
Lagerstroemia	Crape myrtle
Larix	Japanese Larch, American Larch, Tamarack
Ligustrum	Privet
Liquidambar	Sweetgum
Lonicera	Shrubby Honeysuckles
Magnolia stellata	Star Magnolia
Malpighia	
Malus	Apple/Crabapple
Metasequoia	Dawn Redwood
Murraya paniculata	Orange Jasmine, Satinwood

Myrciaria cauliflora	Jaboticaba
Myrtus	Myrtle
Nandina domestica	Sacred or Heavenly Bamboo
Nashia inaguensis	Bahama berry
Neea buxifolia	
Nothofagus	Southern beeches
Olea europaea	Olive, European olive
Parthenocissus	Creeper, Boston Ivy
Phyllostachys	Bamboo
Picea	Spruce
Pieris	Andromeda
Pinus clausa	Sand Pine
Pinus mugo	Mugo Pine, Mountain Pine
Pinus parviflora	Japanese White Pine
Pinus thunbergii	Japanese Black Pine
Pinus virginiana	Virginia Pine
Pinus ponderosa	Western Yellow Pine
Pittosporum	Pittosporum
Podocarpus	Podocarpus
Portulacaria afra	Dwarf jade, elephant food, elephant bush
Prunus serrulata	Japanese Flowering Cherry
Prunus mume	Flowering Apricot
Pseudosasa	Bamboo
Punica granatum	Pomegranate
Pyracantha	Firethorn
Quercus	Oak
Rhaphiolepis indica	Indian Hawthorn
Rhododendron indicum	Satsuki azalea
Rhododendron obtusum amoenum	Kurume azalea
Robinia pseudoacacia	Black Locust
Sageretia	Chinese Sweet Plum, Pauper's Tea
Sasa	Bamboo
Schefflera actinophylla	
Serissa foetida	Snow Rose, Japanese Boxthorn
Sorbus	Rowan and Whitebeam
Syzygium	Brush cherry
Syzygium smithii(=*Acmena smithii*)	Brush cherry
Tamarix	Tamarisk
Taxodium ascendens	Pondcypress

Taxodium distichum	Baldcy press
Taxus	Yew
Thymus	Thyme
Tsuga	Hemlock
Ulmus alata	Winged Elm
Ulmus crassifolia	Cedar Elm
Ulmus minor	Field Elm
Ulmus parvifolia	Chinese Elm
Wisteria floribunda	Japanese Wisteria
Wisteria sinensis	Chinese Wisteria
Zelkova serrata	Japanese Elm
Zanthoxylum	Chinese Pepper

Common styles of Bonsai

The most common styles include formal upright, informal upright, slanting, semi-cascade, cascade, raft, literati and group/forest. Less common forms include windswept, weeping, split-trunk, and drift wood styles.

1. The formal upright style, or *Chokkan*, is characterized by a straight, upright, tapering trunk. Branches progress regularly from the thickest and broadest at the bottom to the finest and shortest at the top.

Fig. 14.1 : Upright/chokkan styles

2. The trunk and branches of the informal upright style, or *Moyogi* incorporate visible curves, but the apex of the informal upright is always located directly above the trunk's entry into the soil line. Similar to the formal upright style, branches generally progress regularly from largest at the bottom to smallest at the top, although this progression may be

broken where the irregular shape of the trunk would make a branch abnormally prominent or obscure.

Fig. 14.2 : Informal upright style or *Moyogi* style

3. Slant-style, or *Shakan*, bonsai possess straight trunks like those of bonsai grown in the formal upright style. However, the slant style trunk emerges from the soil at an angle and the apex of the bonsai will be located to the left or right of the root base.

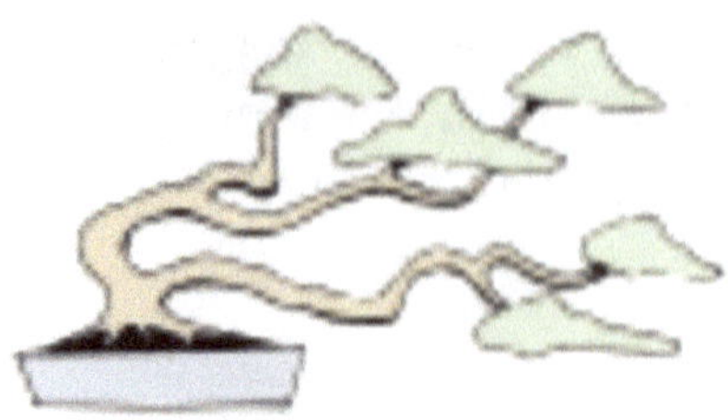

Fig. 14.3 : Slanted or *Shakan* style

4. Cascade-style, or *Kengai*, bonsai are modelled after trees which grow over water or on the sides of mountains. The apex, or tip of the tree in the *Semi-cascade-style*, or *Han Kengai*, bonsai extend just at or beneath the lip of the bonsai pot; the apex of a (full) cascade style falls below the base of the pot.

Fig. 14.4 : Cascade or *Kengai* style & Semi-cascade or *Han Kengai* style

5. Raft-style, or *Netsuranari*, bonsai mimic a natural phenomenon that occurs when a tree topples onto its side, for example, from erosion or another natural force. Branches along the top side of the trunk continue to grow as a group of new trunks. Sometimes, roots will develop from buried portions of the trunk. Raft-style bonsai can have sinuous, straight-line, or slanting trunks, all giving the illusion that they are a group of separate trees—while actually being the branches of a tree planted on its side.

6. The literati style, or *Bunjin-gi*, bonsai is characterized by a generally bare trunk line, with branches reduced to a minimum, and typically placed the top of a long, often contorted trunk. This style derives its name from the Chinese literati who created Chinese brush paintings.

Fig. 14.5 : Literati or *Bunjin-gi* style

7. The group or forest style, or *Yose Ue*, comprises a planting of several or many trees and typically an odd number, in a bonsai pot. The trees are usually the same species, with a variety of heights employed to add visual interest and to reflect the age differences encountered in mature forests.

8. The broom style, or *Hokidachi*, is employed for trees with extensive, fine branching, often with species like elms. The trunk is straight and upright. It branches out in all directions about 1/3 of the way up the entire height of the tree. The branches and leaves form a ball-shaped crown which can also be very beautiful during the winter months.

Fig. 14.6 : Broom or *Hokidachi* style

9. The multi-trunk style, or *Ikadabuki,* has all the trunks growing out of one spot with one root system and it actually is one single tree. Its counterpart in nature is the tree clump formed, for example, where a single pine cone has sprouted a number of seedlings in one spot. All the trunks combine to support one crown of leaves, in which the thickest and most developed trunk forms the top

Fig. 14.7 : Multi-trunk or *Ikadabuki*style

10. The *Shari style,* or *Sharimiki,* style involves portraying a tree in its struggle to live while a significant part of its trunk is bare of bark. In nature, trees in the Sharimiki style are created by lightning or animals eating the bark.
11. The *root-over-rock style,* or *Sekijoju,* is a style in which the roots of the tree are wrapped around a rock. The rock is at the base of the trunk, with the roots exposed to varying degrees as they traverse the rock and then descend into the soil below.
12. The *growing-in-a-rock,* or *Ishizuke,* style means the roots of the tree are growing in soil contained within the cracks and holes of the rock. The rock may serve as a simple container, with the tree escaping the container and forming its own shape. Alternatively, the tree may show a definite relationship to the rock's shape, growing close to the rock and following its contours.

Techniques for Bonsai making

The practice of bonsai development incorporates a number of techniques either unique to bonsai or if used in other forms of cultivation, applied in unusual ways that are particularly suitable to the bonsai domain. These techniques include:

- *Leaf trimming,* the selective removal of leaves (for most varieties of deciduous tree) or needles (for coniferous trees and some others) from a bonsai's trunk and branches.

- *Pruning* the trunk, branches, and roots of the candidate tree.
- *Wiring* branches and trunks allows the bonsai designer to create the desired general form and make detailed branch and leaf placements.
- *Clamping* using mechanical devices for shaping trunks and branches.
- *Grafting* new growing material (typically a bud, branch, or root) into a prepared area on the trunk or under the bark of the tree.
- *Defoliation*, which can provide short-term dwarfing of foliage for certain deciduous species.
- Deadwood bonsai techniques called *jin* and *shari* simulate age and maturity in a bonsai.

Most of these techniques find their origin in Japan, where the art of bonsai got its modern day form. These are very useful in giving bonsai a look of old age. Although plant could be only 10 years old, but it will look like a much older plant.

1. Jinning

This is technique which is used for recreating an image of natural damage and wear on a plant. Jin is a bonsai deadwood technique used on branches or the top of the trunk (the leader). A jin is meant to show age, or show that the tree has had a struggle to survive. Trees might have their branches broken during storms, or bigger branches might block the inner branches from receiving enough sunlight, making these branches die back. A jin requires the complete removal of bark from a given start point to the end of the branch or leader. The remaining wood dies and dries out to form the jin.Many bonsai artists take clue from nature, to create an impression of old age. To practice this on our plant, we need to carefully cut off a branch by making an incision into the branch using a branch cutter, and then use pliers to twist the branch of the plant. After removing the branch apply a good quality wood putty to seal of the exposed wood. This will protect the exposed part from fungal and insect attack.

Creating a jin from the leader (a top jin) can produce a shorter, more visibly tapered bonsai in a single step. The change in proportion can greatly improve the illusion of age in the bonsai specimen. Removing the active leader distributes vigour to lower branches, which will grow more quickly and help increase trunk diameter, reinforcing the illusion of age. A top jin also provides an aesthetic solution for a tree with two leaders, an unaesthetic shape that the designer can modify by turning one of the two into a jin. When used on branches, the jin technique allows the grower to remove some of a bonsai's unwanted branches while increasing the illusion of age. A remaining jin may be long, with a pleasing shape bent or carved into it, or short, like the dead remains of a branch broken off near the trunk.

2. Shari

A *shari* is deadwood on the main trunk of the bonsai.A small shari usually runs vertically on or near the front of the trunk - shari have little aesthetic value at the rear of the trunk, where they are rarely viewed and are obscured by branch growth. In this technique we remove the bark from one side of the plant. This recreates the effect of bark die back which is usually found on old trees. To make a shari on bonsai, first make an outline on the plant surface, clearly demarcating the area of the bark want to cut. Then use a sharp edged knife to cut through the bark along the contours of the line which have already marked. After we have run through the outline, peel of the bark exposing live wood. Apply a good quality cut paste to seal the exposed parts.

The shallow wound exposes a non-living portion of the trunk, which will be surrounded by living bark. The natural causes of shari include a falling branch that has ripped bark from the trunk below it, lightning damage, or trunk damage from another external source. Shari may occur naturally on a bonsai, or may be created by carving the bark.

If the amount of dead trunk (and possibly dead branches situated on the dead trunk area) is large enough, the bonsai is said to be in the *sharamiki* or driftwood style, because a large portion of the tree has the silvery, weathered look of driftwood on a beach, or ancient tree remnants on a harsh mountain landscape.

3. Nebari

The technique of promoting surface root formation, in a circular manner around the base of the plant is called as nebari. Many plants growing in the wild have such root padding and this makes them look very mature and old, many bonsai artists use this technique to give a old age effect to their bonsai plants.

4. Uro

Many a times trees growing in the wild tend to have a hole in their trunk due to rot in the wood, or some other reasons, this make the trees look very old and mature. This is one of the bonsai techniques adopted by artists to recreate the impression of old age. A uro can be seen close to the bottom of the trunk of this bonsai. While jin appear natural on coniferous bonsai, they do not look as appropriate on most deciduous and broadleaf species. In these species, dead branches generally rot and fall off the tree. A small indentation is left where the branch used to be and new wood grows around it forming a small hollow. Bonsai gardeners replicate this hollow as a*uro* by making a small, irregularly-shaped wound in the trunk. For example, when removing a branch from a deciduous or broadleaf species, bonsai growers often make a uro to avoid having an ugly wound healing slowly and scarring without control.

5. Carving

Another popular bonsai technique is carving. Use good quality tools like chisel and grinders and cutters to help you in carving out the bark from the surface of your tree.

Principles in Bonsai Aesthetics

Some key principles in bonsai aesthetics include:

1. *Miniaturization*: By definition, a bonsai is a tree kept small enough to be container-grown while otherwise fostered to have a mature appearance.
2. *Proportion among elements*: The most prized proportions mimic those of a full-grown tree as closely as possible. Small trees with large leaves or needles are out of proportion and are avoided, as is a thin trunk with thick branches.
3. *Asymmetry*: Bonsai aesthetics discourage strict radial or bilateral symmetry in branch and root placement.
4. *No trace of the artist*: The designer touch must not be apparent to the viewer. If a branch is removed in shaping the tree, the scar will be concealed. Similarly, wiring should be removed or at least concealed when the bonsai is shown and must leave no permanent marks on the branch or bark.
5. *Poignancy*: Many of the formal rules of bonsai help the grower create a tree that expresses Wabi-sabi, or portrays an aspect of *mono no aware*. Characteristics of the wabi-sabi aesthetic include asymmetry, asperity, simplicity, economy, austerity, modesty, intimacy and appreciation of the ingenuous integrity of natural objects and processes.The words *wabi* and *sabi* do not translate easily. *Wabi* originally referred to the loneliness of living in nature, remote from society;*sabi* meant "chill", "lean" or "withered".

Containers for Bonsai

A variety of informal containers may house the bonsai during its development and even trees that have been formally planted in a bonsai pot may be returned to growing boxes from time to time. A training box will have a single tree and a smaller volume of soil that helps condition the tree to the eventual size and shape of the formal bonsai container.

Completed trees are grown in formal bonsai containers. These containers are usually ceramic pots, which come in a variety of shapes and colours and may be glazed or unglazed. Unlike many common plant containers, bonsai pots have drainage holes in the bottom surface to complement fast-draining bonsai soil, allowing excess water to escape the pot. Growers cover the holes with a screening to prevent soil from falling out and to hinder pests from entering the pots from below. Pots usually have vertical sides, so that the tree's root mass can easily be removed for inspection, pruning, and replanting, although this is a practical consideration and other container shapes are acceptable.Containers with straight sides and sharp corners are generally used for formally-shaped plants, while oval or round containers are used for plants with informal designs.

Table 14.2 : Common names for bonsai size classes

Large bonsai		
Common name	**Size class**	**Dimensions**
Imperial bonsai	Eight-handed	60–80 in (152–203 cm)
Hachi-uye	Six-handed	40–60 in (102–152 cm)
Dai	Four-handed	30–48 in (76–122 cm)
Omono	Four-handed	30–48 in (76–122 cm)
Medium-sized bonsai		
Common name	**Size class**	**Dimensions**
Chiu	Two-handed	16–36 in (41–91 cm)
Chumono	Two-handed	16–36 in (41–91 cm)
Katade-mochi	One-handed	10–18 in (25–46 cm)
Miniature bonsai		
Common name	**Size class**	**Dimensions**
Komono	One-handed	6–10 in (15–25 cm)
Mame	One-handed	5–8 in (13–20 cm)
Shohin	Palm size	2–6 in (5–15 cm)
Shito	Finger tip size	2–4 in (5–10 cm)
Keshitsubo	Poppy-seed size	1–3 in (3–8 cm)

Great variation exits on the sizes or names bonsai containers. In the very largest size range, a recognized Japanese practice is to name the trees "two-handed", "four-handed", and so on, based on the number of men required to move the tree and pot. These trees will have dozens of branches and can closely simulate a full-sized tree. The very largest size, called "imperial", is named after the enormous potted trees of Japan's Imperial Palace. At the other end of the size spectrum, there are a number of specific techniques and styles associated solely with the smallest common sizes, *shohin* and *shito*. These techniques take advantage of the bonsai's minute dimensions and compensate for the limited number of branches and leaves that can appear on a tree this small.

Materials needed for Bonsai

- *Bonsai Soil*: there are many kinds of it and learning how to mix the bonsai soil can be a little complicated for the beginners. If you are converting your plant into bonsai style, you can use the soil the plant grew up in.
- *Coarse soil*: used to prevent the main soil inside the pot from coming out of the bottom and helps the water to drain. (only needed depending on size and shape of the pot)
- *Wires*: there are two basic wires for bonsai. Aluminum and Copper. Aluminum wire is softer and easier to use.
- *Scissors*: We can use the one we have at home.
- *Pliers/Nippers*: Needs it only to cut and bend the wires.
- *Stick*: something like chopstick.
- *Net*: to cover the hole in the pot.
- *Pot*: any pots are fine. We can use a plate, flat plastic lid of something
- Gloves and newspaper to keep the mess out of hands and house.
- *Green moss*: to make your bonsai have natural feeling. I don't know why bonsai looks really good with it!
- *The plant*: the one you like to have as bonsai style. Without this you can't make anything

Steps of Bonsai Making

Bonsai literally means cultivating trees in a tray. Anyone can make a bonsai. Here are some helpful tips on how you can make a bonsai successfully:

1. Choose a tree species that you want to grow as a bonsai.
2. Know the tree species you have chosen to be propagated as a miniature tree. To make your own bonsai, it is very necessary that you determine what are the specific needs and requirement of the tree species. There are trees that require exposure to direct sunlight and others cannot survive under too much sun exposure. There are also trees that cannot tolerate too much water while other species can.

3. Prepare the tree species you want to grow as a bonsai. You can start to make your bonsai from seeds, layering, grafts, collecting from the wild, or you can purchase a small tree species from a nursery stock or from commercial bonsai growers.
4. Select a good pot for your bonsai. You should make use of a bonsai pot that has drainage holes beneath it. The drainage holes will facilitate draining out of excess water inside the pot. If water cannot be drained out from the bonsai pot, it can lead to water congestion, which will rot the entire root system. Root rotting will eventually lead to the weakness of your bonsai tree.
5. Then, prepare a good soil mixture to your tree. The soil should contain fast-draining components. The soil should be composed of inorganic and organic materials. Organic materials will hold enough moisture for the growing bonsai tree, and as they decay they can provide essential nutrients for the bonsai. Inorganic materials will provide mechanical support for the root system while it grows.
6. You should provide sufficient amounts of water to your miniature tree. You can use a water container that has fine nozzle so that when you pour water, the soil placement will not be disrupted or washing out of the soil will be prevented. Ideally, watering is done early in the morning so that when night comes the soil has already dried up. If you water late in the afternoon, the soil will be moist during night time, and this can lead to the development of fungal infection.
7. Expose your miniature tree to a morning sunshine because its heat is not that intense, while a direct afternoon sunlight is too strong that it can cause burning of its leaves. Sunshine is very necessary because this will help your growing bonsai for the production of their own food, which is important for their growth and survival.
8. Support the growth of your little tree by providing enough fertilizer. You use a high quality fertilizer and this should be given during the season of growth. The fertilizer will aid the proper growth and development of your bonsai tree.

Care and maintenance of Bonsai

Small trees grown in containers, like bonsai, require specialized care. Unlike houseplants and other subjects of container gardening, tree species in the wild generally grow roots up to several meters long and root structures encompassing several thousand liters of soil. In contrast, a typical bonsai container is under 25 centimeters in its largest dimension and 2 to 10 liters in volume. Branch and leaf (or needle) growth in trees is also large-scale in nature.

Wild trees typically grow 5 meters or taller when mature, while the largest bonsai rarely exceed 1 meter and most specimens are significantly smaller. These size differences affect maturation, transpiration, nutrition, pest resistance, and many other aspects of tree biology. Maintaining the long-term health of a tree in a container requires some specialized care techniques:

1. Roots

The roots along with the trunk are fundamental to a bonsai plant. They are the starting point of making a quality bonsai. In root training, a surface rootstructure, called nebari, is important to demonstrating the plant's natural beauty. You need to develop a quality nebari. What is more important than aquality nebari is root pruning, which is intimately involved in the training of the whole bonsai plant.

2. Trunk

The trunk of a bonsai tree is the most appealing part to create the illusion of age. Therefore, as a first step, you should start with developing a well-formed trunk such as good taper (*kokejun*), initial rise (*tachiagari*), smooth curves, etc. The other design elements, including branch location, foliage distribution, leaf reduction and so on, can be established later on in the design process.

3. Wiring

Wiring is a method of bending trunks and branches using wires in order to achieve various impressive shapes. When the plant is still very young, the trunk is bent into the basic shape of a tree. Later on, as the plant grows, you bend the first branch (*ichi-no-eda*), the second branch (*ni-no-eda*), the third branch (*san-no-eda*) and so on until the final branch that determines the shape of the bonsai plant. If you cannot achieve the ideal shape with one wiring, you can do so in stages. By the time you complete the bonsai tree's final shape, you can finesse the details of the small branches with wires.

4. Pruning

Once you have completed the work of tapering (*kokejun*), initial rise (*tachiagari*) and wiring, you can begin the final shaping of the plant. Pruning reduces the volume of the tree in order to achieve the goal of "*keisho-sodai*", literally small size-great similarity. You need to prune the branches, the buds and the leaves.

5. Repotting

Plants that grow in pots must be repotted from time to time. The growing roots of a plant will occupy the limited space in a pot and thereby reduce the free space for the plant to develop. In addition, as time goes by, the soil in the pot will become hard and less permeable for air or water. These conditions will lead to poor health of the plant and the plant may even wither and die. So, it is necessary to repot a bonsai plant. Moreover, repotting provides a good opportunity to prune the roots and shape the plant.

CHAPTER 15

Flower Arrangement

Ikebana

Flowers are generally known as the symbol of beauty around the world. Many people stick flowers in vases to wreath their rooms, houses and other places. Japanese FlowerArrangement is essentially different from the Western style of flower arrangement. The western style involves symmetrical arrangement of flowering plants in a vase whereas, the Ikebana, which means "flowers kept alive", is much more intricate, complex and asymmetric. This style of flower arrangement is one of the famous cultural traditions of Japan known around the world. Generally, all Japanese flower arrangement styles are lumped together as *ikebana* but we cannot say they are all the same. Ikebana is a disciplined art form in which nature and humanity are brought together. Though it is a creative expression, it has certain rules governing its form. The main rule is that all the elements used in construction must be organic, be they branches, leaves, grasses, or flowers. The artist's intention behind each arrangement is shown through a piece's colour combinations, natural shapes, graceful lines and the usually implied meaning of the arrangement. They arrange flowers to recreate nature in miniature to see individual flowers which have been left behind. *Ikebana* is like one of the ways to communicate with nature. Another aspect present in ikebana is its employment of minimalism. That is, an arrangement may consist of only a minimal number of blooms interspersed among stalks and leaves. The structure of a Japanese flower arrangement is based on a scalene triangle delineated by three main points, usually twigs, considered in some schools to symbolize heaven, earth and man and in others sun, moon, love & earth. The container is also a key element of the composition and various styles of pottery may be used in their construction. Over time, the ways and styles have changed but there is still the basic idea of *ikebana* remain same.

Ikebana and Western flower arrangement

1. Symmetry and Asymmetry in Flower Arrangement

One of the most striking differences between Ikebana and Western style flower arrangement is in their shape. While Ikebana is thoroughly asymmetric, Western flowerarrangement stresses symmetry. The contrast is obvious, but the origin and significance of this distinctiveness is not clear.

2. The Ideal Garden

Ikebana, try to mirror the whole of nature incorporating mountain, river, sea, forest and field in to its composition. It endeavours to recreate the harmonious, yet entangled and mysterious world of nature. Artistic appreciation of gardens (Western flower arrangements) in Europe tends to emphasize order and refinement and an untended garden may well give cause for fear. Gardens are usually well ordered, symmetrical and may even be described as disciplined. Colourful flowers in full bloom in perfect array are the highlights of such gardens.

3. Materials

As the final shape of a flower arrangement is necessarily closely connected with the materials from which it is made, it is important to consider the materials used. One of the characteristics of Ikebana is the use of branches and twigs, in addition to flowers. This usage reflects the appreciation for trees in Japanese culture. Trees are paradoxically symbolic of both age and death, as well as of evergreen life. The great symbolic wealth attached to trees and so to their twigs and branches, is certainly a key reason for their use in Ikebana and its attempt to represent the condensed cosmos.

In Western flower arrangement, flowers in bud or in full bloom are the key components and so the youth, love, life and beauty that they symbolize become the theme of the total arrangement.

4. Cycle of Life and Death

The composite Japanese religious context and the aesthetic sense that this has given birth to, have moulded Ikebana during its development.

Certainly the Buddhist concept of the cycle of life, leading from life through death to life again repeatedly, influenced the introduction of 'dead' branches into Ikebana. Ikebana attempts to give artistic expression to this fundamental Buddhist concept. An old twisted limb of a pine tree, a moss covered branch, a spray of maple leaves, a stalk of bamboo or a dried lump of drift wood are all used in Ikebana and are representative of both trees and their symbolic connection with the cycle of life. The art of Ikebana is to recreate a microcosmos, where all of nature coexists in harmony. This unity requires the use of part of a tree to symbolize the cycle of life and death, an essential element in the Buddhist cosmos.

Life is emphasised and celebrated in Western style flower displays, because the flowers that symbolize life, love and beauty are mainly used. Every human being has to go through death to be reborn once and for all. In this created order, life and death are not part of a cycle, but form part of a linear progression that leads us beyond this world. Death does not lead us back to this world, so death does not become the complementary counterpart of life in the great cycle as perceived in Buddhism. Flowers in full bloom signify the full life that God has granted to all creatures and the praise that the creator is due from creation.

Important Features of Ikebana

Ikebana is the time-honoured Japanese art of flower arrangement. This art has been practiced in Japan for more than seven centuries and it shows no sign of losing its appeal.

- Ikebana is not a skill that can be learned overnight. For most, it takes 3-5 years of formal study and practice in order to become proficient at the art.
- Ikebana artists are skilled in all the intricacies of this ancient art. However, you don't have to be a certified artist in order to incorporate Ikebana design elements and techniques into your own flower arrangement designs.
- Ikebana is an art that is practiced by both men and women - in the past; many Samurai practiced this well-respected art.
- The key to Ikebana arrangements is that they must be harmonious with the environment in which they exist. To this goal, Ikebana stresses that indigenous, in-season flowers, should always be used in any flower arrangement so that the arrangement is in harmony with nature. Ikebana also adheres to the idea that the beauty of the individual flowers is paramount. Stems and leaves should only be used when such use would serve to highlight the beauty of the flowers.

Schools of Ikebana

There are many schools, of which the most popular are Ikenobo, Sogetsu and Ohara. Styles in Ikebana depend on the school, the vase and the flowers and plants used. The most popular schools are the following:

1. Ikenobo

The oldest school of Ikebana is Ikenobo and it was founded by the Buddhist priest Ikenobo Senkei in the 15th century. Ikenobo Senkei is often thought to be the creator of the rikka or standing flowers style. This style represents Buddhist expression of the beauty of nature. The school is located in Rokkakudo temple in Kyoto and the present 45th-generation head of the school is Ikenobo Sen'ei. Initially the style of flowering associated with the school of Ikenobo became more and more formal amongst the priests and the elites. But then, during the late 17th century the merchant class practiced a much less complicated style, which is known as seika or shoka.There is one famous episode which illustrates the idea of *ikebana.* It is a story of Rikyu (famous tea master in the *Azuchi* period) and Hideyoshi (the ruler who controlled Japan during the *Azuchi* period). One day, Hideyoshi heard that Rikyu's garden had beautiful *morning glories* which covered the whole garden. He asked Rikyu to have a tea ceremony while looking at those beautiful flowers. Rikyu accepted the order and told him the date to come but Hideyoshi arrived to find all of the flowers cut down and gone. Hideyoshi got angry, but when he entered the teahouse, there was a beautiful flower arranged on the dim floor. This single blossom surprised Hideyoshi and reminded him of hundreds of other *morning glories* which had been cut down. This was not just a single blossom, but one that reminds us the existence of hundreds of others. In this way, Rikyu attempted to make invisible flowers real through the visible.

The flower arrangement got popular with common people in the *Muromachi* period. First, people tried to arrange them for the interior decoration of their house by using beautiful flowers but they also arranged them to show something. At that time, there were many wars. So, many rulers had their own areas, fighting each other to become the leader of Japan. People were tired of those wars and wanted a peaceful life. So, they arranged flowers to show the peaceful life in their design. *Ikenobo* accomplished this new style of flower arrangement. Predecessors of *ikenobo* respected the idea of peace and tried to express peace with flower arrangement. There are many other styles in *ikenobo* flower arrangement but this idea is the basis of them all. They thought that trying to think about the appearances of field flowers and greens is a respectable action. The most popular style under this school is the rikka. This is the standing flower style, which was made as an expression of nature's beauty. This is arranged in a formalized and strict way following its seven elements to represent, among other subjects, valleys, hills and waterfalls. Other styles from this school take intoconsideration the spiritual concepts of heaven, earth and man.

2. Sogetsu

In 1927, Teshigahara Sofu founded this school. This school broke new grounds in Ikebana by incorporating materials such as plastic, plaster and steel.Influence from the artistic movements of the early 20th century led to the development of jiyuka (free-style) arrangement. This was the time when the interest in Ikebana spread across Japan. It attracted all kinds of people from all walks of life. Sogetsu instituted the most modern style, called zen-eibana. This style uses new materials such as steel and plastic. This is a great deviation from the traditional use of flowers and plants.As time went by, flowers were not only a religious offering, but people also began to arrange them for decorating rooms or personal appreciation. Then, people who are good at flower arrangement came out to the public places.

3. Ohara

The first modern school of Ikebana was established in Japan when Ohara Unshin left the Ikenobo school.This is a splinter school of Ikenobo formed in the last part of the Nineteenth Century when Ohara Unshin departed the Ikenobo school.This school uses moribana or pilled-up flowers in flat containers. The style of the school reflects Western influence as the moribana style involves use of Western plants. Its main style is the moribana or the piling up of flowers in a flat and shallow vase. This school has Western influences, since it was formed when Western customs were heavily prominent in Japan. So they allow the use of Western flowers and plants. This school also developed free-style Ikebana called the jiyuka. However, they still maintained formality in their practice.

Styles of Ikebana

There are different styles depending on the school and the plants and vase used. In the late 17th century, the growing merchant class developed a simpler style, called **seika** or **shoka**. Shoka uses only three main branches, known as **ten** (heaven), **chi** (earth) and **jin** (man) and is designed to show the beauty of the plant itself.Flowers can be arranged in several styles. These include Heika, Nageire (flung or thrown-in flowers), Hanaisho, Morimono, and Moribana (piled-up flowers). However,it can be roughly divided into two styles - the moribana shallow vase style and the nageire tall vase style:

1. Heika

Heika is a basic Ikebana style that incorporates the design styles of Rikka (standing flowers), Shoka (living flowers), and Seika. In the Heika style, the emphasis is on strong vertical lines. Tall, narrow, elegant vases are often used to draw the eye upward toward the flower arrangement at the top.

- The Heika style is much formalized and has very explicit rules that must be followed in order to conform to the demands of the style.
- Rikka is one of the earliest forms of Ikebana. It also has the strictest rules.
- Shoka and Seika have less rigid rules.

2. Rikka

Originated from a style that was used in temples and ceremonies. Now a day this style is not popular. Rikka style was established in the Muromachi period, in the 15thcentury. It is said to be the foundation of Ikebana, from which Shoka, Moribana and Nageire styles evolved. There are two varieties of Rikka, the traditional Rikka Shofutai and the more modern, Rikka Shimputai.

Rikka Shofutai

Rikka Shofutai is the most formal style of Ikebana and has a distinctive character, which expresses the beauty of natural landscape. This classical type of Rikka follows strict rules and uses special techniques such as wiring. It consists of nine main stems or parts applying different characters, length, angles, and directions of materials to construct a harmonious, scenic natural landscape. These nine main stems are Shin, Shoshin, Soe, Uke, Nagashi, Mikoshi, Hikae, Do, Maeoki. These are also known as *yakueda*.Each yakueda has a particular function and point of departure from an imagined vertical line running through the center of the arrangement. This point of departure is called *de*. Each yakueda also has a particular insertion position on the kenzan (pin holder). The height, length and position of each yakueda respond to or support another. The harmony of height, width and depth is one of the most important factors to consider. Apart from these nine main stems, additional stems, called *ashirai* are also required for completion. The vase in Rikka usually is 20-30 cm in height and should open out at the top. The stems should appear as a clean, single, vertical line rising from the centre of the kenzan. This is called *Mizugiwa,* literally translated as water's edge.

Rikka Shimputai

This modern Rikka style allows more of the arranger's expression, thoughts and creativity because it is released from the rigid rules and forms seen in Shofutai. For example, Rikka Shimputai does not have fixed points of departure.Instead of having nine main stems seen as in Shofutai, two contrasting main parts construct Rikka Shimputai, these are known as *Shu* and *Yo*. Usually two or three sets of Shu and Yo with ashirai are used in a single arrangement. The important factors to consider are contrasting materials, colour, quantity and texture of the materials. The distinguishing characteristics of Rikka Shimputai are brightness, sharpness and striking or conspicuous appearance. These are similar to Shoka Shimputai, although Rikka Shimputai has a more relaxed feeling and fresh appearance.

2. Nageire (slanted style)

Modern style of flower arrangement that became popular when new flowers were introduced from the West. Nageire and Moribana share the emphasis on the natural beauty and characteristics of flowers and plants with the more traditional Rikka and Shoka styles. However, Nageire and Moribana are arranged much more freely. Vases for Nageire are cylindrical often tall and generally have a narrow mouth. In Nageire, stems are usually inserted into vases without using Kenzan (pinholder). There are various of methods to hold stems at the desired position. Nageire is also composed of three main branches, Shin, Soe and Tai. Shin's height is usually one and a half of the height of the vase plus the width of the vase. Soe is about two-thirds to three-quarters of the Shin. Tai is about one-third to half of the height of the Shin. The basic Nageire composition is leaning style, however, this is also divided into three categories in the same way as Moribana: Chokutai, Shatai, Suitai.

3. Moribana (upright style)

Moribana shares the emphasis on the natural beauty and characteristics of flowers and plants with the more traditional Rikka and Shoka styles. However, Moribana, along with Nageire, are arranged much more freely. Moribana is the simplest style of all Ikebana styles, therefore most suitable for beginners. Moribana means "piling up flowers". This creates beautiful volume which can be viewed from three sides. Vases for Moribana are usually wide mouthed and often flat. There are three sub-styles of Moribana, which are determined by the angle of the Shin. These are called: Chokutai (upright) Shatai (slanting); Suitai (cascading).

The Moribana style of Ikebana is noted for its use of shallow containers as an important design element. This provides a board area in which the flowers can be arranged in a natural like state. Unlike the Heika style, the Moribana style is relatively care free with a minimum of rules that must be followed. The focus of this style are 3 main points: *shin* (center/ truth), *soe* (support) in *hikae* (moderating/overall balance). These points represent sky (heavens) as the leading element, earth (subordinate element) in man (to maintain balance between the heaven and the earth). All these three elements together are called *sansei* which means harmony between nature and human.

Rikka Nageire (slanted style) Moribana (upright style)

Fig. 15.1 : Various styles of Ikebana

If we take a closer look at moribana, we can see that the plants are arranged in a certain composition. As mentioned above, all the elements are positioned to balance each other out. Here's a sketch for easier understanding:

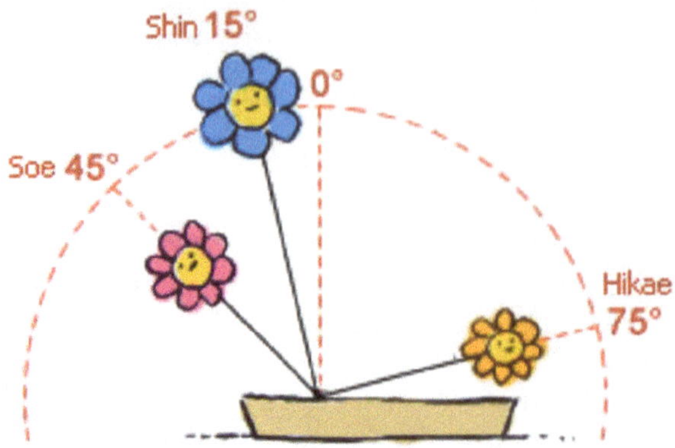

Fig. 15.2 : Making of Ikebana

There are some other rules. The point *soe* lies at the 3/4 length of *shin*, which is positioned at 1.5 times height of the vase size. The vase size itself is measured by vase diameter + vase height. The third point,*hikae*, lies at 1/2 or 3/4 the length of *soe*. The arrangement of these branches and the spiked metal holder are drawn in a simple diagram, called a **kakeizu**. The kakeizu shows a frontal and overhead view of the arrangement. After examining the kakeizu, suitable branches or flowers are chosen for the shushi and trimmed if necessary. The stems are cut to correct lengths according to set formulae. The spiked metal is placed in the vase and just covered with water. The main points are fixed to the spiked metal in order and according to the kakeizu. Short supplementary stems are added to support the main points and give depth to the arrangement. Finally, the composition is examined and any finishing touches applied.

5. Hanaisho

Hanaisho is a very modern Ikebana style that has incorporated a few western elements into it design motifs.

6. Morimono

In addition to flowers and other natural materials, the Morimono style often incorporates in animate objects, such as dolls, sand and folk art in the flower arrangements.

7. Freestyle

Rikka, Shoka, Nageire and Moribana styles have something in common. They emphasise the natural beauty of flowers and plants. However, Free Style, the most recently developed style of Ikebana, is very different. Freestyle seeks creative beauty and it is left entirely up to the arranger what to express and how to express it. It is also entirely up to the arranger what he/she thinks beautiful and what materials to use. The materials in Free Style are unrestricted. The arranger may choose not just floral materials but anything he/she desires. Materials, for example, may be dried, paper and plastic or metals. Various techniques may be used, such as wiring or modifying leaves by cutting them to different shapes from the original, natural shape. The vase is also a very important element of the arrangement. Free style can be divided into two types. The first is naturalistic which expresses but does not recreate nature. The other is abstract, more design-like. To compose a free style arrangement, there are basic principles of formation and some other important factors to consider. Analysis of the elements of the plant's form, shape, colour, texture and quantity should be conducted to decide how it should be used.

Fig. 15.3 : Vases of Ikebana

These decisions will be influenced by such things as desired mass, line, point (focal point) or surface area. The composition or blend of these elements, the sense of proportion, contrasts, rhythm and harmony should be carefully considered.

8. Shoka

Shoka style looks very simple but said to be the most difficult of all Ikebana styles. There are two types of Shoka. One is the classical Shoka Shofutai and more modern style of Shoka Shimputai. Shoka style expresses the feeling of life, growing energy and the natural beauty of the materials. To achieve this it is important that the arranger understands how the plants grow in nature: the natural environment, characteristics and habitat. Each plant species has its own particular characteristics known as"*Shussho*". Shussho is considered to be different from the nature of plants. This nature is something that all plants and flowers share, for example, "growing towards the sun".

A vase for Shoka is more than just a container but represents the source of life. The vase will generally be symmetrical and open at the top, like the one used in this arrangement. This is typical of the shape used for Shoka, as you will see from the other photos on this page. When viewed from the front, the stems should appear as a single vertical line from the water's surface for the first 6-8 cms. This is called *Mizugiwa.*

Shoka Shofutai

Shoka Shofutai consists of three main parts called Shin, Soe and Tai. These yakueda symbolise the cosmos representing heaven, earth and human beings (Ten-Chi-Jin). Shoka Shofutai should bring out the Shussho according to the materials used and create an image of ideal beauty.

In general, Shin is two to three times the height of the vase. Soe is about two thirds the height of the Shin and Tai is about one third. When composed, the concept of sunny side-shady side has to be considered. Soe should always be placed at the sunny side. Shofutai can be arranged using only one kind of material. This is called *Isshu-ike (photo)*. Shofutai Isshu-ike best expresses the characteristics of the plant or flower.Using two kinds (*Nisshu-ike*) emphasises the beauty of contrasting materials. Using three kinds of materials (*Sansyu-ike*) emphasises the beauty of plant combination.

Shoka Shimputai

Shoka Shimputai, is more adapted for our modern life style. Shimputai literally means "new wind style". It is similar in concept to Rikka Shimputai where the arranger can be more expressive because of the release from rigid set of rules and form seen in Shofutai. For example, instead of having three main stems seen as in Shofutai, two contrasting main parts construct Shoka Shimputai.These are known as *Shu* and *Yo*. Usually an additional stem called *ashirai* is used to complete an arrangement. The important factors to consider are contrasting colour, quantity and texture of the materials. The distinguishing characteristics of Shoka Shimputai are brightness and sharpness.

Shoka Shimputai emphasises the unbalanced beauty whereas, Shofutai expresses the well ordered (Shin: Soe: Tai proportion of 7:5:3) beauty.

Equipment to practice Ikebana

In addition to fresh flowers, we will need a few additional items in order to practice Ikebana, these include,

- *Clipers / Scissor*: These are clippers that are used to cut the flower stems. We can also use a pair of scissors.
- *Vases / Kabin*: Kabins are the tall vases used in Heika flower arrangements. We can use any tall, narrow vase that fits our decor and the flowers we will be using.
- *Utsuwa*: This term refers to the shallow containers used for Moribana designs.
- *Kenzan*: This is a small holder with very sharp points that is placed in an Utsuwa container to hold the flowers upright. Kenzans can be obtained at most florist and craft shops. Look for them listed under the name of pin holder, needlepoint holder, or frog. Foam pads can also be used, but are not traditionally acceptable for Ikebana arrangements.

Artificial Flowers

Flowers that are not available naturally, but made artificially from various materials are known as Artificial Flowers. In other words, Artificial Flowers are imitations of natural flowers. Silk Flowers, Soap Flowers, Paper Flowers, Clay Flowers, Plastic Flowers, Porcelain Flowers and Leather Flowers are some examples of artificial flowers. Artificial Flowers are indeed a replica of naturally available flowers. The art of making Artificial Flowers is so specialised that it usually takes more than just a glance to differentiate artificial flowers from natural flowers.

Through the ages people have used different materials to make artificial flowers. The Egyptians used painted linen and shavings of stained horn; the Chinese used rice-paper; the Romans gold and silver. Silkworm cocoons were used in Italy while the South Americans used the plumage of highly colored birds. Modern methods include nylon netting stretched over wire frames, carved or formed soap, ground clay and mass produced injection plastic moldings. The French originally learnt the art of artificial flower production from the Italians. During the 18th century, the French honed their art and towards the end of that century the Parisian manufacturers enjoyed a world-wide reputation. The of manufacturing Artificial flowers was introduced into England by the French refugees around the same time, and soon it spread to

America as well. The modern *Artificial Flower industry* is now a highly specialized one employing specialised processes.

Advantages of Artificial Flowers

- Artificial Flowers last longer.
- It is easy to take care of Artificial Flowers.
- Artificial Flowers are less expensive.
- Artificial Flowers are also colourful.
- Artificial Flowers serve as great gifts.
- Artificial Flowers look like natural flowers.
- Artificial Flowers are just one-of-a-kind.
- Artificial Flowers do not cause allergies.
- Artificial Flowers can be custom designed to suit ones fancy.

Artificial and Dired Flowers Arrangements

Though so similar in use, artificial flowers are an entirely different art form. The line, focal, filler technique is an excellent way to design artificial and dried flower arrangements.

A. Horizontal Arrangements

1. Using a relatively shallow container, anchor foam with a lot of glue or use anchor pins and position sprays of line flowers to establish the shape of the design.
2. Insert focal flowers in the middle so they gently droop over the lip of the container on both sides, reach towards the line material and extend on either side of the middle. Leave room for filler flowers.
3. Fill in and around focal area with filler flowers and foliage.

B. Vertical Arrangements

1. Wedge or secure foam in a vase with hot glue. Cut the stems of the tallest flowers or leaves to reach three or four times the height of the vase.
2. Place the focal flowers vertically within the diameter of the vase.
3. Fill in the areas as needed with filler flowers.

C. Triangular Arrangements

1. Secure floral foam. Determine the vertical height and horizontal width with the smallest line flowers and/or leaves. Make the height higher than the width.

2. Position the largest focal flowers in the heart of the arrangement and slightly lower to give weight and balance.
3. Fill in with the filler flowers and foliage keeping within the triangular shape.

D. Crescent Arrangements

1. Secure foam in container. Determine length of crescent and insert curved line flowers or leaves to follow the crescent form. Angle the shape to balance in the container.
2. Insert the focal flowers fairly low in the bowl to achieve balance, stability and depth.
3. Fill in around the focal flowers with smaller flowers and foliage. Place wisps of filler flowers that gracefully taper off the ends.

E. Oval arrangements

1. Secure the floral foam. Determine the height with line flowers, then frame in the outer edges of the oval shape with light colored flowers and foliage.
2. Place the largest, strongest or brightest flowers in the focal area.
3. Fill in around the larger flowers and leaves with the filler flowers.

F. Minimal Arrangements

1. Adhere foam to container. Insert vertical line flowers to determine the height, and secure the horizontal line flowers to give the basic outline of the arrangement.
2. Place the focal flowers.
3. Fill in with filler flowers as needed.

G. The lazy "S" or "Hogarth's Curve"

1. Anchor the foam securely. Bend the stems gently into graceful curves and insert them in place so they balance.
2. Add the focal flowers following the lines of the upper and lower curves.
3. Cluster filler blossoms and foliage around the central flowers maintaining the rhythm of the 'S'.

H. Free Standing Arrangements

1. Fill a shallow container with foam that extends one inch over the top. Secure the foam with hot glue, floral clay or floral tape. Define the shape of the design with the line flowers and leaves.
2. Place the focal flowers and leaves, turning the vase as you go so all the sides are even.
3. Add filler flowers to integrate the design.

CHAPTER 16

Bio-Aesthetic Planning for India

Plants are important to humankind not only economically, environmentally and industrially but also spiritually, historically and aesthetically, they sustain human life through direct and indirect gains by providing a wide range of products for survival and prosperity. Bio-aesthetic planning to a large extent is a product of the age. With the advancement in agriculture and increase in population, large settlements developed, marking the beginning of urbanization. More and more people moved to urban areas for comfortable life. Increasing human needs resulted in large-scale exploitation of natural resources, great reduction in forest cover and extinction of many species of flora and fauna. The extent of forest cover is a good indication of the health of the land. The large scale deforestation in recent decades has rendered the sensitive catchment areas and hilly areas particularly vulnerable to soil erosion. Professor Lancelot Hogben coined the term 'Bio-aesthetic planning' defined as planning of flora and fauna with the object of beautifying the piece of land (country). Bio-aesthetic planning is an aesthetic branch of Horticulture which deals with planting of ornamental plants in such a way that it creates a picturesque effect. It is a very fascinating and interesting subject. With the greater realization of environmental, ecological and socio-economic benefits of plants, scope of Bio-aesthetic planning has increased further. Bio-aesthetic planning can play important role in environmental amelioration of urban and industrial areas along with their beautification.

Bio-aesthetic planning seeks to create balance between living and natural environments in order to derive the greatest use and enjoyment from outdoor space. Fulfilling recreational needs through ornamental plants and wild animals can be traced back to earliest civilizations and this gave the concept of Bio- aesthetic planning.

Places suitable for Bio-aesthetic Planning

Le Corbusier sums up the ideas of Bio aesthetic planning as: "The sun, space and verdure" are the ancient influences, which have fashioned our body and our spirit. Isolated from their natural environment, all organisms perish, some slowly some quickly and man is no exception to this general rule. Our towns have snatched men from essential conditions, molested them, started them, falsified them, embittered them, crushed them, even sterilized them; the third generation to live in great cities tends to sterility. Fashioned throughout millennia by the conditions of nature, man cannot with impunity disrupt the natural order. Shut up in masonry walls and conditioned to the smell of petrol fumes, men in large towns lead a cramped and unhappy life, deprived of the essential joys of life - sun, space and verdure. Unless the conditions of nature are re-established in man's life, he cannot be healthy in body and spirit. The fundamental elements are: accord with the laws of nature; harmony of actions in the recurrent cycle of the solar day of twenty four hours; experience of the essential joys; an intensity in consecrated work and in consecrated leisure; definition of the metier of the contractor;exploitation of the architectural revolution accomplished in the laboratory by the 19th and 20th centuries; the idea of unity regulating the doctrine of the built domain and its necessary overhauls; the intervention of the law giver reuniting again nature with the built work in the land, in the province, in the region, in the town, determining the scales and types of built volumes and tracing new routes; the whole in conformity with the law of the land."

In search of the sun, space and verdure, man drifts from the ancient town and establishers himself, in garden suburbia. Bio-aesthetic planning should be used for improving environment of all the areas deteriorated by developmental activities. The entire country should be brought under the purview of Bio-aesthetic planning. Priority should be given to public places, which belong to the community as a whole. A large number of persons, especially those who cannot afford gardens of their own, will thus be able to enjoy the splendour of nature. Different places suitable for Bio-aesthetic planning in the country are shown in Fig. 16.1. Public parks and squares, public roads, platforms of railway stations, bus stands, air ports, compounds of courts, hospitals, universities, colleges and schools, ancient historical buildings, government office buildings, dak bungalows are the places in towns which are suitable for bio-aesthetic planning. Industrial areas, various

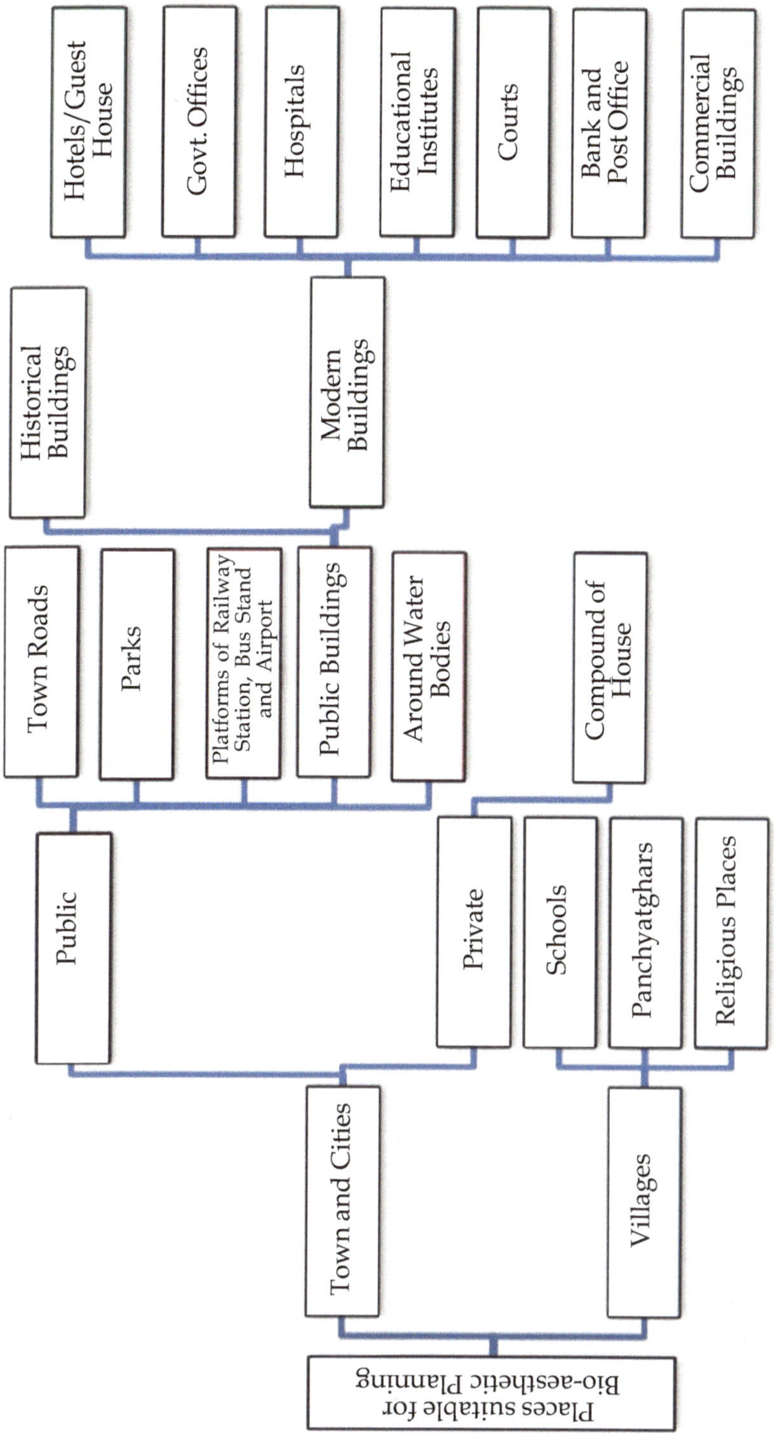

Fig. 16.1 : Places suitable for Bio-aesthetic planning in the country. (Modified from Randhawa, 1978)

categories of roads, banks of rivers and canals are neglected but potential sites for bio-aesthetic planning, which require urgent attention. Beautification of national parks, sanctuaries, biosphere reserves and wetlands will have immense ecological and environmental benefits apart from aesthetic value.

In India tree worship was common in the third or fourth millennium B.C. Among the seals of Mohenjo-daro dating from that period is one depicting a stylised pepal tree with two heads of unicorns emerging from its stem. A seal from Harappa is engraved with the likeness of a weeping willow tree with hanging branches. It seems tree worship was widely prevalent in India before the rise of Buddhism. Mango was their popular fruit tree and shady trees like pipal, banyan, gular and pakur provided welcome shade against the scorching heat of the sun in the summer months. Sal, Asoka, palaksa and pipal trees are associated with Buddha and are thus regarded sacred by Buddhists. Chinese traveller Hiuen Tsiang has described the garden of Lumbini in his travelogue, where Buddha was born under an Asoka tree. On conversion to Buddhism, Asoka (273-326 B.C.) abjured violence and became a messenger of good will and peace. For the first time in history a monarch encouraged arboriculture and adopted it as a state policy. He encouraged planting of trees in gardens and along roads in the form of avenues. (Randhawa, 1964)Trees like pipal, banyan, asoka, siris, sal mango, gular etc. represented in the sculptures of Besnagar, Bharhut, Sanchi, Mathura and many other places signify their importance in the lives of ancient Indians. Gardens had a special place in social and cultural life of people during the Gupta period. Writings of Kalidasa, Ashvaghosha and Vatsyana have vivid description of beautiful flora and gardens of that period. Many festivals were celebrated on the onset of spring and trees were worshiped. Aesthetic considerations lost importance and gardens and tree plantations took a back seat in Mediaeval India, with traces around temples and other religious places. Sher Shah during his brief reign developed many roads, Grand Trunk Road being the most prominent one and promoted planting of shady trees along them. Planned gardens staged a comeback during moghul rule in India who were ardent lovers of plants and Babar, Jahangir, Shah Jahan and Faide Khan developed beautiful formal gardens. However, development by the moghuls was mostly limited to gardens around tombs, palaces and in forts, meant for their personal use. Except the avenues of Chinar along Jehlum River, they did not contribute much to general landscape of the country.

With the advent of British Empire came the era of large-scale introduction of plantsfrom different parts of the world and floral treasure of India was greatly enriched. Many botanical gardens were established during that period, which proved to be important centresfor collection, maintenance and evaluation of flora and education in botany. After independence major problem for the food starved nation was to increase agricultural production and develop basic infrastructure for growth and development of the country. Not much attention was thus paid for aesthetic treatment of environment.M. S. Randhawa was the flag bearer of Bio-aesthetic planning and strived hard for the cause of environmental amelioration by planting of flowering trees. He promoted the idea that rich colourfulness of nature in the form of ornamental trees should be conspicuously noticeable in the form of avenues along streets and roads in towns and even in home gardens. He made great efforts in this direction and made arrangements for planting beautiful trees in the compounds of district courts, tehsil, schools and platforms of railway stations. He advocated garden suburbs around commercial city centres and created garden city of Chandigarh in collaboration with Le Corbusier of France, which has inspired future generations of town planners and architects. Bio-aesthetic planning was included as subject under Ministry of Agriculture and Cooperation in List II of the Seventh Schedule to the Constitution of India.

Advantages of Bio-aesthetic planning

1. *Air quality*: Sulphur Dioxide, Oxides of Nitrogen, Suspended Particulate Matter, Lead, Ammonia and Carbon Monoxide are the major air pollutants, which cause respiratory and other health problems. Plants help in controlling air pollution through acting as biological and physical nets. They cleanse the air by intercepting and slowing dust and other articulate materials causing them to fall out and by absorbing pollutant gases. Plants increase humidity and thus help in settling of particulates. Increased wind turbulence by plants dispenses pollutants. Fragrance of many flowers masks unpleasant odours thereby imparting freshness to air. Interior scape plants can remove organic pollutants from indoor air. An efficient plant air purifier is one, which reduces the concentration of a given pollutant from harmful level to non-injurious level without itself suffering any ill effect. A list of pollution tolerant trees is given in Table 16.1.
2. *Precipitation and humidity*: Plants intercept precipitation and slow it's decent to soil surface thus increasing infiltration and reducing runoff and soil erosion. Pubescence on leaves help in water entrapment. Plants with horizontal branching and rough bark are most effective. Humidity is increased in hot and dry season by means of transpiration.
3. *Noise abatement*: Noise pollution from traffic, loudspeakers, construction & public work and industries affects human system. A noise level above 80 decibels creates tension and increase in blood pressure. Continuous exposure to noise levels of 115-120 db can result in permanent deafness. Along with the engineering methods, plants can be successfully used for noise reduction. Vegetation barriers partially deaden the sound from

traffic, factories and loud neighbours. To be effective, the screen should be dense, tall and wide, planted close to the source of noise. Urban residential areas can be protected by a row of dense shrubs backed by row of tall trees, total 20 feet wide. Vegetation along the highways can screen vehicular noise from reaching the adjoining habitations. Plants also create "background" noise of rustling of leaves and wind through the branches that can help muffle other noises. Plants having thick and fleshy leaves with petioles, which allow the higher degree of flexibility and vibration, are best suited for use as noise screens. Some ofthe trees suitable for reduction of noise pollution are given in table 1.

Table 16.1 : Trees suitable for planting against various types of pollutions

Air Pollution	Dust Pollution	Noise pollution
Tropical Areas	**Tropical Areas**	**Tropical Areas**
Acacia auriculiformis	*Albizzia lebbeck,*	*Alstonia scholaris*
Albizzia lebbeck	*Alstonia scholaris*	*Azadirachta indica*
Ailanthus excelsa	*Bombax malabaricum*	*Anthocephalus cadamba*
Alstonia scholaris	*Cassia seamea* L.	*Butea monosperma*
Anthocephalus cadamba	*Cassia fistula* L.	*Erythrina india*
Azadirachta indica	*Callistemon lanceolatus* Sweet	*Ficus benghalensis* L
Bauhunia purpurea	*Casuarina equisitifolia*Frost	*Ficus infectoria,* Roxb.
Bauhunia variegata	*Erythrina india*	*Ficus religiosa* L.
Butea monosperma	*Ficus benghalensis* L	*Ficus retusa* Linn.
Bombax malabaricum	*Ficus infectoria,* Roxb.	*Ficus cunea,* Ham.,
Cassia seamea L.	*Ficus religiosa* L.	*Ficus glomerata* Roxb.
Cassia fistula L.	*Ficus retusa* Linn.	*Ficus hispida* (L)
Callistemon lanceolatus Sweet	*Ficus cunea,* Ham.,	*Ficus elastica* Roxb.
*Casuarina equisitifolia*Frost	*Ficus glomerata* Roxb.	*Grevillea robusta*
Erythrina india	*Ficus hispida* (L)	*Jacaranda mimosaefolia*
Ficus benghalensis L	*Ficus elastica* Roxb.	*Lagrestroemia flos-reginae*Retz. (syn. L. speciosa Pers.)
Ficus infectoria, Roxb.	*Gliricidia sepium* Walp	*Madhuca indica* Gemel (*Bassia latifolia* Roxb.
Ficus religiosa L.	*Gmelina arborea* Roxb.	*Mimusops elangi* Linn.
Ficus retusa Linn.	*Grevillea robusta*	*Melia azedarach* L.
Ficus cunea, Ham.,	*Ixora coccinea* L	*Millingtonia hortensis*
Ficus glomerata Roxb.	*Jacaranda mimosaefolia*	*Pterospermum acerifolium,* Willd.

Ficus hispida (L	*Leucaena leucocephala* L.	*Pongamia pinnata*, Pierrer, (= *Derris indica* (Lam) Bennett)
Ficus elastica Roxb.	*Lagrestroemia flos-reginae*	*Peltophorum inerme* {=syn.*P. ferrugineum* Benth;*P.pterocarpum* (DC.). Backer}
Gliricidia sepium Walp	*Madhuca indica*	*Saraca indica* Linn. (=*Saraca asoka* Roxb.)
Gmelina arborea Roxb.	*Mimusops elangi* Linn.	*Samanea saman* Jacq.
Grevillea robusta A.	*Melia azedarach* L.	*Spathodea campanulata* Beauv.
Ixora coccinea L	*Millingtonia hortensis*	*Streblus asper* Lour.
*Jacaranda mimosaefolia*D.	*Millingtonia hortensis*	*Spondias pinnata*
Leucaena leucocephala L.	*Nyctanthus arbotristis* Linn (= *Jasminum multiflorum*Andr)	*Swietenia mahogoni* L.
*Lagrestroemia flos-reginae*Retz. (syn. L. speciosa Pers.)	*Polyalthia longifolia*, Var. pendulous Benth and Hook	*Tamarindus indica* Linn.
Madhuca indica Gemel (=*Bassia latifolia* Roxb.	*Pterospermum acerifolium*, Willd.	*Tectona grandis* Linn.
Michelia champaca L	*Pongamia pinnata*, Pierrer, (= *Derris indica* (Lam) Bennett)	*Terminalia arjuna* (Roxb) Weight & Arn
Magnolia stellata		
Mimusops elangi Linn.	*Peltophorum inerme* {=syn.*P. ferrugineum* Benth;*P.pterocarpum* (DC.). Backer}	
Melia azedarach L.	*Saraca indica* Linn. (=*Saraca asoka* Roxb.)	
Millingtonia hortensis	*Spathodea campanulata* Beauv.	
Nyctanthus arbotristis Linn (= *Jasminum multiflorum*Andr)	*Streblus asper* Lour.	
Polyalthia longifolia, Var. pendulous Benth and Hook	*Spondias pinnata*	
Pterospermum acerifolium, Willd.	*Swietenia mahogoni* L.	
Pongamia pinnata, Pierrer, (= *Derris indica* (Lam) Bennett)	*Tamarindus indica* Linn.	
Peltophorum inerme {=syn.*P. ferrugineum* Benth;*P.pterocarpum* (DC.). Backer}	*Tectona grandis* Linn.	

Saraca indica Linn. (=*Saraca asoka* Roxb.)	*Terminalia arjuna* (Roxb) Weight & Arn	
Samanea saman Jacq.	*Thevetia neriifolia* Juss (*T. peruviana* K. Schum)	
Spathodea campanulata Beauv.		
Streblus asper Lour.		
Spondias pinnata		
Swietenia mahogoni L.		
Tamarindus indica Linn.		
Tectona grandis Linn.		
Terminalia arjuna (Roxb) Weight & Arn		
Thevetia neriifolia Juss (*T. peruviana* K. Schum)		
Ziziphus jujuba Lamk.		
Temperate Area	**Temperate Area**	**Temperate Area**
Acer spp.	*Acer spp.*	*Alnus viridis*
Albizia julibrissin	*Ulmus spp.*	*A. crispa*
Betula pendula	*Betula pendula*	*Picia spp.*
Forsythia suspense	*Forsythia suspense*	*Betula pendula*
Ligustrum vulgare	*Crataegus spp.*	*Salix planifolia*
Platanus orientalis	*Quercus palustris*	
Prunus serrulata	*Magnolia stellata*	
Pyracantha coccinea	*Platanus spp.*	
Quercus palustris		
Q.rubra		
Robinia psedocacia		
Malus arnoldiana		
Cornus florida		
Crataegus spp.		
Magnolia stellata		
Ulmus spp.		
Quercus spp.		
Pyrus spp.		
Platanus spp.		
Populus spp.		

4. *Temperature and energy use* : Global surface temperatures have increased by about 0.3-0.6°C since late 19th century. Urban areas in particular behave as heat islands because of concrete structures, decreased wind, increased high-density surfaces and heat generated from human

associated activities. Plants can be successfully used to mitigate heat. Their leaves intercept, reflect, absorb and transmit solar radiations. Trees are called nature's air conditioners because they lower air temperature, shade buildings in the summer and block winter winds. Thus, trees can reduce building energy use and cooling costs. A single large tree can produce the cooling effect of 10 room size air conditioners operating 24 hours/day. Dew and frost are less common under trees because less radiant energy is released from the soil in those areas at night.

5. *Carbon sequestration*: Increase in CO_2 accounts for about 65% of the current direct positive radiative forcing due to greenhouse gases produced as a result of human activities. The atmospheric CO_2 concentration has increased 30% since pre-industrial times, as a result of increasing emissions from fossil fuel combustion, land conversions & cement production and continues to increase by 0.4% per year. Plants remove CO_2 from the atmosphere during photosynthesis to form carbohydrates and return oxygen back as a by-product. Plants therefore, act as carbon sink and remain one of the cheapest, most effective means of drawing excess CO_2 from the atmosphere. One hectare of trees absorbs about 3.7 t of CO_2 from the air generating 2.5 t of life sustaining oxygen.

6. *Water quality/erosion*: Industrial effluents, sewage and domestic waste, excess nutrients and chemicals from agricultural systems are the major pollutants of the water bodies and underground water. Plants prevent harmful pollutants contained in the soil from getting into waterways, reduce top soil erosion, slow down water run-off, and remove nutrients and sediments while increase ground water recharge. Reducing the flow of storm water reduces the amount of pollution that is washed into a drainage area and also prevents sever clogging. Vegetation along rivers, lakes and other water bodies decrease their siltation and reduces cutting of banks during floods.

7. *Wind protection and air movement* : Trees perpendicular to wind direction may reduce the wind speed up to distance of 2-5 times the height of the tallest tree on windward side and 30-40 times on the leeward side. Trees can be planted to funnel or baffle wind away from areas - both vertical and horizontal concentration of foliage can modify air movement patterns. Plants can also be used effectively to control snowdrift.

8. *Glare and reflection control* : Dense foliage absorbs about 70% of the sun's rays, reflect 17% and transmit about 13%. Trees and shrubs help control light scattering, light intensity and modifies predominant wavelengths on a site. Trees block and reflect sunlight and artificial lights to minimize eye strain.

9. *Economic and social benefits*:Besides savings in the form of reduced energy consumption, plants provide timber, pulpwood, fruits, nuts, mulch, composting material, and firewood. Shading by trees extends life of paved surfaces and reduce maintenance cost. Greenery enhances community economic stability by attracting business and tourists, people linger and shop longer along tree lined streets, and apartments/offices in wooded areas rent more quickly and have higher occupancy rates. Property values are higher in well-planned areas lined with trees. Concept of Eco-tourism has gained importance in recent times and many government and private organizations have come up with camping sites and nature tours to boost tourism in many states.
10. *Ecological Advantages*:Bio-aesthetic planning aims at enhancing biodiversity along with beautification of environment. Biological diversity has direct consumptive value in food, agriculture, medicine, and industry. It also has aesthetic and recreational value. Biodiversity maintains ecological balance and continues evolutionary processes. The indirect ecosystem services provided through biodiversity are photosynthesis, pollination, transpiration, chemical cycling, nutrient cycling, soil maintenance, climate regulation, air, water system management, waste treatment and pest control.
11. *Improvement in human behaviour*: People's bodies respond physiologically to viewing and interacting with nature. Studies attribute this response to stress relief, the association of nature with positive emotions through memory and nature's ameliorative effect on mental fatigue. One study showed that hospital patients with a view of trees from their windows had a more positive recovery course than those without. They had shorter hospital stays by 10%. Less violence occurs in urban public housing where there are plants, due to nature-linked decrease in stress and irritability. Parks and open spaces provide opportunities for active and passive recreation to the people, enhancing interaction among them and building a sense of community.
12. *Aesthetic advantages*: Trees and other life forms supported by them have a positive value in scenic quality. Trees change skyline, add variety of form, colour and texture to otherwise dull and monotonous cities. They soften the harshness of concrete structures, complement architectural lines, enframe views and provide background settings to buildings. Dynamic behaviour of vegetation makes the surroundings interesting. Plants can be very effectively used to screen undesirable and disturbing sight lines.
13. *Nature and Wild life Conservation* : To conserve flora and fauna from extinction many botanical gardens,national parks and wild life

sanctuaries have been created.Approximately 1.56 lakh sq. km of area is protected under 89 national parks and 500 sanctuaries. Beauty of these protected areas lies primarily in their wilderness and that is what attracts nature lovers towards them. Wild life has a significant positive impact on scenic quality evaluation. Mere expectation of seeing wildlife significantly increases landscape quality assessment. The wilderness should be given aesthetic treatment with proper planning, which can become an added attraction. The aesthetic and recreational approach places the main emphasis upon preserving the characteristic beauty of the landscape and upon providing ample access to and facilities for open-air recreation and for the enjoyment of beauty in those areas. The major features of the park are made easily accessible by providing roads, tracks bridges and living accommodation in the form of hostels, tourist bungalows, hotels etc. Such utility features can be beautified with ornamental plants bearing attractive flowers and fruits, which will also attract many birds and small animals. Most of these areas are parts of the natural landscapes of that region declared 'protected' by law. Some of these were hunting reserves of former princely states. Although it is impossible to imitate nature but based on understanding of the natural processes one can develop a flexible system that accommodates the natural cycles, rhythms and unpredictability of time as a part of the aesthetics of the place. The landscape of Bharatpur (Keoladeo Ghana) Bird Sanctuary is a wonderful example of such process and aesthetics created through generation of a new wetland for recreational purpose. The once scrub land in the brackish area was developed by the Maharaja of Bharatpur based on the regions bowl shaped topography, potential to collect rainwater and attract wild life. Manipulation of water has been done for nestling of resident birds as well as providing feeding grounds for the migratory birds. Emergent wetland species along with aquatic gardens are designed in some parts of the sanctuary.

14. *Botanical Gardens*: Many botanical gardens have been created for conservation of plant wealth of the country. These gardens serve as centres of collection, maintenance and evaluation of germplasm of important species apart from education in botany. Except the ones of national importance and those established by the British, majority have been planned for scientific purpose only and lack beauty. Botanical gardens and arboreta can be made attractive to general public if aesthetic aspect is integrated with scientific requirement. Proper laid out paths, some benches and a fountain can create the desired effect. Some of the prominent botanical gardens are:

- Lalbagh, Banglore.
- Royal Botanic Garden, Kolkata.
- Royal Agri-Horticultural Society Garden, Kolkata
- Government Botanic Garden, Ootacamund.
- Lloyd Botanic Garden, Darjeeling.
- The National Botanic Garden, Lukhnow.
- The Botanical Garden of Forest Research Institute, Dehradun

A Bio aesthetic planning for Roads / Highways

Trees serve as biological reference points along border of the road and increase driver's perception of orientation. Arranged in varied forms and colours, they can reduce driver's fatigue and boredom. Besides pollution abatement, planting blends the road into the surrounding countryside and enhances the overall beauty. Roads running across all parts of the country are arteries of our economy. Bio-aesthetic planning can be a useful tool in mitigating the environmental impact of road construction and operation activities to a great extent. Problems like soil erosion, siltation of water bodies, loss of vegetation and wild life, aesthetic degradation; air, water and noise pollution can be dealt with to a great extent by careful planning and maintenance of surroundings. Plants should be selected with due regard to soil, rainfall, temperature and water level. The tree species should be either truly evergreen, nearly evergreen or be in leaves at least during height of summer. They must not become too wide to interfere with traffic flow and should be capable of developing clear bole up to a height of 2.5-3.5 m from ground level. They should be fast growing and wind firm and should not be thorny or drop too many leaves. Trees should be deep rooted. Traditional roadside trees like Jamun and seedling mangoes should be preferred which have long life and are well adapted. A general list is presented in Tables 16.3 to 16.5 and suitable tree can be selected according to requirement of site and agro climate.

Bio-aesthetic planning for highways should begin at design stage so that it could be a part of the master plan. Design objectives of highways should be to fit the road into the existing terrain and harmonise it with general landscape of the region. Existing features along the road alignment, like natural wooded areas, streams, ponds, rocky outcrops, scenic vistas, historical buildings, monuments, gardens etc. can contribute significantly to general landscape and every effort should be made to preserve, incorporate and integrate these into the landscape. In the construction stage removal of vegetation and cutting of trees should be done judiciously and limited to minimum possible extent so that least habitat loss occurs. Plantation of ornamental and economically important species should be carried out to a sufficient distance on either side of the road. Functional planting applies to such problems as protection of slopes against erosion, screening of unsightly views, reducing headlight glare, providing shade in summer.

Table 16.2 : Trees for Tropical and Subtropical areas

Botanical name (Family)	Local name	Growth Characteristics, Leaf Canopy characteristics
Acacia auriculiformis A. Cunn (Mimosaeae, Leguminosae)	Australian Wattle; Phyllode Acacia , Akashmoni	Quick growing, drought resistant, hardy, Nitrogen fixer. Not grazzed by cattle; Oblong and dense, Height - 16m; Flowering - June-Jan; Pro.-seed, sapling
Adina cordifolia Hook (Rubiaceae)	Keli-kadam, Haldu	Normally 15-20m, slow growing; Oblong, Leaves are acute at tip and heart-shaped at base; Pro- seeds.
Ailanthus excelsa Roxb (Simaroubaceae)	The tree of Heaven	Fast growing, 12-25m, shade tree.
Alstonia scholaris (Linn) R. Br. (Apocynaceae)	Devil tree, Chatim	Quick growing, hardy, 4-7 leaves in a whorled; produces latex, Round (umbrella shaped); oblong; Height: 10-15m; Flowering :Dec-March
Albizzia lebbeck, Benth. (Mimosae, Leguminosae)	Silk flower or Parrot tree, Siris	Moderate growing; grown in dumps; drought resistant.Round/ spreading; Pro by Seeds. Other two species *A. odoratissima* (Black siris) and *A. procera* also common in ISM campus.
*Anthocephalus cadamba*Miq. (=A. indica A. Reich) (Rubiaceae)	The Kadamba, Kadamba	Quick growing, large; Spreading, Grow rapidly in first 6-8 yr, produce golden ball of flowers
Azadirachta indica A. Juss, (= *Melia azadirachta* L) Meliaceae	Margosa tree , Nim , Neem	Medium size to large tree; Pinnately compound, Semi deciduous; Grow slowly on Mine dumps; Round/ oblong, spreading;. Blossom in the 1st-week of April; Good purifier of air. Seeds do not keep long vitality, sown immediately after collection.
Bauhunia purpurea Linn. (Cesalpineae)	The purple bauhinia or Mountain ebony, Devkanchan	Medium size tree, deep pink-rosy pink flowers; Leaves 10-14 cm long; Open branch9-11 strong nerve radiating from base

Bauhunia variegata Linn (Cesalpineae)	Variegated Bahunia Raktokchan, Kachnar	Small tree, Flowers wwhite with light yellow spot, Open branch 11-15 strong nerve rediating from base;
Butea frondoasa Roxb, (=*Butea monosperma* L) Leguminosae	Flame of forest, Dhak, Palas	Slow growing; Medium size tree; Gives flowers onset of Holi. Oblong/ovoid, In March gets covered with scarlet flowers; extremely resistant to draught
*Bombax malabaricum*D.C. (= *Salmalia malabarica*Schoot; = *Bombax cieba*Linn.) (Bombacaceae)	Silk-cotton tree , Shimool, Simbal	Tall, wide spreading branches, wood is soft, leaves divided into 5-7 leaflets, flowers are bright crimson or red. Spreading. Often reaches 25m height, cuttings or seeds, beauty and brightness flower and shadiness of leaves
Cassia seamea L. (Leguminosae Papilionaceae)	Iron wood tree, Chakundi	Fast growing, pinnately compound, dark green. Grow very well in subsided area, OB dumps. Oblong, Pro. by seeds, sapling; 10-12 m height.
Cassia fistula L. (Leguminosae, Papilionacaea)	Indian laburnum, Amaltash, Bandar lathi	Slow growing; Medium size tree; xerophytic; not eaten by goats or cattle; Fruits pods are long 1-2 ft and dark brown and remain hang for next flowing season.Round/oblong; open branched Height: 10-12m; Pro. by seeds. Easily attacked by caterpillar; Gets covers with golden yellow flowers in April-May
*Callistemon lanceolatus*Sweet. C. linearis L (Myrtaceae)	Lal bottle brash or Bottle brass plant	Handsome evergreen with pendulous branch. Propagate from seed
*Casuarina equisitifolia*Frost. (casuarinaceae)	Beef-wood tree, Australian or Whistling pine, Jau, Casuarina	Tall tree, quick growing, evergreen. Oblong. Multiplied by seeds.
Delonix regia Rafin. (=*Poinciana regia* Bojer.)(Caes alpineae)	The GulMohur, Krisnachura, Gulmohur	Quick growing. Evergreen, Middle size tree, 5-10 m tall. Propagated by seeds or cuttings; Spreading/Flat topped; Open-branched spreading, irregular shaped. Pods are 1-1.5 feet long, black and

Erythrina India Lam. Var. parcelli Papilionaceae (leguminosae).	The Indian Coral Tree	Flowering tree, *E. blakei.* - other flowering tree.
Ficus elastica Roxb. (Moraceae)	Indian rubber tree	Quick growing; Evergreen.
Gliricidia sepium Walp Papilionaceae.	Gliricidia	Quick growing, medium size tree.
Gmelina arborea Roxb. (Verbenaceae)	White teak , Gamar	Moderate growing, wood is valuable, Planted in mining areas.Oblong. Reg by seeds, saplings; 10-12 m height.
Grevillea robusta A. Cunn.(Proteaceae)	Silvery oak	Quick growing, hardy, suitable for mining areas.(Silvery colour in the back of leaves); Oblong. Good for dust and noise attenuation. Reg: by seeds. Very common in West Bokaro Coalfield areas.
Ixora coccinea L (Rubiaceae)	Rangan	Quick growing, Evergreen; Oblong, 6m,By cuttings.
*Jacaranda mimosaefolia*D. Don. (Bignoniaceae)	Nil Gulmohar (E, H, B), The Jacaranda.	Native of Brazil. Quick growing; short lived. Introduced in some dumps but growth is not so attractive. Oblong; Open branched, fruit is flat, disc-like contains numerous papery winged seeds. Feathery leaves and blue colour flowers are added attraction.
Leucaena leucocephala L. Legiminosae (Momosae)	Subabool, Koo - babool	Fast growing and ranked first in mining areas, good fodder; grow very well in dumps. Oblong. Reg: by Seeds, saplings, 7-10m height.
*Lagrestroemia flos-reginae*Retz. (syn. L. speciosa Pers.)	The Queen's flower or pride of India; Jarul	Quick growing; medium size tree, well-known ornamental tree, grown as a avenue tree. Oblong; Shades all leaves in Feb- March, totally looks leafless and dull; Fruit - Woody capsule remain throughout the year.
Madhuca indica Gemel (=*Bassia latifolia* Roxb. (Sapotaceae)	The Butter tree or Mohwa tree, Mauha, Maul	Moderate growth, Oblong/round, Spreading crown

Michelia champaca L. (Magnoliaceae) *M agnolia stellata*	The Golden Champa, Swarnachampa, Champak.	Moderate height tree.
Mangifera indica L. (Anacardiaceae)	Mango, Am	Quick growing after 1yr. Round/ oblong; Height 10-15m; reg - seeds, stem cutting.
Mimusops elangi Linn. (sapotaceae)	Bakul, Bakul	Moderate growth, Round to oblong, By seeds, 4-10m height.
Melia azedarach L. (Meliaceae)	The Persian lilac, Bakayan, Ghoranim	Quick growing, grow very well in OB dumps and subsided area. Medium tree; Oblong to round; Height: 20m; Reg - seeds, saplings; confused with Neem (margosa tree).
Millingtonia hortensis (Bignoniaceae)	The Indian Cork tree akasneem, neem chameli, Jasmine tree	Tall, straight evergreen tree, deep green foliage.
Nyctanthus arbotristis Linn (= *Jasminum multiflorum*Andr) (Oleaceae)	Night jasmine, Sheuli, Harsinghar	Shrub, 4-5m, quick growing, Oblong By seeds, cuttings; FOS- more or less throughout the year.
Plumeria acutifolia Poir (= *P. rubra* Linn) (Apocynaceae)	The temple tree, Gorurchampa, Gulancha, Golainchi	Small deciduous tree, Open. Flowering plant, hardy, pro: by stem cuttings.
Polyalthia longifolia, Var. pendulous Benth and Hook (Anonaceae)	Mast tree, Ashok, Debadaru. (another var is *P. longifolia, erectus*)	Quick growing, tall and majestic tree; branches do not spread out. In erectus var branches are spread out.Conical (Pyramidal); good shade or avenue tree; Height: 10-15m; Flowers in Feb and fruits ripens in Aug - September.
Pterospermum acerifolium, Willd. (Sterculiaceae)	Moochkunda, Kanthachampa	Produce fragrant flower.Round, dense Medium to Big size tree. Dense canopy. Common in lower ground.
Pongamia pinnata, Pierrer, (= *Derris indica* (Lam) Bennett) (Leguminosae, Papilionaceae)	Indian beech, Karanj, Karanja	Quick growing; grow very well in subsided area and OB dumps. Round to oblong.

Peltophorum inerme {=syn.*P. ferrugineum* Benth *P.pterocarpum* (D.C.). Backer} (Cesalpineae, Leguminosae)	The copper Pod; The Rusty Shield-Bearer, Radhachura	Deciduous, Quick growing; Shades all the leaves in Jan-Feb looks totally leafless and dull; Road- side plantation. Oblong;easiest method is through seeds; Blooms twice in a year: March-May and Sep.- Nov.
Saraca indica Linn. (=*Saraca asoka* Roxb.) (Cesalpinaceae)	The asoka tree, Sita asoka, Asoka	Handsome small (5m) dense crown, foliage 7-paris of leaflets +1. Beautiful flowers.Oblong, Propagated from seeds. Beautiful sheltered plant.
Samanea saman Jacq. Mimoseae, Legumonosae	Rain tree	Big Tree, umbrella shaped canopy. Spreading.
Spathodea campanulata Beauv. (Bignoniaceae)	Tulip tree	Dry localities, ornamental flowering plant.
Streblus asper Lour. (Urticaceae)	Siamese rough bush Shaorah, Siora	Medium size tree.
Spondias pinnata	Indian Hogplum, Amra	Big tree.
Swietenia mahogoni L. (Meliaceae)	Mahogony tree	Other sp: are - *S. macrophylla S. microphylla*
Tamarindus indica Linn. (Cesalpineae)	The Tamarind tree, Imli, Tentul	Tree, moderate growing, evergreen.Oblong/ spreading.
Tectona grandis Linn. (Verbenaceae)	Teak tree Segun, Sagwan	Large tree, quick growing.Oblong, 25-30 m tall, by seeds.
Terminalia arjuna (Roxb) Weight & Arn.(Combretaceae)	Arjun	Large evergreen tree with smooth gray barks, quick growing Oblong, By seeds, cuttings.
Thevetia neriifolia Juss (*T. peruviana* K. Schum) (Apocynaceae).	Yellow oleander, Kolkey-phul, Holdey karabi	Small plants.

Ziziphus jujuba Lamk. (Rhamnaceae)	Kul, Ber	Tree, Quick growing (early); evergreenRound to oblong.
Z. oenoplia Mill (Rhamnaceae)	Siakul	Straggler shrub, 5m quick growing, evergreen, round. Other sp: *Z. rugosa* Lamk. *Z. xylopyra* wild

Table 16.3 : Trees for Temperate areas

Scientific Name	Common Name
Acer spp.	Maples
Albizia julibrissin	Silk Tree
Prunus serrulata	Cherry
Malus arnoldiana	Arnold crabapple
Cornus florida	Dogwood
Crataegus spp.	Hawthorn
Magnolia stellata	Magnolia
Ulmus spp.	Ulmus
Quercus spp.	Oak
Pyrus spp.	Pear
Platanus spp.	Plantrees
Populus spp.	Popla

Table 16.4 : Shade Trees for Tropical and Sub-tropical area

Scientific Name	Common Name
Alstonia scholaris	Devil's tree
Azadirachta indica	Neem
Cederella toona	Tun
Eugenia jambolina	Jamun
Ficus benghalensis	Bargad
F. benjamina	Golden fig
F. infectoria	Pilkan/ Pakar
F. religiosa	Pipal
Mangifera indica	Mango
Mimusops elengi	Molsari
Morus sp.	Mulberry
Swietenia macrophylla	Mahagony

Roadside planting of trees may be in the form of avenues, groups or groves. At least 5-6 rows of trees should be planted so as to provide sufficient green cover. Certain restrictions imposed by engineering, traffic and safety requirements like future widening, setback distance, interference with utilities etc. should however, be kept in view while planting. Location of trees and vertical clearance available across the roadway should be checked to ensure that motorists get clear vision of highway signs or signals at all times. Also the foliage of trees should not come in the way of road illumination. Road dividers should be planted with hardy, short height shrubs to cut off headlight glare.

Tree planting along highways

Tree planting along highways can be done according to the following two schemes:

1. *Front row of flowering trees and rear rows of utility trees*: Front row should be planted with flowering trees of different species, each for 3-5 kms and the subsequent rows should be planted with species of economic importance such as timber, fruit and fuel wood.
2. *Mass effect of flowering trees*: Flowering trees should be planted as large groups for about 0.5 kms after every five kms. of plantation of economically important species. Such groups should be arranged alternately on both sides of the road. This will break the monotony of singly planted species.

Roadside amenities like rest areas, truck parks and off strip parking places should be created on all highways for safety, convenience and enjoyment of the travellers. Truck operators in particular have to undertake long journeys, which are often exhausting and boring. Roadside rest areas and truck parks should be planted with shady trees. Such areas should fit into the natural surroundings and become a part of general road landscape.

B. Bio aesthetic planning for City Roads

Roads and streets are lifelines of the urban areas and city dwellers spend considerable part of their outdoor time on them. Without plants, these structures seem to pierce through the body of the city and create restlessness. To make them more enjoyable and safe, they should be lined with shady and flowering trees. An outer row of shady trees and inner row of flowering trees with cycle track/walkway between the two will be a good arrangement for wider roads. This will ensure pedestrian safety and reduction in number of Accidents. Species of flowering trees should be sufficiently varied to avoid monotony. The central verge wherever available should be planted with small to medium sized shrubs to cut off headlight glare.

The road should be planted with green grass and ornamental shrubs, like the bougainvillea. Footpaths, which are provided at the sides for pedestrians, be shaded by four to five rows of trees. On the one hand, it seems useful to demarcate the automobile highway by a border of high trees and on the other to unite with the one glance the entire width of the avenue in question, the shops, pedestrians, parking cars and the localized contacts with the eroded valley and the leisure space. It is equally necessary to cover the pedestrian promenade with shade along the shops. For the car route, a single or double row of trees with high foliage will permit the eye to travel across. This will be with light and evergreen foliage to avoid the need for sweeping. For pedestrians, a multiple row of trees with very heavy deciduous foliage is required so that the sun's rays may pass through in winter. There must also be some evergreen trees with dark and glistening foliage. This arrangement will contrast the height, the thickness, the colours and the permanency of the foliage and will explain the various functions of this essential artery of the country. That is why the trees must be chosen and planted differently. On some of the roads, evergreen trees with large umbrella like crowns like *Ficus infectoria* should be planted and in course of time their crowns would meet, forming a green tunnel. The trees planted as part of avenue landscape shall have clear depth of soil up to 900 mm. It should be planted at an interval of 5000 mm-6000 mm centre to center or as per approved landscape plan. All low height plantations shall be up to a height of 1000 mm maximum wherever, a clear unobstructed vision is required for the road. The plantations shall be separated by a curb of 150 mm to protect plantations from vehicular damage.

Trees selected for roadside plantation should not produce branches at low levels that obscure the view of vehicular traffic, cover the road junctions or the inner sides of curves of the road. Deciduous trees drop a large number of leaves on the road and footpath and render them unsuitable because of uneven surface and the slippery conditions they produce.

Pavement should be kept green with turf grass as far as possible as this ensures proper aeration for trees. If tiling is to be done, porous tiles should be used. Lack of forethought is often seen in the form of tall trees interfering with overhead electricity wires, which are later headed back mercilessly. Due consideration should be given to overhead utilities while selecting plant species and tall shrubs should be preferred to trees in such situations.

Tree planting along City Roads : Tree planting along city roads can be done according to the following two schemes:

1. *One kind of flowering tree on both sides*: In this scheme when the trees come to bloom, there is riot of colour and road looks beautiful for particular period. However, rest of the time road looks dull. Maintenance

is easy in this type of arrangement e.g. *Bauhinia variegata, Cassia nodosa, Lagerstroemia speciosa,Delonix regia, Jacaranda acutifolia, Cassia fistula* etc.

2. *Two kinds of flowering trees blooming at one time on both sides*: Here, the two species are planted alternately on both sides of the road and different colours at same time avoid monotony e.g. *Grevillea robusta* (Yellow) - *Jacaranda acutifolia* (Blue) - *Grevillea robusta*
3. *Two kinds of flowering trees blooming at different time on both sides*: In this case, one tree flowers in one part of the year and other in another part of the year. Roads remain colourful for longer period e.g. *Cassia fistula* (Yellow in May-June) -*Jacaranda acutifolia* (Blue in March-April) - *Cassia fistula.*
4. *Shady trees only on both sides*: This type of arrangement is suitable for roads with heavy pedestrian traffic like approach roads of Bus stops, railway stations and government offices e.g. *Ficus infectoria, Mimuspos elengi, Alstonia scholaris*

A general list is presented in tables - to - and suitable flowering trees and shrubs can be selected according to requirement of site and agro climate.

Table 16.5 : Trees for Tropical and Subtropical areas

Botanical name	Local name	Growth Characteristics
Albizia lebbeck	Woman's Tongue Tree	Flowering Season : April - May This tree is particularly noticeable after the flowering is over and the tree bears the large, light brown flat pods. The pods hanging in good numbers on the tree can be seen during the winter months. These make a loud rustling noise when subject to even a gentle breeze giving the tree its common name. The greenish white flowers of this tree are very pretty, reminiscent of the Rain Tree; they also have a mild fragrance. The tree can be easily propagated by seed.
Anthocephalus cadamba	Common bur-flower	Flowering Season : October to July This tree has large leaves arranged loosely and therefore not forming a dense canopy. The tree grows large with spreading branches and is unfit for planting in small spaces. However,

		it is ideally suited for large parks, gardens and campuses. The flowers somewhat resemble the Badminton Ball tree.
Bauhinia variegata & Bauhinia purpurea	Variegated Bauhinia & Purple Bauhinia	Flowering Season : Feb.-April & June-October. *Bauhinia variegata* and *Bauhinia purpurea* are very similar looking trees. It is very difficult to tell them apart. The former has very pale pink or white coloured flowers with the some petals variegated while the flowers of the latter are pink to purple. Like in all Bauhinias the leaves are split in the middle which is very characteristic. Both the species considered here are indigenous. *B. purpurea* comes into flower when in leaf while *B.variegata* looses most of its leaves when in flower. The trees bear long flat pods which hang and are conspicuous.
Bombax malabaricum	Red Silk Cotton	Flowering Season : January - March The tree is a blaze of colour and completely leafless when in bloom. The numerous, large, cup-shaped, crimson flowers are very attractive. The flowers attract a variety of birds and there is a cacophony of bird calls. Soon after, large green fruits dangle from the tree that turn brown ejecting the soft cotton attached to the seeds.This is an indigenous, fast growing tree and grows in most areas except in the very arid areas.
Brassaia actinophylla	Umbrella Tree or Octopus Tree	Flowering Season : February-March The large palmate leaves are very characteristic and make the tree particularly noticeable. It generally takes a long time before the tree can boast of a substantial canopy. Nevertheless, this evergreen tree looks pretty. Both the common names for the tree are very apt. The name 'Umbrella Tree' is due to the leaf formation while

		the name 'Octopus Tree' can be attributed to the inflorescence.
Butea frondosa	Flame of the Forest	Flowering Season : January - March This tree when in bloom is indeed a sight to behold, particularly so when the forest around it is dry and most trees bereft of leaves. The Flame of the Forest too, like many other trees that share the habitat sheds most of its leaves before putting forth the bright, yet pleasing, orange flowers arranged that are placed in clusters on its branches.It can be propagated from seeds.
Callistemon lanceolatus	Bottle Brush	Flowering Season : February - May; October. The name "Bottle Brush" is very apt for this tree, particularly the flowers.This tree is full activity when in bloom. It attracts nectar feeding insects, particularly bees, in good numbers. The fine leaves and the bright red inflorescence make this tree sought after for landscaping and planting in gardens and parks. As the tree grows very large, it loses its charm. However, when the tree is small it is very attractive. The bright red inflorescence borne amidst fine foliage makes it a good candidate for landscaping. It is a tree suited for planting in small spaces.
Careya arborea	Slow Match Tree	Flowering Season : March - April A nondescript tree with large leathery leaves. However, during the flowering season the tree sheds its leaves and bears large showy flowers. The flowers have several stamens which sometimes have red at their base. Soon after the flowering is over the tree puts forth new leaves. And, the large fruits of this tree remain until much later and well into the rainy months. The fruit is supposed to be edible. The tree can grow in open areas as it requires a lot of sunlight. The tree itself is propagated by seed.

Cassia fistula	Indian Laburnum or Golden Shower	Flowering Season : February - March The inflorescence of golden yellow flowers borne on green stalks resembles a chandelier. Characteristic are the pods of this tree which could be up to two feet in length. Each seed, encapsulated in a sweetish pulp is housed in a separate compartment within the foot-long pod. The pulp is said to have mild laxative properties and is also a favourite of the slothbears. It is propagated from seed with some difficulty. This moderately sized tree is truly spectacular when in bloom.
Cassia javanica	Java Cassia	Flowering Season : April - May A stunningly beautiful tree, the Java Cassia is a medium-sized tree with a spreading crown. It has a straight trunk, horizontal branches and drooping branchlets which bear the foliage. During winter the tree is leafless and is in flush along with the pink flowers during the hot months. The flowers are clustered on the drooping branchlets amidst the pale green leaves. The flowers that are dark pink turn white with age and during the peak season the ground under the tree is covered with the petals. It adds colour to the area where it is planted. Like most other Cassias it bears long, cylindrical, dark brown pods.
Cassia siamea	Siam Cassia	Flowering Season : Most part of the year. This species is planted as an ornamental and well suited for gardens and parks as it is not a very large tree. The yellow flowers and pods are borne in clusters. Pods are flat and occur simultaneously with the flowers.It does very well even in hot, arid conditions and hence of great use in afforesting hot arid areas. This fast growing, evergreen species can be propagated from seed.

Cassia spectabilis	Golden Cassia	Flowering Season : September - November. This is a Cassia with a spreading canopy. It is also short in stature. The large inflorescences are held above the canopy and are produced when the tree has leaves. The combination of green and yellow is quite spectacular when the tree is in bloom. All these features make it well suited for landscaping and also for planting especially in parks and gardens.
Castanospermum australe	Black Bean or the Moreton Bay Chestnut	Flowering Season : February-March This is a handsome tree with glossy dark green leaves and low spreading branches when grown in the open. It bears sprays of red and yellow pea-shaped flowers (about 3 to 4 cm long), which are partly hidden by the dense foliage. The nectar produced by the flowers attracts birds, bats and butterflies in its native land. Large woody pods (measuring 12-20 cm by 4-6 cm) are produced following flowering. The pods look like large groundnuts. These pods split in two, revealing 2-4 large bean-like seeds. The tree is known to contain alkaloids that have been shown to have anti-HIV and anti-cancer properties. It is an ideal shade tree in parks, resorts and gardens. It is known to have a strong root system, which can be used to consolidate stream banks against erosion.
Cochlospermum gossypium	Yellow Silk Cotton	Flowering Season : February - March The Yellow Silk Cotton is a small deciduous tree with a stout trunk. It grows wild in peninsular India, in the dry hilly tracts, which is its habitat. The foliage of the tree is sparse but the large bright yellow flowers borne in terminal clusters are hard to miss;more so since they are in bloom when the

		tree is devoid of leaves. The fruit is a large capsule that turns reddish when dry; the seeds are embedded in short silky wool.The Yellow Silk Cotton is propagated from seed.
Cordia sebestena	Scarlet Sebesten	Flowering Season : Throughout the year; peak between January and March. An evergreen tree with medium sized rough leaves; it can be planted in full sunlight. This tree has a small stature and has a long flowering season. The foliage is not dense. All these features not only make the tree look attractive throughout the year but also make it well suited for large home gardens and parks. It is much sought after for landscaping, particularly where dense foliage is not preferred.
Colvillea racemosa	Colville's Glory	Flowering Season : August - October Although it is very similar to Gulmohur in appearance, a mature individual should pose no problems even to a novice when it comes to recognising the tree.The flowering stalks of this tree are held clear of the foliage mainly on the top. Even after the flowering season these stalks persist for a long time. The dark yellowish-orange flowers are arranged in dense clusters on the drooping stalks. Being a medium sized tree, it is well suited for planting on the roadsides, parks and gardens.
Couroupita guianensis	Cannon Ball Tree	Flowering Season : Almost throughout the year. A truly amazing tree. The tree does not grow branches that reach out from the straight trunk. It bears large, showy flowers, almost through the year, on the trunk and not on branches like most other trees. Though a native of tropical South America, it is considered sacred in India (the structure of the flower is likened to the 'Shiva linga', complete with the snake

		providing shade). The tree also produces globular, brown, woody fruits of an astonishing size, almost the size of a human head. There is also a strong albeit pleasant smell.
Delonix regia	Gulmohur	Gulmohur, though a native of Madagascar, is now common all over India.It is a fast growing species. Truly a spectacular tree when in bloom and otherwise too,owing to its large, flat, spreading canopy. The leaves are feathery and beautiful particularly when young. The tree bursts into bloom with the onset of the hot season when it does,it is bereft of leaves and is transformed into a mass of scarlet (although there is some amount of variation in their hues). The flat, foot long pods of this tree are very characteristic and persist. The tree has conspicuous buttresses that at times spread extensively.
Dolichandrone platycalyx	Nile Tulip Tree	Flowering Season : Most of the year Now it is known scientifically as *Markhamia lutea.* This tree grows tall; and does not branch very much. The tubular yellow flowers are borne in clusters and are very attractive. After the flowering season one can see long pods hanging on the trees. As the pods turn brown, they become twisted. These pods, measuring in excess of a foot and half in length, remain on the tree for a long time. This is a good tree to plant in narrow paces since it tends to grow tall and does not branch out much. It is suitable for parks, gardens, and avenues.
Enterolobium cyclocarpum	Elephant Ear Tree	Flowering Season : February to April This tree grows large and bears flowers that are small and somewhat similar to those of the forget-me-not. It flowers in profusion during the season, to the extent that the ground below the tree

		almost turns white. The pods are large, dark brown in colour and each pod spirals so as to produce a disk like structure. The foliage is fine and feather-like with a spreading crown and is an ideal shade tree.
Erythrina indica	Indian Coral Tree	Flowering Season : February - March This tree, like many others sheds all its leaves when in bloom. The bright red blossoms attract a good lot of birds. The fruits remain on the tree almost through the year. The seed in the pod is a deep reddish-brown. Rub the seed (on the side with the dimple) on a smooth cement floor and put it on your skin; you will realise the amount of heat it can generate. A word of caution, the seed can turn quite hot enough to leave a bad burn on the skin.
Firmiana colorata	Coloured Sterculia	Flowering Season : February-March A large tree with large leaves. It flowers when the tree is leafless. The small, scarlet flowers are borne in erect clusters and when touched leaves behind an orange powder on the fingers. After the flowering season, one can seethe seeds -2 in no.- one each attached to a boat shaped leaf-like structure close to the stalk.
Gliricidia sepium	Gliricidia, Mother of	Flowering Season : February-April A small, deciduous, ornamental tree, planted in gardens and fields as a green manure tree. The tree is leafless when in flower and bears fruits during April and May.The small flowers (barely 2 cm long) are pale pink and they are borne in dense clusters on bare twigs. Flowers fade to white/faint purple with age. It can be used as an ornamental.
Grevillea robusta	Silver Oak	Flowering Season : February - March This tree will be familiar to anyone who has visited coffee/tea estates

		where it is grown as shade trees. The tree perhaps gets it common name due to the pale,silvery undersides of its ornamental leaves. The orange flowers of this tree are borne in clusters.
Jacaranda mimosaefolia	Jacaranda	Flowering Season : February-March *Jacaranda* is a very beautiful tree and spectacular when in flower. It attains a large size. The tree has a canopy consisting of fine, feathery leaves due to which it has a great ornamental value. The leaves are lost during winter and bunches of flowers grow when the tree is leafless. The flowers are mauve coloured and tubular. The seeds are enclosed in a round, woody capsule. They hang from the tree for a long time before dropping off. There is flush even when the tree is in bloom.
Kigelia pinnata	Sausage Tree	Flowering Season : March to June. Bats are known to pollinate the deep liver / maroon coloured flowers of the Sausage Tree. A few of these foul smelling flowers are borne on a thick long stalk which hangs from branches. The fruits, which give the tree its common name, are sausage (cylindrical) shaped, large, woody and hang on long thick stalks. The Sausage tree is evergreen. It also has a shapely crown and ideal for parks and roadside planting.
Lagerstroemia flos-reginae	Queen's Flower	Flowering Season : March - May. This species is also known by the name 'Pride of India'. It certainly merits both the common names attributed to it. The tree, when it puts forth its many inflorescences, each about a foot long, their colour varying from pink to purple. The tree attains majestic proportions in well-watered areas, particularly on river and stream beds. Away from water the growth is stunted making it quite suitable for planting in gardens, parks and on roadsides.

		Never is the tree leafless. The leaves are shed,few at a time and turn red or yellow prior to dropping off, on a scale never equaling the grandeur of temperate trees in autumn. The tree can be easily identified even after the flowering season by the smooth bark (it peels like in guava) and the woody fruits which stay on the tree for a long time - some times even up to the next flowering season.
Michalea champaca	Champak	Flowering Season : May - December. This evergreen tree does not need any introduction. It is largely cultivated for its fragrant flowers. It is very well known to people as they use its flowers for religious ceremonies. Frequently planted in the vicinity of temples, it is considered sacred. The tree requires moderate amount of sunlight and does well in moist climate. This tree has reasonably dense foliage and can grow very large.
Milletia ovalifolia	Moulmein Rose Wood	Flowering Season: January – February. The Moulmein Rosewood is almost leafless when it comes into bloom. The bright mauve flowers are borne on drooping stalks and very similar to that of the familiar Pongam. The flowers attract plenty of bees. Midway during the flowering season the leaves appear. The leaflets of this tree are small and a very pleasant shade of green when new. The seed pods are flat and contain two to three seeds. This being a medium sized tree can be planted is a variety of situations.
Millingtonia hortensis	Indian Cork Tree	Flowering Season: September-October. A glorious tree though it lacks the bright flowers of many ornamental trees. A lofty tree, grows very tall and straight, the drooping branches laden with whitish flowers add to the beauty of the tree. The trumpet-shaped flowers are white and borne inclusters. The

		flowers have a pleasant fragrance that fills the air. They form a carpet on the ground during the flowering season. Though the tree drops most of the leaves for a short period during winter it is never completely bare. The bark of the tree has deep fissures and is used as cork, hence the common name. The flowers are thought to be pollinated by nocturnal insects.
Parkia biglandulosa	Badminton Ball Tree	Flowering Season : October - December This can grow into a large tree indeed with many spreading branches. The leaves are fine and feathery. The inflorescence is similar to a Badminton Ball hanging at the tip of a long thick stalk - the reason for its common name. The fruits that are flat are in clusters at the tip of long thick stalks. Known to grow quite fast. The entire inflorescence has a very soft appearance. However, once the flower is dry the core is pretty hard.
Peltophorum pterocarpum	Copper Pod	Flowering Season :March - September. The Copper Pod is considered to be a semi-evergreen tree. It is seldom leafless though it sheds leaves en masse during the winter months for a very brief period. The name 'Copper Pod' comes from the coppery-red seed cases that are seen in profusion (they turn black with age).A species which is easy to propagate, hardy, and can survive in harsh conditions. A pretty sight when in bloom. It bears clusters of yellow flowers at the end of the branches, together with dark green leaves. The ground under the tree is strewn with a carpet of the yellow petals and is sight to behold.
Plumeria sp.	Frangipani, Temple Tree, Pagoda Tree	Flowering Season : February to October There are 2 species of Temple Trees that are met with often. The Red Frangipani *Plumeria rubra* which is a

		somewhat odd-looking tree while the White Frangipani *Plumeria alba* is the more handsome. Both are of small stature; the former being deciduous and the latter evergreen. They both have leaves that are thick. The leaves and the stem exude a white milky substance when injured. Both, the Red and the White Frangipanis, bear fragrant flowers in clusters at the tips of their branches. The trees are easily propagated from cuttings.
Pterospermum acerifolium	Dinner Plate Tree	Flowering Season : February - March. This tree is also popularly known as Kanaka Champa. The flowers are white with long narrow petals. The sweet-scented flowers open at night and fall by morning; the flowers are pollinated by bats. The name *Pterospermum* means winged seeds. The brown coloured winged seeds fall out of large woody capsules which mature after remaining on the tree for almost a year. This medium sized evergreen tree is a native of the Himalayan foothills. It is planted in many other parts of the country in parks and gardens as it can be propagated from freshseeds.
Samanea saman	Rain Tree	Flowering Season : March-SepFtember The origin of the common name of this species is uncertain. This large tree has an impressive umbrella-like canopy and it can be quite fascinating to watch the tree do its daily exercise of opening and closing its leaves. The flowers are white and pink resembling an open brush contrasting against the green crown.
Saraca asoca	Ashoka Tree	Flowering Season : Throughout the year; mainly November - March. This is a small sized evergreen tree with spreading branches and dense foliage of medium sized leaves. If planted in an open area, it assumes a

		well proportioned, round canopy. The flowers are borne on dry branches in dense clusters that are orange-red that deepens as they age. The tree looks particularly attractive when in bloom. It is propagated by seeds.
Solanum grandiflorum	Potato Tree	Flowering Season : Throughout the year. This species grows into a small tree seldom growing over 15 feet and is very ornamental indeed. It has a very proportionate round crown when allowed to grow unhindered. The tree is in flower most of the year. The flowers are dark purple when theyare young, turning paler with age. They occur in clusters and flowers of different ages can be seen in each cluster. Hence, at any given point of time the tree has flowers of all shades of purple. The large leaves are deeply indented. The young shoots and the undersides of the leaves are covered with prickles.
Spathodea companulata	African Tulip Tree	Flowering Season : August to December. Almost suddenly this species seems to have come into bloom all over town. It is a grand sight particularly where they have been planted at the boundary of parks. A handsome tree bears dark green leaves most of the year. It is in bloom for a long duration providing many birds with nectar from its cup shaped deep-orange flowers. The flowers in clusters held clear off the foliage are in contrast to the green foliage.
Swietenia macrophylla	Large-leaved Mahogany	Flowering Season : March - April. The flowers of this tree are very tiny and hardly noticeable when on the tree. But one can see these pale greenish flowers scattered on the ground beneath the tree. However, the strong yet pleasant fragrance in the vicinity of the tree in flower gives it away. It is quite an experience to be beneath a mahogany in flower on a

		summer morning or in the evening. At this time, the tree also sports fresh leaves which are shed very briefly during December through January. The seeds of mahogany are familiar due to its dispersal method. The seeds are brown and have a long flat wing and are neatly arranged and enclosed in a woody capsule. These seeds can be used for propagation.
Tabebuia argentea	The Tree of Gold	Flowering Season : January - March This tree when in bloom has spectacular yellow blossoms that stand out.It is worth every effort to tend this small tree from South America and wait for it to bloom, though for a short while. It is not very attractive otherwise. The fissured trunk is seldom straight. It has drooping branches that bear long, greyish green leaves in clusters. The tree bears large clusters of bright yellow tubular flowers rendering the tree very conspicuous.The tree is not entirely leafless when in bloom.
Tabebuia avellanedae	Pink Tabebuia	Flowering Season: January – February. A stunningly beautiful tree when in bloom. The Pink Tabebuia is a small statured tree. During most of the year the trees bear very characteristic broad leaflets (five in number) that are pointed at the tip and curved inwards. Prior to bursting into bloom the tree sheds all its leaves. The flowers are deep pink with a pale yellow center and are borne in clusters. Like most other relatives, it can be used as an ornamental very effectively. This species is better suited for this purpose owing to its small stature and they do not attract the axe. All these make it very suitable for planting on avenues.
Tabebuia rosea	Pink Poui	Flowering Season : February-March. It grows large and is very leafy. It

		provides ample shade when in leaf. Leafs resemble the silk cotton leaves and borne on branches that grow at an angle to the main trunk and seem to reach out to the sky (this feature can be used well to differentiate the *T. rosea* from a very similar species *T. pallida* - some botanists can comment on this). Like many flowering trees, it is almost leafless when in bloom. The pale pink flowers grow in clusters similar to the other *Tabebuias*. It is undoubtedly a splendid tree when in bloom. Ideal for parks and gardens where it can be allowed to grow without being subjected to lopping.
Thespesia populnea	Portia tree or Bhendi tree	Flowering Season : Almost throughout the year peaking during winter. Often planted as a roadside tree; particularly common in Tamil Nadu. The Portia tree is a tree that grows to about 30-40 feet in height. Leaves are broad and somewhat reminiscent of the leaves of the Peepul and are heart shaped with prominent veins. The tree, being a relative of the Hibiscus, has flowers similar to it too. They are large crumpled looking flowers with an exterior yellow colour and a deep reddish centre. The flowers turn reddish as they age. The fruit is a capsule with a rounded shape, somewhat like a top. The tree is propagated from cuttings and is known to grow rapidly. It is also easily grown from seed.

C. Bio-aesthetic planning for rural roads

Rural roads are the most neglected ones throughout the country as far as tree plantation is concerned. Here, preference should be given to plant multipurpose trees which besides providing shade and flowers, should also provide economic products such as fuelwood, timber fruits, fodder etc. Roads should be planted with small to medium sized trees preferably on the southern side so that fields are not shaded.

Table 16.6 : Shrubs for Bio-aesthetic planning

Botanical name	Local name
Abelia x grandiflora	Glossy Abelia
Aesculus	Buckeyes
Aucuba japonica 'Variegata'	Gold-dust Plant
Berberis thunbergii var. atropurpurea	Redleaf Japanese Barberry
Buddleia (Buddleja) davidii	Butterfly-bush
Buxus microphylla	Littleleaf Boxwood
Buxus sempervirens	Common Boxwood
Callicarpa americana	American Beautyberry
Calycanthus floridus	Common Sweetshrub
Camellia japonica	Japanese Camellia
Camellia sasanqua	Sasanqua Camellia
Cotinus coggygria	Common Smokebush
Cryptomeria japonica 'Globosa Nana'	Globe Cryptomeria
Cupressus arizonica	Arizona Cypress
Deutzia gracilis	Slender Deutzia
Edgeworthia chrysantha	Paperbush
Euonymus alatus	Winged Euonymus
Euonymus japonicus 'Aureus'	Variegated Euonymus
Euonymus kiautschovicus 'Manhattan	Manhattan Euonymus
Forsythia x intermedia	Border Forsythia
Fothergilla gardenii	Dwarf Fothergilla
Gardenia jasmonoides	Gardenia
Hamamelis virginiana	Witchhazel
Hibiscus syriacus	Althea; Rose-of-Sharon
Hydrangea macrophylla	Bigleaf Hydrangea
Hydrangea paniculata	Panicle Hydrangea
Hydrangea quercifolia	Oakleaf Hydrangea
Ilex cornuta 'Burfordii	Burford Holly
Ilex cornuta 'Carissa'	Carissa Holly
Ilex x meserveae	Blue Holly
Jasminum nudiflorum	Winter Jasmine
Juniperus horizontalis 'Plumosa	Andorra Juniper
Kerria japonica	Japanese Kerria
Ligustrum amurense	Amur Privet
Ligustrum japonicum	Japanese Privet

Lonicera fragrantissima	Winter Honeysuckle
Lonicera maackii	Amur Honeysuckle
Mahonia aquifolium	Oregon Hollygrape
Mahonia bealei	Leatherleaf Mahonia
Nandina domestica	Heavenly Bamboo
Nandina domestica 'Atropurpurea Nana'	Dwarf Nandina
Nerium oleander	Oleander
Philadelphus coronarius	Mockorange
Photinia x fraseri	Fraser Photinia; Redtip Photinia
Pinus mugo	Mugo Pine
Platycladus orientalis (f. Thuja orientalis)	Oriental Arborvitae
Podocarpus macrophyllus	Chinese Podocarpus
Prunus glandulosa	Dwarf Flowering Almond
Pyracantha coccinea	Pyracantha; Firethorn
Rhaphiolepis umbellata	Indian Hawthorn
Rhododendron sp.	Rhododendron
Rhododendron sp.	Rhododendron
Spiraea prunifolia	Bridalwreath Spirea
Spiraea x bumalda	Bumald Spirea
Spiraea x vanhouttei	Vanhoutte Spirea
Syringa vulgaris	Common Lilac
Ternstroemia gymnanthera (f. Cleyera gymna.)	Japanese Ternstroemia
Thuja occidentalis	Eastern Arborvitae
Viburnum plicatum	Japanese Snowball Viburnum
Viburnum trilobum	Cranberrybush viburnum
Viburnum x burkwoodii	Burkwood Viburnum
Weigela florida	Weigela
Yucca filamentosa	Yucca

D. Bio-aesthetic planning for Urban Centres

Cities of today have generally expanded so much due to population pressure that nature, which used to be near at hand previously has receded far away to the country side and citizens have little opportunity to be in contact with its various moods in different seasons. Chirping birds on beautiful trees in fresh morning breeze are like a fairy tale formost of city dwellers. The few public gardens and tree-lined avenues are the only places where the beauty of nature can be appreciated in the towns. Today's citizen lives in a concrete jungle, which cannot inspire or refresh him in any way. Strong structural lines of the urban areas accentuate the stress associated with city life, taking its toll on its inhabitants resulting in irritability, non-sociability and unwillingness to help others in need.Urban greening restores ecosystems that have been displaced or degraded by urban development and other human activities. The creation of healthy natural areas increases biodiversity while providing important habitat for urban wild life. It also contributes to a city stock of "green infrastructure," which includes parks, avenues and green belts.Proper town planning is a pre-requisite for bio-aesthetic treatment of our cities andtowns. This can be seen only in new towns with wide roads and well laid out public parks,rather than in congested old towns with narrow crooked streets. Older parts of our towns like Delhi, Amritsar, Lucknow, Varanasi and Kolkata have ugly, ill ventilated houses joined together along narrow lanes. Deficient in open places, they offer little scope for bio-aesthetic planning. The few existing trees in such areas should be preserved as far aspossible and wherever possible some open spaces should be created and converted into parks. Some open spaces are however, available around religious places and historical buildings, which can be used to give a healing touch by planting flowering trees and shrubs. New townships have greatest potential for proper bio-aesthetic planning. Chandigarh in Punjab provides good example, which can be followed in the development of new population centres throughout the country. Continuous increase in population is putting lot of pressure on limited land available in cities. House with a large garden of its own will soon be a luxury unaffordable, even for the upper class. Gardens near commercial hubs can be very useful for resting and spending time while waiting for public transport. A continuous park system as against a number of small green patches will be much more useful and attractive. Land left by the way of open spaces around buildings for ensuring adequate light and ventilation should be utilized to the best advantage for microclimate improvement by way of plantation. Theme parks like *Bougainvillea* garden, Cacti garden, Seasonal garden, Fragrant garden, Rose garden etc. can be developed with an intention to create awareness among the citizens in respect of different categories of plant species and to encourage academic interest among the students.

E. Bio-aesthetic planning for Compounds of hospitals

Flowering trees, shrubs with fragrant flowers and shady trees should be planted in compounds of hospitals. Medicinal plants like neem should be preferred which also provide good shade. This will improve microclimate of the hospital complex and the beauty of flowers will make the otherwise dull environment, little interesting.

F. Bio-aesthetic planning for Educational institutions

Maximum possible ornamental and economic species should be planted in educational institutions so as to educate the students about importance of plants in everyday life. Students should be involved in maintenance of the plants Courts, office buildings of the government and post offices should be planted with flowering & shady trees and shrubs so that visiting public may rest under the shade and enjoy their beauty while waiting. Railway stations, bus stand and airports, which are gateways to the cities, should be meticulously planted so as to accord welcome to the commuters and make the long waiting period, little interesting. Proprietors of commercial buildings should be encouraged and given all assistance in the planning and planting of ornamental plants.

G. Bio-aesthetic planning forIndustrial areas

Government of India has issued certain guidelines for industrial sitting according to which industries should be located at least 25 km from ecologically sensitive areas such as national parks, sanctuaries, lakes, biosphere reserves; archaeological & and historic places and from major settlements (>3 lakh). Green belts should extend 500 m from the boundary of such areas. However, current scenario of the industrial areas demands strict enforcement of such guidelines for minimizing loss to the environment. Bio-aesthetic plan for industrial areas should include creation of buffer zones around the areas with mixed plantation of economically important species to serve the dual purpose of revenue generation and environmental amelioration. A green belt of suitable trees having a width of 50 to 100 m between residential and industrial zones decrease dust level up to 52% and 40-60% of noise pollution. Five hundred meter belt can mitigate about 76% SO2 and 67% NOx from the atmosphere. Premises of the industrial units should be planted with pollution tolerant flowering plants. This will help in improving working conditions for the industrial workers and increase their output. Deciduous species remove air pollutants in a better way by accumulating them in their leaves, which are all shed during dormant season. Thus mixed plantation of evergreen and deciduous species is recommended. Suitable trees can be selected from above mentioned tables.

H. Bio-aesthetic planning for Historical buildings

We have a rich legacy of historical buildings and heritage sites, which besides their archaeological importance are prime attraction for domestic as well as foreign tourists. Some of these have well planned gardens around them like The Taj while most of them have drab surroundings. With the intended promotion of heritage tourism in new tourism policy, the surroundings of these icons of our glorious past need to be revamped for presenting their true grandeur. Such sites should be beautified keeping in view their significance and type of vegetation historically associated with them should be preferred. Approaches to these sites should also be ornamented with flowering trees and shrubs as a welcome gesture.

I. Bio-aesthetic planning for Banks of rivers

Rivers like Ganga and Yamuna are regarded as sacred and many temples and ghats can be seen along their banks. These should be planted with sacred trees like kadam and aoska, which will greatly enhance their sanctity and also add colour and charm to them. Apart from beautification of banks, plants help in controlling erosion and silting of rivers and also help in reducing cutting of banks during rainy season. However, planting of river banks is not meticulously undertaken and on most of the area general agriculture and forced vegetables are grown. A national level body should be formed to undertake river side plantation and serious efforts should be made for enhancing their beauty and maintain in gecological balance. Many wetlands of international importance have been identified along rivers e.g. Bhoj wetland, Bhopal (M.P.), Deepor Beel, Guwahati (Assam), Harike lake andRopar headworks in Punjab, Pong Dam lake in Himachal Pradesh. Such areas have immense tourist potential which can be exploited by providing proper facilities and aesthetic improvement.

J. Bio-aesthetic planning for Rural areas

Our villages had a tradition of planting trees along the field bunds, panchyat landsand around wells. Although economic considerations always dominated over beauty but there used to be sufficient green cover anyway. Panchyats usually assembled under Pipal or Bargad trees, which used to be centres of village activities. Intensive agriculture has taken its toll and commercial plantations of Eucalyptus and Poplar have replaced long living fruit or timber trees at many places. Village ponds, which used to be prominent features of rural landscape, are fast vanishing or being converted into sewage & waste water dump and every effort should be made not only to conserve them but also to make them attractive. They can be used as water harvest tanks especially in low rainfall areas where they can provide drinking and irrigation water during hot summer months. Planting flowering trees on the approach roads, village schools, panchayatghars and religious places can beautify villages to a great extent. Common lands of the villages usually under the supervision of panchyats provide ample opportunity for beautification with a variety of flowering plants. A scheme was launched in Punjab for planting flowering trees & shrubs and fruit trees near the tube wells in villages and plants were provided to the farmers free of cost.

Bio-aesthetic planning and Vastu Tips for Plants

Vastu Shastra being an ancient science deals in residential buildings, factories and industries. It can be applied on flowers and plants in the house. The plants play significant role in activating positive energy to our day to day life. Here is what Vastu says about plants.

- Tulsi or Basil plant is most powerful and auspicious plant. The plant should be located in the North, East and North-east of the house. Tulsi can be grown at the front or the back of the house.
- Banyan and Peepel tree are sacred trees so they should be planted near temple or sacred place.
- The trees should not be planted in front of the entrance of house.
- Big trees can be planted in the South and West portion of the house. They should not be too close to the building as they block sunlight and roots of the tree can damage foundation. The roots of big trees absorb sunlight and do not allow them to reach the house.
- Small trees can be planted in the East and North side. But no tree should be planted in North-east corner.
- Thorny plants should not be planted in the house as these plants except rose can have negative energy.
- The creepers should be planted near entrance door. It should not be raised on compound wall of the building. It is good to plant creeper in garden.
- Avoid milky trees in the house as it may affect the health of family members.
- Flowering plants should be placed in the South-west portion of the house. Jasmine, Rose, Marigold, Champak, Mogra and Parijat are good around the house as it gives positive energy.
- Money plant should also be planted inside the house as it brings good luck to the house.
- No big tree should be planted in the centre of the house.
- A rock garden should be in southwest portion of the house, not in the northeast portion of the house.

CHAPTER 17

Basics of Landscape Gardening

Landscape Gardening

Landscape gardening is an aesthetic branch of Horticulture which deals with planting of ornamental plants in such a way that it creates a picturesque effect. According to Chambers' dictionary, the definition of landscape is the appearance of that portion of land which the eye can view at once and landscape gardening is the art of laying grounds so as to produce the effect of a picturesque landscape. Landscape gardening can be defined as the decoration of a tract of land with plants and other garden materials so as to produce a picturesque and naturalistic effect in a limited space. According to Bailey, Landscape gardening is the application of garden forms, methods and materials to the improvements of the landscape and the landscape in this connection is any area large or small on which it is desirable to develop a view or design. So landscape may or may not include plants.

Landscape gardening can also be defined as the beautification of a tract of land having a house or other object of interest on it. It is done with a view to create a natural scene by the planting of lawn, trees and shrubs. Landscape gardening is both an art and science of the establishment of a ground in such a way that it gives an effect of a natural landscape. It can be also defined as the imitation of nature in the garden. It can also be defined as improving of total living environment for the people. In light of all the foregoing definition landscape gardening can be defined more appropriately as art of creating gardens, parks and other green areas.

Landscape design seeks to create balance between living and natural environments in order to derive the greatest use and enjoyment from outdoor space. A landscape to a landscape designer is like a slab of marble to a sculptor or the insides of a home to an interior decorator; a landscape is the canvass on which the landscape designer crafts his or her masterpiece. Since the landscape gardening is the making pictures on the ground with plant and other material, landscape designer should be proficient in art, ornamental gardening, ecology and physiology. He should be an architect and engineer to appreciate the relationship between plant form, colours and buildings.

Landscape architecture is based on the use of ever-changing live plants to organize space and the creative combination of natural and man-made elements to form a unified whole. The art includes the creation of gardens and parks; the selection of plants for various climatic and soil conditions; the arrangement of plants along buildings, bodies of water, roads, squares and sculptural works; and the planting and tending of plants. The diverse compositional methods developed by landscape architects may be conventionally reduced to two basic types: formal and natural.

Elements and Principles of Landscape Design

Design is an individualistic as the person who designs. Landscape is an eye of a creator. Abstraction is as good as creativity have been allowed. Elements refer to the basic visual materials used to establish a design. Everything we see has either an inherent or an intended design, or a combination of both. However, the elements and principles of design are means through which a designer monitors the effectiveness and success of a garden. The elements and principles of design are the building blocks used to create a work of art. The elements of design can be thought of as the things that make up a painting, drawing, design etc. Good or bad, all paintings will contain most of if not all, the seven elements of design. These elements and principles serve as structures and guidelines for design development. The design elements are the tools a designer uses to accomplish the design principles in a garden. No consensus exists within the design world on the finite list of elements and principles. Definitions and understanding differ from one person to another. The elements of design are line, form, texture and colour. A designer uses these elements to generate a design. The principles of design are focalization, proportion and scale, balance, order and unity, repetition, rhythm and sequence and interconnection. Use the elements of design to accomplish these design principles. Most artists incorporate the principles of design into their projects. However, some artists break the principles. Experienced artists who break the principles with solid intentions are usually successful. These artists use line, form, texture and colour to complete a composition. However, the artists may or may not incorporate all the principles of design into the composition. Landscape designers are artists. The same elements and principles of design found in works of art also apply to landscape design. These elements and principles of design make up the art and science of landscape design.

Elements of Landscape Design

The elements of design are the visual and physical features of the plant and landscape components that make up a landscape design. Landscape design is defined by manipulating any of its five elements: colour, line, form, texture and scale. A landscape designer uses these features to enhance the design principles.

Line

Line plays an important role in a landscape. This design principle refers to the structural elements in the garden. In design, a line can be a visible, physical mark connecting two or more points or an implied line, drawing the viewer's eye in a particular direction. Line leads the viewer's eyes through the landscaped space. It defines and delineates space. It can mostly be related to the way beds, walkways and entryways move and flow. This design element causes physical and/or visual movement. It may be used to draw attention to an object, divide a space, group related objects together, or separate unrelated objects in landscape design. In landscape design, these can be achieved through the arrangement of plants and borders. Line is some-more constructional beliefs of landscape planning. This can be seen in the way the flowers beds are laid out, the walkways and the entryways that regulates the move and flow of the garden. Line can be considered in two ways. Straight lines are forceful and direct while curvy lines have more natural, gentle, flowing effect.

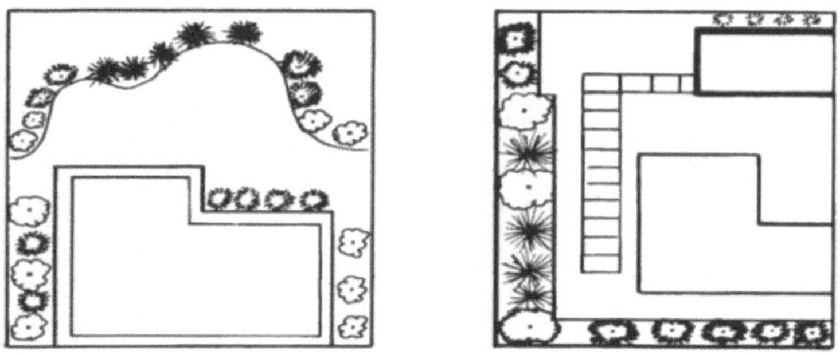

Curved Lines　　Straight Lines

Fig. 17.1 : Various types of lines

Straight lines are used to represent formality or a contemporary concept. Intersecting straight lines suggest hesitation, change of view or direction, or a pause. Meandering or curved lines suggest a more relaxed, slower movement. Use these to create a casual, informal concept.

Form

Form defines the shape and structure of an object. In landscape design, form indicates the shape of a plant and the structure of its branching pattern. Tree forms are defined by branching pattern, while shrub forms are determined by growth pattern. Form is the two or three-dimensional shape and structure of an object or space. Whether it is two or three dimensional, form is line surrounding mass. The shapes of trees and the areas of grass bound by edging are examples of form expressed in a landscape. The air space created by two plant materials set side by side is also an expression of form. All the components in a landscape have a distinctive and natural form. The forms of plants contribute to the total design composition. The basic form of each plant depends on the plant's natural growth habit. Some of the more common forms of landscape plants include round, conical, oval, weeping, horizontal and upright. Most deciduous trees and shrubs have a rounded form. A conical form is characteristic of many evergreen trees. Evergreen shrubs have more of a horizontal form. The concept or theme for a landscape dictates which forms are most appropriate within that concept or theme. Formal concepts suggest the use of much tailored forms of plant material and ground beds. Such a formal landscape would include very straight, crisp, and precise planting beds; topiaries; and other visually clean-lined plants. Informal or woodland concepts mandate much more irregular or natural forms. Casual curving ground beds and loosely branched trees and vines have forms to satisfy this concept. As a designer, incorporate form into a landscape to manipulate a person's emotions. Use vertical forms for strong accents and for adding height. Horizontal or spreading forms add visual width to tall structures. Incorporate weeping or drooping forms to create soft lines and to

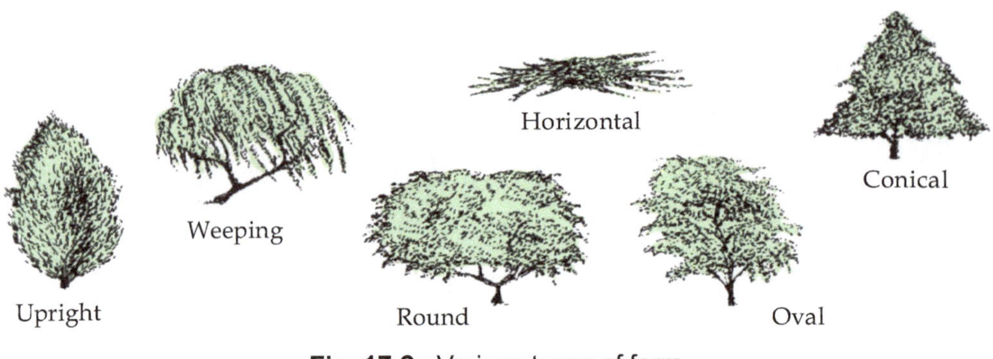

Fig. 17.2 : Various types of form

provide a transition to the ground plane. Rounded plant forms create large masses and are effective as borders and enclosures. Consider the element of form when making design decisions regarding design details for plants, structures and ground patterns.

Texture

Texture is the surface quality of an object. It is the surface quality of any plant material or structure in the landscape. It is how something feels when it is touched or looks like it would feel if touched. Both animate (plants) and inanimate objects (buildings and structures) in landscape design have texture. Texture can be physical (tactile) or visual. It is the feature of a plant or structure's physical surface qualities as determined by form and size.

Texture is also a feature of the aggregation of the minor units that make up the plant or structure. Texture is relative. It must be seen as a comparison. Texture is analyzed by comparison between objects, by association of these objects with each other and by distance.

Fig. 17.3 : Various types of texture

Texture is associated with the senses of touch and sight. Referring to the physical surface of plants (smooth, rough, shiny, or dull), texture is tactile. Texture is also viewed as the organization of the size and arrangement of a plant's component parts (leaves, stems, and branches). In addition to being a physical feeling like rough or soft, texture also describes how one perceives a visual difference. For example, the leaves of one plant are rough and coarse when compared to the smaller leaves of a second plant. However, when compared to the larger leaves of a third plant, the smaller leaves of the first plant appear smooth and fine. As another example, consider coarse-grade pea gravel. Next to fine sand, the pea gravel has a coarse texture. When compared to granite or marble chips, the coarse-grade pea gravel has a finer texture. Visually coarse textures have border impacts. Coarse textures keep the viewer's eyes busy. Progressing from fine textured plant material to coarse textured plant material in a landscape tends to contract the composition.

The coarse textured plant materials appear to advance toward the viewer, making a large space seem smaller and more intimate. Finer textures are easier for the viewer's eyes to span. With finer textures, the eyes continue searching for an accent on which to focus. Progressing from coarse textured plant material to fine textured plant material in a landscape tends to expand the composition. Fine textured plant materials visually recede and appear farther away. They make an area seem larger and more expanded. Texture gives a designer an added dimension for providing greater variety and interest to a landscape. As a designer, proceed with texture changes in a logical and graduated manner. Proceed in sequence so the continuity of the landscape is not disrupted.

Colour

Colour is used to convey emotion and influences the mood and character of the overall landscape design or parts of the design. It adds the dimension of real life and interest to the landscape. Colours can also be used to direct attention to a specific area of the garden. It has three properties: hue or chroma, value and intensity. Hue or chroma refers to the relative purity or strength of the colour. Value determines how light or dark the colour is, whereas, intensity refers to how bright or dull it is.

Contrast can be completed with plants for example fine leaflet verses unenlightened foliage, stimulating leaves spin leaves verses un-stimulating leaves adds tone to contrast. Colour adds a dimension of genuine delay as well as seductiveness to a landscape. Bright colours like reds, yellows and oranges seem to advance toward you and can actually make an object seem closer to you. Cool colours like greens, blues and pastels seem to move away from us and can make an object seem farther from us. A bright display among cooler colours would naturally catch the eye. A cool colour identical to greens resembles sadness. The shades of gray, black and white have been deliberate neutral colours. Grays, blacks and whites are considered neutral colours, are best used in the background with bright colours in the foreground. They have been most befitting with splendid colours used in foreground. However, to increase depth in a landscape, we can use dark and coarse textured plants in the foreground and use fine textured and light colure plants in the background. For enlarging an abyss in a landscape, plants can be dim and for severe hardness it is wise to use excellent make of colour plants in a background.

Light is the source of colour. Light consists of several visible wavelengths. Each wavelength produces a characteristic colour. Colour and light are inseparable. Without light, colour does not exist. White is the accumulation of all light. Black is the absence of light. The colours that people see are reflected wavelengths. For example, the colour perceived as red is the red wavelength reflected from an object. This reflected wavelength is received by the eyes and perceived in the brain. The other visible wavelengths are absorbed by the object and not received by the eyes. Artists and designers use the colour wheel to study colours and their relationships. A position for all colours exists on the colour wheel. Red, blue and yellow are referred to as the primary colours. These colours cannot be made from any other colour. Combining any two of the primary colours produces a secondary colour. The secondary colours are violet, green and orange. Mixing red with blue makes violet. Blue mixed with yellow creates green. Mix yellow with red to get orange.

The primary and secondary colours make up the colour spectrum. The colour spectrum can be seen when visible light passes through a prism. The prism separates, or refracts, the light into the individual colours making up the colour spectrum. Combining equal parts of a primary colour and an adjacent secondary colour produces a tertiary colour. The name of a tertiary colour consists of the names of both colours from which it was created. The primary colour is always named first. For example, red mixed with orange creates red-orange. Six tertiary colours exist on the colour wheel. These colours are red-orange, yellow-orange, yellow-green, blue-green, blue-violet, and red-violet. White, gray, or black added to any of the twelve pure colours (hues), respectively produce a tint (pastel), tone, or shade of the colour. Value describes the lightness or darkness of a hue. A pure hue has full value or intensity. Decrease value by adding white to make a tint. Increase the value by adding black, which creates a shade. Adding gray makes a tone of any hue. Using colour in a landscape is the same as with any other form of art. Knowing the effects of colour is important to the designer when contemplating the landscape design. Designers use colour in the landscape as a background or basic gentle wash to harmonize the view onto the landscape. Colours used in this manner are pleasing and smooth. Another use of colour in the landscape is for accent. In this manner, colour is an "emphasizing factor" for the composition. Colour combinations incorporated in a design influence the moods of those who interact with the design. Bright colours excite or stimulate emotional responses. Subdued or cool colours slow down emotional responses and express a sense of restfulness. Yellows, reds and oranges are warm and advancing colours. These colours are associated with warmth because they are the colours of the sun, fire and heat. Warm colours add a dramatic and excited feeling to an environment. These colours appear to advance and move toward the viewer. Warm colours can infuse a high energy level into those with whom they have contact. Objects or plants with these colours stand out and are the first to be seen. The cool colours of blue and green express restfulness and coolness. These colours are associated with ice, sky and water. Cool colours have a receding visual effect. These colours provide the viewer with a sense of depth. Cool colours make the object of interest appear to be receding into the background. Warm, yellow sunlight and cool, blue moonlight have very different effects on color in the landscape. Artificial lighting also has a dramatic impact on colour.

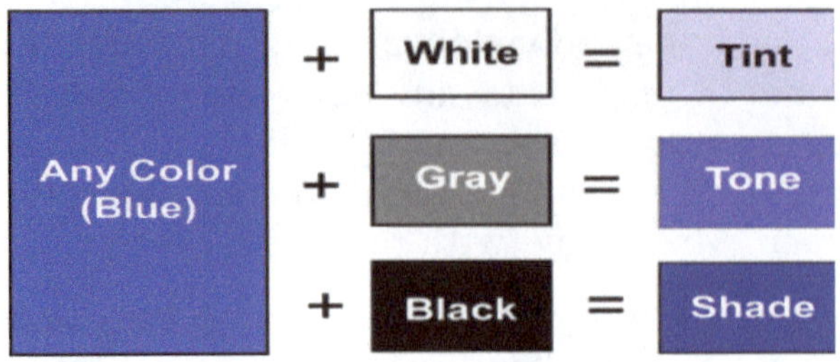

Fig. 17.4 : Various types of colours

Scale

Scale refers to the size of an object in relation to its surroundings. Scale in landscape design is inferred by the size relationship between adjacent objects. As a general rule in landscape design, plants and structures in landscapes should be proportional to the human scale.

As a designer, consider several factors concerning the influences of colour in a landscape design.

1. People have a psychological tendency to "lean toward" light and vivid colours.
2. Bright light and warm colours excite emotional responses. These conditions encourage the viewer's eyes to move throughout the landscape.
3. Subdued light and cool colours are more conducive to moody reflections of thought.
4. Warm colours (red, yellow, orange) appear nearer or advanced. Cool colours (blue, green) appear receded or farther away.
5. Plants or plant masses must blend with their surroundings. If a colour change is desired in a plant mass, proceed in a sequence. A gradual colour change in the plant mass maintains the continuity in the design.
6. Colours and textures relate. Delicate colours (tints and pastels) express fine textures. Earth-tone colours (brown, rust, red) express coarse textures.

Art Principles of Design

Principles refer to standards or prescriptions for working with or arranging various elements to produce the intended landscape design. The principles of design serve as guidelines that govern the organization of the design elements and materials in accordance with the laws of nature. Good landscape design follows a combination of seven principles: unity, balance, proportion, focalization or emphasis, sequence or transition, rhythm and repetition. Simplicity is unequivocally a single of a belief in a pattern as good as art. Landscape designers use these principles of design to create landscape designs that are both functional and aesthetically pleasing.

1. Focalization / Focal Point / Accent / Emphasis

Focalization or Emphasis directs visual attention to a point of interest or prominent part of the landscape design. It's created as a visual break in the sequence and flow of the landscape. The focal point is the point or area of the landscape that attracts the viewer's eyes. Without a point or area on which to focus, the viewer's eyes become lost and confused throughout the landscape. The accent or emphasis is created in the gardens to avoid the monotonous view. It is the method to stress the most important thing. This also serves as the centre of attraction. In English gardens, statues have been used extensively to create such effects. Emphasis in landscape design may be achieved by using a contrasting colour, a different or unusual line, or a plain background space. Paths, walkways and strategically placed plants lead the eye to the focal point of the landscape without distracting from the overall landscape design.

Use the design elements (line, form, texture and colour) to move the viewer's eyes through the landscape to the place of the focal point. In addition to using the design elements to create and draw attention to a focal point, incorporate similar plant forms of varying sizes. Create a focal point with a contrast in spacing between the plants or by grouping several plants together. Most landscapes usually contain visual focal points such as plants and structures. However, running water and rustling leaves are effective as focal points, capturing the sense of sound. The focal points in a landscape may also be focal areas. These focal areas may change throughout the year. For example, the trees with their changing foliage are the focal area during the autumn. In the spring and summer the focal area changes between colourful groups of annual bedding plants and groups of blooming trees and shrubs in the landscape. The focal area attracts the viewer's eyes in the same manner as the focal point. However, the focal area consists of more than a single element or component. Focal areas consist of a group of parts or components. Minor focal points are effective in advancing the viewer's eyes through the landscape to the main attraction or dominant focal point. When incorporating minor focal points into a design, take caution so as not to confuse the viewer. Keep the minor focal points minor, compared to the dominant or major focal point. If colour is the element to be used as the focal point on a site with great depth, use blues or whites as minor focal points. Advance these minor focal points to a dominant red, orange, or strong yellow focal point.

2. Proportion / Percentage / Scale

Proportion describes the size relationship between parts of the landscape design or between a part of the design and the design as a whole. In all the principles of landscape design, this one is quite obvious but still requires a little thought and planning. Most of the elements in landscape design can be intentionally planned to meet the proper proportions. It is the relationship that exists among the components of a landscape. It also describes the relationship between the components of the landscape and the landscape as a whole. Proportion involves the size relationships between and among the components making up the landscape. It describes the mathematical relationships among the dimensions of space and site components making up an area. These mathematical relationships are totally separate from human perception dimensions. In a sense, proportion is similar to a ratio. For example, corner plantings next to a house that are two-thirds the distance from the ground to the eave are proportional to the house. The height of the corner plantings is proportional to the height of the eave. Proportion is relative and elements can be scaled to fit by creating different rooms in the garden. The goal is to create a pleasing relationship among the three dimensions of length, breadth and depth or height. A small water feature can be proportionate if placed in a corner or on the edge of a large area and becomes a focal point of the larger area while creating its own distinct atmosphere.

Scale is the human perception of the size of space and form related to the human dimension. Scale is relative to the perception of the viewer. For a large two-story house, corner plantings that are proportional to the house may appear out of scale to the viewer. Because scale is relative to perception, a designer can manipulate scale psychologically within the site. Although the components of the site may remain proportional, a change in the placing of the components will dramatically affect the scale. A small house surrounded by large trees appears smaller than it actually is. A large house surrounded by small trees appears larger than it actually is. A design is in proper proportion and scale when a pleasing relationship exists among and between each component and the design as a whole.

3. Balance

Balance is concerned with the distribution of visual interest what is wherein a composition.

It is the concept of visual equilibrium and relates to our physical sense of balance. It is a reconciliation of opposing forces in a composition that results in visual stability. It is a design principle defined in terms of weight. Most successful compositions achieve balance in one of two ways: *symmetrically* or *asymmetrically*. Balance in a three dimensional object is easy to understand; if balance isn't achieved, the object tips over. To understand balance in a two dimensional composition, we must use our imaginations to carry this three dimensional analogy forward to the flat surface.

Symmetrical balance can be described as having equal weight on equal sides of a centrally placed fulcrum. It means a mirror image one side is the mirror image of the other. When the elements are arranged equally on either side of a central axis, the result is *Bilateral* symmetry. This axis may be horizontal or vertical. It is also possible to build formal balance by arranging elements equally around a central point, resulting in radial symmetry. There is a variant of symmetrical balance called *approximate symmetry* in which equivalent but not identical forms are arranged around the fulcrum line.

Symmetrical balance is also called formal balance because a form is used. The results look formal, organized and orderly. There is a strong emphasis on the center axis in symmetry since all of the information is reflected from there. This should be taken into consideration when designing with symmetry. Symmetrical balance guarantees left to right balance, which is the most important aspect of balance. But there is more to balance than that. Top to bottom balance is also important. Most images seem more stable if the bottom seems slightly heavier. If the top seems too heavy the composition can look precarious.

Asymmetrical balance, also called *informal balance*, is more complex and difficult to envisage. It involves placement of objects in a way that will allow objects of varying visual weight to balance one another around a fulcrum point. This can be best imagined by envisioning a literal balance scale that can represent the visual "weights" that can be imagined in a two dimensional composition. For example, it is possible to balance a heavy weight with a cluster of lighter weights on equal sides of a fulcrum; in a landscape, this might be a cluster of small objects balanced by a large object. It is also possible to imagine objects of equal weight but different mass (such as a large mass of feathers versus a small mass of stones) on equal sides of a fulcrum. Unequal weights can even be balanced by shifting the fulcrum point on our imaginary scale. Whether the solution is simple or complex, some form of balance can be identified in most successful compositions.

4. Order and Unity

Order and unity are emotional and visual reactions to the overall structure and organization of the design elements. The designer blends the design elements and the design detail decisions of materials with the existing site conditions to establish order and unity. Unity refers to the use of elements to create harmony and consistency with the main theme or idea of the landscape design. It gives the landscape design a sense of oneness and interconnection.

Unity in landscape design can be achieved by using plants, trees or material that have repeating lines or shapes, a common hue, or similar texture. However, too much unity in landscape design can be boring. Therefore, it is important to introduce some variety or contrast into the landscape design. The concept created by order and unity is carried out throughout the design. Order is the overall organization and structure of a design. It is the basic scheme or skeleton of the design. Order is created and carried out through the composition. Examples of order in a design may be symmetrical versus asymmetrical balance or a formal versus naturalistic arrangement. Unity is the harmonious relationship among all elements and characteristics of a design. A unified design is homogeneous and congruous. A design lacking unity appears disorderly and haphazard. Too many components and materials and the complex use of the elements create competitiveness and a lack of integration within a design. To establish unity in a design, stay simple and minimize differences. Limit the complexity of the design elements incorporated into the design. Group the elements in groups of three or uneven numbers. Always remember to simplify diversity and reduce the number of differences between the components in the landscape.It should be one of main goals in design. It may be better understood and applied as consistency and repetition. Repetition creates unity by repeating alike elements like plants, plant groups, or decor throughout the landscape. Consistency creates unity in the sense that some or all of the different elements of the landscape fit together to create a whole.

5. Repetition

Repetition involves repeating or using an element more than once throughout a design. It helps establish and add order and unity to a design. Repetition provides a common feature throughout the design that pulls the design together.

6. Rhythm

Repetition of same object at equidistance is called rhythm. Repeating a colour scheme, shape, texture, line or form evokes rhythm in landscape design. In gardens, generally trees of single species of equal height and shape are planted to create this effect. In Mogul gardens, fountains and water canals have also been extensively used to create such effect. Now-a-days other objects like lights are also used to create the effect of rhythm. Rhythm creates a feeling of motion which leads the eye from one part of the landscape design to another part. Proper expression of rhythm eliminates confusion and monotony from landscape design.

7. Interconnection / Sequence

Rhythm and sequence describe the dynamic unity or the related, orderly movement that implies continuity. They characterize continuity and connection from one part of the design to another part. They group the components together, drawing the design together. This keeps the viewer's eyes busy and allows them to follow easier through the design. Rhythm and sequence lead the viewer's eyes easily, smoothly along a deliberate, dominant and visual path. The viewer's eyes move back and forth with a feeling of smooth motion between the components of the site and the focal point.

Sequence or Transition creates visual movement in landscape design. Sequence in landscape design is achieved by the gradual progression of texture, form, size, or color. Examples of landscape design elements in transition are plants that go from coarse to medium to fine textures or soft scapes that go from large trees to medium trees to shrubs to bedding plants. Transition in landscape design may also be used to create depth or distance or to emphasize a focal point. Natural passing from a single to another can be implemented to equivocate remarkable or in advance changes in a landscape planning. Transition is fundamentally light change. It can be most befitting noticed regulating a plant tallness or color, though can additionally be upon all a elements in a landscape, including, though not used to extent a structure, figure or stretch of a leaves, as good as a stretch as good as figure of assorted elements.

Interconnection is a design principle for producing unity in the design. Various components in the design are physically linked together. Repetition helps in establishing interconnection. A designer may incorporate interconnection into the entire design or into only a small space within the design. Natural transition can be applied to avoid radical or abrupt changes in landscape design. Transition is basically gradual change. It can best be illustrated in terms of plant height or color but can also be applied to all elements in the landscape including but not limited to textures, foliage shape or size and the size and shape of different elements. In other words, transition can be achieved by the gradual, ascending or descending, arrangement of different elements with varying textures, forms, colours, or sizes. An example of a good transition would be a stair step effect from large trees to medium trees to shrubs to bedding plants. This example is where a little knowledge of proper plant selection would come in handy. Transition is one of the principles of landscape design that can be used to create illusions in the landscape. For example a transition from taller to shorter plants can give a sense of depth and distance (like in a painting), making the garden seem larger than it really is. A transition from shorter to taller plants could be used to frame a focal point to make it stand out and seem closer than it really is.

8. Simplicity

Simplicity is actually one of the principles in design and art. It's one of the best guidelines can follow as a beginner. Just keep things simple to begin with. We can do more lately. Simplicity in planting, for instance, would be to pick two or three colors and repeat them throughout the garden or landscape. Keeping decor to a minimum and within a specific theme as well as keeping hardscapes such as boulders consistent is also practicing simplicity.

9. Harmony

It is an overall effect of various features, styles and colour schemes of the total scene. The degree of harmony or unity of various elements of landscape is a measure induced in us and is called as beauty. Therefore, the beauty can be defined as the evident relationship of all parts of a thing observed. When different parts of landscape are correctly placed in right way, produces a harmonious effect. Such landscapes create picturesque effect and appeals to visitors. On the contrary, the absence of harmony or lack of unity is ugliness.

General consideration of landscaping Principles : General considerations of landscaping principles are as follows:

- Simplicity in design should be the key note and undue complexity should be avoided.
- The ground should be so designed that the entire garden is not visible at a glance. Even in a small plot it should not be possible to view the whole garden at a glance. It should be full of surprises, with each turn of the path revealing fresh vistas, or disclosing new interest.
- Colour and contrast in the garden are very much desirable which are of lasting enjoyment and most satisfying means of creating interest in the garden.
- The natural grade of the ground should be taken as a guide.
- Long and straight garden paths should be avoided.
- Judicious use of more number of plants of different varieties should be one of the most important fundamental approaches.
- Garden should be reasonably laid-out for the owners comfort and convenience.
- Beauty and utility should be harmoniously combined.
- The ideal landscape garden is like ideal landscape painting which expresses some single thought or feelings. Its expression may be gay, bold, retired, quiet, etc.
- Variety in a garden gives the greatest pleasure. But attempting too much in a small space is not desirable. Overcrowding of plants and objects should be avoided.

- An ideal landscape should have open space.
- Area should be divided into different parts and plan should be conceived for each area. Overall plan should be such that the observer catches the entire effect and purpose of the plan without stopping to analyse its parts.
- Let the garden and building merge into each other. There should not be stopping abruptly particularly in front of building. The view of garden from the windows and doors should be very attractive. Planting around the building, climbers against wall and on the porch, decoration of verandah and rooms with attractive foliage, flowering plants, hanging baskets serve to unique the building with garden. Every part of the compound should be planned in such a way that it gives surprising effect to visitor.

Factors affecting the landscape design

First of all we should analyze garden site and look at the following factors:

- Climate, sun, rain, temperatures and wind.
- Vehicle and pedestrian access.
- Building laws and boundaries.
- Will garden fit in with the neighbours overhanging plants and trees.
- Does the style fit in with your house style.
- Will you have to remove existing vegetation.
- Maintenance.
- Lighting.
- Allergies.
- Pets.
- Who will be utilizing the garden.
- Budget available.

There are several others factors which affect the making of suitable design for particular site. These factors are:

- *Human choice* : Man's ultimate desire is to make his living pleasurable and surroundings comfortable. His dominance in making designs and selection of plant material is very well evident. Therefore, different styles of gardening have come into existence.

- *Site*: This is an important factor and according to site, suitable design is made. In formal style gardening, the site is selected according to plan. Topography of the site also affects the design.
- *Views*: Distant views of mountains, hills, woods, valley, etc. are preferred from the place of garden.
- *Heritage*: One inherits the knowledge of botany and aesthetic sense and uses accordingly. Our rich heritage teaches us to use flowers and fragrant trees to improve the surroundings.
- *Soil*: According to characteristics of soil types, suitable plants should be selected.

Garden Designing

Garden design is the art and process of designing and creating plans for layout and planting of gardens and landscapes. Garden design may be done by the garden owner themselves, or by professionals of varying levels of experience and expertise. Most professional garden designers are trained in principles of design and in horticulture and have an expert knowledge and experience of using plants. Some professional garden designers are also landscape architects, a more formal level of training that usually requires an advanced degree. Many amateur gardeners also attain a high level of experience from extensive hours working in their own gardens, through casual study, or by joining gardening clubs.

Garden owners have shown an increasing interest in garden design during the late twentieth century, both as enthusiasts of gardening as a hobby, as well as an expansion in the use of professional garden designers.

Styles of Gardening

Garden styles have been changed from time to time with the new ideas and necessities. Broadly, the styles of gardening are grouped into three categories i.e. (I) Formal style, (ii) Informal style and (iii) free style of gardening

1. Formal style of gardening

A formal garden style would match any building with straight boundaries. Geometry and proportions is important. Symmetry, perspective and pattern play a key role in the formal garden design. Symmetry means that the garden is mirrored to look the same on both sides. The formal design plan begins with regularizing the space available. The plan works outward from the house's doors and windows. This governs the location of paths and hedges, and the division of the remaining space. Main features of this style of gardening are: First plan is made on the paper and then land is selected accordingly.

Plan is symmetrical. These types of gardens are of geometric design i.e. squarish or rectangular. Therefore, the roads are cut at right angle. It has some sort of enclosure. Flower beds are also of geometric shapes. The arrangements of tree and shrubs are necessarily geometrical and kept in shape by trimming and training. Other features like fountains, water pools, cascades, etc. are used for further attraction. The examples of such style of gardening are Persian gardens and Mogul gardens.

2. Informal Style of Gardening

The informal garden is a practical, workable garden. Balance is an important design aspect of this style. Informal balance relies heavily on soft curving lines. Undulating levels (grand levels and slopes) are also important. Bold curved lines work well. Too many curves will look too busy and spoil the flow. The curves should lead your eye gently through the garden. A few well placed focal points contribute to the harmony of this landscaping style. This style reflects naturalistic effect of total view and represents natural beauty. This style is just contrast of above formal style. In this, plan is asymmetrical and according to the land available for making garden. Roads, paths are made curvaceous and bending. Water bodies are made of irregular shapes. Hillocks are made to create natural mountainous scenery. Flower beds are made of irregular shapes suiting to surroundings. Plants are allowed to grow in natural form and instead of trimming, annual pruning is done. Japanese gardens are the best example of this style of gardening.

3. Free style of gardening / Picturesque style

This style combines the good points of both formal and informal style of gardening. Rose garden of Ludhiana is an example of this style of gardening.A pleasant and logical modern solution to this confusing choice of styles is to simply incorporate both the formal and informal into your green space. It is uncommon to own a house these days that possesses the architecture and style necessary to display coherently either a completely formal or completely informal styling. Therefore, a happy and harmonious medium can be reached by creating a formal, yet practical area near the house, which addresses both the architectural style of the house yet provides for a space for eating al fresco, then on moving further from the house the garden begins to intertwine with the surrounding countryside, leaving plenty of space for a wild flower, snowdrop or any other plant collecting mania. This can be achieved even in the most limited of spaces, provided we remain within the realms of the scale, which are dictated by the scale of the house itself.

Some important Landscape Gardening terms

Axis: It is an imaginary line, which divides garden into two parts. This also connects two or more points. It is presented in the form of a path, line of fountains or trees, etc. If this axis divides garden into two equal parts, it is called central axis. In formal style, axis is central whereas, in informal style, it is oblique. This axis controls the movement in the garden from the entrance to the terminal. An axis in garden is directional, orderly or dominating.

Circulation in Landscaping: In landscape gardening it means a pathway from entrance to terminal point. Circulation varies with the style of gardening and topography. The more of circulation patterns, it has more points of views and attraction. Common circulation patterns are depicted.

Vista: It is a three dimensional confined view of terminal building or dominant element of feature. It may be natural or man-made. Natural vistas are very common around the lofty mountains and snowy peaks. Overall effect of vista may be of its characteristics. It may be calm or induce motion. There are three different parts of a vista viz.

- It should be subject to a close control.
- It should have a viewing station to see object or objects.
- It should have intermediate ground. These three should be satisfactorily united and thus result into an effect of totality.

Garden

A garden may be defined as place for growing plants and exhibiting various forms of plant life which are consciously directed for ornamental or practical use or both. Now-a-days, we often use the terms like landscape gardening, landscape architecture, landscape design, etc. Arrangement of trees, shrubs, climbers and various other plants together with the building, paths, drives, artificial and natural features for the use of humanity is termed as landscape gardening. Landscape architecture is the art of arranging land and landscape for human use, convenience and enjoyment. Design is the determination of the character of an object to serve a certain purpose known in advance. Design may be of two types i.e. aesthetic and economic. Aesthetic design is purely attractive and pleasing in appearance, while economic design is meant to serve some practical and utilitarian purpose. In most of the cases, landscape architecture should satisfy in its design for both the purposes.

History of Landscape Gardening in India

Early India

- As most gardens and landscapes are usually associated with major buildings, the study of them is intricately bound up with the history of architecture. Most of the buildings surviving from pre-Islamic times are temples because till a fairly late period temples were generally the only buildings built of stone. The remains of some palaces exist, particularly their stone foundations.

- The basis for the construction of a temple is a mandala, (specifically a vaastu purusha mandala). Given the geometric and symmetric layout of mandalas, one would expect the layout of a temple and its compound to follow an equally geometric layout and this is generally so.
- The mandala and the temple are representations of the world or the cosmos. A classic view of the world has Mount Meru at its center, with the mountain standing in the island of Jambudvipa, itself set within an ocean. Temples usually contain representations of these three primary elements, at least in part. Meru or Mount Kailasha is represented by the temple shikara itself, Jambudvipa by the temple and its base or the compound and the ocean by a tank. The basic plan of a temple is a square or rectangle, though this can sometimes be reduced to a linear axis. Where a temple is found within an enclosed space, this is in most cases a rectangular space aligned with the temple. In many cases a water feature is found, often as a tank within the temple compound. The alignment of the temple with the compass directions emphasises its basis in the world. While the symbolism is not always clear, the rectangular layout of the land around a temple is still the rule. There is always water present in some form, for washing and for symbolic purposes.
- Among the best examples of these elements in the Indian subcontinent are the Surya temple at Modhera, the Minaksi temple at Madurai and the Harimandir at Amritsar. The symbolism is probably at its most elaborate at Angkor Wat, Cambodia.
- Temples or temple complexes may contain representations of other landscape elements, although it is not clear that these were always present in pre-Islamic times. Representations of forests occur in the 'thousand pillared halls'. Representations of rivers occur regularly, either on the temple itself, in the shape of goddesses, makaras, etc., or as carvings of rivers on the temple or on its surrounds. Presumably for religious reasons, these representations are more common in Hindu temples than in Buddhist.
- Other important buildings, including palaces, would probably have designed landscapes and gardens associated with them. However, there are few early palaces left, or even the traces of these. The outline of the formal gardens of a palace, from about AD 400 can be seen however, at Sigiriya in Sri Lanka. This is has a strong rectangular layout, based on the axis of the palace. The remains of other palaces can be seen in their foundations, such as in the stone bases to the palace pillars in the royal

center of Vijayanagara. There are references to town, house and palace gardens in the early literature, but this needs further study to try and determine the details of the layouts.

- There are many references to forests, forest glades and flower filled clearings in the passages about life in the forests in the Puranas and the epics. Typically these mention flowering creepers, shading trees, singing birds, fragrant flowers and ponds, often associated with an ashrama or other simple dwelling. They are common in the accounts of the exiles of the principal characters of the Ramayana and the Mahabharata and in the accounts of the lives of rishis. These descriptions occur in such number and detail that it is apparent that an informal garden based on a forest clearing and probably by a river or stream, was seen as an ideal. Of course any such garden or dwelling would have vanished into the forest almost as soon as the gardener gave up, so these accounts will probably have to remain the main evidence. Subsidiary evidence can be seen in the detail of some paintings.
- In some cases, such as the descriptions of early Pataliputra under the Magadhas, accounts of informal gardens are given in relation to cities, or in the immediate neighbourhood of cities.
- It appears from the above, that there were two different traditions of garden and landscape design in pre-Islamic India, which could be called the formal and the informal traditions. The formal is based on the geometric surrounds of a civic building, aligned with the compass directions and is based on the mandala and cosmic order. The informal is that based on the forest clearing and is based on the simple life of a forest dweller living as part of nature.

Islamic and western traditions

- Most of the gardens of the various Islamized cultures are traditionally lumped together under the title 'Islamic gardens'. The common square pattern of the garden or the compound of a tomb probably developed from a fusion of the walled garden, thought to have originated in the Persian paradiaza, with the Semitic concept of the Garden of Eden. The paradiaza is a walled enclosure that shuts out the outside world and encloses a garden. The Garden of Eden is a mythic place from which four rivers flowed out in the four cardinal directions. The fusion of these developed into the 'chahar bagh' (Persian term), the quartered garden. The first known walled tomb garden in India is Sikander Lodi's tomb in Delhi, predating the Mughal tomb gardens.

- In the main, the chahar bagh as seen in India is a square or rectangular enclosure, quartered by water channels that are said to represent the four rivers flowing out of Eden. Examples of these include the principal Mughal tombs - Sikandra, Taj Mahal and Humayun's tomb. These show a layout that could be called the Indian layout of the chahar bagh; the garden is enclosed within walls, is square or nearly so, and has a central reference point, usually the tomb.
- Water is present by mosques for the same reason as by temples; the worshipper is required to be clean before worship.
- Other formal garden layouts exist. A common one is the linear garden (Shalamar bagh, Kashmir and some of the gardens in the Alhambra, Spain). These may blend into the squared layout, as in the chahar bagh of Esfahan. The garden may be part of the surrounding landscape or town. The garden could be divided into numerous squares as in the Aram bagh (Rambagh), Agra.
- In Europe, one line of garden tradition is derived directly from the Islamic interpretations of the Garden of Eden. Later on there were attempts to find ideas in Roman and Greek thoughts and later still in Chinese and Japanese traditions.
- Landscape gardening in Muslim countries greatly influenced the layout and decoration of the gardens of medieval India. In the medieval gardens and parks of the Far East, natural landscaping prevailed. The underlying concept of this form of landscape gardening—the perpetual renewal of nature—was emphasized in Japanese gardens not only by intricately laid-out ponds and paths but also by architectural forms that evoked a feeling of uninterrupted continuity with the natural environment.

Famous landscape Gardens styles

Gardens are categorized into different landscaping styles. Each garden has its own identifying characteristics. The major gardening styles are:-

1. Persian Garden/ Paradise Garden
2. Mughal Garden of India
3. Japanese garden
4. English Garden
5. Italian Gardens
6. French Garden

1. Persian Garden

The Persian Garden gets its name from the Persian word Peiradaeza which means 'enclosed area'. Tree and water stream are the symbols of eternal life for ancient Iranians. Persian gardeners were architects who designed their gardens according to rigorous plans, specific geometric layouts and skilful engineering. The Persian garden is a symbol of paradise.

History of Persian garden design

In the Zoroastrian religion of Iranian Medes and Persians, the cult of trees plays an important role. They brought with them their veneration for trees. The sacred miracle tree was another object for veneration. Tree planting was the sacred occupation for the Persians and a part of education as well. In the ancient time the main purpose of highly planted gardens was to give the army troops the necessary shade and protection mainly during the night times. After few years the Persian gardens were mainly built to give pleasure and relaxation which can only be acquired in contact with the nature.The people of Persia started building formal gardens during Achaemenid era (500-300 B.C.). After the advent of Islam in Persia, in the 600's the basic concept of garden design was developed.

Persian Garden Design

The Persian gardens mainly consist of an enclosing wall, internal canals, rectangular pools, garden pavilions and plantings. The Persian garden designs tend to follow very specific rules, concepts and primarily focus on water and shade. In the typical Persian gardens, the climate is very hot. Therefore, water-efficient elements, shade producing design features are important factors of Persian gardens.

Characteristics of Persian garden design

- The most important factor of Persian gardens is to build shade type structures wherever, necessary in the garden. This helps to cope with hot climates and keep the atmosphere cool.
- These gardens are closed and walled and have very little association with the area outside these walls. As walls is the main feature of the garden design there is an option of fixing existing courtyards in the need of renovation.
- The Persian gardens necessarily have a central water feature or ponds as water is the soul of these gardens. The fountains are also seen in these gardens. Fountains tend to be more of modern Persian garden design. It is a best way to save water and maintain with the original design consisting of a water structure.
- The gardens have crisp colours and clean alignments. The intricate mosaic tiles are occasionally used in these gardens. The gardens have parterre style beds or long beds.

- The garden can be conservative, refined, formal or relaxed and casual. The different garden designs serve different purposes.
- Arches play an important role in Persian garden designs. It shows typical Persian architectural theme. These arches are made from stone work or sometimes wooden support which can be made appear as original.
- Symmetry is very important in these designs. The matching or mirroring elements can give enjoyment and pleasure.

Elements of the Persian Garden

Sunlight and its effects were an important factor of structural design in Persian gardens. Textures and shapes were specifically chosen by architects to harness the light. Elements of the Persian garden, such as the shade and the courtyard style hayat in a public garden in Shiraz. Trees and trellises largely feature as biotic shade; pavilions and walls are also structurally prominent in blocking the sun. Also related to the heat is the importance of water in the gardens. A form of underground tunnel, below the water table, called a qanat is used to irrigate the garden and its environs. Well-like structures then connect to the qanat, enabling the drawing of water.

Alternatively, an animal driven Persian well would be used to draw water to the surface. Such wheel systems could also be used to move water around surface water systems, such as those which exist in the chahar bagh style. Trees were often planted in a ditch called a jub, which prevented water evaporation and allowed the water quick access to the tree roots. The Persian style often attempts to integrate indoors with outdoors through the connection of a surrounding garden with an inner courtyard. Designers often place architectural elements such as vaulted arches between the outer and interior areas to open up the divide between them.

Styles

The six primary styles of the Persian garden are listed below. Gardens are not limited to a particular style, but often integrate different styles, or have areas with different functions and styles.

1. *Hayat (public and private):* Publicly, it is a classical Persian layout with heavy emphasis on aesthetics over function. Man-made structures in the garden are particularly important, with arches and pools (which may be used to bathe). The ground is often covered in gravel flagged with stone. Plantings are typically very simple such as a line of trees, which also provide shade. Privately, these gardens are often pool-centred and again structural. The pool serves as a focus and source of humidity for the surrounding atmosphere. Again, there are few plants - this is often due to the limited water available in urban areas.

2. *Meidan*: This is a public, formal garden that puts more emphasis on the biotic element than the hayat and that minimises structure. Plants range from trees, to shrubs, to bedding plants, to grasses. Again, there are elements such as a pool and gravel pathways which divide the lawn. When structures are used, they are often built, as in the case of pavilions, to provide shade.
3. *Chahar Bagh* : These gardens are private and formal the basic structure consists of four quadrants divided by waterways or pathways. Traditionally, such gardens would be used in work-related functions for the rich (such as entertaining ambassadors). These gardens balance structure with greenery, with the plants often around the periphery of a pool and path based structure.
4. *Park* : Much like many other parks, the Persian park serves a casual public function with emphasis on plant life. They provide pathways and seating, but are otherwise usually limited in terms of structural elements. The purpose of such places is relaxation and socialisation.
5. *Bagh* : Like the other casual garden, the park, bagh emphasizes the natural and green aspect of the garden. Unlike the park it is a private area often affixed to houses and often consisting of lawns, trees and ground plants. The waterways and pathways stand out less than in the more formal counterparts and are largely functional. The primary function of such areas is familial relaxation.

The Persian Garden is normally situated on a slight slope, largely oriented north-south and open on to the surrounding landscape. As the original source of the oriental garden, this model inspired the garden-palaces of central Asia, spreading as far as the Iberian peninsula. In return, it was influenced by European garden culture of the 19th Century, as well as adapting to the urban modernity of the 20th Century.

2. Mughal gardens

Mughal garden history is the best option to study the features and characteristics of the Mughal gardens of the ancient times. Mughals are the origin of Mangolia, who earlier conquered most part of central. Asia spreading their own culture to the territories they ruled. In north India, a great civilization was developed under the rule of Mughals along with garden styles, architectures, etc. It is a traditional garden style, which originated in Central Asia and spread over the South Asia during 16th century up to 18th century.Empire of Mughals was the self-designation of Timurids in Central Asia and Khorasan. Mughal Empire had a largest territorial extent in the Indian subcontinent and Mughals occupied most of the part of India. It was known as Hindustan when Pakistan and Afghanistan was not separated. In 1700, this Empire was extended to a territory of 4 million square km and over 750 million acres of landscape. Some features of Buddhist and Hindu culture had some influences on the gardens of Mughal Emperor. Some other conquerors also followed their earlier invaders. The Emperor decided to build a bridge to link the west and east, which enabled to exchange ideas between China and Europe. After that the conquest of Persia by the Mongols, led to the adoption of Persian culture, art, religion, architecture and language. From the historic records, it is clear that Islamic garden styles were drawn from Persian sources. These sources were drawn from the concepts of Mesopotamia and Egypt.

Babur, the founder of Mughal Empire was very fond of gardens. His most favourite garden type is Charbagh. It brought a new meaning to the development of India, as it explains the importance of flowing streams needed for the Charbagh of Central Asia. The garden of Agra, which is presently known as Ram Bagh, was believed to be the first Charbagh in India. There are numerous number of Mughal gardens in India, Pakistan and Bangladesh, which are very different from the ancient Central-Asian depending on the highly disciplined geometry. The history of Mughal gardens was first studied in the writings of Constance Villiers-Stuart, titled as Gardens of Great Mughals in 1913. As her husband was a colonel in the Indian Army of British, it gave her opportunity to have good contact and traveling. While residing at Pinjore Garden, Villier had the opportunity to maintain an important Mughal garden. Her book had a reference of the Government House at New Delhi. Later, Edwin Lutyens consulted her to know more about the Mughal gardens. This influenced Lutyens to choose the project of Mughal gardens. Some most popular Mughal gardens are:

- Shalimar Gardens (Jammu and Kashmir)
- Shalimar Gardens (Lahore)
- Lalbagh Fort at Dhaka

Babur, constructed gardens in Dholpur and Lahore. However, Humayun, son of Babur, was not seem to be interested in building of the gardens, but he spent lots of time in his father's gardens. After that Akbar built gardens in Delhi and Agra. These gardens were little bit different from the gardens made by his ancestors. Jahangir did not build mush garden, but he played an important role in establishing Shalimar garden, which was very much known for his lover for the flowers. After the trips to Kashmir, it was believed to be the beginning of the natural and floral designs.

After Jahangir, Shah Jahan made an apex in the design, architecture of Mughal gardens. He had also built the Red Fort where Mahtab Bagh was constructed. Here all the flowers are night-blooming with a unique appearance at night. The pavilion is made of white marbles which glows in moonlight of the night.

Shah Jahan's most famous commission was the Taj Mahal, the tomb built for his wife Mumtaz Mahal. The gardens were built between 1632 and 1654. Lesser known and now largely in ruins is the "Moonlight Garden" directly across the river from the Taj Mahal. This garden was built as a mirror image, especially for the enjoyment of the court for evening parties during the full moon from which they could review the reflection of the Taj.

Gardens were also built by noblemen around their houses, creating 'garden cities' such as Mughal Lahore. Hindu rulers in northern India also built Mughal-style gardens. One of the most famous is at Amber in Jaipur (India).

Classic design

The gardens of the Taj Mahal are typical of Mughal gardens, with a formal, symmetrical design and abundant water flowing through channels and pools. Typically, two water channels cross each other, dividing the garden into four quarters. A central pool or pavilion marks the centre of the garden. As well as flowing in channels, water is also used in cascades and fountains and is appreciated for its air-cooling properties. As in earlier gardens, scented flowers, in formal symmetrical beds, was important. Avenues of trees often included cypress trees, with their tall, narrow habit, fruit trees and shade trees such as the plane tree. On forts and hillsides, elaborate terraces were constructed. Paths were usually raised above ground level.

Origins and representation

The roots of Mughal design lie in Persia (present day Iran). Persians are credited with introducing the formal four part walled garden known as "chahar bagh." In fact, the word "paradise" comes down to us today from the Persian *pairi* = around and *daeza* = wall.

The most important element was water, which was channelled in geometric quadrants symbolizing the rivers of paradise mentioned in the Koran. The rectangular Islamic garden, spread to central Asia. Mughal gardens incorporated strong symbolic elements, common to all Islamic gardens. In Mughal India, tombs were often built with gardens, as at the Taj Mahal.

British influence on Mughal garden

The great Mughal tradition of garden design went into decline following the death of the emperor Aurangzib in 1707. After this, the British were to have a huge impact on garden design on the subcontinent. The tradition did not entirely disappear and the President's House (Rashtrapati Bhavan) in New Delhi features a beautiful 'Mughal' garden designed by Sir Edwin Lutyens in 1918-1925.

3. Japanese Garden

The Japanese garden has its roots in the many religions of the area. Many historians consider the Japanese garden to emanate from Shinto, Daoist, Hindu and Buddhist concepts. In earlier age, the features of Japanese garden have large and different influence during the historical periods. Most of people think that typical Japanese garden did not existed, as all the gardens differ largely from each other. The Japanese gardens were build to meet the private needs and sometimes around the temples for a suitable environment for worship. Most of the today's Japanese gardens characters are developed and influence from Zen Buddhism. This idea was brought from China in 13th century and got very much influenced in Japan.

Japanese garden features imply the characteristics of the garden increasing the soul relaxation, meditation, peace and linking with the nature. Japanese gardens are of various types with different characteristics. For instance, the Ryoan-ji Garden, Japan has almost no plants in it except the lakes and seas of fine gray gravels and cluster of rocks and stones. Generally, Japanese gardens are very simple and have lots of impact on the garden visitors. Some of the specific features of Japanese gardens are stone basin and lanterns. Sometimes moss is grown to express Zen esthetic.

Symbolic feature of Japanese garden

Japan is located around ocean and sea forming a cluster of islands. In Japanese garden, water is one of the most important element features. Chisen-style is one of the most famous styles, where ponds are the significant characters in the garden. Water is used as a symbol in this garden to express sea or ocean. The quantity of water present in the garden is not so important. If we vary a very limited space, we can establish a water bucket in a stone basin to enjoy the feeling of sea. It is not required essentially to have a water feature in the garden. If it is a dry garden we can symbolize it by sand or gray gravels with a particular design. Plants and trees used in the Japanese gardens are the portrait of the physical and spiritual life of the people of Japan. Bamboo is also a very common plant found in these gardens. Plum and pine trees are also grown to decorate the garden.

Esthetic feature of Japanese garden

Mostly the esthetic beauty and concepts in Japanese gardens are influenced from Zen Buddhism. The esthetic feature of Japanese garden is unique. It shows elegance, naturalness and simplicity. After the introduction of doctrine, the monks started to build gardens. The elements or features used in these gardens are the way to express the Zen principles and these Zen principles are transferred as the most special esthetic character. The Japanese monks brought most of the Zen teachings to Japan. These Zen philosophies were later followed by the aristocrats, warriors and monks.

In particular, the Japanese garden used four features that distinguished it from others:

- The central feature of water or symbolized water. Streams or their symbolic representation of gravel rivers were likely adopted from Chinese gardens. The focus on water represented the Daoist notion of permanent impermanence always flowing yet always the same. Water was always considered a source of purification and in Japanese gardens specific reverence for water emerged from the Shinto admiration for water.
- The gravel or crushed granite used in Japanese gardens started with Shinto shrines in forest clearings in which the ground was purified and made hospitable to spirits by putting down a layer of washed sand or gravel. They were later used to simulate rivers or seas.
- Trees were a representation or metaphor for life. Often the trees were twisted and bent by Japanese gardeners who carefully trained them to conform to notion of being ancient, noble and wise.
- The use of empty space defines the space around it and is itself defined by the elements surrounding it. This is done in the spirit of in and yo, which Westerners know as ying and yang.

The Garden as Art

The Japanese Garden has evolved from centuries of tradition and a multiplicity of cultural influences synthesized as a uniquely Japanese art form. From Shinto, Japan's first major religion, came the worship of trees, rocks and streams. From Zen Buddhism came the concept of the garden as a symbolic version of nature and an austerity and simplicity that characterize much of Japanese art even today. From both came a reverence for nature. *Seiwa-en* has recreated the tradition, the symbolism and the philosophy of a respected culture in the "garden of pure, clear harmony and peace."

Main principles on the garden's design

Bring the Japanese feeling into garden with these basic steps. First of all, embrace the ideal of nature. That means, keep things in garden as natural as possible, avoiding to include things that could disrupt this natural appearance.

For example, don't include square ponds in design as square ponds are nowhere to be found in nature. Also, a waterfall would be something closer to what exists in nature if we compare it to a fountain. So we also have to consider the Japanese concept of sumi or balance. Because one of Japanese gardening design main purposes is to recreate large landscapes even in the smallest place.

As a miniaturized landscape, the rocks in the garden would represent mountains and the ponds would represent lakes. A space filled with sand would represent an ocean. By that we assume that garden masters were looking to achieve a minimalistic approach, best represented by the phrase less is more.

Basic Arrangements

Despite the fact that certain rules are applied to each individual garden, don't think that there's just one type of garden. There are three basic styles that differ by setting and purpose.

Hill and Pond Garden (Chisen-Kaiyu-skiki)

A China imported classic style. A pond or a space filled with raked gravel fronts a hill (or hills). This style always represents mountainous places and commonly makes use of vegetation indigenous to the mountains. Stroll gardens commonly use this style.

Flat Garden (Hiraniwa)

It derives from the use of open, flat spaces in front of temples and palaces for ceremonies. This is an appropriate style for contemplation and that represents a seashore area (with the use of the right plants). This is a style frequently used in courtyards.

Tea Gardens (Rojiniwa)

Function has a greater importance than form in this type of garden. The Roji or dewy path, is the main point of the garden, along with the pond and the gates. This would be the exception to the rule. The simple and sparse plantings give a rustic feeling to the garden.

Formality has to be taken in consideration

Hill and pond and flat styles may be shin (formal), gyo (intermediate) or so (informal). Formal styles were to be found usually at temples or palaces, intermediate styles were suitable for most residences and the informal style was used in peasant huts and mountain retreats. The tea garden is the one that always fits in the informal style.

The garden components

Rocks (ishi in Japanese) are the main concern of the Japanese garden. If the stones are placed correctly, then the garden shows in a perfect balance. The basic stones are the tall upright stone, the low upright stone, the curved stone, the reclining stone and the horizontal stone. These must be usually set in triads although this doesn't happen always. Two almost identical stones (example, two tall verticals or two reclining stones), one a little quite smaller than the other, can be set together as male and female, but the use of them in threes, fives and sevens is more frequent.

We have to keep away from the Three Bad Stones. These are the Diseased stone (having a withered or misshapen top), the Dead stone (an obviously vertical one used as a horizontal, or vice versa, like the placement of a dead body), and the Pauper Stone (a stone having no connection to the several other ones in the garden). Use only one stone of each of the basic types in any cluster (the rest have to be smaller, modest stones also known as throw away stones). Stones can be placed as sculptures, set against a background in a two-dimensional way, or given a purpose, such as a stepping stone or a bridge.

When used as stepping stones they should be between one and three inches above the soil, yet solid underfoot, as if rooted into the ground. They can be put in straight lines, offset for left foot, right foot (referred as chidori or plover, after the tracks the shore bird leaves), or set in sets of twos, threes, fours, or fives.The pathway stands for the passage through life and even particular stones by the path may have meaning. A much wider stone placed across the path tells us to put two feet here, stopping to enjoy the view. There are numerous stones for specific places. When observing the basic design principles, we can notice the exact character of the Japanese garden.

Water (mizu in Japanese) plays an important part in the composition of the Japanese garden because of Japan's abundant rainfall. Water can be represented even with a raked gravel area instead of water. A rushing stream can be represented by placing flat river stones closely together. In the tea garden, where there isn't any stream or pond, water plays the most important role in the ritual cleansing at the chozubachi, or water basin. As the water fills and empties from the shishi-odoki, or deer scare, the clack of bamboo on rock helps mark the passage of time.

The flow of water, the way it sounds and looks, brings to mind the continual passage of time. A bridge crossing the water stream is often used as a landscaping complement. Bridges denote a journey, just as pathways do. Hashi, in japanese, can mean bridge or edge. Bridges are the symbolic pass from one world into another, a constant theme in Japanese art.

Plants or Shokobutsu may play a secondary role to the stones in the garden, but they are a primary concern in the design too. Stones represent what remains unchanged, so trees, shrubs, and perennials have to represent the passing of seasons. Earlier garden styles used plants to make up poetic connotations or to correct geomantic issues, but these have little meaning today.

As the the Heian style diminished under the Zen influence, perennials and grasses fell out of use. So, for a long time, there were only a few plants that tradition allowed for the garden. However, in modern Japan, designers are again widening the spectrum of materials used. It is highly recommended that native plants are chosen for the garden, because showy exotic plants are not in good taste. Be aware that native plants are used in the garden, because it is in bad taste to use showy exotic plants. Although pines, cherries and bamboo immediatly remind us of Japanese gardens, we encourage to use native plants of locality that can find pleasing. If we choose evergreens as the main plant theme and combine it with deciduous material that may provide seasonal blooms or foliage color we can recreate the look of the Japanese garden.

Now the next thing taken in consideration in a Japanese garden are the ornaments or Tenkebutsu. Stone lanterns are, for westerners, a typical impression of Japanese gardens. Stone lanterns are not important components of the Japanese garden. The reason is that ornaments are subjected to the garden's design. Lanterns, stupas and basins are just architectural complements added when a point of visual interest is necessary to the design.

A good way to finish your garden design could be a well-placed lantern. The three main styles (although with many variations) are: The Kasuga style lantern, is a very formal one featuring a stone base. In the Oribe style lantern, unlike the Kasuga style, the pedestal is underneath the ground. The Yukimi or Snow-Viewing lantern is set on short legs instead of a pedestal. Consider the formality of garden setting to choose the appropiate lantern.

When possible, elements from outside the garden can be included in it. For instance, we can work a far away mountain including the scenery in design, framing it with the stones and plants existing in the garden.

4. English Gardens

For many years, English gardens were based on a formal design as were most European gardens. These were ornate, extravagant and precisely designed in formal mathematical structures primarily based in designs originating from the Paradise or Persian garden. The Italian and French gardens were the foremost exponents of this style. Inspired by critics such as

Richard Addison, English gardens began to change as a reaction against the formal structuring of the landscape. Inspired by art of the time, such as 17th century Dutch art, which strove to impart symbolic meaning, landscape architects such as Lord Burlington began to impart new ways of garden design that were opposed to the old formal look. New strategies to adapt the garden to the external environment demonstrated the following important features:

- Lines were curved and meandering.
- Designs were incorporated to engage the user to enjoy vistas.
- Focal observation points punctuated the landscape garden with sculptures, seats, or temples.
- The garden was opened up to the surrounding land.
- Lawns were extensive.
- Plants of differing heights, shapes and colours were used to create a sense of balance and manoeuvre lines of sight to important views.

Strong design ideas purported mimic nature, to some degree. This can be seen in William Kent's motto that Nature abhors straight lines and, as such, straight lines were to avoided, all water features, even fountains, were to be avoided but nature water features such as a lake were acceptable. This represented a significant contrast from the formal or Persian gardens of the past. Given the strong relationship between art and the English landscape movement, the focus on lighting and shading, were expressed in terms of shrubs and trees. Gardens were to be like pictures. Lawns, which were a predominant feature of English garden, often spread right to the house. Some considered the lawn to be like a carpet in the garden. This more natural style attempted to blend in the garden to the nature landscapes surrounding it. Like most things, the new designs of today's gardens and landscapes borrow heavily from designs of the past. Given our human propensity to merge styles, adopt old concepts and even consider them new by modern standards, there are a few aspects of historical designs in evidence today. These adaptations of historical design qualities include:

- The use of formal design qualities such as the use of geometric design.
- Incorporating the design of the house into the garden.
- The use of individual 'rooms' or defined spaces.
- The use of curved spaces.
- The incorporation of landscape outside of the actual design space.
- Adaptation of features of the land into the garden.

The use of formal design qualities such as geometric design and the matching of the design of the house in the garden design are both examples of adopted techniques of the Paradise or formal garden. The use of garden 'rooms' or places to evoke certain functions or spiritual experiences are examples of design qualities adopted from both the Japanese and formal styles. The incorporation of the external landscape and the adaptation of external features of the land into the garden are examples of using the features of both the English landscape movement and the Japanese garden. Finally, the use of curved lines with an attempt to integrate the shapes and plantings from the surrounding landscapes can be seen as adaptations of the English landscape movement. Examples include:

1. Matching large windows in the house with complementary formal design of patio or garden space.
2. Use of geometric patterns in patio or gardening areas to match a modern house design.
3. Creating rooms or specific areas in the garden for certain activities such as meditating, eating, resting or even certain types of planting such as an orchard or berry patch.
4. Integrating background images from the surrounding areas to broaden the garden footprint.

This may include the limited use of evergreens in the garden to draw in a large forest area beyond the garden itself. The historical gardens of mankind continue to be represented in today's modern garden design whether it is for a small urban space or a larger, more rural, setting. Each of the three styles of gardens discussed herein has had a major influence on garden design and continues to contribute through various aspects of their unique styles to the work of the modern garden and landscape designer.

The 18th century saw a swing from Renaissance formality to a more natural look. One of the prime movers of this style was the art patron, Lord Burlington. William Kent designed an influential garden for Lord Burlington at Chiswick House based on carefully calculated vistas with temples, statues, and classical ornaments punctuating openings in treed parkland.Lines were no longer straight, paths curve and wander, and parterres are replaced by grass. Trees were planted in clusters rather than in straight lines and rounded lakes replaced the rectangular ponds of the earlier style. The garden became open, a park joining the house to the outside world rather than a carefully nurtured refuge from it.This natural style begun by Kent evolved into the "landscape garden" under Kent's pupil and son-in-law, Lancelot "Capability" Brown. Brown, whose curious nickname came from his habit of telling prospective clients that their gardens showed "great capabilities", had an enormous effect upon the course of English gardening and architectural style.The landscape garden made the English country house a part of the fields and farmlands surrounding it. Gone were hedgerows and fences. Gone, too, were formal beds and walks. Grass parkland was brought right up to the doors of the house. The greatest surviving landscape gardens by Capability Brown are at Longleat (Wiltshire), Burghley, Petworth and Blenheim Palace (Oxon).Humphrey Repton carried on Brown's landscape garden mastery, though Repton introduced gravel walks and re-introduced separate flower gardens. Repton replaced the earlier classical ornaments with romantic structure like grottoes and fake ruins.

In the Victorian era the pendulum swung again, to massed beds of flowers (bedding out plants raised in greenhouses), exotic colours and intricate designs. The most influential gardeners of this period were J.C. Loudon and later, Joseph Paxton (Chatsworth House and Kew). The Victorian period also saw a profusion of public gardens and green spaces aimed at bringing culture to the masses. Some of the finest Victorian gardens are public parks, like People's Park in Halifax.Taste in the late Victorian period varied between formal and the "wild" garden advocated by the influential writer William Robinson. Sometimes the formal and informal looks were combined in the same garden, as at Sissinghurst Castle (Kent) and Hidcote (Gloucestershire). Gertrude Jekyll is arguably the most influential gardener of 20th century England. She popularized the herbaceous border and planning a garden based on colour schemes. This built on the tradition of the "Cottage garden", with its profusion of flowers wherever space permits and climbers on trellises and walls. Jekyll saw the house and garden as part of an integral whole, rather than the garden as an afterthought to the building. Her work survives at Marsh Court (Hampshire) and Hestercombe (Somerset).

Gardening has always been a matter of personal taste and often the outstanding works of previous generations are torn down to make way for the style of the next. For that reason it is hard to find unaltered examples of historical gardens in England.

Cottage Garden

Cottage garden is an informal garden style portraying the English countryside and is patterned with perennial flowers. It is a beautiful sight with jumbled, colourful flowers, planted haphazardly in an informal manner designed as a replica of a natural English landscape. They provoke gaiety and joy in a person. The plants that are normally used for this design are rose bushes, perennial flowers, vine-covered arbores etc. It is not an organized and arranged pattern and it is a beauty in itself with bushes and plants tumbling over the walkways. Cottage gardens go back many centuries, but their popularity grew in 1870s England in response to the more structured English estate gardens that used formal designs and massed colours of brilliant greenhouse annuals. They are more casual by design, depending on grace and charm rather than grandeur and formal structure.

The earliest cottage gardens were far more practical than their modern descendants-with an emphasis on vegetables and herbs, along with some fruit trees, perhaps a beehive and even livestock. Flowers were used to fill any spaces in between. Over time, flowers became more dominant. Modern day cottage gardens include countless regional and personal variations of the more traditional English cottage garden.

Residential Garden

A residential or domestic garden is the most common form of garden and is generally found in proximity to a residence, such as the front or back garden. The front garden may be a formal and semi-public space and so subject to the constraints of convention and law. While typically found in the rear of the residence, a garden may also be established on a roof, in an atrium, on a balcony, in window boxes, or on a patio. Residential gardens are typically designed at human scale, as they are most often intended for private use. However, the garden of a great house, castle or a large estate may be larger than a public park in a village and may produce foodstuffs as well. Residential gardens may feature specialized gardens, such as those for exhibiting one particular type of plant, or special features, such as rockery or water features. They are also used for growing herbs and vegetables and are thus an important element of sustainability.

Kitchen Garden

The traditional kitchen gardenis a seasonally used space separate from the rest of the residential garden - the ornamental plants and lawn areas. Most vegetable gardens are still miniature versions of old family farm plots with square or rectangular beds, but the kitchen garden is different not only in its history, but also its design.

The kitchen garden may be a landscape feature that can be the central feature of an ornamental, all-season landscape, but can be little more than a humble vegetable plot. It is a source of herbs, vegetables, fruits and flowers, but it is also a structured garden space, a design based on repetitive geometric patterns.The kitchen garden has year-round visual appeal and can incorporate permanent perennials or woody plantings around (or among) the annual plants.

Rock Garden

A rock garden, also known as a rockery or an alpine garden, is a type of garden that features extensive use of rocks or stones, along with plants native to rocky or alpine environments.

Rock garden plants tend to be small, both because many of the species are naturally small and so as not to cover up the rocks. They may be grown in troughs (containers), or in the ground. The plants will usually be types that prefer well-drained soil and less water.

The usual form of a rock garden is a pile of rocks, large and small, esthetically arranged and with small gaps between, where the plants will be rooted. Some rock gardens incorporate bonsai.

Some rock gardens are designed and built to look like natural outcrops of bedrock. Stones are aligned to suggest a bedding plane and plants are often used to conceal the joints between the stones. This type of rockery was popular in Victorian times, often designed and built by professional landscape architects. The same approach is sometimes used in modern campus or commercial landscaping, but can also be applied in smaller private gardens.

The Japanese rock garden, in the west often referred to as Zen garden, is a special kind of rock garden with hardly any plants. The Rock Garden is a sculpture garden in Chandigarh, India. Spread over an area of forty-acre (160,000 m/square), it is completely built of industrial & home waste and thrown-away items.

Knot Gardens

A Knot garden is symmetrically designed in geometric patterns for a formal garden using herbs. Shrubbery like English boxwood plays an important role in the knot gardens owing to its pruning conformity to the desired precise measurements.

Water Gardens

Water gardens as the name implies features artificial water in your garden. You can bring home the natural gurgling sound of the streams with artificial ponds and fountains. Further more you can beautify your garden pond with floating aquatic plants to add more life to your pond.

Hedges

A hedge is a wall composed of plants. A Hedge not only serves the purpose of security but also privacy and is also a windbreak. When pruned decoratively, they add more charm and beauty to the garden apart from their various functionalities. Holly plants, boxwood shrubs, mountain laurels, English and Japanese yews and deciduous hedge shrubs are some of the examples of hedge plants. You can have the hedges, rock garden and water garden as features in your garden especially featuring a vast expanse of grass. The knot gardens, cottage gardens and the vegetable gardens themselves can provide a stylish garden setting on their own.

5. Italian Gardens

In Italy there has always been a strong tradition for creating formal gardens around the elaborate and romantic architecture, from Sicily to Venice and from Turin to Florence. This style originated in Greek and Arabic garden design, yet continued through Roman architecture into contemporary Italian landscape design. To create such a formal area it is essential to choose a level site that can display the gardens symmetry to the full, an area should be chosen that finds itself within close proximity to the house in order to fully highlight and extend the architecture of the house and propel it into the realm of the garden. Any paths, hedging or axis points in the formal garden should run out parallel or horizontally from the symmetry of the house, and the dimensions of pathways, flower borders etc should be in scale with the principal features of the facade, i.e. the entrance, a large window, or the roof-line. There are many evergreen plants which can be used to create geometric patterns but the most preferred varieties over the centuries have been; Yew, Box, Rosemary, Myrtle, Lavender and Holly, although there are many, many more useful shrubs that can be clipped into small hedges. By far the most versatile is Box (*Buxus sempervirens*), which requires just two simple trims per year to maintain a perfectly complete and tidy geometric shape.

6. French Garden

There is something about a French garden that is charmingly paradoxical. The flowers grow a little wild but in tempered rows. The overall colour is subdued, but is usually dominated by purple, a shade that is rarely found in nature except for in the exotic likes of orchids and lavender. In addition to lavender, other popular flowers include agapanthus, canna and dahlias, to name a few. A French garden accomplishes the cosmetic aesthetic that has vexed women for centuries that is to look the most beautiful. Plant the purple-gry leafed agapanthus in well-drained, rich soil. Know that the cannas abounds throughout Southern France and thrives best under heavy sun and copious amounts of water. Dahlias provide a summer time burst of color and grow hardy with lots of sun and well-drained soil that has a pH of 6.5.Think blending when planting the seeds. There should be a sense of natural abandon in the garden, despite the planning that goes into it. Towering flowers should be near petite ones, while the likes of solitary standing sunflowers should be close to bouquets of dahlias. Do not forget the art of scent when blending the garden. The soothing scent of lavender coupled with the rich aroma of rose creates a perfume a la natural.

Understand that water is an important feature of any French garden, so installing fountains or creating ponds and waterfalls is key.Keep in mind that tucking trellises and arbores throughout the garden will encourage roses and bougainvillea to grow, creating quaint nooks and crannies.

Elements of Garden Design

Whether a garden is designed by a professional or an amateur, certain principles form the basis of effective garden design, resulting in the creation of gardens to meet the needs, goals and desires of the users or owners of the gardens.

Elements of garden design include the layout of hard landscape, such as paths, walls, water features, sitting areas and decking; as well as the plants themselves, with consideration for their horticultural requirements, their season-to-season appearance, life span, growth habit, size, speed of growth, and combinations with other plants and landscape features. Consideration is also given to the maintenance needs of the garden, including the time or funds available for regular maintenance, which can affect the choices of plants regarding speed of growth, spreading or self-seeding of the plants, whether annual or perennial and bloom-time and many other characteristics.

The most important consideration in garden design is how the garden will be used, followed closely by the desired stylistic genres, and the way the garden space will connect to the home or other structures in the surrounding areas. All of these considerations are subject to the limitations of the budgetary concerns for the particular project and time. Budget limitations can be addressed by a simpler more basic garden style with fewer plants and less costly landscape materials, seeds rather than sod for lawns and plants that grow quickly; alternately, garden owners may choose to create their garden over time, area by area, putting more into each section than could be handled all at once.

Location

A garden's location has a substantial influence on the garden design. Many of the great gardens of history and today possess a location that is topographically significant and has a suitable microclimate for plants, a well-designed connection to water and rich soil. However, a good garden design, one that is well-planned and constructed, can increase the value of the garden more than its location.

Soil

The quality of a garden's soil often has a significant influence on the success of the garden. Soil influences the availability of water and nutrients, the activity of beneficial soil organisms and a wide variety of other factors important to plant growth. Traditionally, garden soil is improved by amendment, the process of adding beneficial materials to the excavated native subsoil and topsoil. The materials, which may consist of compost, peat, sand, mineral dust, or manure, among others, are mixed with the excavated soil. The amount and type of amendment may depend on the ratio of clay to humus and on the soil acidity or alkalinity.

Boundaries

The look of the garden can be influenced strongly by the boundary impinges. Planting can be used to modify the boundary line or a line between an area of rough grass and smooth, depending on the size of the plot. Introducing internal boundaries, perhaps in the form of hedges or group of shrubs, can help break up a garden.

Hedges vary their colours throughout the seasons dramatically. Hedges, being strong features in a garden, are often used to divide sections of the garden. However, since they use the moisture and nutrient from the garden soil to grow as well as other plants, they may not be a good choice and may bring a negative effect to the other plants.

Besides the boundaries that are made up of plants like the hedges, walls made up of various materials can be built between regions. There are broadly three types of walling material: stone, either random or coursed, brick, and concrete in its various forms. It is good to determine what colour, size, and texture will be most appropriate for the garden before actually building the wall.

Garden Furniture

Garden furniture may range from a patio set consisting of a table, four or six chairs and a parasol, through benches, swings, various lighting, to stunning artifacts in brutal concrete or weathered oak. A picnic table, is used for the purpose of eating a meal outdoors such as in a garden.The materials used to manufacture modern patio furniture include stones, metals, vinyl, plastics, resins, glass, and treated woods.

Sunlight

While sunlight is not always easily controlled by the gardener, it is an important element of garden design. The amount of available light is a critical factor in determining what plants may be grown. Sunlight will, therefore, have a substantial influence on the character of the garden. For example, a rose garden is generally not successful in full shade, while a garden of host as may not thrive in hot sun. As another example, a vegetable garden may need to be placed in a sunny location, and if that location is not ideal for the overall garden design goals, the designer may need to change other aspects of the garden.

In some cases, the amount of available sunlight can be influenced by the gardener. The location of trees, other shade plants, garden structures, or, when designing an entire property, even buildings, might be selected or changed based on their influence in increasing or reducing the amount of sunlight provided to various areas of the property.

In other cases, the amount of sunlight is not under the gardener's control. Nearby buildings, plants on other properties, or simply the climate of the local area, may limit the available sunlight. Or, substantial changes in the light conditions of the garden may not be within the gardener's means. In this case, it is important to plan a garden that is compatible with the existing light conditions. Light regulates three major plant processes: photosynthesis, phototropism, and photoperiodism.

Lighting

Garden lighting can be an important aspect of garden design. In most cases, various types of lighting techniques may be classified and defined by heights: safety lighting, uplighting, and downlighting. Safety lighting is the most practical application. However, it is more important to determine the type of lamps and fittings needed to create the desired effects.

Features of a formal garden

- Terrace
- Topiary
- Statuary
- Hedge
- Bosquet
- Parterre
- Sylvan theatre
- Pergola
- Pavilion Landscaping

CHAPTER **18**

Low-Maintenance Palms, Cycads and Grasses for Landscape

The term "low-maintenance" refers to a plant that does not require frequent maintenance such as regular watering, pruning, or spraying to remain healthy and to maintain an acceptable aesthetic quality. A low-maintenance plant has low fertilizer requirements and few pest and disease problems. In addition, low-maintenance plants suitable for India must also be adapted to or at least tolerate our poor, alkaline, sand or limestone-based soils. Both native and non-native species are included herein. Also, a variety of plants that are not commonly available in Indian nurseries or garden centres were intentionally included with the hope that increased demand will generate increased supply. Putting the right plant in the right place is of foremost importance in creating a healthy and successful low-maintenance landscape. However, this principle is dependent upon one's ability to accurately select the right plant species for a given location. Common names are often misleading and sometimes more than one plant species may be referred to by the same or similar common name.

Table 18.1 : Low-Maintenance Palms, Cycads and Palm-Like Plants

Common name/ ***Scientific name***	**Size (m)**	**Light preference**	**Salt tolerance**	**Remarks**
Macaw palm, gru-gru palm *Acrocomia aculeata*	9	Full sun	High	Extremely fast-growing palm similar in overall appearance to queen palm, except with denser canopy and 4-6" sharp, black spines on trunk, petiole, and leaf rachis.
Seashore palm *Allagoptera arenaria*	1.5 -3	Full sun to light shade	High	Clumping palm with short prostrate or subterranean stems and graceful pinnate leaves, green above and silver below. Fruit arranged spirally on long stalk; cluster resembles pineapple. Excellent for beach-front. Prefers sandy soils.
Cohune palm *Attalea cohune*	6-15	Full sun	Low	Huge palm with extremely long pinnate leaves (to 33 feet each) that erupt from the trunk in a shuttle-cock shape. Cream-colored flower clusters followed by huge clusters of brownish yellow, woody fruits. Prefers moist soil; can tolerate drought.
Ponytail "palm" *Beaucarnea recurvata*	3-9	Full sun to partial shade	High	Often mistakenly called a palm, this member of the agave family requires excellent drainage. Can tolerate some shifting shade. Also known as *Nolina recurvata*.
Bismarck palm . *Bismarckia nobilis*	6-9	Full sun	None	Enormous palm with huge silvery-blue or green leaves. A native of Madagascar, this is an extraordinarily durable and highly adaptable palm

Blue hesper palm *Brahea armata*	15	Full sun	High	Tall palm with pastel bluish-white, palmate leaves. Trunk slightly swollen at base and up to 1.5 feet wide, bearing persistent leaf scars. Prefers well-drained soil and arid climate.
Sierra Madre palm *Brahea decumbens*	2	Full sun to partial shade	High	Low-growing palm with prostrate trunk and large, blue, fan-shaped leaves. Prefers well-drained soil and arid climate.
Bamboo cycad *Ceratozamia hildae*	1.5-2	Full sun to partial shade	Low	Trunkless cycad with upright, bamboo-like leaves. Extremely cold tolerant. Green or brown emergent leaf forms. Requires well-drained soil.
Cycad *Ceratozimia kuesteriana*	1-1.5	Full sun to partial shade	Low	Trunkless, shrubby cycad with brown emergent leaves. One of the few cycads that is completely spineless, making it a good plant for along sidewalks. Requires well-drained soil.
Cycad *Ceratozamia latifolia*	1.8-3.5	Full sun to partial shade	Low	Large cycad with brown, red, or green emergent leaves. May attain a spread of nine feet. Requires well-drained soil.
Cycad *Ceratozamia robusta*	2-2.5	Full sun to partial sun	Low	Large cycad with green emergent leaves. Cold hardy. Requires well-drained soil.
Stolon chamaedorea *Chamaedorea stolonifera*	1-1.5	Partial shade	None	Sparsely clumping rainforest palm from Central America. Nearly entire leaves and raised annular rings on slender stems quite attractive. Grows best in well-drained soil. Drought tolerant. Forms large stands vegetatively by stolons.

Red leaf palm *Chambeyronia macrocarpa*	6	Full sun to partial shade	None	Rainforest palm from New Caledonia. New leaf is bright red to maroon, fading to dark, glossy green. Requires shade when young and well-drained soil. Two forms: green crownshaft and yellow/cream crownshaft (latter often called *C. hookeri*).
Silver palm *Coccothrinax argentata*	1-3	Full sun	High	Small, slow-growing native palm. Round, palmate leaves with silvery undersides. Numerous small, purplish-black fruit produced nearly year-round.
Old man palm *Coccothrinax crinita*	2.5-3.5	Full sun	High	Slow-growing Cuban palm bearing characteristic dense "beard" of gray or straw-colored fibers on trunk. Leaves large, round, palmate. Fruit glossy purple-black.
Coconut palm *Cocos nucifera*	9	Full sun	High	Popular tropical palm. "Maypan" and "Fiji Dwarf" varieties resistant to Lethal Yellowing.
Bailey copernicia palm *Copernicia baileyana*	9	Full sun	High	Slow-growing palm from Cuba. Huge, stiff, blue-green leaves top massive, slate gray trunk that resembles concrete.
Cuban wax palm *Copernicia hospita*	7.5	Full sun	High	Handsome palm of moderate dimensions. Gray, waxy, fan-shaped leaves on spiny petioles. Prefers well-drained soil and arid climate, but can tolerate south Available in green and blue forms.
Cuban petticoat palm *Copernicia macroglossa*	2.5-4.5	Full sun	High	Another Cuban palm. Old leaves retained on trunk form characteristic skirt or petticoat.

Carnauba wax palm *Copernicia prunifera*	5	Full sun	Moderate	Moderately sized palm with open crown of silvery-white, fan-shaped leaves. Trunk smooth on top third, but bears persistent leaf bases on bottom two-thirds. Wax on leaves used to make heat-resistant carnauba wax. Prefers well-drained soil and arid climate.
Virgin's "palm" *Dioon edule*	1-2	Full sun	High	Medium-sized Mexican cycad. Similar in overall appearance to king sago, but with lighter green leaves. Very cold tolerant. Forms offsets at base that grow into separate trunks. Requires well-drained soil.
Cycad *Dioon mejiae*	1.5	Full sun to partial shade	High	Medium-sized cycad. Larger than *D. edule*, but smaller than *D. spinulosum*. New leaves bear soft, golden hairs. Suitable replacement for queen sago. Requires well-drained soil.
Cycad *Dioon rzedowskii*	1.5-3	Full sun to partial shade	Low	Rare, medium-sized cycad from Mexico. Similar to, but much smaller than, *D. spinulosum*. Use as specimen or accent plant. Requires well-drained soil.
Cycad *Dioon spinulosum*	2.5-9	Full sun to partial shade	Low	Largest American cycad. Similar in overall appearance to queen sago. Grows best in filtered sunlight. Requires well-drained soil.
African oil palm *Elaeis guineensis*	9-15	Full sun to partial shade	Low	Tall palm with long, feathery, pinnate fronds. Trunk smooth on top, rough on bottom from persistent leaf bases. Fruit orange, ripen to black; held tight near trunk;

				high in oil content. Not suitable for small lot.
Cycad *Encephalartos ferox*	2	Partial shade	High	Medium-sized South African cycad with subterranean trunk. Long, glossy, dark green, spiny leaves with leaflets resembling holly leaves. Impressive, large, bright red cones produced on males and females. Offsets form at base of plant. Requires well-drained soil.
Cycad *Encephalartos gratus*	5	Full sun	High	Large, trunk-forming cycad from South African. Long, glossy, dark green, spiny leaves. Offsets form at base of plant. Requires well-drained soil.
Cycad *Encephalartos hildebrandtii*	3.5	Full sun	High	Large South Africa cycad. Long, glossy, dark green, upright, spiny leaves. Smaller than *E. gratus*. Offsets form at base of plant. Requires well-drained soil.
Cycad *Encephalartos villosus*	2-3	Partial to full shade	Moderate	Medium-sized, clumping, trunkless cycad with long, graceful, glossy dark green leaves. Females bear large, orange cones. Not common. Requires well-drained soil.
Cuban belly palm *Gastrococos crispa*	9	Full sun	Moderate	Graceful palm with spiny trunk and leaves. Slow-growing at first, increasing as trunk forms. Trunk distinctly swollen in middle. Orange, golf-ball sized fruit.
Palm *Guihaia argyrata*	0.5-1.5	Partial to full shade	Low	Small, slow-growing, clumping, trunkless palm with fan-shaped leaves, green above and silvery brown below. Can be mistaken for *Rhapis*. Not common.

Sagisi palm *Heterospathe elata*	9	Full sun to partial shade	None	Tall rainforest species from the Philippines and adjacent islands. Slow-growing when young; grows quite fast once trunk forms. Bears crown of gracefully arching leaves. Many small, white fruit produced nearly continuously upon maturation.
Doum palm, gingerbread palm *Hyphaene coriacea*	0.5-1	Full sun	None	Coarse palm with arching, palmate, gray-green leaves that have spiny petioles. Trunk with persistent leaf bases. Unique among palms, in that it branches up to two or three times. Orange, pear-shaped fruit in long clusters.
Cycad *Lepidozamia peroffskyana*	6	Partial to full shade	None	Tall Australian cycad. Elegant, glossy green leaves grow to 10 feet long, with narrow leaflets to 12 inches long. Use as specimen or accent plant. Requires well-drained soil.
Chinese fan palm *Livistona chinensis*	5-12	Full sun to partial shade	None	Medium-sized palm with large, palmate, light green leaves split at the tips. Relatively fast growing. Produces large quantities of olive-shaped blue fruit. Petioles spiny.
Australian ribbon palm *Livistona decipiens*	9	Full sun to partial shade	Low	Tall, slender palm with graceful, deeply segmented, fan-shaped leaves bearing long, thin ribbons. Best when protected from strong winds.
Dwarf livistona palm *Livistona muelleri*	3.5-6	Full sun	Low	Medium-sized palm with stout trunk and crown of large, stiff, erect, bluish-green, deeply segmented, fan-shaped leaves. Striking red, branched inflorescences followed by equally striking

				blue, olive-sized fruit. Not common.
Footstool *Livistona rotundifolia*	9-12	Full sun to partial shade	Low	Tall, slender, fan palm with broad, nearly symmetrical, shallowly segmented leaves. Bright red fruit borne on long, thin, arching, stalk.
Madagascar "palm" *Pachypodium lamerei*	2-5	Full sun	High	Tall-growing, sparsely branching cactus look-alike. Spiny trunks bear showy white flowers upon maturity. Not a palm. Other *Pachypodium* also drought tolerant.
Screw pine, screw palm *Pandanus utilis*	5-7.5	Full sun	High	Many-branched, palm-like plant with saw-toothed leaves that resemble pineapple. Prefers moist conditions, but can tolerate drought. Not a palm. Occasionally susceptible to Lethal Yellowing disease. Other *Pandanus* also drought tolerant.
Canary Island date palm *Phoenix canariensis*	9-15	Full sun	Low	Tall, robust date palm with huge rounded crown of long, arching, pinnate leaves. Best for avenue plantings; not suitable for small lot.
Palm *Pseudophoenix lediniana*	6	Full sun	Moderate	Medium-sized palm with long, graceful, pinnate leaves and smooth, gray trunk distinctly swollen in middle. Related to *P.sargentii*, but much faster growing.
Traveller's tree *Ravenala madagascariensis*	9	Full sun	Low	Bright green, banana-like leaves up to 10 feet long, forming two opposite rows and held on long, tightly overlapping stalks. Spreading form reminiscent of flat fan of foliage. Clusters of white flowers between leaf bases in summer. Not a palm.

Puerto Rican hat palm *Sabal causiarum*	15	Full sun	Moderate	Large palm with massive, smooth gray trunk up to 4 feet in diameter. Large, fan-shaped leaves up to 6 feet long, deep green in color, deeply segmented, drooping at ends. Not for small lots.
Keys thatch palm *Thrinax morrissii*	5	Full sun to partial sun	High	Grows naturally in alkaline soils, sometimes on limestone outcrops. Leaves are bluish-green above and silvery below. Small white fruit produced in the fall.
Veitchia palm *Veitchia* spp.	15	Full sun	None	Fast-growing palms from South Pacific. Widely planted in south Florida. High tolerance of hurricane-force winds. Wide adaptability to varying soil, water, and nutrient conditions.
Washingtonia palm *Washingtonia robusta*	15	Full sun	High	Tall, common fast-growing palm. Leaves large, palmate, with toothed petioles.
Cycad *Zamia herrerae*	1-1.5	Full sun	High	Medium-sized, rare, subterranean cycad from southern Mexico and Central America. Has good landscape potential, with long, stiffly erect, slightly twisting leaves and shrub-like appearance. Requires well-drained soil.
Cycad *Zamia inermis*	1-1.5	Full sun	High	Medium-sized, rare, stem-forming cycad from Veracruz, Mexico. Has good landscape potential, with long, upright leaves and shrub-like appearance. Requires well-drained soil.
Zombie palm *Zombia antillarum*	2.5-6	Full sun	High	Tall, clumping palm with characteristic narrow trunks

				bearing rings of 2-3" yellow-beige spines attached to woven, fabric-like leaf sheathes. Leaves palmate and mid-green. Produces clusters of olive-sized, white fruit.

Table 18.2 : Low-Maintenance Ornamental Grasses

Common name/ ***Scientific name***	**Size (m)**	**Light preference**	**Salt tolerance**	**Remarks**
Big bluestem *Andropogon gerardii*	1.5-2.5	Full sun	Moderate	Perennial, upright, bunch grass forming large clumps. Stems with blue at base. Leaves blue-green in early summer, maturing to copper or orange-brown in fall. Bronze to purplish seed heads, to 3 inches late summer. Tolerates flooding in summer.
Indian wood-oats *Chasmanthium latifolium*	0.5-1.5	Full sun to partial shade	None	Attractive, clump-forming grass with wide leaves and nodding, arching clusters of flat, oat-like seedheads. Resembles small sea-oats. Dried flower heads highly prized.Also grown as ground cover in shaded areas.
Pampas grass *Cortaderia selloana*	2-3	Full sun	Low	Impressive, graceful, fountain-like clumps. Large, 12-18" tall plumes erupt from foliage in mid-summer. Will tolerate poorly drained soil and brief flooding. Dried flower heads highly prized.
Lemongrass *Cymbopogon citratus*	1-2	Full sun to light shade	Low	Elegant, fragrant, clumping grass with thin, strap-like, yellow-green leaves that release citrus aroma when crushed. Prefers moist soil; moderate drought tolerance.

Lilyturf *Liriope muscari*	0.3-0.6	Partial shade	Moderate	Grass-like foliage and trumpet shaped flowers borne on erect stems in blue, purple, pink, or white, depending on species or cultivar. Cold tolerant. Needs freely draining soil. Mulch to control weeds. Variety 'Evergreen Giant' can take full sun.
Pink muhly grass, hairgrass *Muhlenbergia capillaris*	0.6-1	Full sun to partial shade	High	Compact, tuft-forming grass with fine, pink feather-like flower heads held high above the leaves.
Blue muhly grass *Muhlenbergia lindheimeri*	0.3-0.6	Full sun	High	Similar to pink muhly grass, but with bluish-gray foliage and purplish to gray flower heads. Prefers dry, but will tolerate wet soil. Excellent groundcover for poor, sandy soils.
Mondo grass *Ophiopogon japonicus*	2-7.5	Partial to full shade	High	Dark green to blackish, glossy leaves. Small, lavender flowers followed by iridescent, blue-black berries. Good border plant, especially under trees where turfgrass will not grow.
Beach panic grass *Panicum amarum*	0.3-0.6	Full sun	High	Clumping grass with flower stalks rising above foliage. Grows well in sand. Good for beach-front.
Oriental fountain grass *Pennisetum orientale*	0.3-0.6	Full sun to partial shade	High	Low-growing, compact fountain grass with pink or white, fluffy inflorescence in summer and fall. Prefers well-drained soil. The related *P. purpureum* is in invasive category.
Sand cordgrass *Spartina bakeri*	1.2-2.5	Full sun	High	Large, bunch-forming grass with thin, brownish-green, rolled, wire-like, sand-

				papery leaves. Highly adaptable and under-utilized. Good for beach-front.
Fakahatchee grass *Tripsacum dactyloides*	1.2-2.5	Full sun to partial shade	Moderate	Large, native grass. Rich green foliage erupts from fountain-like clumps. Distinctive flowers rise above leaves on slender stems in mid summer. Easy to grow. Virtually free of pests. Prefers moist soil, but has good drought tolerance.

CHAPTER 19

Reference Guide for Landscape Plants

Contd...

Table 19.1 : Reference Guide for Landscape Plants.

Botanical Name	Common Name	Plant Description	Growth	Moisture	Tall/Wide
Abelia X 'Edward Goucher'	Ed Goucher Pink Abelia	Vibrant pink flowers are produced all summer above the dark green leaves of this compact shrub.	Average Growth	Moist	4' x 5'
Abelia X grandiflora	Glossy Abelia	Profuse white flowers appear through the summer above the glossy green leaves of this compact shrub.	Average Growth	Moist	3' x 5'
Abelia X grandiflora 'Francis Mason'	Francis Mason Glossy Abelia	Profuse white flowers top the show y bright yellow and green leaves of this compact shrub.	Average Growth	Moist	6' x 5'
Abelia zanderi 'Sherwoodii'	Sherwood Dwarf Abelia	Showy white flowers during the summer crown the refined dark green leaves of this dense, compact shrub.	Average Growth	Moist	3' x 5'
Abies balsamea 'nana'	Dwarf Balsam Fir	A very dwarf evergreen that grows well in the shade and has the richest dark green color.	Slow Growth	Moist	1' x 3'
Abies concolor	White Fir	A dense pyramidal tree with beautiful silver evergreen folia ge. Prefers open well drained locations.	Slow Growth	Moist	50' x 25'
Abies fraseri	Fraser Fir	A layered deep green pyramidal evergreen tree with soft fragrant foliage.	Slow Growth	Moist	30' x 10'
Acer campestre	Hedge Maple	One tough customer, this small maple thrives in dry, poor soil conditions so common in today's landscape. Leaves turn yellow late in the season.	Slow Growth	Moist	25' x 25'
Acer griseum	Paper Bark Maple	A small tree sought after for its peeling tissue -thin coppery bark that makes this tree a champion of the winter landscape. Scarlet -orange fall color.	Slow Growth	Moist	15' x 15'

Acer palmatum	Japanese Maple	A graceful small tree with delicate foliage and an artistic branch form lending an oriental look to any garden. Fall colors are brilliant.	Slow Growth	Moist	15' x 15'
Acer palmatum 'Atropurpureum'	Red Japanese Maple	This Japanese Maple is noted for its vivid red leaves that light up the summer landscape. The red color intensifies for a brilliant fall display.	Slow Growth	Moist	15' x 15'
Acer palmatum 'Beni Otaki'	Beni Otaki Japanese Maple	A small ornamental tree with beautiful foliage.	Slow Growth	Moist	15' x 15'
Acer palmatum 'Bloodgood'	Bloodgood Japanese Maple	This is the most popular upright Japanese Maple selection because of its very deep red to black-red leaf color and strong growth habit.	Slow Growth	Moist	25' x 25'
Acer palmatum 'Burgundy Lace'	Burgundy Lace Japanese Maple	A spreading small tree with beautiful red finely cut leaves.	Slow Growth	Moist	10' x 10'
Acer palmatum 'Butterfly'	Butterfly Japanese Maple	One of the older variegated forms with emerging leaves green with pink margins.	Slow Growth	Moist	12' x 8'
Acer palmatum 'Crimson Queen'	Crimson Queen Japanese Maple	Known for its fine, delicate deep red leaves that turn extremely bright scarlet in fall; it ages gracefully into a beautiful cascading small tree.	Slow Growth	Moist	15' x 15'
Acer palmatum 'Dissectum'	Cutleaf Japanese Maple	This shade loving small tree is known for its very fine leaves that display blended colors of green, red, white and pink.	Slow Growth	Moist	15' x 25'
Acer palmatum 'Dissectum Seiryu'	Green Cutleaf Japanese Maple	This unique Japanese Maple combines the fine, delicate green leaves with an upright growth habit. Fall colors are spectacular yellow to gold.	Slow Growth	Moist	10' x 15'
Acer palmatum 'Ever Red'	Ever Red Weeping Japanese Maple	A beuatiful weeping mound of finely cut leaves that deep red in the spring and a rich bronze late in the summer.	Slow Growth	Moist	6' x 8'

Acer palmatum 'Inaba Shidare'	Red Weeping Japanese Maple	A beautiful weeping tree with red leaves all summer.	Slow Growth	Moist	10' x 15'
Acer palmatum 'Lion Heart'	Lion Heart Japanese Maple	A compact growing tree with interesting twisted leaves.	Slow Growth	Moist	10' x 6'
Acer palmatum 'Oshio-Beni'	Oshio-Beni Red Japanese Maple	This stout, upright small tree has bright red leaves in spring, changing to maroon red in summer and then turning bright red in fall.	Slow Growth	Moist	20' x 20'
Acer palmatum 'Red Dragon'	Red Dragon Japanese Maple	A small tree with dark red leaves all sumer.	Slow Growth	Moist	15' x 15'
Acer palmatum 'Red Ribbonleaf'	Red Ribbonleaf Japanese Maple	A small tree with dark red leaves all sumer.	Slow Growth	Moist	15' x 15'
Acer palmatum 'Sango Kaku'	Coral Barked Japanese Maple	Brilliant coral-colored bark on young branches is the outstanding feature of this vigorous, medium-sized Japanese Maple. Yellow-gold fall color.	Slow Growth	Moist	15' x 15'
Acer palmatum 'Suminagashi'	Sumininagashi Red Japanese Maple	A small ornamental with beautiful foliage.	Slow Growth	Moist	15' x 15'
Acer palmatum 'Tamugkeyama'	Tamugkeyama Red Japanese Maple	Young foliage is deep crimson-red and matures to purple-red and holds quite well.	Slow Growth	Moist	15' x 15'
Acer palmatum 'virdis'	Green Weeping Japanese Maple	A graceful small weeping tree with small fine textured green leaves that turn yellow in the fall. Grows well in the shade.	Slow Growth	Moist	8' x 10'
Acer platanoides	Norway Maple	A favorite street tree because of its rounded, dense branching habit and it tolerates dry conditions. Dark green leaves turn golden yellow in fall.	Fast Growth	Moist	50' x 35'

Acer platanoides 'Cleveland'	Cleveland Norway Maple	The vigorous, upright growth habit makes this an outstanding shade tree. Excellent golden-yellow fall color.	Fast Growth	Moist	50' x 20'
Acer platanoides 'Columnare'	Columnar Norway Maple	This tree produces a column of dark green leaves where space is limited. A pest-free tree with golden-yellow fall color.	Average Growth	Moist	35' x 10'
Acer platanoides 'Crimson King'	Crimson King Norway Maple	Rich purple leaves provide dense shade throughout the summer under this slow growing tree. Fall color is maroon-red.	Average Growth	Moist	40' x 30'
Acer platanoides 'Crimson Sentry'	Crimson Sentry Norway Maple	A selection of 'Crimson King' with the same dark purple leaves but a very upright growth habit that's ideal for small yards and narrow streets.	Slow Growth	Moist	30' x 10'
Acer platanoides 'Drummondii'	Variegated Norway Maple	Green leaves edged in white sets this Norway Maple apart from the crowd. A slow growing pest free shade tree.	Slow Growth	Moist	40' x 25'
Acer platanoides 'Emerald Lustre'	Emerald Lustre Norway Maple	A vigorous growing shade tree noted for its glossy green leaves and strong branching that can resist wind and snow.	Fast Growth	Moist	50' x 35'
Acer platanoides 'Emerald Queen'	Emerald Queen Norway Maple	Considered the best green-leafed selection of Norway Maple because of its oval shape and tolerance of poor soils. Reliable yellow fall color.	Fast Growth	Moist	40' x 35'
Acer platanoides 'Royal Red'	Royal Red Norway Maple	Excellent maroon-red leaf color all summer crowns this slow-growing, dense shade tree. Considered more cold hardy than 'Crimson King'.	Average Growth	Moist	40' x 25'
Acer platanoides 'Schwedleri'	Schwedler Maple	Leaves emerge bright reddish-purple in spring, turning green in summer and Bronze-yellow in fall. A slow growing shade tree that tolerates poor soils.	Slow Growth	Moist	50' x 35'

Acer rubrum	Red Maple	A rapid growing tree with dense dark green leaves that turn shades of red, orange and yellow in fall. Tolerates poorly drained soil conditions.	Fast Growth	Moist	50' x 40'
Acer rubrum 'Armstrong'	Armstrong Rubrum Maple	A fast-growing upright shade tree for small yards where space demands a tree that won't spread. Light green leaves turn orange-red in fall.	Fast Growth	Moist	50' x 15'
Acer rubrum 'Autumn Flame'	Autumn Flame Maple	The first Red Maple to turn brilliant red in the fall. A slower growing tree with a uniform spreading growth habit. A very cold hardy selection.	Slow Growth	Moist	50' x 35'
Acer rubrum 'Franksred'	Red Sunset Maple	A vigorous growing, uniform tree with lustrous green leaves in summer changing to brilliant red to orange-red in fall.	Fast Growth	Moist	50' x 30'
Acer rubrum 'Northwood'	Northwood Red Maple	A fast growing shade tree with orange fall color. Prefers acid soils and is very hardy.	Fast Growth	Moist	45' x 30'
Acer rubrum 'October Glory'	October Glory Maple	A vigorous growing tree with lustrous green leaves in summer turning crimson red in fall. The best red maple selection for fall color in the South.	Fast Growth	Moist	50' x 35'
Acer saccharinum	Silver Maple	This is a fast growing shade tree that tolerates poor growing conditions where other trees fail. Leave plenty of room for it to grow.	Fast Growth	Moist	75' x 50'
Acer saccharinum 'Silver Queen'	Silver Queen Silver Maple	A more refined selection of Silver Maple with better shape and does not produce seeds that sprout in the flower garden.	Fast Growth	Moist	60' x 50'
Acer saccharum	Sugar Maple	A beautiful shade and street tree that grows at a medium rate and has gold or orange fall color. Sugar Maple needs well drained but fairly moist soil.	Average Growth	Moist	50' x 40'

Acer saccharum 'Bonfire'	Bonfire Sugar Maple	The fastest growing Sugar Maple selection with good heat tolerance and brilliant orange-red fall color.	Fast Growth	Moist	50' x 35'
Acer saccharum 'Green Mountain'	Green Mountain Maple	The most widely planted Sugar Maple because its thick dark green leaves look good all summer and then turn bright reddish-orange every fall.	Average Growth	Moist	50' x 35'
Acer saccharum 'Legacy'	Legacy Sugar Maple	Shiny dark green leaves adorn this Sugar Maple through tough summer conditions and change to dark reddish-orange in fall.	Average Growth	Moist	50' x 40'
Acer tataricum ginnala	Clump Amur Maple	An choice very hardy small tree known for showy red fall color. Use as a alone or as a screen or windbreak.	Average Growth	Semi-Moist	25' x 25'
Acer tataricum ginnala	Amur Maple	A very cold-hardy maple that can be grown as a large shrub or small tree; small green leaves turn orange-red in fall.	Average Growth	Semi-Moist	20' x 20'
Acer tataricum ginnala 'Flame'	Flame Amur Maple	The small leaves of this large shrub to small tree light up the fall landscape with fiery orange-red to deep red color. A very cold hardy Maple.	Average Growth	Semi-Moist	20' x 20'
Acer truncatum	Purple blow Maple	Glossy green leaves turn a brilliant orange-red in fall on this medium-sized Maple that tolerates the poor growing conditions of many urban areas.	Slow Growth	Moist	20' x 20'
Acer x freeman 'Autumn Fantasy'	Autumn Fantasy Maple	A fast growing hybrid shade tree with orange to red fall color.	Fast Growth	Moist	50' x 40'
Acer x freemani 'Autumn Blaze'	Autumn Blaze Maple	A fast growing shade tree with long-lasting red fall color.This tree adapts to most soils and is almost seedless.	Fast Growth	Semi-Moist	50' x 40'

Acer X freemanii 'Morgan'	Indian Summer Maple	A rapid growing Maple selection that has the brightest orange-red to red fall color in mild climates.	Fast Growth	Moist	60' x 35'
Acer X freemanii 'Scarlet Sentenial'	Scarlet Sentenial Maple	A rapid growing upright Maple with rich green leaves that turn yellow-orange to orange-red in fall.	Fast Growth	Moist	40' x 15'
Achillea filipendullina 'Parker's Variety'	Parker's Gold Yarrow	Gold flower heads are produced in mid-summer on sturdy 3 feet tall stems over the gray green feathery leaves. Excellent for cutting and drying.	Fast Growth	Semi-Moist	3' x 4'
Achillea millefolium 'Apple Blossom'	Apple Blossom Yarrow	A compact perennial with gorgeous lilac-pink flower clusters from summer to fall atop dark green deeply cut leaves.	Fast Growth	Semi-Moist	2' x 3'
Achillea millefolium 'Moonshine'	Moonshine Yarrow	Bright yellow flower clusters bloom all summer above bountiful gray feathery leaves. An indispensible perennial for summer gardens.	Fast Growth	Semi-Moist	2' x 2'
Achillea millefolium 'Summer Pastels'	Summer Pastels Yarrow	A blend of subtle pastel-colored flower clusters that bloom all summer.	Fast Growth	Semi-Moist	2' x 3'
Achillea millifolium "Heidi'	Heidi Yarrow	Bright pink flowers in the early summer that last well in the garden and as a cut flower. A tough easy to grow plant.	Fast Growth	Semi-Moist	1' x 2'
Achillea millifolium 'Paprika'	Paprika Yarrow	Rusty red flat clusters of flowers in the early summer and again later if seed heads are removed. A tough easy to grow perennial with a long bloom season.	Fast Growth	Semi-Moist	1' x 2'
Achillea millifolium 'Red Beauty'	Red Beauty Yarrow	Bright red flowers in flat topped clusters in the early summer and again later in the summer if they are cut. A tough easy to grow perennial.	Fast Growth	Semi-Moist	1' x 3'

Aegopodium podagraria 'Variegatum'	Snow on the Mountain	Sometimes grown as a gound cover, its leaflets are neaatlyi edged white.	Fast Growth	Semi-Moist	8" x 3'
Aesculus parviflora	Bottlebrush Buckeye	An unusual shrub with large leaves and flower s that likes open sunny spots.	Slow Growth	Moist	6' x 10'
Aesculus X carnea	Red Horse chestnut	An upright shade tree with rosy red flowers.	Slow Growth	Moist	25' x 25'
Ajuga pyramidalis 'Metallica Crispa'	Bronze Ajuga	Shiny deep bronze-purple evergreen le aves topped with blue flower spikes in spring, makes this an ideal clumping evergreen ground cover.	Slow Growth	Moist	5" x 2'
Ajuga reptans	Green Ajuga	A creeping evergreen groundcover with dark green leaves and blue flower spikes in late spring to early summer.	Slow Growth	Semi-Moist	5" x 2'
Ajuga reptans 'Atropurpurea'	Bronze Ajuga	Dark bronze purple leaves make this an excellent mat-like evergreen perennial for rock gardens. Blue flower spikes in late spring to early summer.	Fast Growth	Moist	5" x 2'
Ajuga reptans 'Burgundy Glow'	Burgundy Glow Bugleweed	White, pink and purple leaves make this low spreading evergreen groundcover a real conversation piece in any garden. Blue flower spikes in spring.	Fast Growth	Moist	4" x 18"
Ajuga reptens 'Silver Beauty'	Silver Beauty Ajuga	A colorful green and white ground cover that can grow in shady or partly shady places. Ajuga is short and neat and will spread to cover the area.	Average Growth	Moist	4" x 18"
Alcea rosea	Hollyhock	An old-time favorite with tall sp ikes of 2" flowers in many colors. Will reseed to stay around in the back of a flower border.	Fast Growth	Semi-Moist	5' x 2'
Alchemilla mollis	Lady's Mantle	A short low maintainance edging plant with nice foliage golden green flowers.	Average Growth	Moist	10" x 2'

Amelanchier canadensis	Shadblow Serviceberry	A native large shrub or small tree with white flowers followed by sweet edible blue fruit and bright colored leaves in the fall.	Average Growth	Moist	10' x 15'
Aquilegia alpina	Blue Alpine Columbine	A long-lived deep blue columbine for part to full shade. This flower will reseed and naturalize in a woodland garden.	Average Growth	Moist	2' x 2'
Aquilegia caerulea	Rocky Mountain Columbine	Huge blue & white flowers cover the plant in the early summer. Native to the Colorado mountain meadows.	Average Growth	Semi-Moist	3' x 2'
Aquilegia caerulea 'Dragonfly Mix'	Dragon fly Columbine	Huge blooms in every color are found in this hybrid mix. Columbines make a good early summer cutflower.	Average Growth	Semi-Moist	3' x 2'
Aquilegia canadensis	Wild Red Columbine	A native northern perennial that explodes with red nodding flowers tinged with yellow in spring. Prefers moist shady areas.	Average Growth	Semi-Moist	3' x 2'
Aquilegia vulgaris 'Biedermeier'	Biedermeier Dwarf Columbine	A compact columbine mix coming with blue, purple, pink and white flowers that look right up at you.	Average Growth	Moist	2' x 2'
Aquilegia vulgaris 'Heidi'	Heidi Columbine	A Columbine with large pink flowers that hummingbirds love.	Average Growth	Moist	3' x 2'
Armeria maritima	Pink Sea Thrift	A low-growing perennial that flowers in early summer. The flowers are a top a slender stalk.	Average Growth	Moist	1' x 1'
Armeria maritima 'Alba'	White Thrift	A low-growing perennial that flowers in early summer. The flowers are a top a slender stalk.	Average Growth	Moist	1' x 1'
Armeria maritima 'Rubra'	Red Splendens Thrift	A low-growing perennial that flowers in early summer. The floweers are atop a slender stalk.	Average Growth	Moist	1' x 1'

Aronia arbutifolia	Red Chokeberry	White flowers in spring are followed by bright red berries in fall and early winter. Popular with birds.	Average Growth	Semi-Moist	6' x 4'
Artemisia arborescens 'Powis Castle'	Powis Castle Wormwood	Outstanding silver-green, lacy leaves and stems make this upright-growing perennial a winner. Excellent contrast plant in the sunny perennial garden.	Average Growth	Semi-Moist	1' x 2'
Artemisia schmidtiana 'Silver Mound'	Silver Mound Artimisia	A sensational low dome of la cy silver leaves throughout the summer. An outstanding perennial for sunny borders or rock gardens.	Slow Growth	Semi-Moist	1' x 2'
Artemisia stelleriana 'Silver Brocade'	Silver Brocade Wormwood	Ghostly silver leaves are finely cut on this compact, mounding perennial. Thrives in full sun with well - drained soils.	Average Growth	Semi-Moist	6" x 18"
Asclepias tuberosa	Butterfly Milkweed	One of the brightest meadow wild flowers that produces broad heads of small orange (sometimes yellow or red) flowers in th e summer.	Average Growth	Semi-Moist	2' x 2'
Aster novi-belgi 'Alert'	Alert Red Aster	One of the most useful plants for a perennial border.	Average Growth	Semi-Moist	2' x 2'
Aster novi-belgi 'Professor Anton Kippenberg	Kippenberg's Blue Aster	One of the m ost useful plants for a perennial border.	Average Growth	Semi-Moist	2' x 2'
Aster x frikartii 'Wunder von Stafa'	Wonder of Staffa Aster	One of the most useful plants for a perennial border.	Average Growth	Moist	2' x 2'
Astilbe chinensis 'Pumila'	Dwarf Purple Astilbe	A small quick spreading groundcover perennial blanketed with lilac-rose flower plumes in mid - summer.	Average Growth	Moist	2' x 2'
Astilbe japonica 'Deutschland'	White Astilbe	Plumes of creamy white flowers that can reach 2 feet high, light up the garden in late spring.	Average Growth	Moist	2' x 2'

Astilbe simplicifolia 'Sprite'	Pink Sprite Astilbe	Delicate pink arching flower plumes cover the dark lacy leaves of this small perennial in mid-summer.	Average Growth	Moist	2' x 2'
Astilbe X arendsii	Mixed Astilbe	Plumes of red, pink, white, or lavender flowers are excellent for cutting or drying and attract butterflies to the shade garden in early summer.	Average Growth	Moist	2' x 2'
Astilbe x arendsii	Assorted Astilbe	A mixture of red, pink and white plumes in the early summer and dark glossy foliage.	Average Growth	Moist	18" x 24"
Astilbe x arendsii	Pink Astilbe	Soft pink plumes in the early summer with glossy dark foliage.	Average Growth	Moist	18" x 24"
Astilbe x arendsii	Red Astilbe	Bright red plumes in the early summer with glossy bronze foliage.	Average Growth	Moist	18" x 24"
Astilbe x arendsii	Purple Astilbe	Deep lavender plumes in the early summer with glossy green foliage.	Average Growth	Moist	18" x 24"
Astilbe x arendsii	White Astilbe	Pure white plumes in the early summer with glossy dark foliage.	Average Growth	Moist	18" x 24"
Astilbe X arendsii 'Fanal'	Fanal Red Astilbe	Deep red flower plumes cover the dark green leaves of this hardy perennial to "fire up" the spring garden.	Average Growth	Moist	2' x 2'
Astilbe X arendsii 'Peach Blossom'	Peach Blossom Astilbe	Delicate pink flower spikes crown the lacy green leaves of this 2 feet high hardy perennial in late spring.	Average Growth	Moist	2' x 2'
Athyrium filix-femina	Lady Fern	A native woodland fern wth finely divided soft green leaves on a compact plant. Does well in the shade and does not spread.	Average Growth	Moist	2' x 2'

Athyrium niponicum 'Pictum'	Japanese Painted Fern	A striking addition to the shade garden because of its burgundy stems and leaves tinged with blue and silver. A very cold hardy fern.	Slow Growth	Moist	1' x 2'
Aucuba japonica 'variegata'	Gold Dust Aucuba	Large green leaves with big gold spots on a compact shrub that grows well in the shade.	Slow Growth	Moist	6' x 5'
Aurinia saxatilis	Basket Of Gold Alyssum	Plant has hairy, gray-green lieaves and is smothered in bright yellow flowers in spring and early summer.	Average Growth	Semi-Moist	6" x 2'
Baptisia australis	Blue False Indigo	Leaves are blue-green and form a loose mound with purplish blue flowers.	Average Growth	Semi-Moist	3' x 2'
Bellis perennis	English Daisy	Forms carpets of crowded rosettes that spread through lawns by short runners.	Average Growth	Moist	6' x 1'
Berberis thunbergi 'auea'	Autumn Blaze Red Maple	A cross that combines the best features of both its parents; the drought tolerance of Silver Maple and brilliant orange red fall color of Red Maple.	Fast Growth	Moist	50' x 35'
Berberis thunbergi 'Bonanza Gold'	Bonanza Gold Barberry	A new dwarf variety that combines golden leaves with a dwarf 2' tall size. The plant is naturally dense without pruning and has orange fall color.	Slow Growth	Moist	2' x 4'
Berberis thunbergi 'Royal Burgundy'	Royal Burgundy Barberry	A new dwarf variety that is better than Crimson Pygmy. Royal Burgundy has smaller darker red leaves on a very dense low mound.	Slow Growth	Moist	18" x 3'
Berberis thunbergii	Green Japanese Barberry	A small, dense shrub, excellent for hedging and mass plantings. Bright green leaves turn orange, scarlet, or reddish-purple in fall.	Average Growth	Semi-Moist	4' x 36"
Berberis thunbergii 'Atropurpurea'	Red Japanese Barberry	A great compact, dense shrub with reddish -purple leaves that turn rich red to redd ish purple in fall.	Average Growth	Semi-Moist	36" x 36"

Berberis thunbergii atropurpurea (var.) 'Crimson Pygmy'	Crimson Pygmy Barberry	One of the most widely planted Barberries, it is a low-growing deciduous shrub.	Slow Growth	Moist	18" x 36"
Berberis thunbergii 'Rose Glow'	Rose Glow Barberry	Leaves are mottled red and white and the new growth has a lively "rosy glow". A densely branched low maintenance shrub for the modern landscape.	Average Growth	Semi-Moist	36" x 36"
Bergenia cordifolia	Heart-Leaf Bergenia	A tough perennial with crinkly edged heart -shaped leaves that produces drooping purple -pink flowers in early spring.	Average Growth	Moist	2' x 2'
Betula nigra	River Birch	A very handsome, tough specimen tree for large areas with an upright growth habit and peeling cinnamon-brown bark on the trunks.	Fast Growth	Moist	25' x 25'
Betula nigra	Clump River Birch	A very handsome, tough specimen tree for large areas with an upright growth habit and peeling cinnamon-brown bark on the trunks.	Fast Growth	Moist	30' x 30'
Betula nigra	2 Stem Clump River Birch	A very handsome, tough specimen tree for large areas with an upright growth habit and peeling cinnamon-brown bark on the trunks.	Fast Growth	Moist	30' x 30'
Betula nigra 'Heritage'	Heritage River Birch	An outstanding tree that'is highlighted by dark, glossy green leaves turning yellow in fall and peeling salmon-white bark.	Fast Growth	Moist	25' x 25'
Betula papyrifera	Paper or White Birch	This upright tree is commonly known as paper or canoe birch. Its white peeling bark had many uses for Native American Indians and early pioneers.	Average Growth	Moist	50' x 25'
Betula pendula	European White Birch	A large upright growing tree with graceful weeping branches. White bark on the trunk and branches	Slow Growth	Moist	25' x 15'

Betula pendula 'Crimson Frost'	Crimson Frost Birch	This new variety of white barked birch has deep red leaves all summer.	Slow Growth	Moist	25' x 15'
Betula platyphylla 'japonica'	Japanese White Birch	A large, upright-growing tree w ith glossy green leaves that turn yellow in fall, and white bark on the trunk and branches.	Fast Growth	Moist	25' x 15'
Betula platyphylla 'japonica'	Clump Japanese White Birch	A large, upright-growing tree with glossy green leaves that turn yellow in fa ll, and white bark on the trunk and branches.	Fast Growth	Moist	30' x 15'
Betula platyphylla 'japonica'	2 Stem Clump Japanese White Birch	A large upright growing tree with graceful weeping branches. White bark on the trunk and branches.	Fast Growth	Moist	30' x 15'
Betula populifolia 'Whitespire'	Whitespire Birch	A large, upright-growing tree with glossy green leaves turning yellow in fall, white bark on trunks and branches. An insect borer -resistant selection.	Fast Growth	Moist	25' x 15'
Betula populifolia 'Whitespire'	Clump Whitespire Birch	A large, upright-growing tree with glossy green leaves turning yellow in fall, white bark on trunks and branches. An insect borer -resistant selection.	Fast Growth	Moist	25' x 15'
Boltonia asteroides 'Snowflake'	Snowflake Boltonia	A large perennial covered with white daisies for 2 months every fall. Tough and easy to grow.	Average Growth	Semi-Moist	3' x 3'
Buddleia davidii	Assorted Butterfly Bush	A hardy summer-blooming shrub in colors of white, pink, red, lavender and purple. As the name implies, butterflies love this vigorous shrub.	Fast Growth	Semi-Moist	6' x 6'
Buddleia davidii 'Black Knight'	Black Knight Butterfly Bush	Deep purple-black flower spikes attract butterflies to the garden all summer long. A hardy , vigorously-growing shrub.	Fast Growth	Semi-Moist	6' x 6'
Buddleia davidii 'Empire Blue'	Empire Blue Butterfly Bush	Rich violet-blue flowers with orange eyes are produced on spikes above the dark green leaves to attract butterflies all summer long.	Fast Growth	Semi-Moist	6' x 6'

Buddleia davidii *'Honeycomb'*	Honeycomb Butterfly Bush	A vigorously-growing variety with large creamy - yellow flower spikes attracting butterflies all summer long.	Fast Growth	Semi-Moist	6' x 6'
Buddleia davidii *'Ile de France'*	Ile de France Butterfly Bush	Dark violet flowers with a yellow -orange throat are produced in spikes above blue -green leaves attracts butterflies to the garden all summer long.	Fast Growth	Semi-Moist	6' x 6'
Buddleia davidii *'Nanho Blue'*	Nanho Blue Butterfly Bush	Mauve-blue flower spikes are abundantly produced over the small gray -green leaves all summer long. A dense, compact butterfly shrub.	Fast Growth	Semi-Moist	36" x 36"
Buddleia davidii *'Nanho Purple'*	Nanho Purple Butterfly Bush	Magenta-purple flower spikes crown the gray - green leaves all summer long on this dense, compact butterfly bush.	Fast Growth	Semi-Moist	6' x 6'
Buddleia davidii *'Pink Delight'*	Pink Delight Butterfly Bush	True pink, fragrant flower spikes produced over gray-green leaves are irresistible to butterflies all summer long.	Fast Growth	Semi-Moist	6' x 6'
Buddleia davidii *'Royal Red'*	Royal Red Butterfly Bush	Rich purple-red flower spikes attract butterflies all summer long to this vigorous, hardy shrub.	Fast Growth	Semi-Moist	6' x 6'
Buddleia davidii *'White Profusion'*	White Profusion Butterfly Bush	Lovely long flower spikes of white blooms with yellow centers attract butterflies all summer long. Narrow light green leaves on an upright shrub.	Fast Growth	Semi-Moist	6' x 6'
Buddleia X *'Lochinch'*	Lochinch Butterfly Bush	Spikes of sweetly scented lavender -blue flowers with orange eyes are produced all summer long above silvery green leaves. A real butterfly magnet.	Fast Growth	Semi-Moist	4' x 4'
Buddleia X *'Sungold'*	Sungold Butterfly Bush	Yellow-orange, fragrant flower spikes attract butterflies to the garden all summer long. A vigorous-growing shrub.	Fast Growth	Semi-Moist	4' x 4'

Buddlia davidii 'Snowbank'	Snowbank Butterfly Bush	Lovely long flower spikes of white blooms with yellow centers attract butterflies all summer long. Narrow light green leaves on an upright shrub.	Fast Growth	Semi-Moist	5' x 5'
Buddlia x 'Nanho White'	Nanho White Butterfly Bush	Produces hundreds of white flowers all summer long on a naturally dense compact pl ant that butterflies cannot resist.	Fast Growth	Semi-Moist	4' x 4'
Buxus microphylla insularis (var.) 'Wintergreen'	Wintergreen Boxwood	Dense, light green, small leaves grace this compact evergreen shrub through all seasons. A great landcape shrub for nor thern climates.	Slow Growth	Moist	24" x 12"
Buxus microphylla japonica (var.)	Japanese Boxwood	Lustrous dark green leaves grace this dense, compact, slow-growing low maintenance shrub all year long.	Slow Growth	Moist	24" x 12"
Buxus microphylla koreana (var.)	Korean Boxwood	Dark green, oval-shaped leaves grace this dense, compact, low maintenance shrub all year long.	Slow Growth	Moist	24" x 36"
Buxus microphylla 'Welleri'	Weller's Boxwood	A dense plant that is green throughout the winter.	Slow Growth	Moist	3' x 3'
Buxus microphylla 'Winter Gem'	Winter Gem Boxwood	Rich deep green leaves grace this dense compact shrub all year long. A low maintenance plant that grows to 2 feet high.	Slow Growth	Moist	24" x 12"
Buxus sempervirens	English Boxwood	The aristocrat of hedging plants. Dense, compact, slow-growing shrubs are covered with small, dark green leaves all year long.	Slow Growth	Moist	6' x 10'
Buxus X 'Green Mountain'	Green Mountain Boxwood	An upright growing boxwood that forms a pyramidal oval 5 fee t high by 3 feet wide covered with dark green leaves all year long.	Slow Growth	Moist	6' x 24"
Buxus X 'Green Velvet'	Green Velvet Boxwood	An outstanding selection blanketed with velvety green leaves all year long on a low maintenance shrub that grows to 4 feet high by 4 feet wide.	Slow Growth	Moist	36" x 36"

Caladium bicolor	Assorted Caladium	Large, multi-colored leaves lend a semi -tropical look to any garden or patio container. Prefers light shade.	Fast Growth	Moist	18" x 18"
Calamagrostis X acutiflora 'Karl Foerster'	Karl Foerster Feather Reed Grass	A compact, clumping ornamental grass with dark green foliage crowned with stiff, tawny flower heads from spring through winter.	Average Growth	Moist	30" x 15"
Callicarpa americana	White Beautyberry	A rather coarse open shrub that produces lavender - pink flowers on new growth from June into August.	Average Growth	Moist	4' x 4'
Callicarpa dichotoma	Purple Beautyberry	The most graceful of the Beautyberries and produces small pinkish lavender flowers from June to August.	Average Growth	Moist	4' x 4'
Campanula carpatica 'Blue Clips'	Blue Clips Bellfower	Well suited for use as an edging or rock garden plant. Flowers from late spring through summer.	Average Growth	Moist	10" x 18"
Campanula carpatica 'White Clips'	White Clips Bellflower	Well suited for use as an edging or rock garden plant. Flowers from late spring through summer.	Average Growth	Moist	10" x 18"
Campsis radicans	Orange Trumpet Vine	The narrow flowers are open at the throat. Flowers well against a warm wall.	Fast Growth	Semi-Moist	10' x 20'
Campsis radicans 'Flava'	Yellow Trumpet Vine	The narrow flowers are open at the throat. Flowers well against a warm wall.	Fast Growth	Semi-Moist	10' x 20'
Canna X generalis 'Primrose Yellow'	Yellow Canna	A mid to large size upright perennial with stalks of pure yellow flowers from mid -summer through fall. Loves wet soils.	Fast Growth	Moist	5' x 3'
Carex morrowii 'Evergold'	Evergold Sedge	Bright creamy yellow -striped, narrow leaves makes this a wonderful ac cent plant in rock gardens and near water gardens. Needs some shade.	Slow Growth	Moist	1' x 18"

Carex morrowii 'variegatum'	Variegated Japanese Sedge	Bright creamy white-striped, narrow leaves makes this a wonderful accent plant in rock gardens and near water gardens. Needs some shade.	Slow Growth	Moist	1' x 18"
Carpinus betulus 'Fastigiata'	Columnar European Hornbeam	Serves as an effective screen in winter because of the dense, compact, close-knit nature of the branches.	Slow Growth	Moist	25' x 15'
Carya illinoensis 'Choctaw'	Coctaw Pecan	Large nut with a very thin, soft hull. Richly flavored kernel is very smooth and high in oil content.	Average Growth	Semi-Moist	40' x 40'
Carya illinoensis 'Desirable'	Desirable Pecan	Large nut with a very thin, soft hull. Richly flavored kernel is very smooth and high in oil content.	Average Growth	Semi-Moist	40' x 40'
Carya illinoensis 'Meramec'	Meramec Pecan	Large nut with a very thin, soft hull. Richly flavored kernel is very smooth and high in oil content.	Average Growth	Semi-Moist	40' x 40'
Carya illinoensis 'Pawnee'	Pawnee Pecan	Produces a large, soft shelled nut of excellent quality. Ripens early.	Average Growth	Semi-Moist	40' x 40'
Carya illinoensis 'Stuart'	Stuart Pecan	Large nut with a very thin, soft hull. Richly flavored kernel is very smooth and high in oil content.	Average Growth	Semi-Moist	40' x 40'
Carya illinoensis 'Western'	Western Pecan	Large nut with a very thin, soft hull. Richly flavored kernel is very smooth and high in oil content.	Average Growth	Semi-Moist	40' x 40'
Carya illinoinensis	Northern Pecan	Grows well as far north as northeastern Nebraska. Has a smaller nut than southern varieties.	Average Growth	Semi-Moist	75' x 50'
Carya illnoensis 'Wichita'	Wichita Pecan	Produces a medium-size nut, thin shelled but high quality.	Average Growth	Semi-Moist	40' x 40'

Caryopteris clandonesis 'Dark Knight'	Dark Knight Blue Mist Spirea	Produces deep blue-purple flowers on new growth.	Average Growth	Semi-Moist	3' x 4'
Caryopteris x clandonensis "Blue Mist'	Blue Mist Spirea	Produces powder-blue flowrers on new growth.	Average Growth	Semi-Moist	3' x 4'
Caryopteris X clandonensis 'Longwood Blue'	Longwood Blue Mist Spirea	A low mounding shrub with silvery blue-green leaves and covered with powder blue flowers in late summer through fall.	Average Growth	Semi-Moist	3' x 4'
Caryopteris X clandonensis 'Worcester Gold'	Golden Blue Mist Spirea	Produces yellow foliage and blue flowers.	Average Growth	Moist	3' x 4'
Castanea mollissima	Chinese Chestnut	A stately tree with lustrous dark green leaves and produces edible chestnuts in fall, if you can beat the squirrels to them.	Average Growth	Moist	25' x 35'
Catalpa speciosa	Northern Catalpa	A tree with a narrow, open, irregular, oval crown that can be quite striking in winter with bold rugged outline.	Fast Growth	Moist	35' x 35'
Cedrus deodara	Deodar Cedar	An excellent specimen evergreen because of extremely graceful habit.	Fast Growth	Moist	50' x 25'
Cedrus deodara 'Aurea'	Golden Deodar Cedar	A graceful open evergreen with bright golden tipped branches.	Fast Growth	Moist	50' x 25'
Cedrus deodara 'Pendula'	Weeping Deodar Cedar	A graceful weeping tree with soft green foliage.	Fast Growth	Semi-Moist	50' x 25'
Cedrus libani atlantica	Atlas Cedar	A graceful open tree with intense silver foliage.	Average Growth	Semi-Moist	50' x 35'

Cedrus libani atlantica 'Glauca Pendula'	Weeping Blue Atlas Cedar	A graceful weeping tree with intense silve-blue foliage.	Average Growth	Semi-Moist	10' x 15'
Celtis laevigata	Sugarbush Hackberry	A valued street tree. Grows fast when young and has deep green foliage.	Average Growth	Semi-Moist	50' x 50'
Celtis occidentalis	Hackberry	Performs admirably under adverst conditions. Has the ability to grow in dry soils and windy conditions.	Fast Growth	Semi-Moist	50' x 35'
Centaurea montana	Mountain Bluet	Produces wide violet flowerheads in early summer.	Fast Growth	Semi-Moist	2' x 2'
Cerastium tomentosum	Snow in Summer	A vigorous, fast-growing ground cover that produces star-shaped white flowers in late spring and summer.	Fast Growth	Semi-Moist	6" x 30"
Cercidiphyllum japonicum	Katsura Tree	Valued chiefly for its foliage which is reddish when first showing in spring then changing to a mixture of yellow, pink, orange and red in the fall.	Average Growth	Moist	35' x 25'
Cercis canadensis	Eastern Redbud	A wonderful small, native tree whose branches are covered with light pinkish purple flowers in early spring before the leaves emerge.	Average Growth	Semi-Moist	15' x 15'
Cercis canadensis	Clump Eastern Redbud	A wonderful small, native tree whose branches are covered with light pinkish purple flowers in early spring before the leaves emerge.	Average Growth	Semi-Moist	15' X 15'
Cercis canadensis	2 Stem Clump Eastern Redbud	A wonderful small, native tree whose branches are covered with light pinkish purple flowers in early spring before the leaves emerge.	Average Growth	Semi-Moist	15' x 15'
Cercis canadensis alba (var.)	Whitebud	A small, native tree whose branches are covered with clusters of small white flowers in early spring, followed by green leaves.	Average Growth	Semi-Moist	25' x 15'

Cercis canadensis 'Flame'	Flame Redbud	Has an interesting double rose-pink flower.	Average Growth	Semi-Moist	15' x 15'
Cercis canadensis 'Forest Pansy'	Forest Pansy Redbud	A handsome small native tree with rosy pink flowers in spring followed by shimmering red-purple new leaves. Leaf color becomes more subdued in summer.	Slow Growth	Semi-Moist	15' x 15'
Cercis chinensis 'Avondale'	Avondale Redbud	A beautiful small tree with clean glossy foliage and red flwers.	Average Growth	Semi-Moist	15' x 15'
Cercis reniformis 'Oklahoma'	Oklahoma Redbud	An excellent small native tree with branches that are covered with clusters of magenta flowers in early spring, followed by lustrous green leaves.	Average Growth	Semi-Moist	15' x 15'
Cercis reniformis 'Texas White'	Texas White Redbud	A great native small tree with branches covered with clusters of pure white flowers, followed by lustrous wavy green leaves.	Average Growth	Semi-Moist	15' x 15'
Chaenomeles japonica 'rubra'	Red Flowering Quince	An early bloomer with red flowers, this plant is used often for a hedge, shrub border, massing or a grouping.	Average Growth	Semi-Moist	4' x 6'
Chaenomeles japonica 'Spitfire'	Spitfire Flowering Quince	An early bloomer with red flowers, this plant is used often for a hedge, shrub border, massing or a grouping.	Average Growth	Semi-Moist	4' x 6'
Chaenomeles speciosa 'Nivalis'	White Flowering Quince	An upright form of Quince with pure white flowers.	Average Growth	Semi-Moist	4' x 6'
Chaenomeles speciosa 'Texas Scarlet'	Texas Scarlet Flowering Quince	A profusion of tomato-red flower clusters open in late winter on this compact, spreading shrub.	Average Growth	Moist	4' x 36"
Chaenomeles X superba 'Cameo'	Cameo Flowering Quince	Considered the best quince selection because of its profuse, fluffy apricot pink flower clusters in late winter, followed by bright green leaves.	Average Growth	Moist	24" x 36"

Chamaecyparis nootkatensis 'Pendula'	Weeping Alaskan Cedar	Has beautiful, graceful, elegantly arranged branches and rich green foliage.	Average Growth	Moist	25' x 15'
Chamaecyparis obtusa	Hinoki Cypress	Tree has richly textured, deep green foliage and branches touch the ground.	Slow Growth	Moist	50' x 15'
Chamaecyparis obtusa 'aurea'	Golden Hinoki Cypress	An oriental looking evergreen with twisting gold-tipped foliage. Fits well in small spaces since it grows slowly.	Slow Growth	Moist	4' x 2'
Chamaecyparis obtusa 'nana gracilis'	Dwarf Hinoki Cypress	The rich green foliage twists and undulates into interesting shapes. Great in rock gardens and tight places.	Slow Growth	Moist	4' x 2'
Chamaecyparis obtusa 'Nana Lutea'	Dwarf Golden Hinoki Cypress	An oriental looking evergreen with twisting gold-tipped foliage. Fits well in small spaces since it grows slowly.	Slow Growth	Moist	6' x 36"
Chamaecyparis obtusa 'Torulosa'	Torulosa Hinoki Cypress	The rich green foliage twists and undulates into interesting shapes. Great in rock gardens and tight places.	Slow Growth	Moist	50' x 15'
Chamaecyparis pisifera 'Boulevard'	Boulevard Cypress	A dense upright mound of twisted gray foliage. Great in rock gardens and narrow places.	Slow Growth	Moist	20' x 15'
Chamaecyparis pisifera 'Filifera Aurea'	Golden Threadbranch Cypress	An arching mound of golden foliage for sun or partial shade.	Slow Growth	Moist	25' x 15'
Chamaecyparis pisifera 'Sungold'	Sungold False Cypress	An arching mound of golden foliage for sun or partial shade.	Slow Growth	Moist	20' x 12'
Chasmanthium latifolium	Northern Sea Oats	Grown for its unusual seedheads which can be used as dried decorations.	Average Growth	Moist	2' x 2'

Chelone lyonii	Pink Turtlehead	Has erect, angled stems and dark green leaves and produces rosy purple summer flowers.	Average Growth	Moist	3' x 3'
Chionanthus virginicus	White Fringetree	A very beautiful specimen shrub that is excellent in groups, borders, or near buildings.	Slow Growth	Moist	10' x 8'
Clematis viticella 'Etoile Violette'	Etoile Violette Clematis	This hybrid form is very vigorous and free flowering. Flowers are deep purple with cream stamens.	Average Growth	Moist	8' x 5'
Clematis x Duchess of 'Edinburgh'	Duchess of Edinburgh Clematis	Produces double white flowers with yellow anthers and green outer petals in spring and summer.	Average Growth	Moist	8' x 3'
Clematis x 'Ernest Markham'	Ernest Markham Clematis	One of the most vigorous and widely grown. Flowers are bright magenta in color. Blooms in July to September.	Average Growth	Moist	8' x 3'
Clematis X 'General Sikorski'	General Sikorski Clematis	Flowers are mid-blue with a reddish flush at the base of the petals. Blooms in June and September.	Average Growth	Moist	8' x 3'
Clematis X 'Hagley Hybrid'	Pink Chiffon Clematis	Has medium-size righ pink flowers with ruffly edges and reddish anthers. Blooms in July and September.	Average Growth	Moist	8' x 3'
Clematis X 'Henryi'	Henryi White Clematis	Produces very large pure white flowers with brown anthers. Blooms June and September.	Average Growth	Moist	8' x 3'
Clematis x hybrids	Assorted Clematis	These showy vines come in almost any color during the early summer. They can be trained to climb on a trellis or fence.	Average Growth	Moist	8' x 3'
Clematis X jackmanii	Jackman's Purple Clematis	The most profuse blooming dark purple Clematis. Blooms in June, July and September.	Slow Growth	Moist	8' x 3'
Clematis X 'Nelly Moser'	Nelly Moser Clematis	Has pale pink with red bar down the middle of each flower petal. Blooms May, June and September.	Average Growth	Moist	8' x 3'

Clematis X 'Niobe'	Niobe Red Clematis	Has very dark ruby-red pointed petals with gold anthers. Blooms June to September.	Average Growth	Moist	8' x 3'
Clethera alnifolia 'Ruby Spice'	Ruby Spice Summer sweet	Fragrant, rich rose-colored flower spikes of this upright shrub light up the summer garden. Lustrous dark green leaves turn golden yellow in fall.	Fast Growth	Moist	36" x 36"
Clethra alnifolia 'Hummingbird'	Humingbird Summer sweet	Fragrant white flower spikes blanket this compact (30-40 inches high) shrub in summer, lustrous black-green leaves that turn golden yellow.	Fast Growth	Moist	18" x 12"
Clethra alnifolia 'Rosea'	Pink Summer sweet	A tall upright shrub producing spikes of fragrant pinkish white flowers in summer, glossy dark green leaves that turn golden yellow in fall.	Fast Growth	Moist	36" x 36"
Convallaria majalis	Lily of the Valley	Renowned for its glorious perfune, the dainty white bell-shaped flowers appear in spring.	Average Growth	Moist	10" x 2'
Coreopsis auriculata 'Nana'	Dwarf Coreopsis	A compact perennial with yellow daisy-like flowers in late spring. The plant grows 6-9 inches high and is excellent for borders or edging in full sun.	Average Growth	Moist	6" x 15"
Coreopsis lanceolata 'Baby Sun'	Baby Sun Coreopsis	Single golden yellow flowers are produced on long stalks throughout the summer making this an excellent cut flower or naturalizing selection.	Fast Growth	Moist	2' x 2'
Coreopsis lanceolata 'Early Sunrise'	Early Sunrise Coreopsis	Double yellow flowers crown this compact perennial throughout the summer. An excellent cut flower plant.	Fast Growth	Moist	3' x 2'
Coreopsis lanceolata 'Sunburst'	Sunburst Coreopsis	Double yellow flowers crown this compact perennial throughout the summer. An excellent cut flower plant.	Fast Growth	Moist	3' x 3'

Coreopsis lanceolata 'Sunray'	Sunray Coreopsis	Large double golden yellow flowers rise above the compact dark green leaves throughout the summer. An excellent cut flower or naturalizing perennial.	Fast Growth	Moist	3' x 2'
Coreopsis rosea	Pink Coreopsis	A compact, fine-textured perennial that turns into a pink cloud of small flowers in summer. An excellent plant for naturalizing or edging.	Slow Growth	Moist	8" x 15"
Coreopsis verticillata 'Moonbeam'	Moonbeam	Creamy yellow daisy-like flowers are produced throughout the summer above dark green, fine-textured leaves.	Fast Growth	Moist	1' x 2'
Coreopsis verticillata 'Zagreb'	Zagreb Coreopsis	Bright yellow daisy-like flowers blanket the dark green leaves all summer long on this compact perennial. Excellent for borders or naturalizing.	Fast Growth	Moist	1' x 2'
Cornus alba 'Argenteo-marginata'	Variegated Red Twig Dogwood	Shiny red branches and twigs that are the brightest in winter or late fall. Gray green leaves are white on their margins.	Average Growth	Moist	6' x 6'
Cornus alba 'Elagantissima'	Variegated Red Twig Dogwood	Shiny red branches and twigs that are the brightest in winter or late fall. Gray green leaves are white on their margins.	Average Growth	Moist	8' x 8'
Cornus alba 'Gouchetti'	Golden Variegated Dogwood	Showy yellow edged leaves in the summer and red twigs in the winter make for year-round color.	Average Growth	Moist	8' x 8'
Cornus alternifolia	Pagoda Dogwood	An attractive deciduous tree that in late s[pring bears small clusters of cream flowers followed by buhches of blue-black berries.	Average Growth	Moist	15' x 15'
Cornus baileyi	Bailey Red Twig Dogwood	An upright shrub with dark red twigs in the winter. Makes a good fast growing screen or windbreak on rich moist soils	Fast Growth	Moist	8' x 5'

Cornus florida	Frowering Dogwood	The aristocrat of native flowering trees, in spring abundant, large white flowers are followed by bright red berries; reddish-purple fall leaf color.	Average Growth	Moist	20' x 20'
Cornus florida 'Cherokee Chief'	Cherokee Chief Flowering Dogwood	this small tree is covered with rich ruby-red flowers in spring. Shiny red berries and blazing reddish-purple leaves in fall.	Average Growth	Moist	20' x 20'
Cornus florida 'Cherokee Daybreak'	Cherokee Daybreak Flowering Dogwood	A small tree producing white flowers in spring followed by green leaves with white edges that turn deep red in fall.	Average Growth	Moist	20' x 20'
Cornus florida 'Cherokee Princess'	Cherokee Princess Flowering Dogwood	Abundant large white flowers are produced consistently every spring by this cold hardy tree. Considered by many as one of the best dogwood selections.	Average Growth	Moist	20' x 20'
Cornus florida 'Cloud 9'	Cloud 9 Flowering Dogwood	A cold hardy Dogwood that produces abundant, showy white flowers in spring, even on young trees. Considered one of the best dogwood selections.	Average Growth	Moist	25' x 20'
Cornus florida 'Rainbow'	Rainbow Flowering Dogwood	Showy white flowers in sprin, followed by deep green leaves edged in yellow that turn carmine red in fall. A rainbow of colors throughout the summer.	Average Growth	Moist	20' x 20'
Cornus florida 'Red Cloud'	Red Cloud Flowering Dogwood	Large rosey flowers are showy in the spring. The foliage is slightly crinkled and colors well in the fall.	Average Growth	Moist	20' x 20'
Cornus florida rubra (var.)	Red Flowering Dogwood	An upright-growing Dogwood with pinkish red flowers in spring and dark green leaves that turn dark reddish-purple in fall.	Average Growth	Moist	15' x 15'

Cornus florida rubra (var.) 'Cherokee Brave'	Cherokee Brave Flowering Dogwood	A vigorous disease-resistant Dogwood with reddish-pink flowers with a white center. Brilliant dark reddish-purple fall leaf color.	Average Growth	Moist	15' x 25'
Cornus florida rubra (var.) 'Cherokee Sunset'	Cherokee Sunset Flowering Dogwood	This vigorous, disease-resistant Dogwood has bright red spring flowers followed by green leaves with a band of yellow that turn pinkish red in fall.	Average Growth	Moist	15' x 15'
Cornus florida rubra rosea	Pink Flowering Dogwood	An upright-growing Dogwood with pinkish pink flowers in spring and dark green leaves that turn dark reddish-purple in fall.	Average Growth	Moist	20' x 20'
Cornus florida rubra rosea	Clump Pink Dogwood	An upright-growing Dogwood with pinkish pink flowers in spring and dark green leaves that turn dark reddish-purple in fall.	Average Growth	Moist	20' x 20'
Cornus kousa	Kousa White Flowering Dogwood	This vase-shaped Dogwood produces large white flowers 2 weeks after Flowering Dogwood that are followed with large, raspberry-like berries.	Slow Growth	Moist	15' x 20'
Cornus kousa	Clump Kousa Dogwood	This vase-shaped Dogwood produces large white flowers 2 weeks after Flowering Dogwood that are followed with large, raspberry-like berries.	Slow Growth	Moist	20' x 20'
Cornus kousa 'Milky Way'	Milky Way Flowering Dogwood	A short spreading Dogwood that's covered with an abundance of large white flowers in late spring followed by numerous, large red berries.	Slow Growth	Moist	20' x 20'
Cornus kousa X C. florida(n) 'Constellation'	Constellation Flowering Dogwood	Upright, vigorous hybrid Dogwood with large white flowers that bloom after Flowering Dogwood, but before Kousa Dogwood. Excellent red fall leaf color.	Average Growth	Moist	15' x 15'
Cornus sericea	Red Twig Dogwood	A fast growing shrub on rich moist soil with showy red twigs in the winter.	Fast Growth	Moist	6' x 8'

Cornus sericea 'flaviramea'	Yellow Twig Dogwood	A fast growing shrub with golden twigs in the winter that are showy against dark backgrounds.	Fast Growth	Moist	6' x 8'
Cornus sericea 'Isanti'	Isanti Red Twig Dogwood	A dense fast growing shrub with red twigs in the winter. Makes a fast hedge on good soil and is good cover for wildlife.	Fast Growth	Moist	5' x 5'
Cornus stolonifera	Red Twig Dogwood	A great plant for the winter garden because of its red stem color that looks great with snow. Rich green leaves that turn reddish purple in fall.	Average Growth	Moist	8' x 8'
Cornus stolonifera 'Cardinal'	Cardinal Red Twig Dogwood	Bright cherry red twig color in winter is dazzling in the winter garden where there's snowfall, or mixed with evergreen plants. A cold hardy shrub.	Average Growth	Moist	8' x 8'
Cornus stolonifera 'Silver and Gold'	Variegated Yellow Twig Dogwood	Showy green and white leaves in the summer and bright golden twigs in the winter is colorful year-round.	Average Growth	Moist	8' x 8'
Cortaderia selloana	Pampas Grass	A tropical-looking upright ornamental grass grows to 10 feet high and has beautiful, bushy white flower plumes in late summer to fall.	Fast Growth	Semi-Moist	8' x 4'
Cortaderia selloana 'Gold Band'	Gold Band Pampas Grass	Showy gold-edged leaves and tall silver plumes for the fall and winter.	Average Growth	Semi-Moist	5' x 4'
Cortaderia selloana 'Pumila'	Dwarf Pampas Grass	A compact ornamental grass that grows to 6 feet high with thick green leaves and numerous beautiful, bushy white flower plumes in late summer to fall.	Average Growth	Semi-Moist	5' x 4'
Cortaderia selloana 'Rosea'	Pink Pampas Grass	A tropical-looking upright ornamental grass grows to 10 feet high and has beautiful, bushy pink flower plumes in late summer to fall.	Fast Growth	Semi-Moist	8' x 4'
Corylus americana	American Hazelnut	Probably not good for the small yard. Squirrels love the nuts.	Average Growth	Moist	6' x 6'

Corylus avellana 'Contorta'	Harry Lauder's Walkingstick	Stems are curled and twisted as well as the leaves. Best for winter effect.	Slow Growth	Semi-Moist	6' x 6'
Cotinus coggygria	Smoketree	A large upright multi-stemmed shrub that has blue-green leaves and hairy flowers in early summer that look like large puffs of smoke.	Slow Growth	Semi-Moist	10' x 15'
Cotinus coggygria 'Purpureus'	Purple Smoketree	An outstanding upright multi-stemmed shrub with rich maroon-red leaves that persists all summer and smoky purplish red flowers in early summer.	Slow Growth	Semi-Moist	10' x 10'
Cotinus coggygria 'Royal Purple'	Royal Purple Smoketree	An outstanding upright multi-stemmed shrub with rich maroon-red leaves that persists all summer and smoky purplish red flowers in early summer.	Slow Growth	Semi-Moist	10' x 15'
Cotoneaster apiculatus	Cranberry Cotoneaster	A low, spreading ground cover shrub with glossy dark green leaves that create a perfect background for the cranberry red berries produced in the fall.	Slow Growth	Semi-Moist	24" x 36"
Cotoneaster dammeri 'Coral Beauty'	Coral Beauty Cotoneaster	A low, spreading, evergreen groundcover with glossy, dark green leaves and coral red berries produced in late summer.	Slow Growth	Semi-Moist	12" x 36"
Cotoneaster divaricatus	Spreading Cotoneaster	Can be used in foundation plantings, hedges, groups, borders or masses. Blends well with other plants.	Fast Growth	Moist	6' x 5'
Cotoneaster horizontalis	Rockspray Cotoneaster	A low, dense shrub with horizontal branches covered with glossy, dark green leaves that turn reddish purple in fall and red berries in late summer.	Slow Growth	Semi-Moist	24" x 5'
Cotoneaster lucidus	Hedge Cotoneaster	Primarily used as a hedge because of its upright branching habit. Excellent for screens or groupings.	Average Growth	Semi-Moist	6' x 4'

Cotoneaster salicifolius	Willowleaf Cotoneaster	A large evergreen shrub valued for fruits especially in European gardens.	Average Growth	Semi-Moist	10' x 12'
Cotoneaster salicifolius 'Scarlet Leader'	Scarlet Leader Cotoneaster	An excellent low, spreading groundcover with glossy dark green leaves that turn reddish-purple in fall and has red berries in late summer.	Fast Growth	Semi-Moist	36" x 10'
Crataegus laevigata 'Crimson Cloud'	Crimson Cloud Hawthorn	Has good red, single flowers and does well under city conditions.	Slow Growth	Moist	12' x 8'
Crataegus phaenopyrum	Washington Hawthorn	A dense, upright, medium-sized tree covered with white flowers in spring followed by bright red glossy fruit. Excellent as a specimen tree.	Average Growth	Moist	15' x 15'
Crataegus viridis 'Winter King'	Winter King Green Hawthorn	A lovely, rounded, medium-sized tree with white flowers in spring followed by bright red fruit and green leaves that turn scarlet-purple in fall.	Average Growth	Moist	15' x 25'
Crocosmia X crocosmiiflora 'Lucifer'	Lucifer Crocrosmia	Has bright red flowers and is frost hardy but needs full sun in cold climates.	Average Growth	Moist	3' x 2'
Cryptomeria japonica 'Globosa Nana'	Dwarf Globe Japanese Cedar	A globe of interesting twisted light green foliage.	Slow Growth	Semi-Moist	24" x 24"
Cryptomeria japonica 'Yoshino'	Yoshino Japanese Cedar	A choice pyramidal evergreen that grows well in the shade of big trees for screening.	Slow Growth	Semi-Moist	25' x 12'
Cuphea hyssopifola	Mexican Heather	A dense, compact plant with small, glossy green leaves providing the perfect background for lavender flowers that appear from spring through fall.	Average Growth	Moist	1' x 2'

Cupressocyparis leylandii	Leyland Cypress	A very picturesque vigorous, upright -growing evergreen with a pyramid al form and pleasant green color. Excellent for screening and mass plantings.	Fast Growth	Moist	25' x 15'
Cupressus arizonica arizonica (var.)	Arizona Cypress	Grown as both a specimen tree and as a hedge. Mature foliage is gray -green.	Average Growth	Semi-Moist	25' x 15'
Cyrtomium falcatum 'Rochfordianum'	Holly Fern	Reserved strictly for the southern United States. Has lustrous dark green leaves.	Average Growth	Moist	1' x 2'
Cytisus scoparius	Scotch Broom	A very good plant for poor soils. There is a tendancy touse the plant inmasses especiallyh for color splashes.	Fast Growth	Semi-Moist	8' x 6'
Daphne odora 'Aureo-marginata'	Vairegated Daphne	Has yellow-edged leaves and flowers marked with reddish purple.	Slow Growth	Semi-Moist	36" x 36"
Delosperma cooperi	Pink Hardy Iceplant	A good ground cover shrub for a sunny rock garden.	Fast Growth	Semi-Moist	4" x 12"
Delphinium elatum 'Magic Fountains'	Magic Fountains Delphinium	Flowers bloom from summer to fall and are ideal for picking.	Average Growth	Moist	3' x 2'
Delphinium grandiflora 'Blue Butterfly'	Blue Butterfly Delphinium	Produces large bright blue flowers over a long period in summer.	Average Growth	Moist	18" x 18"
Dendranthema nipponicum 'Montauk'	Maontauk Japanese Daisymum	Produces loads of big wh ite daisies in mid -summer.	Average Growth	Semi-Moist	18" x 18"
Dendranthema X grandiflorum	Assorted Garden Mums	Nothing beats a mum for fall blooms in every color except blue.	Average Growth	Moist	2' x 2'

Dendranthema x 'Mary Stoker'	Mary Stoker Daisymum	Straw yellow daisies cover this plant in the late summer and fall. A very hardy long -lived perennial.	Average Growth	Moist	2' x 2'
Dendranthema zawadskii 'Clara Curtis'	Clara Curtis Daisymum	Loads of pink daisies cover this plant in the late summer and fall. A hardy long-lived perennial.	Average Growth	Moist	18" x 18"
Deschampsia caespitosa	Tufted Hairgrass	A mound forming evergreen perennial whose flowers are borne in open panicles.	Average Growth	Semi-Moist	18" x 18"
Deutzia gracilis 'Nikko'	Nikko Slender Deutzia	A graceful low spreading gem with abundant white flowers in late spring and attractive burgundy fall color. An excellent groundcover for full sun.	Average Growth	Moist	12" x 6'
Deutzia scabra 'Godsall Pink'	Pink Deutzia	The longest established and most robust of the species that produces large panicles of bell -shaped flowers from mid-spring to early summer.	Average Growth	Moist	4' x 4'
Dianthus barbatus	Sweet William	A wonderful mix of red, pink and white flowers in fragrant clusters. A biennial that will reseed to come back every year that makes a good cutflower.	Average Growth	Moist	1' x 2'
Dianthus barbatus 'Indian Carpet'	Dwarf Sweet William	A wonderful mix of red, pink and white flowers in fragrant clusters. A biennial that will reseed to come back every year that makes a good cutflower.	Average Growth	Moist	8" x 15"
Dryopteris marginalis	Leatherwood Fern	Dark evergreen leaves are clustered into an arching vase-like arrangement 18 to 24 inches high. A cold - hardy, shade- and moisture-loving fern.	Average Growth	Moist	1' x 2'
Echinacea purpurea	Purple Coneflower	Large purple to pink to white flower petals surround the brownish cone on flowers that are produced in mid-summer by this tall perennial.	Average Growth	Semi-Moist	2' x 2'
Echinacea purpurea 'Bravo'	Bravo Purple Coneflower	Large purple to pink to white flower petals surround the brownish cone on flowers that are produced in mid-summer by this tall perennial.	Average Growth	Semi-Moist	2' x 2'

Echinacea purpurea 'Magnus'	Magnus Purple Coneflower	The flowers on this compact perennial are rose - pink petals surrounding the maroon cone.	Average Growth	Semi-Moist	2' x 2'
Echinacea purpurea 'White Swan'	White Swan Purple Coneflower	This coneflower has bountiful pure white flowers that begin in mid-summer and continue to fall. An excellent perennial for cut flowers.	Average Growth	Semi-Moist	2' x 2'
Elaeagnus angustifolia	Russian Olive	Silvery gray-green leaves and branches set this large shrub, small tree apart from the crowd. A cold-hardy, vigorous plant for dry climates.	Fast Growth	Semi-Moist	20' x 20'
Elaeagnus pungens	Thorny Olive	This vigorous large shrub has glossy dark green leaves with a silvery underside. Excellent plant for hedges, screening and bank stabilization.	Fast Growth	Moist	15' x 10'
Elaegnus pungens 'Fruitlandi'	Fruitland Eleagnus	A tough fast growing dense shrub for screening with silvery leaves and small fragrant flowers.	Fast Growth	Semi-Moist	8' x 8'
Eleagnus x ebbingii 'Gilt Edge'	Gilt Edge Eleagnus	A dense shrub yellow edged silver leaves.	Fast Growth	Semi-Moist	6' x 6'
Euonymus alatus 'Compactus'	Burning Bush	A compact, dense, slow -growing shrub with dark green leaves that turn brilliant red in fall. Attractive twigs with wings of corky bark.	Average Growth	Moist	6' x 6'
Euonymus fortunei 'Canadale Gold'	Canadale Gold Euonymus	Large, light green leaves are bordered with golden yellow. Forms a sturdy compact plant.	Average Growth	Moist	2' x 5'
Euonymus fortunei 'Coloratus'	Purple Winter creeper	An excellent vigorous groundcover with glossy dark green leaves that turn plum -purple to purple-red in fall and persists through the winter.	Average Growth	Moist	12" x 6'
Euonymus fortunei 'Emerald Cushion'	Emerald Cushion Euonymus	A dwarf, mounded form with dense branching habit that holds rich green foliage.	Slow Growth	Moist	3' x 3'

Euonymus fortunei 'Emerald Gaiety'	Emerald Gaity Euonymus	A compact, evergreen shrub with dark green leaves edged in white. Excellent as a groundcove r or accent plant; will climb if planted by a structure.	Average Growth	Moist	4' x 24"
Euonymus fortunei 'Emerald 'n Gold'	Emerald 'n Gold Euonymus	A low mounding evergreen shrub covered by dark green leaves with golden yellow edges turning pinkish-red in cold weather. Ideal for mass plantings.	Average Growth	Moist	4' x 24"
Euonymus fortunei 'Emerald Pride'	Emerald Pride Euonymus	A small erect form with lustrous dark green foliage and a close branching habit.	Average Growth	Moist	3' x 5'
Euonymus fortunei 'Sarcoxie'	Sarcoxie Euonymus	An upright form plant with glossy polished dark green leaves.	Fast Growth	Moist	3' x 5'
Euonymus fortunei 'Sheridan Gold'	Sheridan Gold Euonymus	Has deep green foliage with sunshineyellow coloring in a full sun situation.	Average Growth	Moist	6" x 3'
Euonymus fortunei 'Sun Spot'	Sun Spot Euonymus	A mounding evergreen shrub covered by dark green leaves with bright yellow centers. Ideal for specimen, mass, or foundation plantings.	Average Growth	Moist	24" x 4'
Euonymus japonica ''aureo-variegata'	Gold Spot Euonymus	Has dark green foliage with bright gold centers.	Average Growth	Moist	6' x 4'
Euonymus japonica 'Bekkomasaki'	Bekkomasaki Euonymus	Ground cover with small dark green leaves. Grows in sun or shade.	Average Growth	Moist	1' x 4'
Euonymus japonicus	Japanese Green Euonymus	Very dense, oval evergreen shrub with lustrous, waxy dark green leaves. Excellent plant for hedges or mass plantings.	Average Growth	Moist	6' x 36"
Euonymus japonicus 'Aureo-marginatus'	Golden Euonymus	A dense, oval evergreen shrub covered by lustrous waxy green leaves with wide, irregular bright gold edges. Extremely showy for specimen plantings.	Average Growth	Moist	4' x 4'

Euonymus japonicus 'Microphyllus'	Boxleaf Euonymus	A compact, dwarf evergr een shrub with small green leaves set close together. A nice selection for a low formal effect.	Average Growth	Moist	12" x 12"
Euonymus japonicus 'Silver King'	Silver King Euonymus	A fast-growing, upright shrub covered by glossy green leaves with silvery white irregular margins. Good for large hedges or specimen plantings.	Average Growth	Moist	4' x 36"
Euonymus kiautschovicus 'Manhattan'	Manhattan Euonymus	Has excellent glossy dark green foliage.	Average Growth	Moist	6' x 4'
Euonymus kiautschovicus 'Pauli'	Pauli Euonymus	This plant has dark green leaves and is an upright grower.	Fast Growth	Moist	2' x 5'
Fagus sylvatica 'Asplenifolia'	Fernleaf European Beech	Gracefully cut leaves offer a fern -like appearance. Dark green lieaves turn golden brown in fall .	Slow Growth	Moist	50' x 50'
Fagus sylvatica 'Pendula'	Green Weeping European Beech	A most beautiful weeping form.	Average Growth	Moist	25' x 35'
Fagus sylvatica 'Purpurea Pendula'	Purple Weeping European Beech	A broad mushroom-shapping weeping, purple - leaved tree.	Slow Growth	Moist	10' x 10'
Fagus sylvatica 'Riversii'	Rivers European Beech	A deep purple tree that holds its color into summer.	Slow Growth	Moist	50' x 50'
Fagus sylvatica 'Tricolor'	Tricolor European Beech	Beautiful pink edged purple leav es grace this pyramidal medium sized tree.	Slow Growth	Moist	50' x 50'
Fallopia Aubertii	Silver Lace Vine	A deciduous, vigorous, twining climber. Has white to green flowers, maturing to pink in summer.	Fast Growth	Moist	10' x 10'
Festuca ovina glauca	Blue Fescue	Clumps of dense blue leaves makes this a favorite ornamental grass for edging or planting in masses for color. Thrives in well -drained soils.	Average Growth	Semi-Moist	1' x 1'

Festuca ovina glauca 'Elijah Blue'	Elijah Blue Fescue	This outstanding ornamental grass has fine powder blue leaves that form short, dense clumps. Excellent as an edging plant or mass plantings for color.	Average Growth	Semi-Moist	10" x 10"
Ficus carica 'Brown Turkey'	Brown Turkey Fig	A rather coarse, broad-rounded shrub that produces purple-grown fruit.	Fast Growth	Moist	8' x 6'
Forsythia suspensa	Weeping Forsythia	Growth habit is upright, arching, almost fountain-like and flowers are yellow.	Average Growth	Moist	7' x 5'
Forsythia viridissima 'Bronxensis'	Bronxensis Dwarf Greenstem Forsythia	Primrose yellow flowers blanket this short, compact Forsythia selection about 2 weeks later than other Forsythia. Excellent groundcover or specimen.	Slow Growth	Moist	12" x 24"
Forsythia X intermedia 'Lynwood'	Lynwood Border Forsythia	Brilliant golden-yellow flowers that are evenly distributed along the long slender branches of this shrub proclaim that spring has arrived.	Fast Growth	Moist	8' x 6'
Forsythia X intermedia 'Spectabilis'	Showy Border Forsythia	This is the forsythia most commonly seen blanketed in bright yellow flowers in early spring. An excellent large shrub for mass plantings.	Fast Growth	Moist	8' x 8'
Forsythia X intermedia 'Spring Glory'	Spring Glory Border Forsythia	Bountiful sulfur-yellow flowers cover the slender branches of this large shrub in early spring. One of the earliest-blooming Forsythia varieties.	Fast Growth	Moist	5' x 6'
Forsythia x ovata 'Meadowlark'	Meadowlark Forsythia	Sometimes called Early Forsythia, this variety blooms yellow and is very flower bud hardy.	Fast Growth	Moist	8' x 8'
Fothergilla gardenii 'Blue Mist'	Blue Mist Dwarf Fothergilla	A wonderful compact shrub for shade that has fragrant, white flowers in spring and powdery blue leaves all summer. Prefers moist, well-drained soils.	Slow Growth		4' x 36"

Fragaria x 'Ozark Beauty'	Ozark Beauty Everbearing Strawberry	Everbearing, large, firm fruit. Has a high sugar content. Self-pollinating. Excellent for eating fresh, canning, freezing and preserves.	Fast Growth	Semi-Moist	10" x 2'
Fraxinus americana	White Ash	A handsome native tree that grows rapidly to a rounded form. Dark green leaves turn yellow to deep purple in fall. Grows 50 to 80 feet high and wide.	Fast Growth	Moist	50' x 40'
Fraxinus americana 'Autumn Applause'	Autumn Applause White Ash	A smaller, more densely-branched White Ash with small dark green leaves that turn maroon every fall. Grows 40 feet high by 25 feet wide.	Fast Growth	Moist	50' x 25'
Fraxinus americana 'Autumn Purple'	Autumn Purple White Ash	A fast-growing, rounded White Ash that has glossy, dark green leaves that turn a rich dark reddish-purple in fall. Grows 45 feet high by 60 feet wide.	Fast Growth	Moist	50' x 35'
Fraxinus americana 'Rosehill'	Rosehill White Ash	A vigorous-growing shade tree that tolerates poor soils with dark green leaves that turn bronsze-red in fall. Grows 50 feet high by 30 feet wide.	Fast Growth	Moist	50' x 40'
Fraxinus angustifolia 'Raywood'	Raywood Ash	A vigorous-growing ash for the South with lustrous dark green leaves that turn rich plum purple in fall. Grows 50 feet high by 30 feet wide.	Fast Growth	Semi-Moist	45' x 35'
Fraxinus pennsylvanica	Green Ash	A rapid-growing tree that tolerates poor soil conditions. Has shiny green leaves that turn yellow in fall. Grows 50 feet high by 25 feet wide.	Fast Growth	Moist	50' x 35'
Fraxinus pennsylvanica 'Cimmzam'	Cimmaron Red Ash	A fast growing upright oval shaped tree that is seedless and salt tolerant.The thick green foliage turns a brick red in the fall.	Fast Growth	Semi-Moist	50' x 25'

Fraxinus pennsylvanica 'Dakota Centennial'	Dakota Centennial Green Ash	A fast growing upright oval shaped tree that is seedless. Yellow fall color.	Fast Growth	Semi-Moist	50' x 30'
Fraxinus pennsylvanica lanceolata 'Marshall's Seedless'	Marshall's Seedless Ash	A fast growing seedless rounded shade tree. The glossy green foliage turns gold in the fall.	Fast Growth	Semi-Moist	50' x 40'
Fraxinus pennsylvanica 'Patmore'	Patmore Green Ash	A rapid-growing, uniformly oval, cold hardy Green Ash with glossy dark green leaves that turn yellow in fall. Grows 50 feet high by 35 feet wide.	Fast Growth	Moist	50' x 35'
Fraxinus pennsylvanica 'Summit'	Summit Green Ash	A rapid-growing, upright pyramidal Green Ash with refined green leaves that turn a pleasing golden-yellow in fall. Grows 45 feet high by 25 feet wide.	Fast Growth	Moist	50' x 35'
Fraxinus pennsylvanica 'Urbanite'	Urbanite Green Ash	A fast growing seedless tree with glossy green leaves with golden fall color.	Fast Growth	Semi-Moist	50' x 35'
Fraxinus velutina	Arizona Ash	A fast growing shade tree with yellow fall color. This Ash will take heat and wind well.	Fast Growth	Semi-Moist	50' x 40'
Gaillardia aristata 'Kobald'	Goblin Blanketflower	Has a compact growth and produces rich red flowers with yellow tips.	Average Growth	Semi-Moist	1' x 1 1/2'
Gardinia jasmoides 'August Beauty'	August Beauty Gardenia	Large, fragrant white flowers bloom in spring above lustrous dark green leaves. Should be planted near a patio where people will notice the fragrance.	Average Growth	Semi-Moist	4' x 36"

Gardinia jasmoides 'prostrata'	Dwarf Gardinia	Fragrant white flowers bloom all spring above lustrous green leaves on this compact shrub. The best selection for fragrance in today's landscapes.	Average Growth	Moist	12" x 36"
Gazania rigens	Gazania Mix	A mat forming plant with leaves that are green above and whitish beneath.	Fast Growth	Semi-Moist	
Gelsemium sempervirens	Carolina Jessamine	Profuse, fragrant yellow flowers bloom in early spring above the glossy dark green leaves of this twining evergreen vine. Excellent for naturalizing.	Fast Growth	Moist	10' x 10'
Geranium X 'Johnson's Blue'	Johnson's Blue Geranium	Clear blue, cup-shaped flowers blanket this sprawling perennial in late spring. Great in rock gardens or as a border a long the garden path.	Average Growth	Moist	
Ginkgo biloba	Ginkgo Tree	Tree's green, fan-shaped leaves turn golden yellow in fall.	Fast Growth	Moist	50' x 40'
Gleditaia tricanthos 'inermis aurea'	Golden Honeylocust	This medium-sized shade tree has golden yellow leaves in spring, eventually turning bright green in summer. Grass grows well under this tree.	Fast Growth	Semi-Moist	35' x 35'
Gleditsia triacanthos inermis (var.)	Thornless Honeylocust	A rapid-growing shade tree which has a short trunk and open-spreading branches. Leaves are glossy, bright green and turn a clear yellow in fall.	Fast Growth	Semi-Moist	50' x 35'
Gleditsia triacanthos inermis (var.) 'Halka'	Halka Honeylocust	A rapid-growing shade tree which has a short trunk and open-spreading branc hes. Leaves are glossy, bright green and turn a clear yellow in fall.	Fast Growth	Semi-Moist	40' x 35'
Gleditsia triacanthos inermis (var.) 'Imperial'	Imperial Honeylocust	A medium-sized shade tree with fine, glossy dark green leaves that turn clear yello w in fall. Open branching allows grass to grow under the tree.	Fast Growth	Semi-Moist	35' x 35'
Gleditsia triacanthos inermis (var.) 'Moraine'	Moraine Honeylocust	A medium-sized shade tree with fine-textured glossy green leaves that turn golden yellow in fall. Open branching allows grass to grow under the tree.	Fast Growth	Semi-Moist	60' x 50'

Gleditsia triacanthos inermis (var.) 'Shademaster'	Shademaster Honeylocust	An upright-growing, medium-sized shade tree, perfect for today's yards. Glossy green leav es turn yellow in fall. Grass grows well under this tree.	Fast Growth	Semi-Moist	50' x 45'
Gleditsia triacanthos inermis (var.) 'Skyline'	Skyline Honeylocust	A pyramidal, compact shade tree fits well in today's yards. Glossy green leaves turn golden yellow in fall. Grass grows well under this tree.	Fast Growth	Semi-Moist	50' x 45'
Gleditsia triacanthos inermis (var.) 'Sunburst'	Sunburst Honeylocust	This medium-sized shade tree has golden yellow leaves in spring, eventually turning bright green in summer. Grass grows well under this tree.	Fast Growth	Semi-Moist	25' x 35'
Gleditsia tricanthos 'Green Glory'	Green Glory Honeylocust	Retains its foliage later than other types. Dark green fern-like leaves.		Moist	
Gymnocladus dioica	Kentucky Coffee Tree	A choice tree for parks, goldf courses and toehr large areas.	Average Growth		50' x 45'
Halesia carolina	Silverbell	A handsome lawn tree. Flowers are normally white but rarely pale rose and bell -shaped.	Average Growth	Moist	25' x 25'
Hamamelis virginiana	Common Witchhazel	Valuable in a naturalized situation, shrub border, or near large buildings. Flowers are yellow and fragrant.	Slow Growth	Moist	15' x 15'
Hedera helix	English Ivy	A shade-loving, groundcover with lustrous dark green leaves. Will spread along the ground or climb trees and structures with root-like holdfasts.	Fast Growth	Semi-Moist	
Hemerocallis flava	Lemon Daylily	Large yellow flowers in the early summer.	Average Growth	Moist	3' x 2'
Hibiscus moscheutos 'Moy Grande'	Moy Grande Red Hibiscus	A large perennial that produces huge 10" red flowers in the late summer and fall.	Fast Growth	Moist	5' x 4'

Hibiscus rosa-sinensis	Tropical Hibiscus Mix	A compact semi-tropical shrub covered with glossy dark green leaves that provide the perfect background for the red, pink, white and yellow flowers.	Fast Growth	Semi-Moist	6' x 6'
Hibiscus syriacus	Rose of Sharon Mix	A dense upright shrub with numerous white, pink, red, or purple flowers which bloom in late summer. Works best in mass plantings and hedges.	Slow Growth	Moist	10' x 6'
Hibiscus syriacus 'Aphrodite'	Aphrodite Rose of Sharon	A U. S. National Arboretum selection that has dark pink, single flowers with dark red centers in late summer. An upright multi -stemmed shrub.	Slow Growth	Moist	7' x 6'
Hibiscus syriacus 'Ardens'	Double Purple Althea	2" double lavender purple flowers cover the plant from mid-summer until frost. Althea are also called Rose of Sharon and like warm sunny locations.	Fast Growth	Semi-Moist	10'x 5'
Hibiscus syriacus 'Diana'	Diana Rose of Sharon	A U. S. National Arboretum selection with large, single, pure white flowers that bloom from mid - summer to fall. A great upright multi -stemmed shrub.	Slow Growth	Moist	8' x 6'
Hibiscus syriacus 'Freedom'	Red Freedom Rose of Sharon	2" double rosey flowers cover the plant from mid - summer until frost. Althea is also called Rose of Sharon and like warm sunny spots.	Slow Growth	Moist	10' x 6'
Hibiscus syriacus 'Jeanne d'Arc'	Double White Rose of Sharon	Double white flowers cover this upright, den se multi-stemmed shrub in late summer. Works great in mass plantings and for hedging.	Slow Growth	Moist	10' x 6'
Hibiscus syriacus 'Lucy'	Double Red Althea	2" double rosey flowers cover the plant from mid - summer until frost. Althea is also called Rose of Sharon and like warm sunny spots.	Fast Growth	Semi-Moist	10' x 5'
Hibiscus syriacus 'Lucy'	Double Red Althea Tree	A small tree with 2" double rosey flower covering the plant from mid -summer until frost.	Fast Growth	Semi-Moist	10' x 5'

Hibiscus syriacus 'Lucy'	Lucy Rose of Sharon	2" double rosey flowers cover the plant from mid - summer until frost. Althea is also called Rose of Sharon and like warm sunny spots.	Fast Growth	Semi-Moist	10' x 5'
Hibiscus syriacus 'Minerva'	Minerva Rose of Sharon	A U. S. National Arboretum selection with single purple flowers with red centers throughout the summer. A wonderful upright, multi -stemmed shrub.	Slow Growth	Moist	10' x 6'
Hibiscus syriacus 'Peonyflorus'	Double Pink Althea	2" double pink flowers cover the plant fro m mid-summer until frost. Althea are also called Rose of Sharon and like warm sunny loca locations.	Fast Growth	Semi-Moist	10' x 5'
Hibiscus syriacus 'Peonyflorus'	Double Pink Althea Tree	A small tree with 2" double pink flowers from mid - summer until fro st.	Fast Growth	Semi-Moist	10' x 5'
Hibiscus syriacus 'Red Heart'	Red Heart Rose of Sharon	Single white flowers with red centers cover this upright multi-stemmed shrub all summer long. Works best in mass plantings.	Slow Growth	Moist	10' x 6'
Hibscus syriacus "Ardens'	Double Purple Althea Tree	A small tree with 2" double lavender purple flowers from mid-summer until frost. Althea are also callled Rose of Sharon and like warm sunny locstions.	Fast Growth	Semi-Moist	10' x 5'
Hibuscus syriacus 'Double White'	White Tree Althea	Double white flowers cover this upright, dense multi-stemmed shrub in late summer. Works great in mass plantings and for hedging.	Fast Growth	Semi-Moist	10' x5'
Hosta fortunei 'Aureo Marginata'	Gold Edge Hosta	A medium-sized Hosta with dark green leaves framed in rich, gold edges. Lavender flowers bloom in mid-summer. Excellent edging plant.	Average Growth	Moist	
Hydrangea arborescens 'Annabelle'	Annabelle White Hydrangea	Big, bold snowball-shaped flowers change from green to white, lasting up to 2 months on this light shade-loving shrub. Great for mass plantings.	Average Growth	Moist	36" x 36"

Hydrangea macrophylla	Pink Hydrangea	A compact shrub with pink snowball flower heads that begin to bloom in early summer and turn rose - red in cool weather. Great specimen plant.	Average Growth	Moist	4' x 4'
Hydrangea macrophylla 'All Summer Beauty'	All Summer Beauty Hydrangea	An outstanding small shrub with rich blue snowball-shaped flowers, in acid soils, in mid - summer. Great for foundation , mass, or specimen plantings.	Average Growth	Moist	4' x 6'
Hydrangea macrophylla 'All Summer Beauty'	All Summer Beauty Hydrangea	An outstanding small shrub with rich blue snowball-shaped flowers, in acid soils, in mid - summer. Great for foundation, mass, or specimen plantings.	Average Growth	Moist	4' x 4'
Hydrangea macrophylla 'Alpenglow'	Alpenglow Hydrangea	Hot pink large flowers are real traffic stoppers on a compact plant.	Average Growth	Moist	4' x 6'
Hydrangea macrophylla 'Blue Wave'	Blue Wave Hydrangea	Produces rich blue flowers in acid soils, otherwise pink or lilac..	Average Growth	Moist	4' x 6'
Hydrangea macrophylla 'Cardinal Red'	Cardinal Red Hydrangea	A great small, summer -blooming shrub with cardinal red snowball -shaped flowers that make a dazzling display over the dark green leaves.	Average Growth	Moist	4' x 6'
Hydrangea macrophylla 'Forever Pink'	Forever Pink Hydrangea	A compact shrub with pink snowball flower heads that begin to bloom in early summer and turn rose - red in cool weather. Great specimen plant.	Average Growth	Moist	36" x 6'
Hydrangea macrophylla 'Glory Blue'	Glory Blue Hydrangea	An old form that is quite vigorous producing large rounded medium blue flowers in acid soil.	Average Growth	Moist	4' x 6'
Hydrangea macrophylla 'Glowing Embers'	Glowing Embers Hydrangea	Hot pink large flowers are real traffic stoppers on a compact plant.	Average Growth	Moist	4' x 6'

Hydrangea macrophylla 'Merritt Supreme'	Merritt Supreame Hydrangea	An old form that is quite vigorous producing large rounded medium blue flowers in acid soil.	Average Growth	Moist	4' x 6'
Hydrangea macrophylla 'Nikko Blue'	Nikko Blue Hydrangea	An old form that is quite vigorous producing large rounded medium blue flowers in acid soil.	Average Growth	Moist	4' x 4'
Hydrangea macrophylla 'Pia (Pink Elf)'	Pink Pia Hydrangea	Has pink flowers that are 3" to 4" in diameter.	Average Growth	Moist	24" x 24"
Hydrangea macrophylla 'Pink Beauty'	Pink Beauty Hydrangea	A compact shrub with pink snowball flower heads that begin to bloom in early summer and turn rose-red in cool weather. Great specimen plant.	Average Growth	Moist	4' x 4'
Hydrangea macrophylla 'Sister Therese'	Sister Theresa Hydrangea	A compact shrub that's covered with white snowball flowers over dark green leaves in mid-summer. Great for foundation, mass, or specimen plantings.	Average Growth	Moist	4' x 6'
Hydrangea macrophylla 'Variegata'	Variegated Blue Hydrangea	An excellent companion shrub with hosta because of its dark green leaves with white edges and mauve flowers in mid-summer.	Average Growth	Moist	6' x 6'
Hydrangea paniculata 'Grandiflora'	Pee Gee Hydrangea	A large shrub or small tree with spectacular, large white flower heads from mid-summer to fall. Great for border, mass, or specimen plantings.	Average Growth	Moist	8' x 10'
Hydrangea paniculata 'grandiflora'	Pee Gee Tree Hydrangea	A small tree with huge white flowers that turn to pink as they age in August.	Average Growth	Moist	12' x 6'
Hydrangea paniculata 'Pink Diamond'	Pink Diamond Hydrangea	A large shrub or small tree with spectacular, large pink flower heads from mid-summer to fall. Great for border, mass, or specimen plantings.	Average Growth	Moist	5' x 6'

Hydrangea paniculata 'Tardiva'	Late Panicle Hydrangea	A stout large shrub with large, white flower heads in late summer to fall that are great for drying. A great foundation or specimen plant.	Average Growth	Moist	6' x 8'
Hydrangea paniculata 'Webbs'	Webb's Pee Gee Hydrangea	A compact growing Pee Gee type plant with 6" white flowers in the late summer.	Slow Growth	Moist	4' x 4'
Hydrangea quercifolia	Oakleaf Hydrangea	A great shade plant with large white flowers and huge leaves that turn burgundy in the fall.	Fast Growth	Moist	6' x 6'
Hydrangea quercifolia 'Alice'	Alice Oak Leaf Hydrangea	A great shade plant with large white flowers and huge leaves that turn burgundy in the fall.	Fast Growth	Moist	8' x 4'
Hydrangea quercifolia 'Snow Queen'	Snow Queen Oakleaf Hydrangea	A large shrub with masses of white flower heads held erect above green leaves that turn bright burgundy in fall. Excellent specimen plant.	Fast Growth	Moist	6' x 6'
Hydrangea quercifolia 'Snowflake'	Snowflake Oakleaf Hydrangea	A large shrub with masses of white flower heads that create a weeping habit in mid-summer. Dark green leaves turn burgundy-red in fall.	Fast Growth	Moist	6' x 6'
Hydrangea quercifolia 'Syke's Dwarf'	Syke's Dwarf Oakleaf Hydrangea	A great shade plant with large white flowers and huge leaves that turn burgundy in the fall.	Average Growth	Moist	36" x 6'
Hydrangea serrata 'Coerulea Lace'	Blue Lace Hydrangea	Dainty circles of baby blue flowers cover the plant for much of the summer.	Average Growth	Moist	4' x 4'
Hylotelephium X spectabile 'Autumn Joy'	Autumn Joy Sedum	Showy bright pink flowers cover the plants for two months in the fall.	Average Growth	Semi-Moist	3' x 3'
Hypericum frondosum 'Sunburst'	Sunburst St John's Wort	Stems are upright and ensely covered with oblong leaves. Bright yellow flowers appear in summer.	Fast Growth	Moist	24" x 36"

Hypericum patulum 'Hidcote'	Hidcote St John's Wort	Has lance-shaped, dark green leaves and produces golden-yellow flowers from mid-summer to early fall.	Average Growth	Moist	18" x 30"
Iberis sempervirens	White Candytuft	Ideal for rock gardens. Produces rounded heads of unscented white flowers in spring and early summer.	Average Growth	Moist	1' x 2'
Ilex cornuta	Chinese Holly	A bushy, dense evergreen shrub that can be trained into a small tree.	Average Growth	Moist	
Ilex cornuta 'Burfordii'	Burford Holly	A vigorous, dense evergreen shrub with glossy green leaves and large red berries. Excellent large hedge plant. Tolerates poor soil conditions.	Average Growth	Moist	15' x 15'
Ilex cornuta 'Burfordii Nana'	Dwarf Burford Holly	Glossy, dark green leaves provide a perfect background for red berries produced by this dense, evergreen shrub. Excellent for hedging.	Average Growth		10' x 6'
Ilex cornuta 'Carissa'	Carissa Holly	A compact, dense evergreen shrub with glossy, dark green leaves. Excellent for a low hedge or in mass plantings. No berry production.	Slow Growth		24" x 36"
Ilex cornuta 'Needlepoint'	Needlepoint Holly	One of the best hollies for screening, massing and grouping.	Average Growth	Moist	10' x 10'
Ilex cornuta 'Rotunda'	Dwarf Chinese Holly	A very tough, durable plant, akin to a living pin cushion.	Slow Growth	Moist	6' x 6'
Ilex crenata 'Compacta'	Compact Japanese Holly	A compact mounding evergreen shrub with lustrous dark green leaves and purplish twigs. Excellent for foundation, hedge, or mass plantings.	Slow Growth	Moist	6' x 6'
Ilex crenata 'Convexa'	Convexa Japanese Holly	A hardy, dense, evergreen shrub with dark green leaves and black berries. An excellent selection for hedges or mass plantings.	Slow Growth	Moist	6' x 6'

Ilex crenata 'Green Dragon'	Green Dragon Japanese Holly	A slow-growing, rounded evergreen shrub with small dark green leaves. A unique small accent plant for the garden. No berry production.	Slow Growth	Moist	24" x 24"
Ilex crenata 'Green Lustre'	Green Lustre Holly	A low, spreading evergr een shrub with dense, lustrous dark green leaves. Excellent for hedges or mass plantings. Produces some black berries.	Slow Growth	Moist	5' x 8'
Ilex crenata 'Helleri'	Heller Japanese Holly	A dense, compact, mounded evergreen shrub with small dark green leaves. Outstanding plant for border plantings, hedges, accent or mass plantings.	Slow Growth	Moist	36" x 6'
Ilex crenata 'Hetzii'	Hetzi Japanese Holly	A vigorous, evergreen shrub with large, lustrous dark green leaves and large black berries. Excellent for hedges or foundation plantings.	Slow Growth	Moist	6' x 6'
Ilex crenata 'Rotundiolia'	Japanese Holly	A dense shrub with dark evergreen leaves. Can be pruned into many shapes.	Slow Growth	Moist	4' x 4'
Ilex crenata 'Soft Touch'	Soft Touch Holly	A dense, compact shrub with lustrous dark green leaves. Flexible branches and solf foliage.			
Ilex crenata 'Soft Touch'	Soft Touch Holly	A dense, compact shrub with lustrous dark green leaves. Flexible branches and solf foliage.			
Ilex decidua 'Warren Red'	Warren Red Possum Haw	A native shrub with showy red berries in the fall	Slow Growth	Moist	15' x 10'
Ilex glabra	Inkberry	Excellent for foundation, hedges, masses and as an accent plant.	Slow Growth		6' x 6'
Ilex glabra 'Compacta'	Compact Inkberry	A dense, compact evergreen shrub with dark green leaves and jet black berries that persist through the winter. Thrives in wet soil conditions.	Slow Growth		4' x 6'
Ilex glabra 'Nordic'	Nordic Compact inkberry	A compact rounded form with lustrous deep green leaves. Quite cold hardy.			

Ilex glabra 'Shamrock'	Shamrock Inkberry	A dense, compact evergreen shrub with glistening, bright green leaves that turn dark green with age. Outstanding for mass plantings.	Slow Growth		36" x 36"
Ilex verticillata	Winterberry Holly	An upright multi-stemmed shrub with bright red berries in fall and persisting into winter. A tough shrub that thrives in wet soil conditions.	Slow Growth	Moist	6' x 6'
Ilex vomitoria	Yaupon Holly	An upright, multi-stemmed evergreen shrub or small tree with lustrous dark green leaves and scarlet berries in fall that persist through winter.	Slow Growth	Moist	15' x 15'
Ilex vomitoria 'nana'	Dwarf Yaupon Holly	A dwarf compact form with slightlyi smaller leaves than the standard species. New growht is ric h yellow-green.			
Ilex vomitoria 'Pendula'	Weeping Yaupon Holly	A graceful multi-stemmed evergreen shrub or small tree with gray weeping branches, lustrous dark green leaves and scarlet berries in fall.	Average Growth	Moist	15' x 10'
Ilex X aquipernyi 'San Jose'	San Jose Holly	A narrow, upright, pyramidal evergreen shrub with lustrous dark green leaves and bright red berries. Needs a male for good berry production.	Slow Growth		25' x 15'
Ilex X attenuata 'Fosteri '	Foster Holly	An excellent upright, pyr amidal evergreen shrub to 20 feet high with dark green and large red berries in fall. Needs male for berry production.	Slow Growth	Moist	20' x 6'
Ilex X attenuata 'Savannah'	Savannah Holly	An upright, pyramidal evergreen that grows 20 to 30 feet high by 15 feet wide with fluorescent red berries in fall. Needs male for berry production.	Average Growth	Moist	25' x 6'
Ilex X 'Emily Brunner'	Emily Brunner Holly	An upright, pyramidal evergreen shrub that has dark green leaves and large red berries. Needs a male for berry production. Great accent plant.	Average Growth	Moist	15' x 6'

Ilex X 'Mary Nell'	Mary Nell Holly	An upright, pyramidal evergreen shrub with glossy, dark green leaves and bright red berries. Excellent for southern landscapes.	Slow Growth	Moist	25' x 15'
Ilex X meserveae 'Blue Prince'	Blue Prince Holly	Bright glossy blue-green leaves highlight this dense, rounded, cold-hardy evergreen shrub. The male pollinator for female Blue Holly selections.	Slow Growth	Moist	8' x 4'
Ilex X meserveae 'Blue Princess'	Blue Princess Holly	Lush, glossy blue-green leaves provide the perfect background for the prolific red berries produced by this dense, rounded, evergreen shrub.	Slow Growth	Moist	8' x 4'
Ilex x meserveae 'Mesdob'	China Boy Holly	Makes a good hedge. A good pollinator			
Ilex X meserveae 'Mesgolg(China Girl)'	China Girl Holly	Rich, glossy dark green leaves form the perfect background for the profusion of red berries produced by this dense, evergreen shrub in fall.		Moist	5' x 4'
Ilex X 'Nellie R. Stevens'	Nellie Stevens Holly	A pyramidal evergreen shrub with lustrous dark green leaves and red berries. A great holly for southern landscapes. Needs a male for berry production.	Average Growth	Moist	15' x 6'
Imperata cylindrica 'Red Baron'	Japanese Bloodgrass	Has erect mid-green leaves and in mid-summer their ends turn the color of blood.			
Incarvillea delavayi	Hardy Gloxinia	Useful for rock gardens and borders. Fern-like foliage and erect stems bears trumpet-shaped rosey purple flowers in summer.	Average Growth	Moist	
Iris germanica	Tall Bearded Iris Mix	An upright, hardy perennial with large, fragrant flowers of many colors in late spring. Excellent for cut flowers. Great for naturalizing.	Average Growth	Semi-Moist	

Iris sibirica	Siberian Iris	Multi-color flowers crown the narrow green leaves of this hardy perennial in mid - to late spring. Outstanding for mass plantings and naturalizing.	Fast Growth	Moist	
Iris sibirica 'Butter and Sugar'	Butter and Sugar Siberian Iris	Delicate white and yellow flowers bloom in late spring above the narrow green leaves of this hardy perennial. Great for cut flowers.	Fast Growth	Moist	
Iris sibirica 'Caesar's Brother'	Caesar's Brother Siberian Iris	Purple flowers bloom above the narrow green leaves of this hardy perennial in late spring to early summer. Great for cut flowers and naturalizing.	Fast Growth	Moist	
Iris x Pumila	Blue Dwarf Bearded Iris	A dainty little bearded iris. Prefers slightly alkaline soil.	Average Growth	Semi-Moist	
Itea virginica	Virginia Sweetspire	A deciduous shrub of upright, slender form. In summer bears fragrant, creamy white flowers.	Average Growth	Moist	4' x 6'
Itea virginica 'Henry's Garnet'	Henry's Garnet Virginia Sweetspire	A compact, rounded shrub with fragrant white flowers in spring and brilliant garnet red fall color that lasts into winter. Great for mass plantings.	Average Growth	Moist	4' x 6'
Itea virginica 'Sprich (Little Henry)'	Little Henry Virginia Sweetspire	Sweetly scented white flowers cover this compact shrub in spring and brilliant red fall color lasts into winter. Great for edging or mass plantings.	Slow Growth		36" x 4'
Jasminum nudiflorum	Winter Jasmine	Clusters of yellow flowers liven up the landscape in winter and early spring. A fast -growing shrub that grows into an arching mound with green twigs.	Fast Growth		24" x 4'
Juglans cinerea	Butternut	Seeds are sweet, edible and very oily.	Slow Growth	Moist	40' x 45'
Juglans nigra	Black Walnut	Wood is highly prized and has been used for many furniture pieces.	Average Growth	Moist	50' x 50'

Juglans regia	English or Carpathian Walnut	Nuts are thin-shelled and widely available in stores.	Average Growth	Moist	40' x 40'
Juniperous chinensis 'Admiral'	Admiral Juniper	A bright green upright juniper.			
Juniperous chinensis 'Sea Green'	Espallier Sea Green Juniper	A form created by careful pruning and tieing to the trellis			
Juniperous sabina 'tamariscifolia'	Tami Juniper	A low spreader with fine textured rich green foliage. Grows well in dry climates.	Average Growth	Dry	2' x 4'
Juniperus chinensis 'Blue Vase'	Blue Vase Juniper	Steel blue foliage covers the branches of this attractive, naturally vase-shaped evergreen. Excellent for low hedges or foundation plantings.	Slow Growth	Semi-Moist	10' x 6'
Juniperus chinensis 'Columnaris'	Columnar Juniper	Deep green foliage covers the branches of this upright, narrow, pyramidal evergreen. An excel lent specimen plant or as a living fence.	Average Growth	Semi-Moist	25' x 8'
Juniperus chinensis 'Nicks Compact'	Nick's Compact Juniper	A flat-topped, wide-spreading evergreen with dark bluish-green foliage. Excellent for mass plantings in difficult sites, such as slopes.	Slow Growth	Semi-Moist	24" x 8'
Juniperus chinensis 'Old Gold'	Old Gold Juniper	One of the finest golden spreading evergreens. Golden foliage color intensifies in spring. A compact shrub that works great in mass plantings.	Slow Growth	Semi-Moist	24" x 6'
Juniperus chinensis 'Pfitzeriana'	Pfitzer Juniper	Feathery gray-green foliage blanket the arching branches of this spreading evergreen. Excellent for mass plantings in large areas.	Slow Growth	Semi-Moist	5' x 10'
Juniperus chinensis 'Pfitzeriana Aurea'	Gold Tipped Pfitzer Juniper	Mint green foliage with bright golden yellow tips cover the arching branches of this evergreen. Excellent as an accent or as a foundation plant.	Slow Growth	Semi-Moist	36" x 8'

Juniperus chinensis 'Pfitzeriana Aurea'	Goldtip Pfitzer Pom Pom Juniper	A form created by years of careful pruning. To maintain this form just touch up the pruning twice a year.			
Juniperus chinensis 'Pfitzeriana Compacta'	Compact Pfitzer Juniper	A flat-topped, wide-spreading evergreen with gray-green foliage. Excellent as a foundation plant or in mass plantings.	Slow Growth	Semi-Moist	12" x 36"
Kniphofia uvaria	Red Hot Poker	Carries brightly colored tubular flowers on spikes in the summer.			24"
Koelreuteria paniculata	Golden Rain Tree	A wonderful dense, rounded tree with bright yellow flower clusters in late summer followed by showy fruit capsules. Excellent small lawn tree.	Average Growth	Moist	25' x 35'
Kolkwitzia amabilis	Beautybush	Masses of large, clear pink flowers blanket this graceful arching shrub in spring. Works best in group plantings to announce the arrival of spring.	Fast Growth		6' x 6'
Laburnum X watereri	Golden Chain Tree	Good in a shrub border, near buildings or corners of houses. Makes a tremendous show in flower.	Average Growth		15' x 10'
Lagerstoemia indica 'Byers Standard Red'	Byers Red Crape Myrtle	Flowers are dark red and bloom in mid -july to September.		Semi-Moist	
Lagerstroemia indica	Crape Myrtle	Handsome and very beautifyl specimen shrub or tree. Some smaller types are used for hedges, screens and in masses.	Average Growth	Semi-Moist	15' x 15'
Lagerstroemia indica 'Bayou Marie'	Bayou Marie Crape Myrtle	An upright spreading shrub with clear pink flowers in mid-summer atop dark green leaves. Makes a beautiful specimen shrub or small tree.	Average Growth	Semi-Moist	12' x 15'
Lagerstroemia indica 'Bourbon Street'	Bourbon Street Juniper	A miniature shrub covered with watermelon -red flowers in mid-summer. Makes an attractive flowering pot plant and color bord ers.	Average Growth	Semi-Moist	4' x 4'

Lagerstroemia indica 'Byers Regal Red'	Byers Regal Red Crape Myrtle	Flowers are dark red and bloom in mid -july to September.		Semi-Moist	
Lagerstroemia indica 'Candystripe'	Candystripe Crape Myrtle	Deep pink white e dged flowers produce a candy strip effect.		Semi-Moist	
Lagerstroemia indica 'Carolina Beauty'	Carolina Beauty Crape Myrtle	Flowers are dark red and bloom in mid -july to September.		Semi-Moist	20' x 12'
Lagerstroemia indica 'Catawba'	Catawba Crape Myrtle	Has glossy dark green foliage and produces dark purple flowers from late July to September.		Semi-Moist	10' x 10'
Lagerstroemia indica 'Centenial'	Centenial Crape Myrtle	A dwarf compact shrub that produced bright lavender flowers blooming mid -June and later.		Semi-Moist	3' x 4'
Lagerstroemia indica 'Chickasaw'	Chickasaw Dwarf Crape Myrtle	A true genetic dwarf producing pink -lavender flowers.		Semi-Moist	2' x 3 1/2'
Lagerstroemia indica 'Choctaw'	Choctaw Crape Myrtle	A large multi-stemmed shrub or small tree that produced bright pink flowers.		Semi-Moist	18'
Lagerstroemia indica 'Christmas Time'	Christmas Time Crape Myrtle	An upright grower that produces white blooms.		Semi-Moist	12' x 10'
Lagerstroemia indica 'Cordon Bleu'	Cordon Bleu Crape Myrtle	Lavender flower clusters cover this miniature shrub in mid-summer. Perfect flowering pot plant and for color borders.	Average Growth	Semi-Moist	15' x 15'
Lagerstroemia indica 'Hope'	Hope Crape Myrtle	An open-growing dwarf that produced white flowers blooming mid-June.		Semi-Moist	4' x 4'

Lagerstroemia indica 'Hopi'	Hopi Crape Myrtle	A spreading, semi-dwarf, multi-stemmed variety that produces pink blooms from late June to late September.		Semi-Moist	10' x 6'
Lagerstroemia indica 'Miami'	Miami Crape Myrtle	An upright, milki-stemmed small tree that produces dark pink blooms from early July to September.		Semi-Moist	16' x 8'
Lagerstroemia indica 'Muskogee'	Muskogee Crape Myrtle	A large tree or shrub with light lavender -pink flowers in July to September.		Semi-Moist	18' x 14'
Lagerstroemia indica 'Natchez'	Natchez Crape Myrtle	Can be a large tree or shrub that produces pure white flowers in late June-July into September.		Semi-Moist	20' x 18'
Lagerstroemia indica 'Osage'	Osage Crape Myrtle	A multi-stemmed large shrub or small tree producing light pink flowers .		Semi-Moist	12' x 10'
Lagerstroemia indica 'Ozark Spring'	Ozark Spring Crape Myrtle	A compact dwarf variety with lavender flowers.		Semi-Moist	16' x 12'
Lagerstroemia indica 'Peppermint'	Peppermint Crape Myrtle	Deep pink white edged flowers produce a candy strip effect.		Semi-Moist	15' x 15'
Lagerstroemia indica 'Potomac'	Potomac Crape Myrtle	Has an upright growth habit with medium pink flowers mid-July into October.		Semi-Moist	10' x 8'
Lagerstroemia indica 'Powhatan'	Powhatan Crape Myrtle	Has a compact, medium growth havig producing medium purple flowers in late July until late September.		Semi-Moist	10' x 10'
Lagerstroemia indica 'Prairie Lace'	Prairie Lace Crape Myrtle	A compact, upright semi-dwarf shrub that produces blooms with medium pink petals with white outer margins in mid-June to September.		Semi-Moist	6' x 6'
Lagerstroemia indica 'Royalty'	Royalty Dwarf Crape Myrtle	Lavender-purple flower clusters bloom in mid-summer on this dwarf shrub. Great as an accent plant or when used in mass plantings.	Average Growth	Semi-Moist	4' x 36"

Lagerstroemia indica 'Sarah's Favorite'	Sarah's Favorite Crape Myrtle	A large shrub or small tree with white flowers.		Semi-Moist	20' x 18'
Lagerstroemia indica 'Seminole'	Seminole Crape Myrtle	Clear pink flower clusters bloom in mid -summer on this dense shrub. Glossy, dark green, disease -resistant leaves turn bright yellow in fall.	Slow Growth	Semi-Moist	10' x 10'
Lagerstroemia indica 'Tonto'	Tonto Crape Myrtle	A semi-dwarf shrub with fuchsia red flowers produced in July to September.		Semi-Moist	10' x 10'
Lagerstroemia indica 'Victor'	Victor Dwarf Crape Myrtle	A dwarf compact variety producing dark red flowers that bloom in July to August.		Semi-Moist	3' x 4'
Lagerstroemia indica 'Wichita'	Wichita Crape Myrtle	An upright multii-stemmed small tree whose blooms are light magenta to lavender showing in earlyi July to October.		Semi-Moist	16' x 9'
Lagerstroemia indica 'Wonderful White'	Wonderful White Crape Myrtle	An upright form with clear white flowers blooming in July.		Semi-Moist	20' x 14'
Lagerstroemia X 'Acoma'	Acoma Crape Mytle	Clusters of white flowers bloom in mid -summer above the dark green leaves on the upright, weeping branches of this dense, disease -resistant shrub.	Average Growth	Semi-Moist	12' x 15'
Lagerstroemia X 'Biloxi'	Biloxi Crape Myrtle	Pale pink flowers are produced in mid -summer on this tall, upright, multi-stemmed shrub. Cinnamon - brown trunks add all season interest.	Average Growth	Semi-Moist	15' x 15'
Lagerstroemia X 'Miami'	Miami Crape Myrtle	Dark pink flower clusters bloom in mid -summer on this multi-stemmed tall shrub or small tree. Showy chestnut brown bark adds all season interest.	Fast Growth	Semi-Moist	15' x 10'
Lagerstroemia X 'Muskogee'	Muskogee Crape Mytle	Light lavender flowers bloom in mid -summer on this multi-stemmed tall shrub or small tree. Glossy green, disease-resistant leaves turn red in fall.	Average Growth	Semi-Moist	15' x 15'

Lagerstroemia X 'Natchez'	Natchez Crape Myrtle	Pure white flowers bloom in mid -summer on this multi-stemmed shrub or small tree. Cinnamon brown bark adds interest all season long.	Average Growth	Semi-Moist	25' x 20'
Lagerstroemia X 'Tuscarora'	Tuscarora Crape Myrtle	Dark coral-pink flower clusters bloom in mid-summer on this multi-stemmed large shrub or small tree. Dark green leaves turn orange -red in fall.	Average Growth	Semi-Moist	15' x 10'
Lagerstroemia X 'Tuskegee'	Tuskegee Crape Myrtle	A multi-stemmed small tree producing dark pink to almost red flowers from early July to September.	Average Growth	Semi-Moist	12' x 16'
Lagerstroemia X 'Yuma'	Yuma Crape Myrtle	Lavender flower clusters bloom in mid -summer on this large, multi-stemmed shrub. Dark green, disease-resistant leaves turn yellow -orange in fall.	Slow Growth	Semi-Moist	12' x 10'
Lagerstroemia x 'Zuni'	Zuni Crape Myrtle	Lavender flower clusters bloom in mid -summer on this large, multi-stemmed shrub. Dark green, disease-resistant leaves turn yellow -orange in fall.		Semi-Moist	
Lagerstromeia x 'Centenial Spirit'	Centenial Spirit Crape Myrtle	Lavender-purple flower clusters bloom in mid -summer on this dwarf shrub. Great as an accent plant or when used in mass plantings.		Semi-Moist	
Lagerstromeia x 'Natchez'	Natchez Crape Myrtle	Can be a large tree or shrub that produces pure white flowers in late June-July into September.		Semi-Moist	
Lamium maculatum 'Beacon's Silver'	Beacon Silver Lamium	A semi-evergreen perennial that produces purplish flowers with silvery green leaves edged dark gre en	Fast Growth	Moist	8" x 12"
Lantana camara	Lantana Mix	Masses of small flowers are produced all summer long above fragrant dark green leaves and attract butterflies. Great for borders and mass plantings.	Fast Growth	Moist	
Lantana camara 'New Gold'	New Gold Lantana	Masses of yellow flowers bloom from spring through fall, attracting butterflies to the garden. A great plant for borders and mass plantings.	Fast Growth	Moist	

Lantana camara 'Radiation'	Radiation Lantana	Masses of rich orange-red flowers bloom from spring through fall attracting butterflies to the garden. Great for borders and mass plantings.	Fast Growth	Semi-Moist	
Larix decidua 'Pendula'	Weeping European Larch	Very effective for park and large area use.	Average Growth	Moist	15' x 8'
Lavandula angustifolia 'Munstead'	Munstead English Lavender	Grown mainly for the long-stemmed heads of purple sented flowers that appear in summer and throughout the warm season.	Slow Growth	Moist	
Leucothoe fontanesiana 'Girard's Rainbow'	Rainbow Leucothoe	New growth may emerge white, pinkish or coppery. A showy evergreen when new leaves are emerging.	Average Growth	Moist	4' x 24"
Liatris spicata 'Kobold'	Dwarf Purple Gayfeather	A dwarf cultivar produces bright purple flowers.	Average Growth	Moist	15"
Ligustrum lucidum	Waxleaf Ligustrum	A large evergreen shrub with glossy green leaves that is very tolerant of poor soil conditions. Excellent choice for screening and hedges.	Average Growth	Moist	25' x 25'
Ligustrum sinense 'Variegatum'	Variegated Chinese Privet	A dense evergreen shrub with green leaves with creamy white edges. Works well for hedges and foundation plantings.	Average Growth	Moist	10' x 10'
Ligustrum x ibolium 'Variegatum'	Variegated Privet	A dense evergreen shrub with green leaves with creamy white edges. Works well for hedges and foundation plantings.	Average Growth	Moist	10' x 10'
Ligustrum X vicaryi	Golden Vicary Privet	A dense, upright-growing, evergreen shrub with bright golden yellow leaves. Plant this cold-hardy shrub in full sun for best leaf color.	Average Growth	Moist	10' x 10'
Liquidambar styraciflua	Sweetgum	A rapid-growing, pyramidal shade tree with star-shaped leaves that turn to brilliant yellow, red and orange in fall.	Average Growth	Moist	50' x 35'

Liquidambar styraciflua 'Rotundiloba'	Fruitless Sweetgum	A rapid-growing upright shade tree with star-shaped leaves with rounded lobes. Sets no fruit, so it's a perfect tree for the home landscape.	Average Growth	Moist	50' x 35'
Liriodendron tulipifera	Tulip Tree	A large,pyramidal tree with yellow tulip-shaped flowers in spring and bright green leaves that turn brilliant golden yellow in fall.	Average Growth	Moist	65' x 35'
Liriope muscari 'Big Blue'	Big Blue Lilyturf	Has grass-like shining, dark green leaves and bears erect spikes of rounded, bell-shaped flowers.			
Lobelia cardinalis	Cardinal Flower	Useful for growing in wet places and produces spikes of scarlet-red flowers from late summer to mid-fall.	Average Growth		
Lonicera sempervirens	Trumpet Honeysuckle	A great twining vine that bursts into bloom with orange-red flowers in spring followed by red fruit. Hummingbirds find the flowers irresistable.	Fast Growth		
Lonicera tatarica	Tatarian Honeysuckle	Foliage is bluish-green and flowers are pink to white.	Fast Growth	Moist	10' x 10'
Lonicera X heckrottii	Goldflame Honeysuckle	A rapid-growing vine with clusters of fragrant, brilliant red-yellow flowers from spring through summer that attract hummingbirds and butterflies.	Fast Growth	Moist	10' x 10'
Lonicera xylosteum 'Clavey's Dwarf'	Clavey's Dwarf Honeysuckle	Foliage is grayish-green. Flowers are white or yellowish white often tinged reddish.	Fast Growth	Moist	4' x 8'
Lonicera xylosteum 'Emerald Mound	Emerald Mound Honeysuckle	Hardier and more compact than other varieties but fruits and flowers are not conspicuous.	Fast Growth	Moist	4' x 4'
Loropetalum chinense 'Burgundy'	Burgundy Loropetalum	A unique evergreen shrub with reddish -purple leaves and hot pink flowers that bloom in late winter through early spring at the branc h tips.	Average Growth	Moist	12' x 10'
Lychnis chalcedonica	Maltese Cross	A favorite of gardeners since the 17th century. Color is a dazzling orange-red.	Average Growth	Moist	4'

Maclura pomifera	Osage Orange	Used for hedgerows in the plains states. Go od for rugged and polluted areas.	Fast Growth	Moist	25' x 25'
Magnolia grandiflora	Southern Magnolia	This stately evergreen tree is known for its large, fragrant white flowers in early summer followed by red berries and glossy, dark green leaves.	Average Growth	Moist	60' x 35'
Magnolia grandiflora 'D.D. Blanchard'	Blanchard Southern Magnolia	Pyramidal with lustrous dark green leaves and rich orangish brown undersides.	Average Growth	Moist	25' x 10'
Magnolia grandiflora 'Little Gem'	Little Gem Southern Magnolia	This small tree, large shrub produces fragrant white flowers throughout the growing season. An excellent Magnolia for today's smaller landscapes.	Slow Growth	Moist	20' x 10'
Magnolia stellata	Star Magnolia	Fragrant white flowers bloom in early spri ng on this dense, rounded large shrub. Excellent specimen or accent plant for spring color.	Slow Growth	Moist	10' x 10'
Magnolia stellata 'Royal Star'	Royal Star Magnolia	Pink buds open to fragrant, pure white flowers with many petals in early spring. Thi s dense, upright shrub is great for specimen or accent plantings.	Slow Growth	Moist	8' x 6'
Magnolia virginiana	Sweetbay Magnolia	Lemon-scented creamy white flowers bloom from early summer to fall on this small evergreen tree. Leaves are dark green with s ilvery undersides.	Average Growth	Moist	25' x 15'
Magnolia X 'Ann'	Ann Magnolia	Large, deep purplish-pink flowers bloom in late spring, after most frosts.	Average Growth	Moist	10' x 10'
Magnolia X 'Betty'	Betty Magnolia	Large deep purple flowers bloom in late spring after most frosts. A vigorous, upright -growing shrub that's a U. S. National Arboretum selection.	Average Growth	Moist	10' x 10'
Magnolia X loebneri 'Merrill'	Dr Merrill Magnolia	Masses of fragrant flowers turn this vigorous, spreading deciduous tree into a white cloud in late spring. Dark green leaves turn yellow in fall.	Average Growth	Moist	25' x 25'

Magnolia X soulangiana	Saucer Magnolia	Large white flowers with pink t o purplish tones bloom in early spring on this vigorous, upright, multi-stemmed shrub. A great accent plant.	Slow Growth	Moist	15' x 15'
Magnolia X soulangiana 'Alexandrina'	Alexandrina Saucer Magnolia	Large rose-purple flowers bloom in early spring on this vigorous, upright, multi -stemmed shrub. A great accent plant.	Average Growth	Moist	20' x 15'
Mahonia aquifolium	Oregon Grapeholly	Bright yellow flowers bloom in mid -spring amid the lustrous dark green leaves on this upright, evergreen shrub. A great fo undation plant.	Slow Growth	Moist	4' x 36"
Mahonia aquifolium 'Compacta'	Oregon Grapeholly	Dwarf form with very glossy leaves and bronze winter color.	Slow Growth	Moist	36" x 36"
Mahonia bealei	Leatherleaf Mahonia	Golden yellow flowers bloom in spring, f ollowed by blue-black clusters of grape -like berries on this upright, evergreen shrub.	Slow Growth	Moist	4' x 36"
Mahonia repens	Creeping Grape Holly	Deep yellow flowers bloom in spring followed by black grape-like berries on this stiff, spreading evergreen groundcover.	Average Growth		6" x 12"
Malus domestica "Anna'	Anna Semi-dwarf Apple	A low chill apple that bears well in the Southern USA. The sweet red and green fruit ripen in mid to late summer.	Average Growth	Semi-Moist	15' x 15'
Malus domestica 'Braeburn'	Braeburn Semi-dwarf Apple	A tart red and green striped apple that ripens in late October and keeps well.	Average Growth	Semi-Moist	18' x 18'
Mandevilla X amoena 'Alice du Pont'	Alice du Pont Pink Mandevilla	Clusters of pure pink flowers grace the glossy, dark green leaves of this twining vine all summer long. Great for posts, arbors and trellises.	Fast Growth	Moist	
Mentha suaveolens 'Variegata'	Variegated Pineapple Mint	A vigorous perennials with felted variegated apple - scented leaves.	Fast Growth	Moist	3'
Metasequoia glyptostroboides	Dawn Redwood	Well suited to parks, gold courses and other laerge areas.	Fast Growth	Moist	60' x 20'

Microbiota decussata	Siberian Cypress	A very cold-hardy, spreading evergreen with rich green leaves that turn bronze in winter. A great groundcover that does well in shade.	Slow Growth		6" x 4'
Miscanthus sinensis 'Silberfeder'	Silver Feather Grass	Produces a generous crop of light, open, silvery pink flowers on erect stems.			9'
Miscanthus sinensis 'Strictus'	Porcupine Silvergrass	A tall upright ornamental grass known for its wide green leaves with horizontal yellow bands. Reddish-silver flowers in fall. A great accent plant.	Fast Growth	Moist	
Miscanthus sinensis 'Variegatus'	Variegated Silvergrass	A tall ornamental grass having wide green leaves with bold white stripes. Red-tinted flowers bloom in fall. A great arching accent plant.	Fast Growth		
Miscanthus sinensis 'Zebrinus'	Zebra Silvergrass	A tall, spreading ornamental grass known for its wide green leaves with horizontal yellow bands. Copper flowers in fall. A great accent plant.	Fast Growth		
Monarda didyma	Bee Balm	Masses of dark green leaves are crowned with scarlet-red flowers al summer long on this upright perennial. A great plant for naturalizing in sun.	Average Growth	Moist	
Morus alba 'Chaparral'	Chaparral Weeping Mulberry	A non-fruiting, weeping form that is grafted at a certain height and weeps to the ground from there. Bright green leaves turn yellow in fall.	Average Growth	Moist	10' x 6'
Myrica cerifera	Southern Wax Myrtle	Makes an excellent pruned screen and can be limbed to make a small tree.			20' x 20'
Myrica pensylvanica	Northern Bayberry	Lustrous dark green leaves blanket this dense, upright shrub. An excellent selection for foundation and mass plantings. Thrives in poor, salty soils.	Slow Growth	Moist	10' x 6'
Nandina domestica	Heavenly Nandina	An upright, evergreen shrub with fine-textured blue-green leaves that turn bright red in fall and are complimented by masses of red berries..	Average Growth	Moist	6' x 12"

Nandina domestica 'Compacta'	Compact Nandina	A compact, dense mounding, evergreen shrub with fine-textured blue-green leaves that turn red in fall. Excellent selection for mass plantings.	Slow Growth	Moist	24" x 12"
Nandina domestica 'Firepower'	Firepower Nandina	A dense, upright evergreen shrub with purplish-green leaves that turn fluorescent red in fall. Works great for foundation and mass plantings.	Slow Growth	Moist	12" x 12"
Nandina domestica 'Harbor Dwarf'	Harbor Dwarf Nandina	A compact, dense mounding, evergreen shrub with bluish-green leaves that turn reddish-purple in fall. Great for mass plantings.	Slow Growth	Moist	24" x 12"
Nandina domestica 'Nana Purpurea'	Dwarf Purple Nandina	A compact, dense mounding, evergreen shrub with reddish-green, fine-textured leaves that turn bright red in fall and winter. Great for mass plantings.	Slow Growth	Moist	12" x 12"
Nandina domestica 'Royal Princess'	Royal Princess Nandina	Beautiful fine textured leaves that turn red in the fall and winter.	Average Growth	Moist	5' x 5'
Nerium oleander	Oleander Mix	Flowers of white, pink, red, yellow and orange blanket this low maintenance, rounded, evergreen shrub throughout the summer.	Fast Growth		10' x 10'
Nyssa sylvatica	Black Gum	A picturesque, native shade tree with nice form and lustrous dark green leaves that turn to fluorescent red to red-orange in early fall.	Slow Growth	Moist	50' x 25'
Oenothera fruticosa	Sundrops	Has a reddish, erect, branching stem and narrow leaves. Produces deep yellow flowers.			3'
Oenothera macrocarpa	Missouri Evening Primrose	Usually almost stemless. Flowers are large, lemon yellow in color and open in the evening and summer.	Average Growth	Semi-Moist	6" x 24"
Oenothera speciosa	Pink Evening Primrose	Has spikes of profuse, fragrant, saucer-shaped, pink-tinted white flowers.	Fast Growth		24" x 18"

Ophiopogan planiscapus 'nigrescens'	Black Mondo Grass	Grown particularly for its distinctive purple-black leaves. Lilac flowers appear in the summer.			10" x 12"
Ophiopogon japonicus	Dwarf Mondo Grass	A clumping, evergreen perennial with thin, dark green, grass-like leaves. A slow grower that makes and excellent edging plant or groundcover.	Slow Growth	Moist	
Ophiopogon japonicus 'Ebony Knight'	Ebony Knight Lilyturf	Grown particularly for its distinctive purple-black leaves. Lilac flowers appear in the summer.			
Osmunda cinnamonea	Cinnamon Fern	Fronds are bluish green from spring to summer.	Average Growth	Moist	4'
Osmunda regalis	Regal Fern	Commonly found in swamps and other boggy areas.	Average Growth	Moist	6'
Pachysandra terminalis	Green Pachysandra	A great semi-evergreen groundcover for shady areas with white flowers blooming in early spring above the dark green leaves.	Average Growth	Moist	
Paeonia lactiflora	Peony Mix	Among the great garden plants, prized for its handsome foliage, longevity and huge blooms in late spring.	Average Growth	Semi-Moist	3'
Paeonia lactiflora	Pink Peony	Among the great garden plants, prized for its handsome foliage, longevity and huge pink blooms in late spring.	Average Growth	Semi-Moist	3'
Paeonia lactiflora	Red Peony	Among the great garden plants, prized for its handsome foliage, longevity and huge red blooms in late spring.	Average Growth	Semi-Moist	3'
Paeonia lactiflora	White Peony	Among the great garden plants, prized for its handsome foliage, longevity and huge white blooms in late spring.	Average Growth	Semi-Moist	3'

Papaver nudicaule	Iceland Poppy	Good for rock gardens and for cutting. Large scented flowers are borne in winter and spring.	Average Growth	Moist	24" x 8"
Papaver orientale	Oriental Poppy	Bears spectacular flowers in summer.	Average Growth	Semi-Moist	18"
Parrotia persica	Persian Parrotia	A short trunked tree whose leaves turn shades of yellow, orange and crimson in fall.	Slow Growth		25' x 25'
Parthenocissus tricuspidata	Boston Ivy	A deciduous vine that makes a dense groundcover and climbs with adhesive tendrils. The dark green leaves turn vivid red in fall.	Fast Growth		40' x 40'
Pennisetum alopecuroides	Fountain Grass	A wonderful ornamental grass with tufts of enormous flower tassels in fall above the dark green leaves. Great for mass plantings.	Fast Growth	Moist	
Pennisetum alopecuroides 'Hamelin'	Hamelin Dwarf Fountain Grass	A compact ornamental grass with tufts of flower tassels in early fall above the dark green leaves. Works great as a border or in mass plantings.	Average Growth	Moist	
Pennisetum alopecuroides 'Moudry'	Black Fountain Grass	Dark brown to black flower tassels appear in mid-fall above the dark green leaves of this compact ornamental grass. Good for cut flowers.	Fast Growth	Moist	
Pennisetum orientale	Oriental Fountain Grass	Delicate pink flowers bloom from early summer through fall above thin, glossy blue-green leaves. Great as a specimen plant or in mass plantings.	Average Growth	Semi-Moist	
Pennisetum setaceum 'Burgundy Giant'	Burgundy Giant Fountain Grass	Makes an attractive tall ground cover and has purple-red foliage.	Average Growth	Semi-Moist	3' x 8"
Pennisetum setaceum 'Rubrum'	Purple Fountain Grass	Stunning purple leaves and dark red flower plumes from summer through fall make this an outstanding plant for containers or mass plantings.	Fast Growth	Moist	

Penstemon digitalis 'Husker Red'	Husker Red Pentstemon	Spikes of white bell-shaped flowers bloom in summer above the deep bronze-purple leaves.	Average Growth	Moist	
Perovskia atriplicifolia	Russian Sage	This tall tough species produces soft, gray-green foliage that compliments the pale lavender-blue flowers in late smmer and fall.	Average Growth	Semi-Moist	4' x 3'
Phalaris arundinacea picta (var.)	Variegated Ribbongrass	Bears reed-like leaves with white stripes. A clump form perennial grass that is easily grown.	Fast Growth		5'
Philadelphus coronarius	Sweet Mockorange	An old favorite for its sweet-scented flowers.	Average Growth	Moist	10' x 8'
Philadelphus x 'Natchez'	Natchez Mockorange	In flower, perhaps the handsomest of all mockoranges with 1 1/2" diameter, slightly fragrant, pure white flowers.			
Philadelphus X virginalis 'Minnesota Snowflake'	Minnesota Snowflake Mockorange	Fragrant white flowers bloom in spring among the green leaves of this upright shrub. Plant it where the wind will carry the fragrance into the garden.	Average Growth	Moist	10' x 8'
Phlox divaricata	Woodland Phlox	A great clumping perennial with white to blue flowers in late spring. Makes an excellent groundcover or mass planting in light shade.	Average Growth	Moist	
Phlox paniculata 'Starfire'	Starfire Garden Phlox	True, bright red flower clusters bloom in summer on this upright perennial. A great plant for the summer sun garden.	Average Growth	Moist	
Phlox subulata	Creeping Phlox	Fragrant star-shaped blooms completely cover mounds of green moss-like foliage in early spring.	Average Growth	Semi-Moist	
Photinia X fraseri	Red Tip Photinia	New leaves and branches emerge a brilliant red, later turning deep green on this vigorous, evergreen shrub. Makes a great screen.	Average Growth	Semi-Moist	15' x 6'

Physocarpus opulus 'Darts Gold'	Darts Golden Ninebark	A good choise for spring and early summer foliage color.	Average Growth	Semi-Moist	10' x 8'
Picea abies	Norway Spruce	Native to Scandanavia. The traditional Christmas tree in Europe.	Slow Growth	Moist	60' x 25'
Picea abies 'nidiformis'	Bird's Nest Spruce	A spreading, dense, broad plant with usually a depression in the center.	Slow Growth	Moist	3'
Picea abies 'Pendula'	Weeping Norway Spruce	Has a weeping branching structure.	Slow Growth	Moist	25' x 10'
Picea glauca	White Spruce	Useful as a specimen, mass, hedge or windbreak.	Average Growth	Moist	35' x 15'
Picea glauca 'Conica'	Dwarf Alberta Spruce	A natural dwarf it can become broadly conical in time.	Slow Growth	Semi-Moist	8' x 36"
Picea mariana 'Nana'	Blue Nest Spruce	A small to medium-size tree which develops a limited spread.	Average Growth	Moist	24" x 6'
Picea pungens	Colorado Blue Spruce	A frost hardy species tha t has a pyramid of bluish green foliage.	Average Growth	Moist	35' x 15'
Picea pungens 'Bakeri'	Bakeri Blue Spruce	Has deep blue foliage.	Average Growth	Moist	35' x 15'
Picea pungens 'Fat Albert'	Fat Albert Blue Spruce	Dense upright pyramidal form with go od blue needle color.	Average Growth	Moist	35' x 15'
Picea pungens 'Glauca Pendula'	Weeping Blue Spruce	Makes a somewhat gaudy rock garden plant.	Slow Growth	Moist	24" x 4'
Picea pungens 'Hoopsii'	Hoopsi Blue Spruce	Dense, pyramidal form with spreading b ranches. Foliage is blue-white.	Average Growth	Moist	35' x 15'

Picea pungens 'Montgomery'	Globe Blue Spruce		Slow Growth	Moist	10' x 10'
Pieris japonica	Japanese Pieris	Arching white flower clusters bloom in spring over the bronze-pink new leaves of this medium-sized evergreen shrub. Brightens a shady area in spring.	Slow Growth		9' x 6'
Pinus aristata	Bristlecone Pine	A very slow-growing, irregularly-shaped tree, as you'd expect of the oldest living plant on earth. Does best in a rock garden or as a bonsai tree.	Slow Growth	Moist	20' x 20'
Pinus cembra	Swiss Stone Pine	A picturesque and hardy tree useful as a specimen or in mass. Foliage is dark green outside and bluish white innerside.	Slow Growth		35' x 10'
Pinus densiflora 'Umbraculifera'	Tanyosho Japanese Red Pine	A unique evergreen noted for the dense, spreading top of the tree as it forms an umbrella over the reddish brown trunk. A great accent tree.	Average Growth	Moist	15' x 35'
Pinus elliottii	Slash Pine	Fast growing pyramidal pine that yields turpentine, rosin, timber and wood pulp.	Fast Growth	Moist	75' x 35'
Pinus flexalis 'Vanderwolf'	Vanderwolf Limber Pine	A handsome specimen that is adapted for planting on rocky slopes.	Average Growth	Moist	
Pinus mugho	Mugho Pine	A low-spreading, bushy evergreen with dark green needles that's especially suited to foundation or mass plantings.	Slow Growth	Semi-Moist	10' x 10'
Pinus nigra	Austrian Pine	A graceful, pyramidal evergreen with beautiful, long green needles that fill the tree branches all the way to the ground. A great landscape tree.	Average Growth		25' x 25'
Pinus nigra 'Globosa'	Dwarf Austrian Pine	A unique evergreen with a dense, rounded top filled with stiff, dark green needles. Makes a wonderful accent plant.	Average Growth		25' x 25'

Pinus parviflora 'Glauca'	Blue Japanese Pine	A slow-growing, upright evergreen with gray-blue needles. A graceful addition to any landscape as a specimen plant.	Slow Growth	Moist	25' x 25'
Pinus ponderosa	Ponderosa Pine	A large, pyramidal evergreen with dark green needles that thrives in well-drained soils. Best used in mass plantings.	Fast Growth		50' x 30'
Pinus strobus	Eastern White Pine	A graceful, tall evergreen with soft, fine-textured blue-green needles. Best suited for use as a specimen in large areas.	Average Growth	Moist	50' x 25'
Pinus strobus 'Contorta'	Contorted White Pine	Has an open, irregular, pyramidal form with slightly twisted branches.	Slow Growth	Moist	20' x 25'
Pinus strobus 'Fastigiata'	Columnar White Pine	When young, narrowly upright and columnar but develops a wider character with age.	Average Growth	Moist	50' x 15'
Pinus strobus 'Nana'	Dwarf White Pine	A dwarf, mounded form of this popular evergreen with dense, blue-green needles covering the branches. A good accent plant.	Slow Growth	Moist	15' x 8'
Pinus strobus 'nana'	Dwarf White Pine	A dwarf, mounded form of this popular evergreen with dense, blue-green needles covering the branches. A good accent plant.			
Pinus strobus 'Pendula'	Weeping White Pine	Weeping type with long branches that sweep the ground.	Slow Growth	Moist	10' x 25'
Pinus sylvestris	Scotch Pine	A rapid-growing, adaptable evergreen with blue-green needles and bright orange, papery bark. A great selection for mass plantings.	Slow Growth	Moist	45' x 25'
Pinus sylvestris 'Fastigiata'	Columnar Scotch Pine	One of the narrowest of pines.	Slow Growth	Moist	35' x 5'

Pinus taeda	Loblolly Pine	A rapid-growing, adaptable evergreen with dark green needles. An excellent choice for screening and mass plantings.	Slow Growth	Moist	50' x 25'
Pinus thunbergii	Japanese Black pine	A medium-sized evergreen with dark green needles that thrives in poor sandy soils. Can be used on shorelines because it's very tolerant of salt spray.	Average Growth	Moist	15' x 15'
Pinus virginiana	Virginia Pine	Valuable as a cover for dry and barren soils.	Slow Growth	Semi-Moist	30' x 25'
Pistacia chinensis	Chinese Pistachio	A medium-sized tree with bright green leaves that turn brilliant orange and red in fall. A great shade tree for today's landscapes.	Slow Growth	Moist	25' x 30'
Pittosporum tobira	Green Pittossporum	Fragrant creamy white flowers bloom in spring among the lustrous, dark green leaves of this dense, evergreen shrub. Works great in mass plantings.	Average Growth	Semi-Moist	6' x 10'
Pittosporum tobira 'Variegata'	Variegated Pittosporum	Very fragrant creamy white flowers bloom in spring among the dark green leaves edged with white, making a stunning display. Great for mass plantings.	Average Growth		6' x 10'
Pittosporum tobira 'Wheeleri'	Dwarf Pittosprorum	Fragrant, creamy white flowers bloom in spring among the lustrous dark green leaves on this compact evergreen shrub. Works great in mass plantings.	Average Growth	Semi-Moist	24" x 36"
Platanus occidentalis	Sycamore	A rapid-growing, wide-spreading shade tree that works best in large areas. Has cream and olive bark that peels adding interest in winter and summer.	Fast Growth	Moist	70' x 70'
Platanus orientalis	European Planetree	A rapid-growing, disease-resistant shade tree that has cream and olive bark that peels making an interesting mosaic in summer and winter.	Fast Growth	Moist	70' x 70'

Platanus X acerifolia 'Bloodgood'	Bloodgood London Planetree	A wide-spreading shade tree with beautiful cream and olive bark that looks good in winter and summer, and dark green leaves that turn yellow in fall.	Fast Growth	Moist	70' x 70'
Platycladus orientalis 'Aurea Nana'	Berkman's Golden Arborvitae	Dwarf, dense, globular to ovid form with golden yellow foliage. Slightly browish in winter.	Average Growth		4' x 36"
Polygonatum odoratum 'Variegatum'	Variegated Solomon's Seal	This perennial originates from Europe, Russia, Japan and the Caucasus.	Average Growth	Moist	
Populus alba	White Poplar	Two-tone leaves with Dark green above and silvery white below look stunning fluttering in the wind on this rapid-growing, wide-spreading tree.	Fast Growth	Moist	50' x 35'
Populus canadensis 'Nor'Easter'	Nor'Easter Cottonwood	A seedless canker resistant cottonwood.	Fast Growth	Semi-Moist	
Populus nigra 'Italica'	Lombardy Poplar	An upright, fast-growing tree planted for wind breaks.	Fast Growth	Moist	25' x 6'
Populus tremuloides	Quaking Aspen	The most widely distributed tree of North America. A fast-growing tree whose leaves flutter in the breeze.	Fast Growth	Moist	50' x 20'
Populus X canadensis 'Siouxland'	Siouxland Cottonless Cottonwood	A rapid-growing, pyramidal tree with lustrous light green leaves that flutter in the wind.	Fast Growth	Moist	50' x 35'
Populus x 'Robusta'	Robusta Poplar	Averages 4'1" of growth per year under optimum conditions.	Fast Growth	Semi-Moist	
Potentilla fruiticosa 'Red Sunset'	Red Potentilla	Has orange flowers fading to yellow.	Fast Growth	Semi-Moist	

Potentilla fruiticosa 'Tangerine'	Tangerine Potentilla	Flowers are yellow in full sun. Medium yellow, pale copper orange or orange-red under ideal conditions.	Fast Growth	Semi-Moist	4' x 2'
Potentilla fruticosa 'Abbotswood'	Abbotswood Potentilla	White flowers bloom from spring through fall on this compact, cold-hardy plant with bluish leaves. Works great as a low hedge or in mass plantings.	Slow Growth	Moist	36" x 36"
Potentilla fruticosa 'Coronation Triumph'	Coronation Triumph Potentilla	Brilliant yellow flowers bloom from early summer through fall on this compact, cold-hardy shrub. Works great as a low hedge or in mass plantings.	Slow Growth	Moist	24" x 24"
Potentilla fruticosa 'Gold Drop'	Gold Drop Potentilla	Large, intense yellow flowers bloom from early summer through fall on this compact, cold-hardy shrub. Works well as a low hedge or in mass plantings.			24" x 36"
Potentilla fruticosa 'Goldfinger'	Goldfinger Potentilla	Large, deep yellow flowers bloom from early summer through fall on this compact, cold-hardy shrub. A winner as a low hedge or in mass plantings.		Moist	36" x 36"
Potentilla fruticosa 'Jackmanii'	Jackman Potentilla	Deep yellow flowers bloom from early summer through fall on this rounded, cold-hardy shrub. Works great as a hedge or in mass plantings.	Slow Growth	Moist	36" x 24"
Potentilla fruticosa 'Katherine Dykes'	Katherine Dykes Potentilla	Soft yellow flowers bloom from early summer through fall on this low-spreading, cold-hardy shrub. Ideal for mass plantings.	Slow Growth	Moist	12" x 24"
Potentilla fruticosa 'McKay's White'	McKay's White Potentilla	Creamy white flowers bloom from early summer through fall on this low-spreading, cold-hardy shrub. Works great in mass plantings.	Slow Growth	Moist	36" x 4'
Potentilla fruticosa 'Primrose Beauty'	Primrose Beauty Potentilla	Creamy yellow flowers bloom from early summer through fall on this upright, cold-hardy shrub. Works great as a hedge or in mass plantings.	Slow Growth	Moist	4' x 5'

Potentilla fruticosa 'Sunset'	Sunset Potentilla	Pale yellow flowers with tints of orange grace this upright, cold-hardy shrub from early summer through fall. A great choice for hedges.	Slow Growth	Moist	4' x 4'
Potentilla nepalensis 'Miss Willmott'	Miss Willmott Potentilla	Cherry pink flowers bloom from late spring through summer on this trailing perennial. Looks great in the rock garden where it sprawls over rocks.	Average Growth	Semi-Moist	
Prunus americana	American Plum	Red and yellow fruit. Excellent food for wildlife but suitable for humans as well.	Average Growth	Moist	15' x 10'
Prunus americana 'Toka'	Toka Semi-dwarf Plum	Reddish bronze fruit. Firm, aromatic, freestone flesh. Rich, spicy, sweet flavot. Excellent for fresh eating.			
Prunus armeniaca 'Chinese'	Chinese Semi-dwarf Apricot	Sweet golden fruit are good for canning and drying.The heavy bearing tree is self pollinating.The pit are edible an rich in natural nutrients.	Average Growth	Semi-Moist	15' x 15'
Prunus Armeniaca mandshurica	Mandchurian Apricot	A very hardy ornamental tree with bland semi-sweet yellow fruit for jam.The vase shaped tree has showy flowers and fall color.	Average Growth	Semi-Moist	15' x 15'
Prunus Armeniaca mandshurica 'Moongold'	Moongold Semi-dwarf Apricot	Moongold has amber yellow semi-sweet fruit for jams and canning. The tree is vase shaped and ornamental and bears better with a pollinator.	Average Growth	Semi-Moist	15' x 15'
Prunus Armeniaca mandshurica 'Scout'	Scout Semi-dwarf Apricot	A very hardy ornamental tree with bland semi-sweet yellow fruit for jam.The vase shaped tree has showy flowers and fall color.	Average Growth	Semi-Moist	20' x 20'
Prunus armeniaca 'Moorpark'	Moorpark Semi-dwarf Apricot	A great home variety that is self pollinating and has good fruit quality. Red skin covers orange full favored flesh for fresh eating and canning.	Average Growth	Semi-Moist	18' x 18'

Prunus armeniaca 'Sunglow'	Sunglow Semi-dwarf Apricot	Sweet golden fruit are good for canning and eating fresh.	Average Growth	Semi-Moist	15' x 15'
Prunus cerasifera 'Krauter Vesuvius'	K V Purple Plum	Light pink flowers bloom in early spring, followed by the dark purple leaves on this upright, rounded small tree.	Average Growth	Moist	20' x 15'
Prunus cerasifera 'Mount Saint Helens'	Mt St Helens Purple Plum	Light pink to white flowers bloom in early spring, followed by large rich purple leaves on this upright, rounded small tree.	Slow Growth	Moist	20' x 20'
Prunus cerasifera 'Newport'	Newport Purple Plum	Light pink to white flowers bloom in early spring, followed by dark purple leaves on this upright, rounded small tree.	Slow Growth	Moist	15' x 15'
Prunus cerasifera 'Thundcloud'	Thundercloud Purple Leaf Plum	Sincle pionk flowers appear before the leaves and are fragrant. Tree retains its deep purple foliage through growing season.	Slow Growth	Moist	15' x 15'
Prunus glandulosa 'rosea'	Pink Flowering Almond	The stems are covered with double pink flowers at the first sign of spring.	Average Growth	Semi-Moist	3' x 2'
Prunus laurocerasus	Cherry Laurel	Creamy white flowers bloom in spring among the lustrous dark green leaves of this spreading, evergreen shrub. Makes a handsome hedge plant.	Fast Growth	Moist	12' x 12'
Prunus laurocerasus 'Otto Luyken'	Otto Luyken Cherry Laurel	Profuse white flowers bloom in early spring among the dark green leaves of this compact, evergreen shrub. Excellent for hedges and mass plantings.	Average Growth	Moist	4' x 6'
Prunus laurocerasus 'Schipkaensis'	Skip Cherry Laurel	White flowers bloom in early spring among the glossy, dark green leaves of this compact, evergreen shrub. Works well as a hedge or in mass plantings.	Average Growth		4' x 8'
Prunus maackii	Amur Bird Cherry	White flowers bloom in early spring on this dense, rounded, small tree. Wonderful cinnamon-brown bark peels back for great winter interest.	Average Growth	Moist	25' x 25'

Prunus padus	European Birdcherry	Fragrant white flowers announce the arrival of spring by this medium-sized, rounded tree. The small black fruit attract birds when fully ripe.	Average Growth	Moist	20' x 15'
Prunus persica	Peach	The Peach tree has pink flowers in early spring are followed by luscious fruit. A wonderful small fruit tree for the backyard gardener.	Fast Growth	Moist	15' x 15'
Prunus serrulata 'Kwanzan'	Kwanzan Flowering Cherry	The large deep pink flowers that bloom in mid-spring on this medium-sized Cherry are the true indication that spring has arrived to many.	Average Growth	Moist	15' x 15'
Prunus subhirtella 'Autumnalis'	Autumn Higan Cherry	Abundant reddish-pink flowers bloom in early spring on this rounded, medium-sized tree.	Fast Growth	Moist	25' x 15'
Prunus subhirtella 'Pendula'	Weeping Higan Cherry	A gracefully weeping tree that becomes a waterfall of single pink flowers in early spring. An outstanding specimen plant.	Fast Growth	Moist	25' x 15'
Prunus tomentosa	Nanking Cherry	Fragrant white flowers bloom in early spring, followed by edible scarlet fruit on this dense, upright-spreading shrub. Looks great in mass plantings.	Fast Growth	Moist	6' x 10'
Prunus virginiana 'Schubert'	Canada Red Cherry	Bright red leaves whose fruit is red to dark purple 1/3" across.	Fast Growth		20' x 15'
Prunus X cistena	Purpleleaf Sand Cherry	A profusion of fragrant, white flowers bloom in early spring before the deep purple leaves emerge on this upright shrub.	Average Growth	Moist	6' x 6'

Prunus X 'Hally Jolivette'	Hally Jolivette Flowering Cherry	Pinkish white flowers bloom in mid-spring on this dense, rounded tree. A fruitless, ornamental Cherry tree.	Average Growth	Moist	10' x 10'
Prunus X incamp 'Okame'	Okame Flowering Cherry	A great Cherry selection with striking pinkish-red flowers in early spring, followed by reddish-green leaves that turn orange-red in fall.	Average Growth	Moist	15' x 15'
Prunus x 'Superior'	Superior Semi-dwarf Plum	Large golden fruit turns pink then a deep red blush. Flesh is clingstone. Good fresh and for preserves.			
Prunus X yedoensis	Yoshino Flowering Cherry	Known for its beauty and fragrant flowrs.	Average Growth	Moist	15' x 15'
Prunus X yedoensis 'Afterglow'	Afterglow Flowering Cherry	Abundant pink flowers bloom in spring are followed by glossy, bright green leaves that turn yellow in fall. A vigorous, rounded, medium-sized tree.	Average Growth	Moist	25' x 25'
Prunus X yedoensis 'Akebono'	Akebono Yoshino Cherry	Soft pink flowers bloom in mid-spring, followed by glossy, bright green leaves that turn yellow in fall. A vigorous, rounded, medium-sized tree.	Average Growth	Moist	25' x 25'
Pseudotsuga menziesii	Douglas Fir	A noble forest tree that is an excellent planted as a specimen, in groupings or in mass.	Average Growth	Semi-Moist	40' x 25'
Punica granatum 'Nana'	Dwarf Pomegranate	Vivid orange-red flowers bloom in summer followed by red fruit on this small, densely-branched shrub. Great as a specimen or for mass plantings.	Average Growth		5' x 5'
Pyracantha angustifolia 'Gnome'	Gnome Dwarf Pyrocantha	A medium sized shrub of compact, densely branched growth.	Fast Growth	Moist	6' x 10'

Pyracantha coccinea "Kasan'	Kasan Pyracantha	Produces orange-red fruits. Originated in Russia.			
Pyracantha coccinea 'Mohave'	Mohave Pyrocantha	Abundant white flowers in spring become masses of bright orange-red berries in fall among glossy dark green leaves on this upright, evergreen shrub.	Fast Growth	Moist	6' x 10'
Pyrus calleryana 'Aristocrat'	Aristocrat Flowering Pear	A great medium-sized tree for all seasons with white flowers in early spring, followed by shimmering dark gree leaves that turn shades of red in fall.	Fast Growth	Moist	25' x 25'
Pyrus calleryana 'Autumn Blaze'	Autumn Blaze Flowering Pear	A rounded, medium-sized tree with white flowers in early spring, followed by glossy dark green leaves that turn blazing reddish-purple in fall.	Fast Growth	Moist	35' x 20'
Pyrus calleryana 'Bradford'	Bradford Flowering Pear	A dense, upright tree with white flowers in early spring, followed by glossy, dark green leaves that turn vibrant red in fall.	Fast Growth	Moist	40' x 35'
Pyrus calleryana 'Capital'	Capital Flowering Pear	A columnar tree with white flowers in early spring, followed by lustrous, dark green leaves that turn coppery-red in fall.	Fast Growth	Moist	35' x 12'
Pyrus calleryana 'Chanticleer'	Chanticleer Flowering Pear	A fine upright-pyramidal tree with abundant white flowers in early spring, followed by glossy, dark green leaves that turn reddish-purple in fall.	Fast Growth	Moist	40' x 15'
Pyrus calleryana 'Cleveland Select'	Cleveland Flowering Pear	Much narrower than Bradford. Flowers heavily with a good reddish purple fall color.	Average Growth	Moist	
Pyrus calleryana 'Redspire'	Redspire Flowering Pear	A medium-sized pyramidal tree with white flowers in early spring, followed by shiny, dark green leaves that turn bright yellow in fall.	Fast Growth	Moist	35' x 25'

Pyrus calleryana 'Stonehill'	Stonehill Flowering Pear	Narrow growth habit that flowers heavily. Good fall color.	Fast Growth	Moist	40' x 20'
Pyrus calleryana 'Trinity'	Trinity Flowering Pear	Tightly rounded head with profuse single white flowers. Leaves turn orange-red in fall.			
Pyrus calleryana 'Whitehouse'	Whitehouse Flowering Pear	An upright-columnar, medium-sized tree with white flowers in early spring, followed by glossy dark green leaves that turn reddish-purple in fall.	Fast Growth	Moist	40' x 10'
Pyrus communis 'Ayers'	Ayers Semi-dwarf Pear	Small to medium sized high quality yellow fruit. Excellent fresh eating quality. Needs a pollinator.			
Pyrus communis 'Bartlett'	Bartlett Pear	White flowers bloom in mid-spring and are followed by large, flavorful, golden yellow pears. Needs another pear for best fruit production.	Fast Growth	Moist	35' x 35'
Pyrus communis 'Bosc'	Bosc Pear	White flowers bloom in mid-spring and are followed by large, yellowish-tan fruit. Needs another pear for best fruit production.	Fast Growth	Moist	35' x 35'
Pyrus communis 'Clapp's Favorite'	Clapp's Favorite Flowering Pear	Lemon-yellow fruit whose white flesh is high quality, very sweet and highlyi flavored. Good for desserts. Needs a pollinator.	Average Growth	Semi-Moist	30' x 20'
Pyrus communis 'Kieffer'	Kieffer Semi-dwarf Pear	Golden yellow fruit, white flesh that crisp and juicy. Exccellent for canning and baking. Self pollinating.	Average Growth	Semi-Moist	30' x 20'
Pyrus communis 'Max Red Bartlett'	Red Bartlett Semi-dwarf Pear	Large juicy pears with red skin. High yielding tree that is good fresh and canned.	Average Growth	Semi-Moist	30' x 20'
Pyrus communis 'Moonglow'	Moonglow Semi-dwarf Pear	Yellow fruit with mildly juicy, soft, white flesh. Mild flavor. Good for canning. Requires 700 chilling hours. Needs a pollinator.	Average Growth	Semi-Moist	30' x 20'

Pyrus communis 'Seckel'	Seckel Semi-dwarf Pear	Small yellowish brown fruit. Flursh is extremely sweet and very juicy. Required 500-800 hours of chilling. Self fertile.	Average Growth	Semi-Moist	25' x 15'
Pyrus pyrifolia '20th Century'	20th Century Asian Pear	Best known Asian pear. Extremely juicy, crisp flesh. Required 450-500 hours of chilling. Needs a pollinator.	Fast Growth	Semi-Moist	35' x 20'
Pyrus pyrifolia 'Orient'	Orient Asian Pear	Yellow fruit with juicy, creamy white flesh. Used mainly for canning. Requires 350 hours of chilling. Needs a pollinator.	Fast Growth	Semi-Moist	30' x 20'
Quercus acutissima	Sawtooth Oak	A dense, rounded tree that grows moderately fast and has clean, shiny green leaves with saw-like teeth along the edge that turn yellow in fall.	Average Growth	Moist	40' x 50'
Quercus alba	White Oak	A large spreading, stately tree with dark green leaves. A beautiful tree for parks, golf courses and other large areas.	Slow Growth	Moist	50' x 50'
Quercus bicolor	Swamp White Oak	A large, spreading tree with dark green leaves that turn yellow in fall. Prefers moist, well-drained soils.	Average Growth		50' x 50'
Quercus coccinea	Scarlet Oak	A large, rounded tree with glossy dark green leaves that turn scarlet red in late fall.	Average Growth	Moist	50' x 40'
Quercus falcata	Southern Red Oak	A large, rounded tree found throughout the South that tolerates dry, poor soils.	Average Growth	Moist	75' x 50'
Quercus gambelii	Gambel Oak	An important tree for the mountain regions of the western USA. Can be pruned into a small tree form.	Average Growth	Moist	30' x 40'
Quercus macrocarpa	Burr Oak	A broad-spreading tree with large glossy green leaves and large acorns. Makes an excellent tree for parks or other large areas.	Average Growth	Moist	50' x 50'

Quercus muehlenbergii	Chinkapin Oak	Quite an attractive tree especially in old age.	Fast Growth	Moist	50' x 50'
Quercus myrsinifolia	Chinese Evergreen Oak	Superb small street or lawn tree. New folilage is purple-bronze.	Slow Growth	Moist	20' x 25'
Quercus nigra	Water Oak	A moderately-fast growing, rounded tree with deep green leaves that is popular as a shade tree in the South.	Fast Growth	Moist	50' x 50'
Quercus nuttallii	Nuttall Oak	A pyramidal tree with dark green leaves that turn reddish-purple in fall, tolerates a wide range of soil conditions, making this a great shade tree.	Average Growth	Moist	25' x 35'
Quercus palustris	Pin Oak	A moderately-fast growing, pyramidal tree with shiny, dark green leaves that turn red to red-orange in fall. A great shade tree.	Average Growth	Moist	50' x 35'
Quercus phellos	Willow Oak	An upright-pyramidal tree that grows moderately fast with strap-like, glossy dark green leaves that turn golden yellow in fall. A great shade tree.	Fast Growth	Moist	50' x 35'
Quercus prinus	Chestnut Oak	A dense, pyramidal tree with lustrous dark green leaves in summer changing to orange-yellow in fall. Wildlife love the acorns of this tree.	Average Growth	Moist	50' x 50'
Quercus robur	English Oak	A large, rounded tree with long, dark blue-green leaves. Best suited for parks and other large areas.	Average Growth		50' x 50'
Quercus rubra	Northern Red Oak	A moderately-fast growing, rounded tree with shiny dark green leaves in summer, turning bright red in fall. A great shade tree for the North.	Fast Growth	Moist	50' x 50'
Quercus shumardii	Shumard Red Oak	A moderately-fast growing, pyramidal tree with lustrous, dark green leaves that turn red-orange in fall. Makes a great shade tree.	Fast Growth	Moist	75' x 35'

Quercus virginiana	Live Oak	The magnificent evergreen shade and mansion tree of the South. A large tree with dark green leaves that tolerates poor soil conditions.	Average Growth	Moist	50' x 50'
Rhamnus frangula 'Asplenifolia'	Fernleaf Buckthorn	A unique upright shrub with long, narrow light green leaves gives the plant a fern-like look. Black berries and yellow leaves in fall.	Average Growth	Semi-Moist	10' x 10'
Rheum rhubarum	Garden Rhubarb	Victoria Rhubarb has broad crimson stalks. A heavy commercial producer.	Average Growth	Semi-Moist	
Rhododendron catawbiense	Catawba Rhododrendron	Buds are lilac opening to pure white flowers with greenish yellow spotting.	Slow Growth	Moist	6' x 6'
Rhododendron catawbiense 'Album'	White Catawba Rhododendron	Large, pure white flowers bloom in mid-spring above the large, leathery, dark green leaves of this upright, evergreen Rhododendron.	Slow Growth	Moist	6' x 6'
Rhododendron catawbiense 'America'	America Rhododendron	Brillinat deep red.	Slow Growth	Moist	6' x 6'
Rhododendron catawbiense 'Blue Ensign'	Blue Ensign Rhododendron	Lavender-blue with a prominent dark blotch.	Slow Growth	Moist	6' x 6'
Rhododendron catawbiense 'Dora Amateis'	Dora Amateis Rhododendron	White flowers lightly spotted with green. Slightly aromatic.	Slow Growth	Moist	24" x 4'
Rhododendron catawbiense 'Lee's Dark Purple'	Lee's Dark Purple Rhododendron	Buds very dark purple opening to dark purple.	Slow Growth	Moist	4' x 6'
Rhododendron catawbiense 'Roseum Elegans'	Roseum Elegans Rhododendron	Rosy-lilac flowers bloom in mid-spring above the large, leathery, dark green leaves of this cold-hardy, evergreen Rhododendron.	Average Growth	Moist	6' x 6'

Rhododendron catawbiense 'Vulcan's Flame'	Vulcan's Flame Rhododendron	Bright red flowers bloom in mid-spring above the large leathery, dark green leaves of this upright, evergreen Rhododendron.	Slow Growth	Moist	6' x 6'
Rhododendron eriocarpum 'Pink Gumpo'	Pink Gumpo Azalea	Large ruffled pink flowers cover a compact evergreen shrub.	Slow Growth	Moist	3' x 6'
Rhododendron eriocarpum 'White Gumpo'	White Gumpo Azalea	Large ruffled white flowers cover a compact evergreen shrub.	Slow Growth	Moist	3' x 6'
Rhododendron indica 'George l. Taber'	George Taber Azalea	A profusion of very large, light orchid-pink flowers blanket the dark green leaves of this sun-tolerant, evergreen Azalea in early spring.	Slow Growth	Moist	6' x 5'
Rhododendron 'Nova Zembla'	Nova Zembla Red Rhododendron	Red fowers and quite cold tolerant and heat resistant.	Slow Growth	Moist	
Rhododendron X 'A Bedford'	Bedford Rhododendron	Lavender-blue with dark blotch.	Slow Growth	Moist	6' x 6'
Rhododendron x Assorted Azalea	Assorted Hardy Azaleas	Compact evergreen shrubs that are covered with showy red, white, pink or lavender flowers in the spring.	Slow Growth	Moist	3' x 6'
Rhododendron X 'Ben Morrison'	Ben Morrison Rhododendron	Masses of large, salmon-pink flowers with white edges cover this evergreen Azalea in mid-spring. Great for mass plantings.	Slow Growth	Moist	24" x 36"
Rhododendron x Glenn Dale 'Fashion'	Fashion Azalea	Masses of large orange-red flowers grace this medium-sized, evergreen Azalea in mid-spring. Great for mass plantings.	Slow Growth	Moist	24" x 36"
Rhododendron x Kurume 'Hino-Crimson'	Hino-Crimson Azalea	Masses of brilliant, crimson-red flowers bloom in early spring above the glossy, dark green leaves of this compact, evergreen Azalea.	Slow Growth	Moist	24" x 36"

Rhododendron X 'PJM'	PJM Rhododendron	Vivid, bright lavender flowers bloom in mid-spring above the small, dark green leaves of this compact, cold-hardy, evergreen Rhododendron.	Slow Growth	Moist	36" x 4'
Rhododendron X 'Scintillation'	Scintillation Rhododedron	Fragrant, luminous, light pink flowers with an amber throat bloom in mid-spring above the large, lustrous green leaves of this evergreen Rhododendron.	Slow Growth	Moist	4' x 6'
Rhus aromatica 'Gro-low'	Gro-Low Fragrant Sumac	Fragrant yellow flowers bloom in spring, followed by red, hairy fruit on this low,spreading shrub that looks great in mass plantings.	Slow Growth		24" x 4'
Rhus copallina	Flameleaf Sumac	A thick-branched, upright shrub with lustrous dark green leaves that turn flaming scarlet in fall. A great naturalizing plant.	Average Growth	Semi-Moist	15' x 15'
Rhus typhina	Staghorn Sumac	An irregular, upright shrub with glossy, dark green leaves that turn bright orange-red in fall and flattened branches that resemble deer antlers.	Average Growth	Semi-Moist	15' x 15'
Ribes uva-crispa 'Pixwell'	Pixwell Gooseberry	Light green berries turn pink when ripe. Soft, juicy and high in sugar. Excellent for pies and desserts.	Average Growth	Moist	36" x 4'
Robinia pseudoacacia	Black Locust	Can survive under the toughest conditions. Flowers are fragrant from which bees produce a delicious honey.	Fast Growth	Semi-Moist	50' x 30'
Robinia pseudoacacia 'Purple Robe'	Purple Robe Black Locust	Fragrant dark rose-pink flowers bloom in late spring among the bronze-red new leaves of this medium-sized tree that survives tough growing conditions.	Fast Growth	Semi-Moist	25' x 25'
Robinia X ambigua 'Idahoensis'	Pink Idaho Locust	Flowers are rose-pink and fragrant.	Average Growth	Semi-Moist	25' x 12'

Rosa banksiae	Yellow Lady Banks Rose	A vigorous climbing rose with slightly fragrant, small yellow flowers in spring. Looks great on fences, trellises and trained against walls.	Fast Growth	Moist	20'
Rosa foetida 'Bicolor'	Austrian Copper Rose	Has copper-red petals, yellow on the back side.	Average Growth	Semi-Moist	36" x 36"
Rosa rugosa	Rugosa Rose	A very durable, disease-free shrub rose that produces fragrant white or pink flowers in late spring, and tolerates salty conditions.	Average Growth		4' x 4'
Rosa rugosa 'Alba'	White Rugosa Rose	Large, fragrant pure white flowers begin to bloom in spring above the glossy, dark green, disease-free leaves of this tough shrub rose.	Average Growth		4' x 4'
Rosa rugosa 'F. J. Grootendorst'	F. J. Grootendorst Rugosa Rose	Fright red, slightly fragrant flowers begin blooming in spring above the leathery, dark green, disease-free leaves of this tough shrub rose.	Average Growth		4' x 4'
Rosa rugosa 'rubra'	Red Rugosa Shrub Rose	Valuable for difficult sites such as banks, cuts, fills, or sandy soils. Very beautiful in foliage.	Average Growth	Semi-Moist	
Rosa X 'Angel Face'	Angle Face Floribunda Rose	Clusters of lavender flowers edged with purple bloom from late spring through summer on this low, floribunda rose. All American Rose Selection winner.	Average Growth	Moist	36" x 24"
Rosa X 'Augusta'	Augusta Rose	Clusters of deep peach flowers bloom from late spring through summer above the glossy, dark green leaves of this compact shrub rose.	Average Growth	Moist	36" x 24"
Rosa X 'Betty Prior'	Betty Prior Rose	Clusters of large single pink flowers bloom from late spring through summer on this vigorous-growing, upright, floribunda rose.	Average Growth	Moist	36" x 36"

Rosa X 'Charisma'	Charisma Floribunda Rose	Clusters of fragrant orange flowers bloom from late spring through summer above the dark green leaves of this shrub rose.	Average Growth	Moist	36" x 24"
Rosa X 'Circus'	Circus Floribunda Rose	This floribunda rose provides a riot of color with its fragrant, rich red and yellow to orange and pink flowers that bloom all summer.	Average Growth	Moist	36" x 24"
Rubus 'Cumberland'	Cumberland Black Raspberry	Large black berries that are firm and juicy. Good for freezing, canning, jellies, sauce and pies.			
Salix caprea	French Pink Pussywillow	The showy flowers that appear in early spring of this small, upright tree are the reason for its name "Pussy Willow". An excellent accent plant.	Average Growth		20' x 15'
Salix caprea 'Pendula'	Weeping Pussywillow	A graceful weeping tree with the characteristic flowers appearing in early spring leading to its name "Pussy Willow". A great accent plant.	Average Growth		20' x 15'
Salix integra 'Hakuro Nishiki'	Variegated Hakuro Nishiki Willow	An upright shrub that has purplish branches with white-mottles leaves, pink at the branch tips.	Average Growth		20' x 15'
Salix matsudana 'Tortuosa'	Corkscrew Willow	An unusual small tree with twisted, spiral branches and bright green leaves. Grows very rapidly. A great accent plant.	Fast Growth	Moist	25' x 10'
Salix melanostachys	Black Pussy Willow	Stems assume a rich purple-black color in winter.	Average Growth	Semi-Moist	10' x 6'
Salix purpurea 'Nana'	Dwarf Artic Willow	Has blue-green leaves. Good for wet areas such as banks and ponds.	Average Growth	Moist	4' x 4'
Salix x blanda	Wisconsin Weeping Willow	Branches are pendulous and has a weeping look.	Fast Growth	Moist	40' x 35'

Salvia nemerosa 'May Night'	May Night Blue Salvia	Indigo blue flower spikes bloom in late spring to summer above the ruffled dark green leaves of this perennial.	Average Growth	Semi-Moist	
Salvia superba 'Blue Queen'	Blue Salvia	A clump-forming erect perennial that bears tender flowers from mid-summer to early fall.	Average Growth	Semi-Moist	3'
Salvia superba 'East Friesland'	Blue Salvia	A clump-forming erect perennial that bears tender flowers from mid-summer to early fall.	Average Growth	Semi-Moist	3'
Sambucus canadensis	American Elderberry	Fruit is good for jellies, pies, juice and wine. Attracts birds.	Average Growth	Moist	10' x 6'
Sambucus canadensis 'Aurea'	Golden American Elder	Produces cherry red fruit and golden yellow leaves.	Average Growth	Moist	10' x 6'
Scabiosa caucasica 'Fama'	Fama Blue Pincushion Flower	Produces flat, many-petalled flowerheads with pincushion-like centers. Popular in borders or as cut flowers.			24'
Scabiosa caucasica 'Pink Mist'	Pink Mist Pincushion Flower	Produces flat, many-petalled flowerheads with pincushion-like centers. Popular in borders or as cut flowers.	Average Growth	Moist	24'
Schizachyrium scoparium	Little Bluestem Grass	Forms a perennial tuft with upright stems and green to blue-green leaves often shaded bronze or purplish.			3'
Sedum kamtschaticum 'Variegatum'	Variegated Gold Sedum	Yellow flowers bloom in late summer above the green leaves with white margins of this low-growing groundcover. Looks great in rock gardens.	Average Growth	Semi-Moist	
Sedum spurium 'Dragon's Blood'	Dragon's Blood Sedum	Blood red flowers bloom in mid-summer above the bronze red leaves of this low-growing groundcover. Looks great in rock gardens.	Average Growth	Semi-Moist	

Sedum ternatum 'Vera Jameson'	Vera Jameson Sedum	Pink flowers bloom in fall, crowning the purple - green leaves of this upright, co mpact perennial. A great accent plant for fall color.	Slow Growth	Semi-Moist	3" x 24"
Sophora japonica	Japanese Pagodatree	Good tree for city conditions, parks and golf courses.	Fast Growth	Moist	50' x 40'
Sorbus alnifolia	Korean Mountain Ash	Good specimen tree for lawns.	Fast Growth	Moist	25' x 15'
Sorbus aucuparia	European Mountain Ash	Excessive summer heat may cause problems.	Average Growth	Moist	25' x 15'
Sorbus aucuparia 'Cardinal Royal'	Cardinal Royal European Mountain Ash	A vigorous grower. Leaves are dark green above and silvery beneath.	Average Growth	Moist	25' x 15'
Spiraea cantoniensis	Reeves Spirea	Mounded with arching branches. White 1/2" diameter flowers appear in early April.	Average Growth	Moist	4' x 36"
Spiraea japonica 'Alpina'	Alpine Japanese Spirea	Normal flower color is rosey pink to carmine.	Average Growth	Moist	12" x 24"
Spiraea japonica 'Anthony Waterer'	A W Spirea	Carmine-red flowers cover the blue -green leaves of this dense, compact shrub all summer. Leaves turn wine-red in fall.			36" x 4'
Spiraea japonica 'Crispa'	Crispa Spirea	Pink flowers bloom in summer above the crinkled and twisted leaves, giving the plant a fine -textured appearance. Leaves turn red in fall.	Average Growth	Moist	24" x 24"
Spiraea japonica 'Dart's Red'	Dart's Red Spirea	Deep carmine-red flowers bloom in late spring through summer above the blue -green leaves of this compact shrub. Leaves turn red in fall.	Average Growth	Moist	36" x 36"
Spiraea japonica 'Frobelii'	Frobel Spirea	Clusters of bright pink flowers bloom in late spring above the blue-green leaves of this rounded shrub. Leaves turn red in fall.	Average Growth	Moist	36" x 36"

Spiraea japonica 'Goldflame'	Goldflame Spirea	In spring, the leaves of this mounding shrub are a blazing bronze-red turning yellow and topped with clusters of crimson-red flowers in summer.	Average Growth	Moist	24" x 36"
Spiraea japonica 'Goldmound'	Goldmound Spirea	A compact mound of bright yellow leaves topped with clusters of light pink flowers in summer. An excellent accent plant.	Average Growth	Moist	36" x 36"
Spiraea japonica 'Limemound'	Limemound Spirea	In spring, the leaves of this compact, dense shrub emerge lemon-yellow but turn lime green in summer when crowned with clusters of pink flowers.			24" x 36"
Spiraea japonica 'Little Princess'	Little Princess Spirea	A low, spreading shrub with covered with clusters of dark pink flowers in summer. Excellent as a groundcover or as a dwarf specimen plant.	Average Growth	Moist	36" x 36"
Spiraea japonica 'Neon Flash'	Neaon Flash Spirea	Leaves emerge bronze-red in spring, then fade to dark green when clusters of rich red flowers bloom in summer. A wonderful, compact shrub.	Average Growth	Moist	36" x 36"
Spiraea japonica 'Norman'	Little Prince Spirea	A small, compact shrub with rosy pink flower clusters in summer and long-lasting raspberry-purple to red leaf color in fall.	Average Growth	Moist	36" x 36"
Spiraea japonica 'Shirobana'	Shirobana Spirea	Clusters of white, pink and red flowers bloom in summer above glossy, dark green leaves of this dense, mounding shrub.	Average Growth	Moist	36" x 36"
Spiraea nipponica 'Snowmound'	Snowmound Spirea	Has small dark blue-green lieaves and white flowers that appear in late May.	Average Growth	Moist	36" x 4'
Spiraea X vanhouttei	Briadal Wreath Spirea	Clusters of white flowers burst into bloom in early spring on the upright, arching branches of this fountain-like shrub.	Average Growth	Moist	6' x 5'

Stachys byzantina	Lamb's Ears	Leaves are lance-shaped and have the same white downy feel of a lamb.	Fast Growth	Semi-Moist	18" x 24"
Stokesia laevis	Stoke's Aster	An attractive late flowering perennial that flowers from late summer to fall.	Average Growth	Semi-Moist	
Syringa josikaea 'Royalty'	Royalty Lilac	Violet-purple flower clusters bloom in late spring among the dark green leaves on this upright, arching, multi-stemmed shrub.	Average Growth	Moist	8' x 8'
Syringa meyeri 'Palibin'	Dwarf Korean Lilac	A slow-growing dwarf cultivar with violet to rose-pink flowers in small dense clusters.	Average Growth	Moist	5' x 8'
Syringa patula 'Miss Kim'	Miss Kim Lilac	Dark purple buds open to fragrant light purple flower clusters in spring among dark green leaves that turn burgundy-red in fall on this compact shrub.	Average Growth	Moist	36" x 36"
Syringa pekinensis	Pekin Tree Lilac	Large clusters of creamy white flowers bloom in spring on this large, multi-stemmed shrub with an upright, arching shape.	Average Growth	Moist	25' x 20'
Syringa reticulata 'Ivory Silk'	Ivory Silk Japanese Tree Lilac	Fragrant creamy white flower clusters bloom in late spring on this handsome, care-free, small tree. Makes an excellent specimen or street tree.	Average Growth	Moist	14' x 10'
Syringa villosa	Late Lilac	A nice lilac for the shrub border with rosy lilac to white flowers.	Average Growth	Moist	6' x 6'

Syringa vulgaris 'Charles Joly'	Charles Joly French Lilac	Clusters of fragrant, magenta flowers bloom in spring on this upright, arching, multi -stemmed shrub.			8' x 8'
Syringa vulgaris 'Ellen Willmott'	Ellen Willmott French Lilac	Hybrid lilac with a double white bloom.			8' x 8'
Syringa vulgaris 'Mme. Lemoine'	Madam Lemoine French Lilac	Clusters of pure white flowers bloom in spring on this vigorous, upright, arching, multi -stemmed shrub.			5' x 4'
Syringa vulgaris 'Pres. Grevy'	President Grevey French Lilac	Clusters of lavender-blue flowers bloom in spring on this upright, arching, multi -stemmed shrub.			8' x 8'
Syringa X hyacinthiflora 'Pocahontas'	Pocahontas Hybrid Lilac	Clusters of deep maroon buds open to fragrant, deep violet flowers in early spring before the dark green leaves emerge on this u pright shrub.	Average Growth	Moist	8' x 8'
Syringa X persica	Persian Lilac	A graceful shrub with upright branches. Flowers are pale lilac.	Average Growth	Moist	5' x 5'
Syringa X prestoniae 'James Macfarlane'	James MacFarlane Preston Lilac	Clusters of bright pink, fragrant flowers bloom in late spring among the dark green leaves of this upright, multi-stemmed shrub.	Fast Growth	Moist	8' x 8'
Tamarix ramosissima 'Summer Glow'	Summer Glow Tamarix	Has dense, feathery, blue-tinged foliage with bright pink flowers.	Fast Growth	Moist	10' x 10'
Tanacetum coccineum	Painted Daisy	A frost-hardy perennial with dark green scented leaves. Flowers may be pink, red, purple or white.	Average Growth	Semi-Moist	
Taxodium ascendens	Pond Cypress	A large, pyramidal tree wi th bright green leaves that turn orangish-brown in fall. Makes a fabulous accent plant near water features.	Fast Growth	Moist	50' x 10'

Taxodium distichum	Bald Cypress	A large, pyramidal tree with bright green leaves that turn russet-brown in fall. Makes a fabulous accent tree around water features.	Fast Growth	Moist	50' x 25'
Taxus baccata 'Repandens'	Repandens English Yew	A dwarf, wide-spreading yew with sickle-shaped needles.	Slow Growth	Moist	36" x 5'
Taxus cuspidata	Japanese Yew	An upright spreading evergreen small tree or large shrub with dark green needles. Makes a great hedge or screen.	Slow Growth	Moist	15' x 15'
Taxus cuspidata 'Capitata'	Upright Japanese Yew	An upright, pyramidal evergreen shrub with dark green needles that responds well to pruning. Makes a great accent plant in today's landscapes.	Slow Growth	Moist	25' x 15'
Taxus cuspidata 'Intermedia'	Intermedia Japanese Yew	A dwarf, rounded evergreen shrub with shiny, dark green leaves. Makes an excellent foundation or border plant.	Slow Growth	Moist	6' x 8'
Taxus X media 'Brownii'	Browni Globe Yew	A dense, rounded evergreen shrub with dark green needles that responds well to pruning. A great foundation or border shrub.	Slow Growth	Moist	10' x 8'
Taxus X media 'Densiformis'	Densiformis Yew	The standard Yew, forms a dense, rounded evergreen shrub with dark green needles and responds well to pruning. A great foundation or border plant.	Slow Growth	Moist	36" x 4'
Taxus X media 'Hicksii'	Hick's Yew	A columnar evergreen shrub with dark green needles that becomes pyramidal with age. A great accent plant.	Slow Growth	Moist	10' x 8'
Taxus X media 'Wardii'	Wardi Globe Yew	A low, spreading evergreen shrub with dense, dark green needles. Responds well to pruning, making it ideal for foundation and border plantings.	Slow Growth	Moist	6' x 15'

Thuja occidentalis	American Arborvitae	A dense, upright evergreen shrub with bright green needles in summer, turning yellow -green in winter. Great for hedges and as a specimen plant.	Slow Growth	Moist	25' x 10'
Thuja occidentalis 'Globosa'	Globe Arborvitae	A dwart, globe-shaped evergreen shrub that is great in mass plantings and as an accent plant.	Slow Growth	Moist	20' x 5'
Thuja occidentalis 'Golden Globe'	Glolden Globe Arborvitae	Has golden yellow foliage an d a broad globe-shaped growth habit.	Slow Growth	Moist	25' x 10'
Thuja occidentalis 'Hetz's Midget'	Hetz Midget Globe Arborvitae	A small, globe-shaped evergreen with dense, dark green needles. Excellent plant for formal gardens or patio containers.	Slow Growth	Moist	24" x 4'
Thuja occidentalis 'Little Giant'	Little Giant Globe Arborvitae	A slow-growing, globe-shaped shrub with rich green foliage.	Slow Growth	Moist	4' x 4'
Thuja occidentalis 'nigra'	Dark Green Arborvitae	A dense, upright evergreen that th rives in all locations. Very dark green needles, even in winter. Ideal for screens or as a specimen plant.	Slow Growth	Moist	20' x 10'
Thuja occidentalis 'Pyramidalis'	Pyramidal Arborvitae	A narrow, pyramidal evergreen with bright green, soft textured nee dles. Makes a great screen or accent plant.	Slow Growth	Moist	20' x 5'
Thuja occidentalis 'Rheingold'	Rheingold Arborvitae	A slow-growing, oval evergreen shrub with dense, rich gold needles. Makes a wonderful small hedge or accent plant.	Slow Growth	Moist	5' x 4'
Thuja occidentalis 'Smaragd'	Emerald Arborvitae	One of the finest pyramidal evergreens. Rich, bright green, fine-textured needles that hold their color in winter. A great hedge or accent plant	Slow Growth	Moist	12' x 36"
Thuja occidentalis 'Techny'	Techny Arborvitae	A broad, pyramidal evergreen with dark green needles that hold their color all year. Makes a great hedge plant.	Slow Growth	Moist	15' x 5'

Thuja occidentalis 'Tiny Tim'	Tiny Tim Globe Arborvitae	A dense, rounded globe -shaped evergeen with dark green needles that turn bronze in winter. Makes a great low hedge plant.	Slow Growth	Moist	36" x 5'
Thuja occidentalis 'Woodwardii'	Woodward Globe Arborvitae	A dwarf, globe-shaped evergreen with very dark green needles that turn bronze in winte r. Makes a great low hedge or accent plant.	Slow Growth	Moist	36" x 36"
Thuja occidentalis 'Yellow Ribbon'	Yellow Ribbon Arborvitae	Yellow tipped foliage on a dense upright evergreen shrub that makes a great accent or screen.	Average Growth	Semi-Moist	15' x 6'
Thuja orientalis 'Aurea Nana'	Berkman's Golden Arborvitae	A naturally dense cone of golden tipped evergreen foliage.Holds up well in the sun and to dry air.	Average Growth	Semi-Moist	8' x 4'
Tilia americana	American Linden	A tall statley tree with leaves that are dark green above and a paler green beneath.	Average Growth	Moist	75' x 35'
Tilia americana 'Redmond'	Redmond Linden	A dens pyramidal tree with large leaves.	Average Growth		50' x 35'
Tilia cordata	Little Leaf Linden	An excellent shade tr ee for lawns, large areas, streets, planters, etc.	Average Growth	Moist	50' x 35'
Tilia cordata 'Glenleven'	Glenleven Linden	Fast growing with a straight trunk.	Fast Growth	Moist	50' x 35'
Tilia cordata 'Greenspire'	Greenspire Little Leaf Linden	Widely used as a street tree and does well under difficult conditions.	Average Growth	Moist	40' x 30'
Trachelospermum asiaticum	Japanese Star Jasmine	A vigorous growing, evergreen vine with fragrant white flowers in spring among the small dark green leaves. Makes a great groundcover.	Fast Growth	Moist	12" x 6'
Trachelospermum asiaticum 'Variegatum'	Variegated Star Jamine	A vigorous-growing evergreen vine with fragrant, white flowers in spring among the lustrous dark green leaves with white edges. A great groundcover	Fast Growth	Moist	6" x 10'

Trachelospermum jasminoides	Confederate Jasmine	A strong-growing, evergreen vine with masses of fragrant, star-shaped, white flowers in spring among the shiny dark green leaves.	Fast Growth	Moist	12" x 10'
Trachelospermum jasminoides 'Variegatum'	Variegated Conferate Jasmine	A twining evergreen vine with leaves bordered with white.	Fast Growth	Moist	12" x 10'
Tsuga canadensis	Canadian Hemlock	Makes an excellent graceful evergreen hedge	Slow Growth	Moist	50' x 25'
Tsuga canadensis 'Sargentii'	Sargent Weeping Hemlock	A broadly spreading weeping form.	Slow Growth	Moist	10' x 15'
Ulmus parvifolia	Lace Bark Elm	A wonderful, disease-resistant shade tree with glossy dark green leaves and beautiful peeling bark mottled with gray, green, orange and brown.	Average Growth	Moist	40' x 35'
Ulmus parvifolia 'Drake'	Drake Lace Bark Elm	Has rich dark green foliage on upright-growing branches.	Average Growth	Moist	25' x 40'
Ulmus parvifolia 'Dynasty'	Dynasty Lace Bark Elm	A vigorous, rounded shade tree with glossy dark green leaves that turn red in fall. A U. S. National Arboretum selection.	Average Growth	Moist	40' x 40'
Vaccinium corymbosum	Highbush Blueberry	Fruit is bluish black. Ripens from June to August.	Average Growth		6' x 6'
Verbena canadensis "Homestead Purple'	Homestead Purple Verbena	A trailing or sprawling perennial bearing flowers from summer through fall.			
Veronica longifolia 'Sunny Border Blue'	Sunny Border Blue Veronica	Dark blue flowers bloom in late spring to summ er on this compact, hardy perennial. The 1993 Perennial Plant of the Year.	Average Growth	Moist	

Viburnum carlesii	Koreanspice Viburnum	Clusters of fragrant white flowers open from pink buds in spring. Red fruit turns black in fall on this dense, rounded shrub.	Average Growth		4' x 5'
Viburnum dentatum	Arrowwood Viburnum	A large, spreading shrub with white flowers that bloom in spring over the glossy, dark green leaves. Bluish-black berries are a favorite of birds.	Average Growth		5' x 5'
Viburnum dentatum 'Synnestvedt'	Chicago Lustre Arrowood Viburnum	Glossy dark green foliage turns reddish purple in fall.	Average Growth		5' x 5'
Viburnum dilatatum 'Erie'	Erie Linden Viburnum	A dense, rounded shrub that's covered with clusters of white flowers in spring , followed by coral-red berries in fall.	Average Growth		6' x 6'
Viburnum lantana	Wayfaring Viburnum	Used for hedges, screens, masings and a shrub border.	Average Growth		10' x 10'
Viburnum lantana 'Mohican'	Mohican Wayfaring Viburnum	A great compact, rounded shrub with clusters of white flowers in spring, followed by orange -red berries. A U. S. National Arboretum Selection.	Average Growth	Moist	10' x 8'
Viburnum opulus	European Highbush Cranberry	An upright spreading shrub producing white flowers. Excellent for its showy flowers and fruit display.	Average Growth	Moist	8' x 10'
Viburnum opulus 'Compactum'	Compact European Highbush Cranberry	A very dense, rounded shrub with masses of white flowers in spring followed by bright red berries in fall. Makes a great mass planting.	Average Growth	Moist	6' x 6'
Viburnum opulus 'Nanum'	Dwarf European Highbush Cranberry	A very dense, low, globe-shaped shrub that requires little maintenance to maintain its shape. Very cold hardy and makes a great low hedge.	Average Growth	Moist	36" x 36"

Viburnum opulus 'Roseum'	Snowball Viburnum	A showy, cold-hardy, upright shrub that's blanketed with clusters of white flowers in spring that fade to pink.	Average Growth	Moist	8' x 10'
Viburnum plicatum	Japanese Snowball Viburnum	A fabulous rounded shrub with horizontal branches that are accentuated by clusters of white flowers in spring. Birds love the reddish -black berries.	Average Growth	Moist	10' x 15'
Viburnum plicatum 'Newzam'	Newport Doublefile Viburnum	Extremely dense with small dark green leaves that turn burgundy in fall.	Average Growth	Moist	10' x 15'
Viburnum plicatum tomentosum (var.)	Doublefile Viburnum	Produces snow white flowers. Excellent use as screens, borders, and in mass.	Average Growth	Moist	10' x 15'
Viburnum plicatum tomentosum (var.) 'Mariesii'	Mariesii Doublefile Viburnum	A horizontal growth habit and a magnificant sight when in full flower.	Average Growth	Moist	8' x 10'
Viburnum plicatum tomentosum (var.) 'Shasta'	Shasta Doublefile Viburnum	The broad, horizontal branches of this shrub are smothered with clusters of white flowers in spring.	Average Growth		5' x 10'
Viburnum rhytidophyllum	Leatherleaf Viburnum	Excellent foliage and can be used for screens, as a foundation plant or blended with other broadleaf evergreens.	Average Growth	Moist	8' x 6'
Viburnum setigerum	Tea Viburnum	Use in a shrub border because of its leggy habit. Produces white flowers.	Average Growth	Moist	6' x 4'
Viburnum trilobum	American Highbush Cranberry	A medium-sized, rounded shrub with clusters of white flowers in spring, followed by bright red, edible berries amid the glossy dark green leaves.	Average Growth	Moist	8' x 8'

Viburnum trilobum 'Compactum'	Compact American Highbush Cranberry	Dwarf form with good flowering habit.	Average Growth	Moist	5' x 4'
Viburnum X juddii	Juddii Spice Viburnum	Fragrant clusters of white flowers bloom in spring on this full, rounded shrub. A great addition to any landscape because of its wonderful fragranc e.	Average Growth		6' x 10'
Viburnum X pragense	Prague Viburnum	Clusters of creamy white flowers bloom in spring above the glossy, dark green leaves of this cold - hardy, evergreen shrub.	Average Growth	Moist	8' x 6'
Vinca minor	Periwinkle	An excellent eve rgreen groundcover for shade with dainty blue flowers in spring and small, glossy dark green leaves.	Average Growth	Moist	
Weigela florida	Pink Weigela	Used in the shrub border, grouping or massing.			
Weigela florida 'Alexandra'	Wine & Roses Weigela	Dark burgundy-purple leaves shimmer like a fine wine. Hot rose-pink flowers bloom against the glossy dark leaves in spring. Attracts butterflies.	Average Growth	Moist	4' x 4'
Weigela florida 'Bristol Snowflake'	Snowflake Weigela	Flowers are white but wi th some pink.			
Weigela florida 'Java Red'	Java Red Weigela	Carmine-red buds open to deep pink flowers in spring above the dark purplish -green leaves on this compact, mounding shrub.	Average Growth	Moist	36" x 4'
Weigela florida 'Minuet'	Minuet Weigela	Dark ruby-red, slightly fragrant flowers bloom in spring above the dark purplish -green leaves of this dwarf, compact shrub.	Slow Growth		24" x 24"
Weigela florida 'Pink Princess'	Pink Princess Weigela	Masses of lavender-pink flowers bloom in spring on this rounded shrub. Attracts butterflies and hummingbirds into the garden.	Average Growth	Moist	5' x 5'

Weigela florida 'Red Prince'	Red Prince Weigela	Masses of red trumpet-shaped flowers bloom in spring on this upright, arching shrub. Attracts butterflies and hummingbirds into the garden.	Average Growth	Moist	5' x 5'
Weigela florida 'Rumba'	Rumba Weigela	Yellow-green leaves has purple edges. Flowers are dark red with a yellow throat.	Slow Growth	Moist	30" x 30"
Weigela florida 'Samba'	Samba Weigela	Leaves are dark green with purple tips and edges. Flowers are red with a yellow throat.	Slow Growth	Moist	24" x 24"
Weigela florida 'Tango'	Tango Weigela	Foliage is purple on the upper surface and dark green below. Flowers are red on the outside, yellowish in the throat.	Slow Growth	Moist	24" x 36"
Weigela florida 'Variegata'	Variegated Pink Weigela	Deep rose flowers bloom in spring above the green leaves with creamy white edges on this compact shrub. Attracts butterflies and hummingbirds.	Average Growth	Moist	4' x 5'
Weigela florida 'Variegata Nana'	Dwarf Variegated Weigela	Deep rose flowers bloom in spring above the green leaves with creamy white edges on this dense, dwarf shrub. Attracts butterflies and hummingbirds.	Slow Growth		36" x 4'
Wisteria floribunda	Purple Japanese Wisteria	An excellent flowering vine used over patios, large structures, trellises or trained into a tree form. Purple flowers.	Fast Growth	Moist	15' x 40'
Yucca filamentosa	Adam's Needle Yucca	Tall spikes of creamy white flowers bloom in late spring above the stiff, upright, evergreen leaves of this low, clumping shrub. Loves hot weather.	Slow Growth	Semi-Moist	4' x 3'
Yucca glauca	Plains Yucca	A tall spike of creamy white flowers blooms in late summer above the cluster of narrow, evergreen leaves.	Average Growth	Semi-Moist	3' x 3'

Zelkova serrata	Japanese Zelkova	A very handsome shade tree with clean, dark green leaves that turn yellowish-orange in fall and cinnamon-red bark. Tolerates poor soil conditions.	Average Growth	Moist	75' x 50'
Juniperous chinensis 'Sea Green'	Espallier Sea Green Juniper	The bright green foliage is trained on trellis for walls and fences. Regular pruning is needed to maintain this effect.	Average Growth	Semi-Moist	5' x 6'

□□□

Selected Readings

Abbs, B. (2003). Choosing and Using Climbing Plants. San Diego, California : Laurel Glen Publishing. 96 p.

Abellanosa, A.L. and H.M. Pava.(1987). An Introduction to Crop Science. CMU, Musuan, Bukidnon : Publications Office. 245 p.

Aldrich, R. A. and John,W. B. (1994). Greenhouse Engineering. Storrs : University and Applications, 403-408.

Allard, R.W. (1960). Principles of Plant Breeding. New York : John Wiley & Sons, Inc. 485 p.

Arditti J. and Ernst R. (1993). Micropropagation of Orchids, John Wiley and Sons, New York.

Arditti, J. (1992). Fundamentals of Orchid Biology, John Wiley and Sons, New York.

Arnon, D.I. (1943). Mineral Nutrition of Plants, *Annual Review of Biochemistry*, 12: 493-528.

Arora, J.S. (1998). *Introductory Ornamental Horticulture.* Kalyani Publishers, Ludhiana. pp. 204.

Bagust, H. (2003). The Firefly Dictionary of Plant Names : Common & Botanical. Ontario, Canada : Firefly Books Ltd. 440p.

Bechtel, H., Cribb, P., and Launert, E. (1992). The Manual of Cultivated Orchid Species, Third Edition. The MIT Press, Cambridge, MA.

Bhattacharya, U.C. (1969). New distributional records of orchids for West Himalaya. *Bull. Bot Soc.* Bengal, 23: 161-165.

Boden, T.A., Kanciruk, P. and Farrell, P. (1990). *Trends 90, A compendium of Data on Global Change*, Carbon Dioxide Information Analysis Center, Oak Ridge, Tennessee, U.S.A. pp. 104.

Bremness, L. (1988). The Complete Book of Herbs : A Practical Guide to Growing & Using Herbs. London: Dorling Kindersley Ltd. 288 p.

Brown J.H., Lynch D.V., Thompson J.E. (1987). Molecular species specificity of phospholipid breakdown in microsomal membranes of senescing carnation flowers. *Plant Physiol*, 85: 679-683.

Brundrett, M.C. (2002). Tansley review, Co evolution of roots and mycorrhizas of land plants, *New Phytologist*,154 : 275–304.

Burns, R. (1979). A Red, Red Rose, in *The Literature of England* (eds. G. K. Anderson *et al.*) 3rd Ed., Scott, Foresman and Company, London, p. 630.

Byung, D.L., Wan, G.K. Weon, D.C. and Jae, M.S. (2002). Occurrence of dry rot on *Cymbidium* Orchids Caused by *Fusarium* spp., *Plant Pathol. J.* 18(3): 156-160.

Chadha, K.L. (2000). Post harvest management of flowers. Hand Book of Horticulture, I.C.A.R, New Delhi, 969-981.

Chauhan N.S. (1999). Medicinal and Aromatic Plants of Himachal Pradesh. Indus Publishing Company, New Dehli.

Culquhoun, J.B. (2006). Allelopathy in weeds and crops : myths and facts. Proc. Of the 2006 Wisconsin Fertilizer, Aglime & Pest Management Conference. 45 : 318-320. Retrieved June 4, 2010 from http://www.soils.wisc.edu/extension/wcmc/2006/pap/Colquhoun2.pdf.

D'Urzo, M.P., Alkire, B. and T. Aillo. (1998). Capers. New Crop Factsheet. Retrieved July 11, 2010 from http://www.hort.purdue.edu/newcrop/cropfactsheets/caper.html#Origin.

Douglas P.C. and Barry J.J. (2002). Optimizing a Bacillus subtilis isolate for biological control of sugar beet *Cercospora* leaf spot. *Biological Control,* 26(2): 153-161.

Downie, D.G. (1957). *Corticium solani*-an orchid endophyte. *Nature*, 179: 180.

Dube, H.C. (2007). An Introduction to fungi : 3rd edition, Vikash Publishing House Pvt. Ltd. Delhi.

Duff. J. and Daly, A. (2002). Orchid diseases in the Northern Territory. Agnote, 13: 1-5.

Dunne, J.A. and Harte, J. (2001). Greenhouse effect. In *Encyclopedia of Biodiversity,*

E. Moritz, Greifswald.G.A. Stuart (1984). *Chinese Materia Medica.*Taipei. Southern Materials Centre. A translation of an ancient Chinese herbal, 31: 751-759.

Ferguson, J.J. and B. Rathinasabapathi, (2003). Allelopathy : how plants suppress other plants. University of Florida IFAS Extension Publication #HS944. Retrieved July 10, 2010 from http://edis.ifas.ufl.edu/hs186.

Food and Agriculture Organization (2010). Crops statistics - concepts, definitions and classifications. Retrieved May 29, 2010 fromhttp://www.fao.org/economic/ess/methodology/methodology-systems/crops-statistics-concepts-definitions-and-classifications/en/.

Ge, S., Li, A, Lu, BR., Zhang, SZ. and DY. Hong. (2002). A phylogeny of the rice tribe Oryzeae (Poaceae) based on *matK* sequence data. American Journal of Botany. 89: 1967-1972. Retrieved July 6, 2010 from http://www.amjbot.org/cgi /content/ full/89/12/1967.

Ghanaksh, A. and P. Kaushik (1999). Antibacterial effect of*Aerides multiflora*Roxb. A study *in vitro, Journ. Orchid Soc.India,*13 : 65-68.

Griesbach, R.J. (1985). An orchid in every pot.*Florists' Rev.* 176(4548): 26–30.

H. Kizu, E.I. Kaneko E-I., T. Tomimori (1999). Studies on Nepalese crude drugs. XXVI. Chemical constituents of Panch aunle, the roots of *Dactylorhiza hatagirea*D. Don. *Chem. Pharm. Bull.,* 47(11): 1618-1625.

Hagelberg, G.B. (2003). Sugar crops and natural sweeteners. Food & Culture Encyclopedia. Retrieved July 18, 2010 from http://www.answers.com/topic/sugar-crops-and-natural-sweeteners.

Hajra, P.K. (1983). New Species of Lister a from Nandadevi National Park, Chamoli, District, Uttar Pradesh. Bull.*Bot-Surv India,* 25: 181-182.

Harley, J.I. (1965). Mycorrhiza. *In* : Ecology of soil born plant pathogens. K.F Waker and W C Synder (eds.). University California Pres, Berkeley, Los Angles. 218-230.

Hew C.S., Yong J.W.H. (2004). Recent advances in orchid tissue culture. In : *The Physiology of Tropical Orchids in Relation to the Industry* (2nd Edn) World Scientific Publishing, Singapore, 288-317.

Hijner, J.A., Arditti J. (1973). Orchid mycorrhiza; vitamin requirements and production by the symbionts. *Amer.J.Bot.,* 60: 829-835.

How to Use Bulbous Plants | eHow.com http : //www.ehow.com/how_4478382_use-bulbous-plants.html#ixzz1G6b4XBFI\

http : //chemistry.about.com/od/chemistryhowtoguide/ht/makephpaper.htm .Retrieved on 2010-07-1.

http : //faq.gardenweb.com/faq/lists/orchids/2003011119000930.html. Retrieved on 2010-07-26.

http : //forum.terrorchid.org/viewtopic.php?t=900. Retrieved on 2010-08-28.

http : //fpdn.ifas.ufl.edu/. Retrieved on 2010-08-28.

http : //hrdc.pcarrd.dost.gov.ph/phil-organic/Technologies/techgen-pestmgt.pdf, accessed June 13, 2010.

http : //mycology.cornell.edu/listings.aspx?findex=Discussion. Retrieved on 2010-08-28.

http : //www.angelfire.com/on4/angelorchids/PropagatingOrchidsnew. Retrieved on 2010-08-9.

http : //www.ehow.com/about_6617456_do-orchid-flowers-adapt-pollination. html#ixzz0zEk5XFeA retrieved on 2010-06-17.

http : //www.indexfungorum.org/Names/NamesRecord.asp?RecordID. Retrieved on 2010-08-30.

http : //www.indg.in/agriculture/ruralemployment. Retrieved on 2010-08-28.

http : //www.ionopsis.com/sainsburyp.htm. Retrieved on 2010-098-28.

http : //www.orchidsusa.com/3Lightlevels.htm. Retrieved on 2010-09-27.

http : //www.pfc.cfs.nrcan.gc.ca/biodiversity/bcern/manual/index_e.html. Retrieved on 2010-08-3.

Huang F.Y., Philosoph-Hadas S., Meir S., Callahan D.A., Sabato R., Zelcer A., Hepler P.K. Ichikawa, K. and Aoki, T. (2000). New leaf spot disease of *Cymbidium* species caused by *Fusarium subglutinans* and *F. proliferatum. J.G.Pl. Path.*, 66(3): 213-218.

Ichikawa, S., Kawasaki, S., Tanaka, C. and Tsuda, M. (2003). Induced restance against Fusarium diseases of *Cymbidium* species by weakly virulent Strain HPF-1(*Fusarium* sp.).*Gen. Plant. Pathol.* 69: 400-405.

Janick, J. (2005). Horticultural Plant Breeding : Past Accomplishments, Future Directions. In : Proc. IS on Hort. in Asian-Pacific Region. (R. Drew, ed.). Acta Hort. 694. Retrieved May 30, 2010 from http : //www.hort.purdue.edu/newcrop/acta 694.pdf.

Janick, J. (1972). Hortiultural Science. 2nd ed. San Francisco : W. H. Freeman and Co. 586 p.

Khory, N. (1982). Materia Medica of India and their Therapeutics. Neeraj Publishing House.

Knapp J.E., Kausch A.P., Chandlee J.M. (2000). Transformation of three genera of orchidusing the *bar* gene as a selectable marker. *Plant Cell Reports* 19,893-898.

Knudson L. (1946). A new nutrient solution for germination of orchid seed. *American Orchid Society Bulletin* 15: 214-217.

Kudoh, M. (1985). *Ikebana no Michi* (*The Road to Ikebana*), Shufu no tomo, Tokyo.

Kuehnle A.R. (1997). Molecular biology of orchids. In : Arditti J, Pridgeon AM (Eds) *Orchid Biology: Reviews and Perspectives, VII*, Kluwer Academic Publishers,The Netherlands 75-115.

Kuehnle A.R., Sugii N. (1992). Transformation of *Dendrobium* orchid using particle bombardment of protocorms. *Plant Cell Reports,* 11: 484-488.

Kusuyama, M. (1985). Hanasaka Jijii, in *Nihon no Shinwa to Juhdai Mukashi Banashi* (*Japanese Mythology and Ten Folktales*), Kohdansha, Tokyo, pp. 153–160.

Lam E., Kato N., Lawton M. (2001). Programmed cell death, mitochondria and the plant hypersensitive response. *Nature,* 411: 848-853.

Lantican, R.M. (2001). The Science and Practice of Crop Production. College, Los Baños, Laguna, Phils. : SEAMEO SEARCA and UPLB. 330 p.

Leist, N. and A. Jonitz. (2009). Tools for seed identification species classification and nomenclature. ISTA Purity Sem. June 15, 2009, Zurich. Retrieved June 30, 2010 from https://www.seedtest.org/upload/cms/user/6Toolsforseed Identification_NorbertKompatibilittsmodus.pdf.

Leonardi, C. and D. Romano. (2004). Recent issues on vegetable grafting. Proc. XXVI IHC Transplant Production and Stand Establishment. Acta Hort. 631 : 163-174. Retrieved July 9, 2010 from www.ecoponics.de.

Lin C.C., Huang P.C., Lin J.M. (1992). Antioxidant and hepatoprotective effects of*Anoectochilus formosanus* and *Gynostemma pentaphyllum.AmJ. Chin. Med.* 28(1): 87-96.

M.R. Uniyal, M. Uniyal, M. (1977). Medicinal Plants and Minerals of Uttarakhand Himalaya. Baidyanath, Ayurved Shodh Sanstha, Patna 26.

Mahawer, L.N. and Mishra, R.L. (1996). Trees for combating dust and noise pollution. In*Eco-friendly trees for urban beautification,*Khosla, P.K. Ed., pp. 285-290. Indian Society of Tree Scientists, Solan.

Majumdar P.L., Majumder S. and Sen S. (2003). Triperpenoids from the orchids *Agrostophyllum breirps* and *Agrostophyllum callosum.Phytochemistry*, 62: 591.

Majumder, P.L. Banerjee, S. Sen, S. (1996). Three stilbenoids from the orchid Agrostophyllum callosum.*Phytochemistry,* 42(3): 847-853.

Mayak, S. Dilley, D. (1976). Effect of sucrose on response of cut carnation flowers to kinetin, ethylene and abscisic acid. *J Am Soc Hort Sci.*, 101: 583-585.

McKinley, M. (ed.) (2005). Ortho Complete Guide to Orchids. Des Moines, IA : Meredith Books. 224 p.

Mcneill, J., Barrie, F.R., Burdet, H.M., Demoulin. V., Hawksworth, D.L., Marhold, K., Nicolson, D.H., Prado, J., Silva, P.C., Skog, J.E., Wiersema, J.H. and N.J. Turland. (eds.) (2006). International Code of Botanical Nomenclature (Vienna Code), adopted by the Seventeenth International Botanical Congress, Vienna, Austria, July 2005. (electronic version). Retrieved July 5, 2010 from http://ibot.sav.sk/icbn/main.htm.

Men S., Ming X., Wang Y., Liu R., Wei C., Li Y. (2003a). Genetic transformat of two species of orchid by biolistic bombardment. *Plant Cell Reports,* 21: 592-598.

Merrill, E.D. (1912). A Flora of Manila. Manila, Phils. : Bureau of Printing. 491 p.

Mishiba K., Chin D.P., Mii M. (2005). *Agrobacterium*-mediated transformation of *Phalaenopsis* by targeting protocorms at an early stage after germination.*Plant Cell Reports,* 24: 297-303.

Mishra, R.L. and Mahawer, L.N. (1996). Trees for combating dust and noise pollution. In *Eco-friendly trees for urban beautification,*Khosla, P.K. Ed., pp. 285-290. Indian Society of Tree Scientists, Solan.

Murasaki S. (1965). *Genji Monogatari* (*The Tale of Genji*) (trans. J. Tanizaki) Vol. 2, Chuokohron, Tokyo, pp. 3–38.

Murashige T., Skoog F. (1962). A revised medium for rapid growth and bioassays with tobacco tissue cultures. *Physiologia Plantarum,* 15: 473-497.

Nicholas D.J.D. (1961). Minor Mineral Elements, *Annual Review of Plant Physiology,* 12: 63-90.

Nichols R. (1973). Senescence and sugar status of the cut flower. *Acta Hortic.,* 41: 21-27.

North Carolina State University.(2010). New technique improves efficiency of biofuel production. ScienceDaily. July 1, 2010. Retrieved July 4, 2010 fromhttp : //www.sciencedaily.com/releases/2010/06/100630115143.htm.

Ornamental Bulbous Plants | eHow.comhttp : //www.ehow.com/list_7349655_ornamental-bulbous-plants.html#ixzz1G6bfTw2A.

P.K. Kasera, J.K. Shukla (2001). Bio-medicinal properties and cultivation of *Leptadaenia reticualta* (Jivanti)-an endangered plant of the Thar Desert, India. *Curr. Sci.,* 84(7): 877-869.

Peters, C.M. (1994). Sustainable harvest of non-timber plant resources in tropical moist forest : an edological primer. Retrieved June 13, 2010 from. http : // www.worldwildlife.org /bsp/bcn/learning/primer/eng1.htm

Philippine Coconut Authority (PCA) (2003). Information bulletin on biological control of 'hagonoy'. Primer on Biological Control of Hagonoy (Chromolaena odorata). CPD Technoguide No. 8. Retrieved July 10, 2010 from http : //www.bar.gov.ph/biofuelsinfo /downloads/hagonoy.pdf.

Pisto, J. (1997). Cooking with Mushrooms : A Culinary Guide to Chef Pisto's Favorite Fungi. Pacific Grove, CA : Pisto,s Kitchen. 137 p.

Pittendrigh, S. and M. Davis. (1977). Lawns and Groundcovers. Sydney, Australia: Ure Smith. 80 p.

Poehlman, J.M. (1977). Breeding Field Crops. Connecticut : AVI Publishing Co., Inc. 427 p.

Poincelot, R.P. (1980). Horticulture : Principles and Practical Applications. New Jersey : Prentice-Hall, Inc. 652 p.

Priyadarshan, P.M. and P. de S. Goncalves. (2003). Hevea gene pool for breeding. Genetic Resources and Crop Evolution. 50 : 11-114. Retrieved June 13 from http: //www.iac.br/Centros/centro_cafe/seringueira/PubPDF/Hevea_gene_pool.pdf.

Quimio, T.H. (1996). Fungi : Mycologists' Delight. UPLB, College, Laguna, Phils. : The Museum of Natural History. 19 p. (booklet).

Randhawa, M.S. (1964). *The cult of trees and tree-worship in Buddhist-Hindu sculptures.* All India Fine Arts & Crafts Society, New Delhi. pp. 66.

Randhawa, M.S. (1978). *Beautifying cities of India.* Indian Council of Agricultural Research, New Delhi. pp 15.

Robinson, P. (2003). The Practical Rock & Water Garden. London : Hermes House. 256 p.

S.M. Khasin, P.R. Mohana Rao (1999). Medicinal Importance of Orchids.*The Botanica*, 49: 86-91.

Saigyoh (1988). *Sanka-shu* (*Collected Poems of Sanka*), Iwanami, Tokyo, p. 31.

Saint Exupéry, A. de (1971). *The Little Prince* (trans. K. Woods), Harcourt Brace Jovanovich, New York.

Sakcali, M.S., Bahadir, H. and M. Ozturk. (2008). Eco-physiology of *Capparis spinosa* L.: a plant suitable for combating desertification. Pak. J. Bot. 40(4) : 1481-1486. Retrieved July 11, 2010 from http: //www.pakbs.org/pjbot/PDFs/40%284%29/PJB40%284%291481.pdf.

Shakespeare, W. (1982). *Hamlet* (ed. H. Jenkins), The Arden Shakespeare, Methuen, London, p. 188.

Shih, C.C., Y.W. Wu and W.C. Lin. (2002). Antihyperglycemic and antioxidant properties of *Anoectochilus formosanus*in diabetic rats.*Clinical and Experimental Pharmacology and Physiology*, 29(8): 684-688.

Singh R.S. (1994). Plant Disease. 7th edition, Oxford and IBH publishing Co. Pvt. New Delhi.

Singh R.S. (1994). Plant Pathogens.The fungi. Third edition, Oxford and IBH publishing Co. Pvt. New Delhi.

Smith S.E., Read D.J. (1997). Mycorrhizal Symbiosis, 2nd edn. San Diego, CA, USA : Academic Press.

Steiner, M.L. (1986). Philippine Ornamental Plants. Las Piñas, Metro Manila : M&L Licudine Enterprises. 233.

Swartz O. (1799). Dianome Epidendri generis Linn. *Nova ActaRegiae Societatia Scientiarum Upsaliensia,* 6: 61–88.

Talkovski, A. (1986). *Offret* (*The Sacrifice*), Argos films, Paris.

Tee C.S., Marziah M., Tan C.S., Abdullah M.P. (2003). Evaluation of different promoters driving the GFP reporter gene and selected target tissues for particle bombardment of *Dendrobium* Sonia-17. *Plant Cell Reports,* 21: 452-458.

Teixeira da Silva J.A., Chan M.T., Sanjaya, Chai M.L., Tanaka M. (2006a). Priming abiotic factors for optimal hybrid *Cymbidium* (Orchidaceae) PLB and callus induction, plantlet formation, and their subsequent cytogenetic stability analysis. *Scientia Horticulturae* 109: 368-378.

Than, P.P., Haryudian, P. Sitthisack, P., Paul W.J.T. and Kevin D.H. (2008). Chilli anthracnose disease caused by *Colletotrichum* species.*J Zhejiang Univ Sci B.* 9(10): 764-778.

Thomas, M. (1937). Plant Hormones and their Possible Importance in Horticulture. *Scient. Hort.,* 5: 141.

Thompson, H.C. (1940). Effect of Temperature and Length of Day on Growth of Vegetables. *Minnesota Hort.* 68: 163.

Thornberry, H.H. Thompson, M.R., Izadpanah, K. and Canares, P.H. (1968). Orchid viruses : causality of *Cattleya* infection blossom necrosis. *J. Phyto. Path.,* 62 : 305-310.

Training and Pruning Apple Trees in Intensive Orchards, Virginia Cooperative Extension publication 422-024, *http : //pubs.ext.vt.edu/422-024/*

Training and Pruning Apples Trees, Virginia Cooperative Extension publication 422-021, *http : //pubs.ext.vt.edu/422-021/*

Vacin E.F., Went F.W. (1949). Some pH changes in nutrient solutions. *Botanical Gazette,*110: 605-613.

Valpuesta V., Lange N.E., Guerrero C., Reid M.S. (1995). Up-regulation of a cysteine protease accompanies the ethylene-insensitive senescence of daylily (*Hemerocallis*) flowers. *Plant Mol Biol* 28: 575-582.

Van Doorn W.G. (2001). Categories of petal senescence and abscission : a re-evaluation. *Ann Bot,* 87: 447-456.

van Doorn W.G., Stead A.D. (1997). Abscission of flowers and floral parts. *J Exp Bot,* 309: 821-837.

Verheij, E.W.M. and R.E. Coronel (eds.) (1992). Edible fruits and nuts. Plant Resources of South East Asia No. 2. Bogor, Indonesia : Prosea. 447 p.

Verma, N., MacDonald, L. and Punja, Z.K. (2006). Inoculum prevalence, host infection and biological control of *Colletotrichum acutatum* causal agent of blueberry anthracnose in British Columbia. *Plant Pathology,* 55: 442–450.

Wan G.K., Byung D.L., Woo S.K. and Weon D.C. (2002). Root Rot of Moth Orchid Caused by *Fusarium* spp. *Plant Pathol. J.,*18(4): 225-227.

Warcup J.H. and Talbot P.H.B. (1971). Perfect statesof*Rhizoctonias* associated with orchids I-III. New Phytologist, 66: 631-641.

Washington State University (2010). Agriculture's next revolution- perennial grain-within sight. *Science Daily.* June 27, 2010. Retrieved July 6, 2010 from http : //www.sciencedaily.com/releases/2010/06/100624144111.htm.

White J. (1990). Watering and Fertilizing by - American Orchid Society Bulletin.

Wikipedia (2010). Liana. Retrieved July 10, 2010 from http : //en.wikipedia.org/wiki/Liana.

Wikipedia (2010). Orchidaceae. Retrieved July 10, 2010 from http : //en.wikipedia.org /wiki/Orchidaceae.

Wikipedia (2010). Plant defense against herbivory. Retrieved March 17 , 2010 from http : //en.wikipedia.org/wiki/Plant_defense_against_herbivory.

Winter K., Wallace B.J., Stocker G.C., Rocksandic Z. (1983). Crassulacean acid metabolism in Australian vascular epiphytes and some related species. *Oecologia (Berlin)* 57: 129–141.

Zorkaya, N. (1995). Owari (An end), in *Tarkovski no Sekai* (*The World of Tarkovski*) (ed. A.M. Sandrell), Kinema Junpo, Tokyo, pp. 227–251.http : //www.ncdc.noaa.gov /img/climate /online/doe/indiaanntemp.gif

Index

A

B

C

D

E

F

G

H

I

J

K

L

M

Q

R

S

T

U

□□□